Sugarcane-Based Biofuels and Bioproducts

Sugarcane-Based Biofuels and Bioproducts

EDITED BY

Ian M. O'Hara

and

Sagadevan G. Mundree

WILEY Blackwell

Published by John Wiley & Sons, Inc., Hoboken, New Jersey
Published simultaneously in Canada

Library of Congress Cataloging-in-Publication Data

Names: O'Hara, Ian M., editor. | Mundree, Sagadevan G., 1946- editor.
Title: Sugarcane-based biofuels and bioproducts / edited by Ian M. O'Hara and
Sagadevan G. Mundree.
Description: Hoboken, New Jersey : John Wiley & Sons, 2016. | Includes index.
Identifiers: LCCN 2016007511| ISBN 9781118719916 (cloth) | ISBN 9781118719923
(epub)
Subjects: LCSH: Biomass energy. | Sugarcane–Biotechnology.
Classification: LCC TP339 .S84 2016 | DDC 662/.88–dc23 LC record available at
http://lccn.loc.gov/2016007511

Cover image: Getty/Felipex

Printed in Singapore by C.O.S. Printers Pte Ltd

10 9 8 7 6 5 4 3 2 1

Contents

Preface

As a society we are faced with significant issues. There is an urgent need to address the challenge of climate change while continuing to promote development in the world's poorest countries. From an agricultural perspective, our land, water, energy, and food systems are inextricably linked. New technologies are needed to provide sustainable energy solutions and at the same time enhance food availability and distribution.

Sugarcane is one of the world's most important agricultural crops with a long history of use for the production of food, energy, and coproducts. Growing across many countries in tropical and subtropical regions, sugarcane has a significant global footprint. The high photosynthetic efficiency and high biomass production makes sugarcane an ideal feedstock for both food production and the coproduction of non-fossil-based chemicals, polymers, and energy products.

While the opportunities for the use of sugarcane for ethanol production are well-known, there are many other potential products of similar or higher value that can be produced from the crop. Technology developments, most particularly in the fields of agricultural and industrial biotechnology, are providing new opportunities to diversify the revenue base for sugar producers. Not only does the application of this technology promote economic viability of sugarcane producers and their regional communities, it also helps to address our over-reliance on products from fossil-based resources, and hence contributes to global decarbonization activities. These economic, social, and environmental benefits, however, will only be achieved where technologies are adopted in an appropriate manner

This book provides a comprehensive overview of current and future opportunities for the production of biofuels and bioproducts from sugarcane. The first section of the book (Chapters 1 and 2) provides an overview of the sugarcane industry and presents the opportunities and challenges in this area. This section also examines the sugarcane crop biotechnology and the opportunities that this field presents in enhancing opportunities for the production of bioproducts. The second section of the book (Chapters 3–12) provides detailed overviews of the current state-of-the-art relating to a variety of biofuel and bioproduct opportunities from sugarcane. These opportunities include more traditional products such as ethanol production, pulp and paper, animal feed products and cogeneration to future opportunities such as the production of fermentable sugars from bagasse and their subsequent conversion into specialty chemical products. The final section of the book addresses aspects relating to sugarcane biofuel and bioproduct sustainability, techno-economics, and whole-of-system process integration.

The editors are very grateful to the many authors who contributed to this book. All of the authors are recognized as leading experts in their fields and provide unique perspectives as a result of their many decades of experience in sugar, biofuels, and bioproducts research. Without their contributions, this book would not have been possible and we appreciate their insights and highly value the contributions that they have made.

Ian M. O'Hara
Sagadevan G. Mundree
9 July 2015
Brisbane, Australia

List of contributors

Sébastien Bonnet
Life Cycle Sustainability Assessment Laboratory, The Joint Graduate School of Energy and Environment (JGSEE), King Mongkut's University of Technology Thonburi (KMUTT), Bangkok, Thailand; Center of Excellence on Energy Technology and Environment, PERDO, Bangkok, Thailand

Antonio Bonomi
Laboratório Nacional de Ciência e Tecnologia do Bioetanol (CTBE), Centro Nacional de Pesquisa em Energia e Materiais (CNPEM), Campinas, Brazil; Faculdade de Engenharia Química, Universidade Estadual de Campinas (FEQ/UNICAMP), Campinas, Brazil

Anthony K. Brinin
Centre for Tropical Crops and Biocommodities, Queensland University of Technology (QUT), Brisbane, Australia

Otávio Cavalett
Laboratório Nacional de Ciência e Tecnologia do Bioetanol (CTBE), Centro Nacional de Pesquisa em Energia e Materiais (CNPEM), Campinas, Brazil

Geoff Covey
Covey Consulting, Melbourne, Australia

Sudipta S. Das Bhowmik
Centre for Tropical Crops and Biocommodities, Queensland University of Technology (QUT), Brisbane, Australia

Marina O. de Souza Dias
Instituto de Ciência e Tecnologia (ICT), Universidade Federal de São Paulo (UNIFESP), São Paulo, Brazil; Laboratório Nacional de Ciência e Tecnologia do Bioetanol (CTBE), Centro Nacional de Pesquisa em Energia e Materiais (CNPEM), Campinas, Brazil

William O.S. Doherty
Centre for Tropical Crops and Biocommodities, Queensland University of Technology (QUT), Brisbane, Australia

Kameron G. Dunn
Centre for Tropical Crops and Biocommodities, Queensland University of Technology (QUT), Brisbane, Australia

Troy Farrell
Mathematical Sciences, Queensland University of Technology (QUT), Brisbane, Australia

Rubens M. Filho
Laboratório Nacional de Ciência e Tecnologia do Bioetanol (CTBE), Centro Nacional de Pesquisa em Energia e Materiais (CNPEM), Campinas, Brazil; Faculdade de Engenharia Química, Universidade Estadual de Campinas (FEQ/UNICAMP), Campinas, Brazil

Shabbir H. Gheewala
Life Cycle Sustainability Assessment Laboratory, The Joint Graduate School of Energy and Environment (JGSEE), King Mongkut's University of Technology Thonburi (KMUTT), Bangkok, Thailand; Center of Excellence on Energy Technology and Environment, PERDO, Bangkok, Thailand

Ava Greenwood
Mathematical Sciences, Queensland University of Technology (QUT), Brisbane, Australia

Mark D. Harrison
Centre for Tropical Crops and Biocommodities, Queensland University of Technology (QUT), Brisbane, Australia

Philip A. Hobson
Centre for Tropical Crops and Biocommodities, Queensland University of Technology (QUT), Brisbane, Australia

Cecília Laluce
Biochemistry and Chemical Technology Department, Institute of Chemistry, Univ Estadual Paulista (UNESP), São Paulo, Brazil

Guilherme R. Leite
Biochemistry and Chemical Technology Department, Institute of Chemistry, Univ Estadual Paulista (UNESP), São Paulo, Brazil

Anthony P. Mann
Centre for Tropical Crops and Biocommodities, Queensland University of Technology (QUT), Brisbane, Australia

Sagadevan G. Mundree
Centre for Tropical Crops and Biocommodities, Queensland University of Technology (QUT), Brisbane, Australia

Ian M. O'Hara
Centre for Tropical Crops and Biocommodities, Queensland University of Technology (QUT), Brisbane, Australia

Darryn W. Rackemann
Centre for Tropical Crops and Biocommodities, Queensland University of Technology (QUT), Brisbane, Australia

Thomas J. Rainey
School of Chemistry, Physics and Mechanical Engineering, Science and Engineering Faculty, Queensland University of Technology (QUT), Brisbane, Australia

Marguerite A. Renouf
School of Geography, Planning and Environmental Management, Faculty of Science, University of Queensland, St. Lucia, Brisbane, Australia

Karen T. Robins
Sustain Biotech Pty Ltd, Sydney, Australia

Thapat Silalertruksa
Life Cycle Sustainability Assessment Laboratory, The Joint Graduate School of Energy and Environment (JGSEE), King Mongkut's University of Technology Thonburi (KMUTT), Bangkok, Thailand; Center of Excellence on Energy Technology and Environment, PERDO, Bangkok, Thailand

Robert E. Speight
School of Chemistry, Physics and Mechanical Engineering, Science and Engineering Faculty, Queensland University of Technology (QUT), Brisbane, Australia

Ricardo Ventura
Integra Consultoria Química LTDA, Ribeirão Preto, Brazil

Brett Williams
Centre for Tropical Crops and Biocommodities, Queensland University of Technology (QUT), Brisbane, Australia

Thamires T. Zamai
Biochemistry and Chemical Technology Department, Institute of Chemistry, Univ Estadual Paulista (UNESP), São Paulo, Brazil

Bruna Z. Zavitoski
Biochemistry and Chemical Technology Department, Institute of Chemistry, Univ Estadual Paulista (UNESP), São Paulo, Brazil

Zhanying Zhang
Centre for Tropical Crops and Biocommodities, Queensland University of Technology (QUT), Brisbane, Australia

PART I
Sugarcane for biofuels and bioproducts

CHAPTER 1

The sugarcane industry, biofuel, and bioproduct perspectives

Ian M. O'Hara
Centre for Tropical Crops and Biocommodities, Queensland University of Technology (QUT), Brisbane, Australia

1.1 Sugarcane – a global bioindustrial crop

Sugar (or more specifically sucrose) is one of the major food carbohydrate energy sources in the world. It is used as a sweetener, preservative, and colorant in baked and processed foods and beverages and is one of lowest cost energy sources for human metabolism.

On an industrial scale, sucrose is produced from two major crops – sugarcane, grown in tropical and subtropical regions of the world, and sugar beet, grown in more temperate climates. Sugarcane, however, accounts for the vast majority of global sugar production.

For much of the history of sugarcane production, sugar was a scarce and highly valued commodity. Sugarcane processing focused on extracting sucrose as efficiently as possible for the lucrative markets in the United Kingdom and Europe. The potential for the production of alternative products from sugarcane, however, has long been recognized. The key process by-products including bagasse, molasses, mud, and ash have all been investigated as a basis for the production of alternative products (Rao 1997, Taupier and Bugallo 2000).

Sugarcane is believed to have originated in southern Asia, and migrated in several waves following trade routes through the Pacific to Oceania and Hawaii and through India into Europe. Sugarcane was introduced and spread through the Americas following the expansion by British, Spanish, and Portuguese colonies in the 15th and 16th centuries (Barnes 1964).

While various methods of juice extraction and sugar production have been used over centuries to produce sugar, substantial innovations in sugar chemistry and processing technologies throughout the 18th and 19th centuries have formed the basis of modern sugar production methods (Bruhns *et al.* 1998). Dramatic improvements in processing efficiency, sugar quality, and automation and control characterized sugar processing throughout the 20th century.

Sugarcane-Based Biofuels and Bioproducts, First Edition. Edited by Ian M. O'Hara and Sagadevan G. Mundree.

While the production of alcoholic liquors from sugarcane juice and molasses has been known since ancient times, the production of rum has been associated with industrial sugar production since the introduction of sugarcane to the Caribbean in the 17th century. More recently, further coproducts started being produced including paper products, cardboard, compressed fiber board, and furfural from bagasse; ethanol, butanol, acetone, and acetates from molasses; and cane wax extracted from filter mud (Barnes 1964).

Perhaps the most significant development in sugarcane coproducts, however, occurred in 1975 when the Brazilian Government established the National Alcohol Program (the ProÁlcool program) in response to high oil prices and increasing costs of oil imports to Brazil. This program established a large domestic demand for ethanol, which resulted in the rapid expansion of the sugarcane industry in Brazil, enhancing technical capability, increasing the scale of factories, and lowering production costs of sugar and ethanol (Bajay *et al.* 2002).

The impact on global sugar and ethanol markets of ProÁlcool was profound, and this impact is still being felt today with Brazil being the undisputed global powerhouse of sugarcane production. The ProÁlcool program demonstrated the viability of sugarcane as a truly industrial crop, not just for food markets but also as a large-scale feedstock for the coproduction of energy products in integrated factories.

The period of the 1980s and 1990s saw sustained periods of low world sugar prices, in part the result of lower crude oil prices and increased Brazilian sugar exports, and increasing electricity prices in many countries. These factors focused the attention of the sugar industry on diversification opportunities and, in particular, the utilization of the surplus energy from bagasse to produce electricity for export into electrical distribution networks.

The past two decades have seen the emergence into the public consciousness of global challenges of climate change and increasing crude oil prices. Both these factors have enhanced human desires to find more renewable feedstocks for fuels, chemicals, and other products currently manufactured from fossil-based resources leading to direct consumer demand for more sustainable consumer products.

At the same time, human achievements and growth in our understanding of biotechnology have resulted in a suite of new tools that allow us to more readily convert renewable feedstocks into everyday products.

Sugarcane is widely acknowledged to be one of the best feedstocks for early-stage and large-scale commercialization of biomass into biofuels and bioproducts. As such, the sugarcane industry, with its abundant agricultural resource, is poised to benefit as a key participant in the growth of biofuel and bioproduct industries throughout the 21st century.

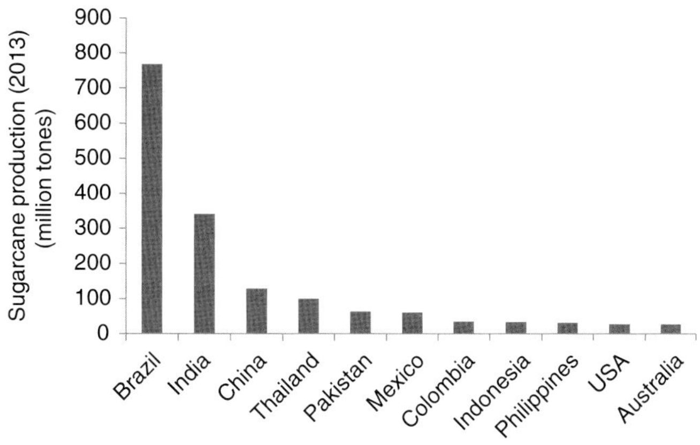

Figure 1.1 Leading sugarcane-producing countries (FAO 2015).

1.2 The global sugarcane industry

In 2013, more than 1.9 billion tons of sugarcane was grown globally at an average yield of 70.9 t/ha dominated by production in Brazil and India. Sugar beet production in 2013 was 247 million tons at an average yield of 56.4 t/ha (FAO 2015). The leading sugarcane-producing countries are shown in Figure 1.1.

Sugarcane is the largest agricultural crop by volume globally and the fifth largest by value with a production value in 2012 of US$103.5 billion (FAO 2015).

The principal use of sugarcane throughout the world is for crystal sugar production for human consumption. In several countries, including Brazil, a sizable portion of the crop is also used for ethanol production from both sugarcane juice and molasses. Many other countries produce lesser quantities of ethanol from sugarcane juice or molasses.

Over the past decade, global sugarcane production has increased by 35%, driven by a doubling in sugarcane production in Brazil (FAO 2015). This increased sugarcane production has resulted in both increased crystal sugar production and increased ethanol production, and has had a significant impact on the world price of raw sugar. Land-use change enabling this global expansion of sugarcane production has both direct and indirect sustainability implications, and the factors relating to these implications are diverse and complex (Martinelli and Filoso 2008, Sparovek *et al.* 2009, Martinelli *et al.* 2010).

1.2.1 Sugarcane

Sugarcane is a C4 monocotyledonous perennial grass grown in tropical and subtropical regions of the world. Modern sugarcane varieties are complex hybrids derived through intensive selective breeding between the species *Saccharum officinarum* and *Saccharum spontaneum* (Cox *et al.* 2000).

Globally, the 1.9 billion tons of sugarcane produced annually is grown on about 26.9 million hectares (FAO 2015) in tropical and subtropical regions. Modern sugarcane varieties are capable of producing more than 55 t/ha/y of biomass (dry weight). The development of high biomass sugarcane (often referred to as energy cane) has the potential to significantly increase the amount of biomass available.

1.2.2 Sugarcane harvesting and transport

Sugarcane harvesting and transport practices vary around the world, principally depending upon the degree of mechanization of the process. Sugarcane may be burnt before harvesting or cut in a green state without burning. The burning of sugarcane is becoming less prevalent with the introduction and enforcement of environmental air quality guidelines and this is increasing the amount of sugarcane leaf material available for coproducts.

In some countries, hand cutting of sugarcane is still widely practiced, although this has been completely replaced by mechanical harvesting in many countries. The transition to mechanized harvesting has often been driven by the difficulty in attracting labor to the very physically demanding work of hand cutting. This transition has not been without significant challenges in ensuring the delivery of both the optimum sugarcane weight and a quality product low in dirt, leaves, and low-sucrose sugarcane tops, which are collectively referred to as extraneous matter.

Traditional sugarcane-harvesting processes cut the stalk around ground level and discard tops and leaf materials. Only the clean stalk (either as a whole stalk or cut into billets) is transported into the factory for the extraction of the juice and production of sugar. Tops and leaf material separated in harvesting (trash) are generally left in the field to decompose, acting as mulch and providing organic matter and nutrient for the soil, or raked and burnt depending upon farming practices.

Some proportion of this leaf material is of value in the agricultural system, improving the soil condition. The remainder of this extraneous matter is potentially available as a feedstock for biomass value-adding processes such as bioethanol production. The impacts of harvesting and transporting extraneous matter on the sugar milling process, and the economics of the industry, are complex and integrated modeling approaches have been developed to analyze these effects (Thorburn *et al.* 2006).

Transport of sugarcane to the factory in a timely manner is important to ensure that little sucrose is lost through degradation processes. Not only is this a requirement to ensure maximum recovery of the sugar product, but a significant presence of one of the key degradation products, dextran, has a major impact on sugar quality. Minimizing the formation of this polysaccharide is crucial to efficient sugar production.

In order to maximize the availability of biomass for cogeneration or coproducts production, some movement has been made toward whole-of-crop harvesting. In this harvesting approach, the entire crop including the field trash may be collected and transported to the mill. Ideally, this trash is separated before processing, as there are significant efficiency, sugar recovery, and sugar quality challenges associated with processing sugarcane trash in a conventional sugar factory.

1.2.3 The raw sugar production process

Sugarcane is processed in factories generally located close to sugarcane farming areas to minimize the cost of sugarcane transportation. The factories are constructed to crush the sugarcane to extract the juice and produce non-food-grade raw sugar as the primary product. Raw sugar from these factories is generally transported to sugar refineries where the sugar is further decolorized and purified to produce the high-quality white "refined" sugar that is used as table sugar and in industrial sugar applications.

Sugarcane factories do not typically operate year round, but only during the period in which sugarcane harvesting is done. This period, which varies throughout the world from around 5 to 9 months, is largely determined by climate and economic factors associated with the period of peak sugar content of the sugarcane.

In the raw sugar production process (Figure 1.2), sugarcane is first shredded to produce a fibrous material and the sugarcane juice extracted from the fiber through a process of milling and/or diffusion. Water is used to assist in washing the sugar from the fiber. The fibrous residue of this process is known as bagasse, and this bagasse is burnt in suspension in bagasse-fired water tube boilers to produce steam. The steam is used to provide energy to drive mill machinery, to produce electricity in turbo-alternators, and to provide heat for the process. The quantity of ash residue from the combustion process, known as boiler ash, varies depending upon the incoming dirt levels of the sugarcane.

The sugarcane juice is heated, limed, and clarified to separate the dirt and other insoluble impurities from the juice. The clean juice, generally known as clarified juice (CJ) or evaporator supply juice (ESJ), is fed into multiple effect vacuum evaporators where the juice is concentrated to around 65–70 brix to produce a concentrated syrup. The syrup is then passed to the panstage where the sugar crystallization occurs in a series of product and recovery sugar strikes.

High-grade (product) sugar from the panstage is centrifuged to produce sugar crystals of the target polarization and the molasses from these centrifugals is recycled to the panstage for further processing. The wet sugar from the centrifugals is passed to the sugar drier that dries the sugar to the target moisture specification, and this product is shipped to a refinery for further decolorization and impurity removal.

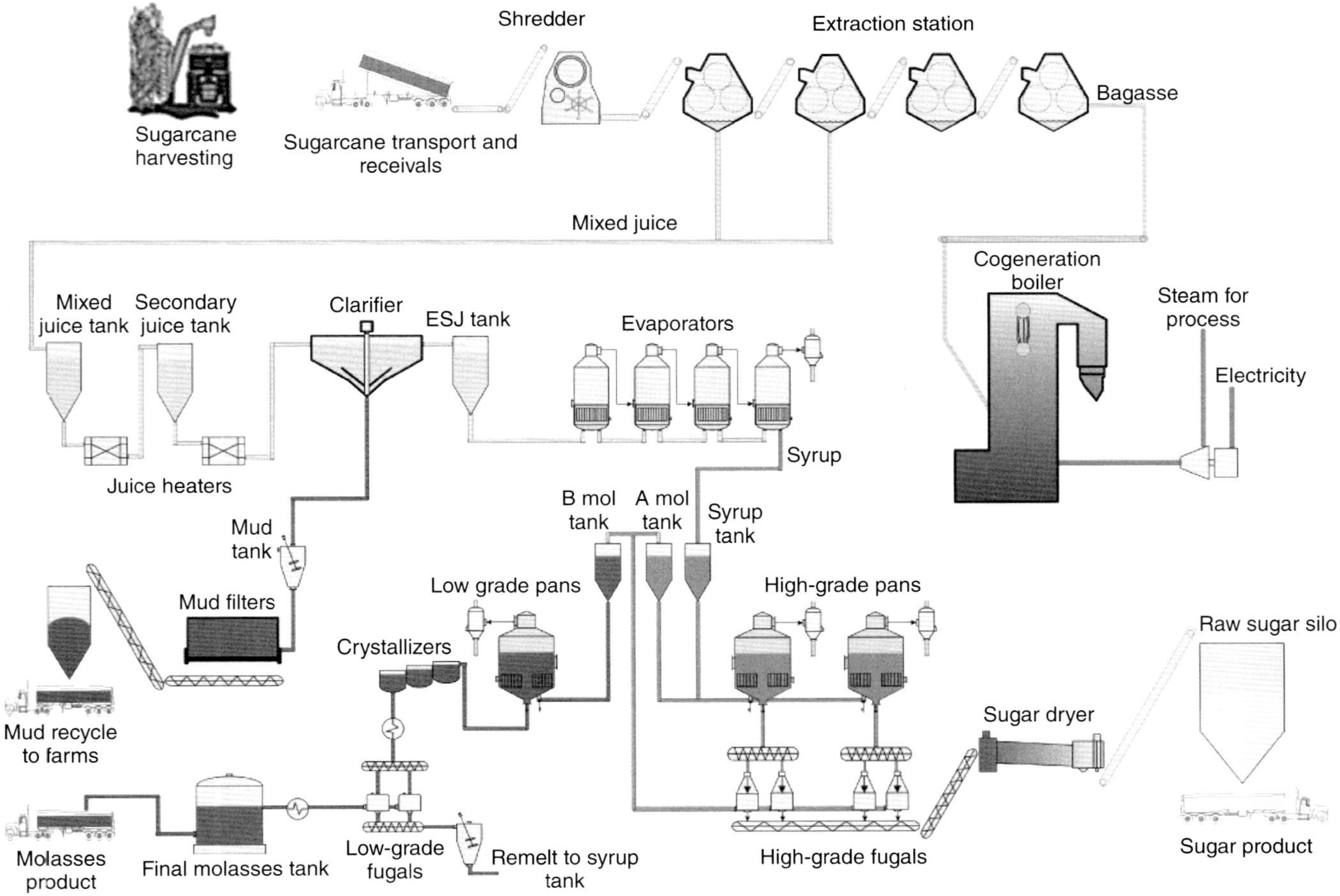

Figure 1.2 Typical schematic of the raw sugar production process.

Low-grade massecuite from the panstage is further processed to recover as much of the remaining sugar as possible from the molasses. This involves a process of cooling crystallization of the low-grade massecuite, followed by centrifugation to separate the recovered sugar from the final molasses. The quantity of final molasses produced depends on the quantity and types of impurities present in the sugarcane but is generally around 3–5% (w/w) of the sugarcane processed.

1.2.4 The refined sugar production process

The process for the conversion of raw sugar to refined sugar (Figure 1.3) is principally designed to achieve decolorization to a desired product specification. A series of processes are used to remove impurities while maximizing the yield of refined sugar. Several processing options exist and the number of decolorization stages required is determined by the purity and color of the initial sugar and the required color standard of the refined sugar product.

In the typical refined sugar process, raw sugar is initially processed through an affination station in which the raw sugar is mixed with affination centrifugal syrup (known as raw wash) and centrifuged to remove impurities contained in the highly colored molasses layer surrounding the sugar crystal. After the affination station, the affined sugar is remelted using water and steam to create melt liquor.

The melt liquor is processed through a primary decolorization stage using either a carbonatation process or a phosphatation clarification process. In carbonatation, the melt liquor is limed to a high pH, and carbon dioxide is bubbled through the liquor in a carbonatation column. The resultant calcium carbonate precipitate that is formed in this process removes impurities, and this precipitate is then filtered from the clarified liquor. In the phosphatation process, the melt liquor undergoes a clarification process with the addition of lime and phosphoric acid. In this case, the calcium phosphate complex adsorbs impurities, and the precipitate is skimmed off the surface of a flotation clarifier.

The clarified liquor then enters the second major decolorization process, and again there are several process options. These options include the use of activated carbon or ion-exchange resins to adsorb impurities from the clarified liquor. Both processes are highly effective at color removal from clarified liquor and the processes generate fine liquor suitable for crystallization.

The final stage of the refinery process is crystallization of the fine liquor to produce refined sugar massecuite, which is then centrifuged to separate the refined sugar crystals from the refined molasses. Several refined sugar strikes can be boiled and the number of product strikes is determined by the color specification of the product sugar. The refined sugar is dried and packaged for transport to retail and industrial customers.

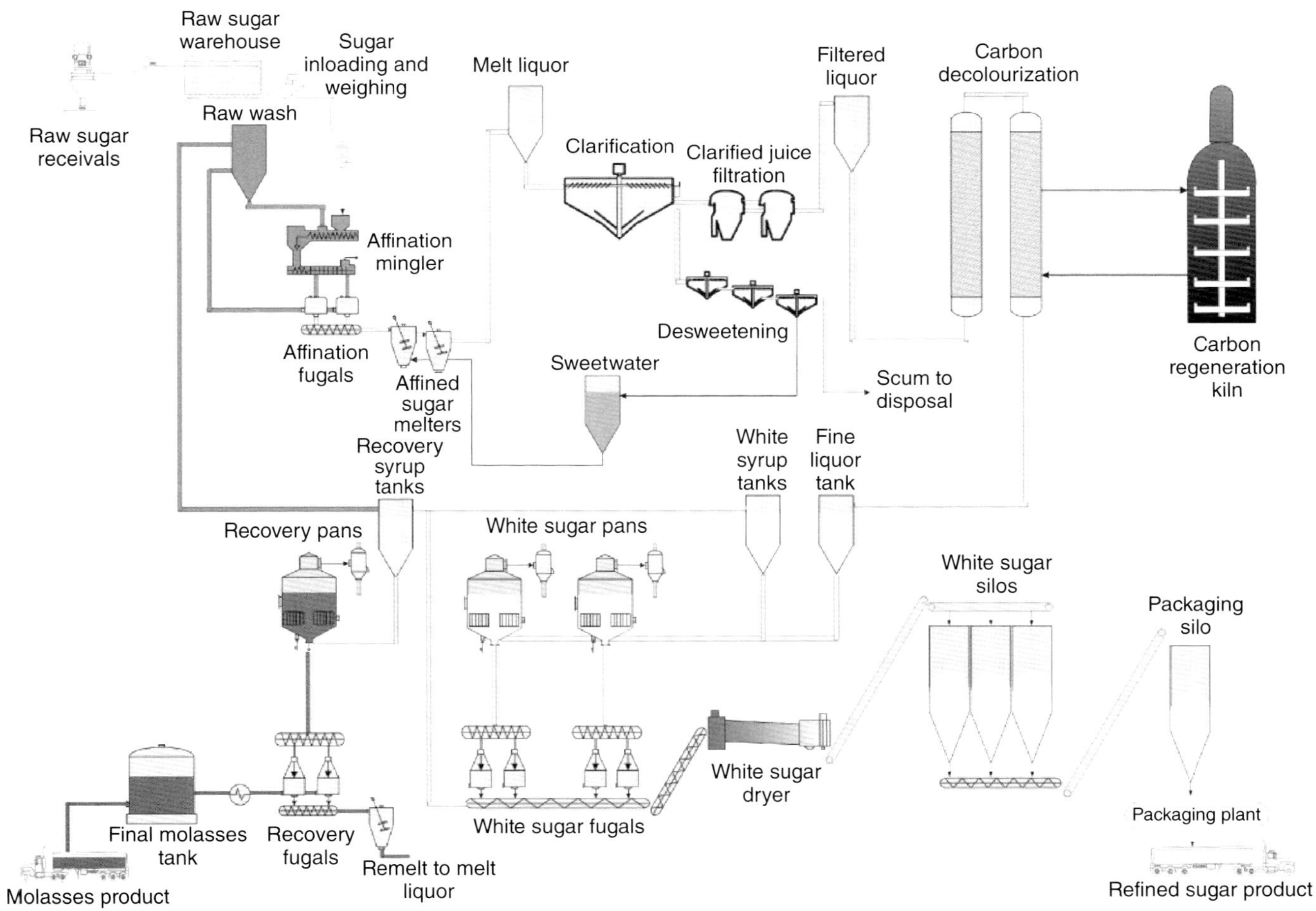

Figure 1.3 Schematic of a typical refined sugar production process showing phosphatation clarification and ion exchange resin decolorization processes.

1.2.5 The sugar market

While raw sugar physically flows from raw sugar manufacturers to refineries, the price of sugar is generally determined with reference to a futures price and a basis price. The futures market allows for price discovery in a transparent market and provides risk management tools for sugar suppliers and purchasers. The basis price accounts for variation in the sugar quality between producers and freight costs differentials between sugars of varying countries of origin.

Raw sugar futures and options on futures are traded globally through the Intercontinental Exchange (known as ICE Futures US), which also trades futures of other soft commodities including cocoa, frozen concentrated orange juice, and cotton. Internationally, raw sugar is traded with reference to the Sugar No. 11 contract (US c/lb), which is for the physical delivery of lots of 112,000 lb of raw cane sugar, free on board the receiver's vessel at a port within the country of origin (Intercontinental Exchange Inc 2012).

There is a separate futures contract (Sugar No. 16) for the physical delivery of cane sugar of the United States or duty-free origin into US destinations. This is the result of the high import tariffs into the United States, which create a distinct market for US destination sugar and typically trades 35–50% higher than the Sugar No. 11 price (Intercontinental Exchange Inc 2012).

White sugar futures and options on futures are traded through the NYSE Euronext London International Financial Futures Exchange (LIFFE) White Sugar Futures Contract. This contract (in US dollars per ton) is for the delivery of 50 tonnes of white or refined beet or cane crystal sugar with a minimum polarization of 99.8° and maximum color of 45 ICUMSA units at the time of delivery to the vessel in the port of origin (NYSE Euronext 2013).

The raw sugar (ICE Futures US Sugar No. 11) to white sugar (LIFFE White Sugar Futures) differential is typically between 2 and 4.5 US c/lb (Intercontinental Exchange Inc 2012).

In a highly volatile market, the raw and white sugar futures markets allow sugar producers and their customers to manage price and currency risks using sophisticated tools in a transparent market. For raw sugar producers, this ability to manage price risk is particularly important given the inherent production risks associated with weather, pests, and diseases experienced in agricultural systems. Despite these markets, however, many sugarcane-processing factories are highly exposed to the revenue generated from sugar. This has led producers to seek alternative revenue streams to produce a more diversified revenue base from sugarcane.

1.3 Why biofuels and bioproducts?

1.3.1 The search for new revenue

Sucrose accounts for about 40% of the dry matter produced by the sugarcane plant but for conventional sugarcane factories producing raw sugar as the primary product, raw sugar revenue accounts for more than 95% of the total revenue. Profitability in these factories is directly linked to the prevailing price of sugar on the volatile global market and the ability of the factory to limit production costs. For this reason, there is a strong interest among the global sugar community to diversify the revenue streams from sugarcane.

The process of revenue diversification seeks to create additional revenue streams such that there are multiple revenue streams contributing in a substantive way to the overall profitability of the facility. Ideally, at least some of these additional revenue streams have price profiles that are countercyclical to sugar. In this way, a downturn in the market price of one product has a lower impact on profitability resulting in less volatile revenue base. This can directly impact the investment attractiveness for current and potential shareholders, a more stable sugarcane price for suppliers, and better access to debt and equity markets at a lower price.

1.3.2 Sugar, ethanol, and cogeneration

The most common diversification strategies for sugarcane industries globally are for the coproduction of ethanol and large-scale cogeneration.

In diversifying into ethanol production, a portion of the sugarcane juice or molasses is directed to a distillery producing ethanol from the sugars contained in that material. For the utilization of sugarcane juice, A molasses or B molasses for the production of ethanol, there is a decrease in crystal sugar production and hence sugar revenue. The utilization of the C or final molasses for ethanol production does not come at the expense of crystal sugar production but much smaller ethanol production quantities can be achieved.

In sugarcane factories, bagasse is burnt to produce heat and power for the process. There is, however, much more energy in bagasse than is required for the process and, historically, sugarcane factories and combustion equipment were designed to be energy inefficient to ensure complete disposal of the bagasse, which had little value for alternate uses.

Increasing electricity prices, carbon pricing mechanisms, and renewable energy incentive schemes in many countries have resulted in a greater focus on increasing the energy efficiency of the sugar production process and equipment to produce large amounts of surplus electricity. This electricity can be fed into local transmission or distribution networks to provide renewable electricity to the local community and local industries.

The electricity that can be produced from bagasse can be increased by the utilization of other supplementary fiber sources including sugarcane trash or other local fiber crops.

While the technology for producing electricity from bagasse via combustion in water tube boilers and steam-driven turbo-alternators is well established, the potential revenue able to be generated from electricity sales (even including green credits) is quite moderate. With the fiber proportion of sugarcane (including trash) being about two-thirds of the total above-ground component of the sugarcane crop (dry matter basis), there is significant interest in turning this high-volume, low-value resource into higher value products.

1.3.3 Fiber-based biofuels and bioproducts

Bagasse is an attractive feedstock for the production of fiber-based products. Bagasse has been used to commercially produce energy products (electricity via combustion or gasification), fuels, fiber products (paper and carton board), structural building materials, animal feed products, and chemicals such as furfural.

While the quality of many of these products is high, few of these products (other than electricity via combustion) are being produced in large quantities globally. One of the key challenges is for bagasse to compete with the best alternative feedstocks for the corresponding products, such as *Eucalypt* pulp for paper products and crude oil for industrial chemicals. Ensuring the availability of surplus bagasse in sufficient quantities for a world-scale chemicals or other manufacturing plant can also be a challenge and must be considered when entering competitive markets.

The rapid improvements in technology for the production of bioproducts is driving down the cost of production and decreasing the economically viable scale of production facilities. Further improvements in technology over the coming decade are expected to further enhance the opportunities for global sugar industries to add value to bagasse.

1.3.4 Climate change and renewable products

In 2006, the Stern Review on the Economics of Climate Change (Stern 2006) concluded that the scientific evidence on climate change is overwhelming, a serious and urgent issue and that the benefits of strong, early action considerably outweigh the costs of action. Independent reviews from many sources now recognize the majority scientific opinion that the climate is changing as a result of anthropogenic greenhouse gas emissions (Stern 2006, IPCC 2007, Garnaut 2008, The Royal Society 2008) and that the energy future we are creating is unsustainable (IEA 2006).

In general, these reports conclude that it is economically advantageous to undertake early action, and that deep cuts in carbon emissions in the first half of the 21st century are not only essential but achievable and affordable. It is generally recognized that there is no single solution for the challenges that climate change will bring through the 21st century and beyond, and that multiple strategies are required to both reduce carbon emissions and to adapt to the climate change effects that will inevitably occur.

The production of biofuels and bioproducts from renewable feedstocks such as sugarcane bagasse rather than equivalent products from nonrenewable fossil-based feedstocks is one path to reducing the intensity of emissions in modern human society. This provides a compelling incentive for increased government investment in research and development that aims to fast-track the commercial release of biobased products and their broad-scale manufacture. The success of the Brazilian sugarcane ethanol industry and the US corn ethanol industry are good examples of how government policy can drive rapid change in investment in biobased technologies and drive down the cost for new capital investment.

1.3.5 New industries for sustainable regional communities

Many countries are becoming increasingly concerned with ensuring the security of their future energy resources and seeking to ensure continued scope for a proportion of domestic production. Renewable energy technologies have the potential to play a significant role in enhancing energy security (IEA 2007) through diversifying energy sources.

In addition, domestic production of biofuels reduces (to some degree) exposure to the price volatility in international energy markets, stimulates rural development, creates jobs, and saves foreign exchange (Kojima and Johnson 2005).

As an agricultural industry, the sugarcane industry is regionally based and central to the economic viability of rural and regional communities. The industry provides employment, economic growth, development, and in many cases essential services to the local communities in which they exist. As sugarcane is a rapidly perishable product, sugarcane-processing infrastructure must be located close to the sugarcane-growing region which ensures the ongoing regional nature of the industry.

The conversion of bagasse into biofuels and bioproducts offers the opportunity to significantly increase the value from sugarcane supplementing the revenue from sugar. Bagasse to bioproducts converts the lowest value component of the crop, the fiber component, into revenue sources that in the future could be at least as valuable, or potentially more valuable, than sucrose.

The development of new biofuel and bioproduct industries throughout regional sugarcane growing areas will, therefore, enhance regional development, provide employment opportunities in construction and operational phases, and provide revenue that will flow back through the communities to retail, services, and support industries. This offers the opportunity to reinvigorate rural and regional communities based around low-carbon industries and enhance economic and social sustainability of these communities.

1.4 Sugarcane biorefinery perspectives

1.4.1 The sugarcane biorefinery

The production of multiple coproducts from sugarcane biomass in integrated processing facilities is known as biorefining, and these facilities can be considered sugarcane biorefineries. Several assessments of sugarcane biorefineries have been previously described (Godshall 2005, Pye 2005, Edye *et al.* 2006, Peterson 2006, Erickson 2007, Day *et al.* 2008).

Sugarcane bagasse is widely considered to be one of the best feedstocks for early-stage commercialization of biorefining technologies. Sugarcane bagasse has many key advantages as a biorefinery feedstock including the following (O'Hara *et al.* 2013):

1. Sugarcane is a highly efficient C4 photosynthetic crop producing high yields of biomass on an annual basis.
2. The sugarcane resource is massive and globally distributed.
3. Sugarcane is an established industrial crop with well-understood farming practices, pest and disease profiles, and well-established and sophisticated varietal development programs.
4. In terms of potential economic value, the biomass component of the crop (bagasse and trash) is vastly underutilized.
5. The major biomass residue from the crop (bagasse) is already at a centralized processing facility (the sugarcane factory).

As a result, sugarcane bagasse has a much lower feedstock risk profile and often a lower feedstock price than many other potential biorefinery feedstocks. The commercialization of any new biorefining technology is subject to significant technical and commercial risk, and the ability to reduce feedstock supply cost and risk is a key advantage of sugarcane bagasse as a biorefinery feedstock.

In centralized infrastructure, sugarcane factories process sugarcane into products. For this purpose, they require essential infrastructure including boilers, electrical generation and distribution equipment, cooling water, effluent treatment, maintenance, and other support services.

In biorefineries, sugarcane factories not only integrate sugarcane processing, sugar production, and renewable energy production, but in addition produce biotechnology products from biomass. Further to the emergence of sugarcane biorefineries is the opportunity for these facilities to be the catalyst for new regional renewable energy and biotechnology hubs attracting related industries and innovation enterprises able to make use of the central infrastructure, energy availability, and coproduct streams as inputs to their processes (Figure 1.4).

Most organic chemicals produced from fossil-based resources can also be produced from biomass (Bridgwater *et al.* 2010). Several studies have assessed the range of potential chemical products from biomass and more than 300 potential products have been identified (Werpy *et al.* 2004, Bridgwater *et al.* 2010).

Products that are able to be produced in biorefineries include alcohols (methanol, ethanol, and butanol), macromolecules, and other compounds derived from lignin, specialty sugars, organic acids, fermentation products, and energy products including biodiesel, hydrogen, gasoline, and diesel replacements (Table 1.1).

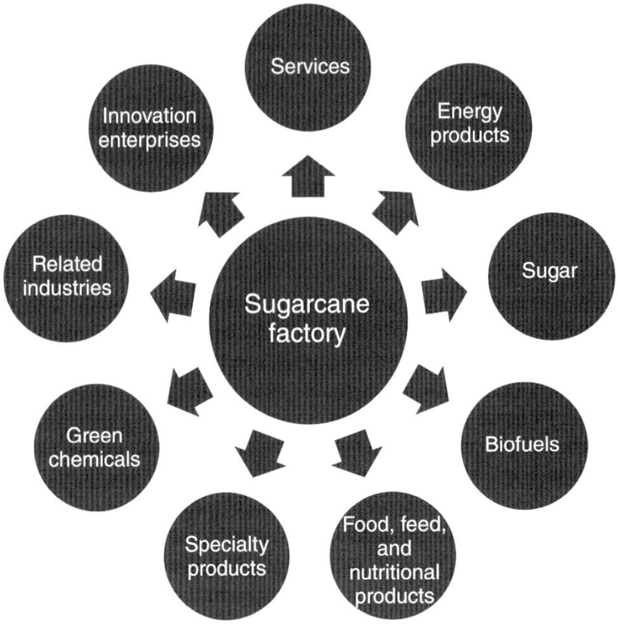

Figure 1.4 Conceptual model of a sugarcane biorefinery with the sugarcane factory as a hub for renewable energy and bioproduct technologies and services (O'Hara *et al.* 2013).

Table 1.1 Potential chemicals and bioproducts from biomass (O'Hara *et al.* 2013).

Products from biomass (Bridgwater *et al.* 2010)	**Chemicals from sugars (Werpy *et al.* 2004)**	**Chemicals from lignin (Holladay *et al.* 2007)**	**Chemicals from syngas (Spath and Dayton 2003)**
1,2-Propanediol	1,4-Succinic, fumaric and	Macromolecules	Hydrogen
Epichlorohydrin	malic acids	Carbon fiber	Ammonia
Lactic acid	2,5-Furan	Polymer modifiers	Methanol and derivatives
Diesel	dicarboxylic acid	Thermoset resins	*di*-methyl ether (DME)
Gasoline	3-Hydroxy propionic acid	Aromatic chemicals	Acetic acid
Kerosene	Aspartic acid	BTX (benzene, toluene,	Formaldehyde
Ethanol	Glucaric acid	xylene) derivatives	Methyl *tert*-butyl ether
Methanol	Glutamic acid	Phenol	(MTBE)
DME	Itaconic acid	Lignin monomers	Methanol to olefins
Char	Levulinic acid	Propylphenol	Methanol to gasoline
Wood pellets	3-Hydroxybutyrolactone	Eugenol	Ethanol
Animal feed	Alcohols (e.g., glycerol,	Syringol,	Mixed higher alcohols
1,3-Propanediol	sorbitol, xylitol/arabinitol)	Oxidized lignin monomers	Oxosynthesis products
Carbon dioxide		Syringaldehyde	(C_3–C_{15} aldehydes)
		Vanillin	Isosynthesis products
		Vanillic acid	(isobutene, isobutane)

1.4.2 The sustainability imperative

While there is significant consumer demand for renewable and sustainable products that demonstrate green credentials, consumers have generally shown an unwillingness to pay more for green products than their fossil-fuel-derived counterparts. It is critical, therefore, that biofuel and bioproduct technologies continue to develop to be cost-competitive with their fossil fuel equivalents.

However, it is critical as we move toward large-scale change from fossil-based products to biobased products that the industry demonstrates its advantage in environmental sustainability over alternative production systems.

The production of sugarcane-based biofuels and bioproducts has the potential to result in both positive and negative environmental outcomes. Indeed, these outcomes may vary based on the location or even the way the technology is implemented.

Sugarcane production requires the use of land, water, fertilizer, agricultural chemicals, fuels, and other inputs. The implications of land-use change, which can impact directly on forestation, biodiversity, food crop production, and competition for constrained resources, can have profound implications for regional and global communities. The challenges associated with measuring and assessing indirect land-use change are very complex but important.

Sugar production also has potential environmental impacts associated with emissions from fossil fuel combustion, chemicals utilization, and waste water treatment and discharge.

However, sugarcane also contributes to positive environmental outcomes through the production of electricity, bioproducts, and fuels from a renewable feedstock. The growth of sugarcane fixes carbon dioxide into plant biomass resulting in sugarcane being a contributor to the low-carbon manufacturing economy.

The assessment of the environmental credentials of production systems is undertaken through life cycle assessment (LCA). The license to operate for future production systems will require demonstration of their environmental credentials using these tools.

LCA considers the production system from cradle-to-grave within defined system boundaries. Many LCA techniques consider not just the environmental impacts but social impacts as well. Carbon footprint analysis is one of the critical components of LCA but many other factors are also identified as important in the development of global standards and assessment methodologies, such as ISO 14040:2006 and the Roundtable for Sustainable Biomaterials (RSB) standards.

Public debate throughout the past several years has also focused on the potential for bioproduct systems (in particular biofuels) to negatively impact on food production with particular implications for food prices on the poorest people in society. While this is a potential consequence of certain biofuels and bioproducts systems, the challenge for human society is to deliver both food and energy in an adequate, sustainable, and affordable manner. Modern human society is critically dependent upon both food and energy, and in fact the food and energy systems are inextricably linked with around 30% of total primary energy consumption in the "paddock-to-plate" food supply chain (FAO 2011).

In today's global society, the public perception of the economic, social, and environmental sustainability of biofuels and bioproducts from sugarcane will be influenced as much or more by international reputation than regional sugarcane production standards. It is critically important, therefore, that all sugarcane industries around the world contribute to continual improvement in sustainability of their domestic production systems to ensure their future license to operate and ensure that sugarcane continues to be considered by the international community as a highly desired feedstock for the production of biofuels and bioproducts.

1.4.3 Future developments in biotechnology for sugarcane biorefineries

Biotechnology is causing rapid changes in many areas of human endeavor including medicine, health, environmental remediation, agriculture, and manufacturing. This change is leading to significant increases in yield and productivity of agricultural crops and biotechnology processes and hence a reduction in the cost of bioproducts. A good example of this is the dramatic decrease in cellulase enzyme cost that has been reported over the last decade (Stephen *et al.* 2012).

Biotechnology offers significant opportunities for the future development of sugarcane biorefineries. These future developments will improve productivity and yields of sugarcane feedstocks and biorefinery products and further improve the sustainability outcomes. Most remarkable are the opportunities in agricultural biotechnology to improve sugarcane as a feedstock and industrial biotechnology to improve the biorefinery process.

While sugarcane is inherently a good feedstock for biorefineries, biotechnology offers the opportunity to improve agricultural yields with reduced crop inputs. Key opportunities in agricultural biotechnology to improve sugarcane as a biorefinery feedstock include

- more biomass through increased sugarcane yields per hectare;
- increased sucrose and total fermentable sugar contents of sugarcane;
- improved sugarcane resilience to abiotic and biotic stresses including drought, salinity, frost, pest, and disease;
- modified sugarcane fiber composition or morphology targeted at more efficient processing (e.g., lower lignin contents or higher cellulose contents); and
- more value embedded in the sugarcane such as through the *in planta* production of proteins, enzymes, specialty sugars, chemicals, or plastics.

The growing field of industrial biotechnology also offers opportunities to enhance value-creation from sugarcane processing through

- cost-effective processes for creating value-added products from sucrose and fermentable sugars;

- the production of value-added products from sugarcane by-products including bagasse, trash, molasses, vinasse, and filter mud;
- increased focus on clean technology production processes reducing energy requirements and environmental impacts from sugarcane processing; and
- enhanced processes for wastewater treatment.

While the sugar production process is considered by many to be technologically mature, biotechnology will play an important role in the next generation of sugarcane production and sugar-processing improvements. In particular, biotechnology improvements will be necessary to ensure that sugarcane remains amongst the lowest cost feedstocks for biorefinery processes and that the profitability of the production of biofuels and bioproducts from sugarcane carbohydrates can match and exceed that of their current fossil fuel equivalents.

1.5 Concluding remarks

Sugarcane is an important global agricultural crop that has made a major contribution to the development of communities and nations throughout tropical and subtropical regions of the world over the past few centuries. It remains the fifth largest crop (by production volume) and is a major contributor to gross national product in many tropical countries.

Sugarcane has been principally used for the production of crystal sugar, although increasingly ethanol and cogeneration are contributing to total sugarcane revenue.

The production of biofuels and bioproducts offers significant opportunities to enhance the revenue from sugarcane and contribute to more economically, environmentally, and socially sustainable sugarcane production around the world. The transition of sugarcane-processing factories into biorefineries coproducing food, feed, biofuels, and bioproducts in integrated facilities will be one of the most important changes to impact the future viability of the industry. These changes will generate new industries for regional communities in low emission manufacturing technologies.

Biotechnology is poised to bring significant new developments that will further position sugarcane as a leading feedstock for new biorefinery industries. However, the sugarcane industry needs to place sustainability at the core of its operations and continue to build and reinforce its social license to operate.

Indeed, the ongoing social license to operate requires the sugarcane industry globally to further improve its triple bottom line performance, and the production of biofuels and bioproducts can assist in furthering this aim. This will assist in ensuring a vibrant and sustainable future for sugarcane production globally and place sugarcane production as a major contributor to sustainable human societies over the next century.

References

Bajay, S. V., de Carvalho, E. B. and Ferreira, A. L. (2002). Energy from biomass in Brazil. In *Industrial Uses of Biomass Energy: the Example of Brazil*. F. Rosillo-Calle, S. V. Bajay and H. Rothman, eds. Taylor & Francis: London.

Barnes, A. C. (1964). *The Sugar Cane*. Leonard Hill/Interscience Publishers: London/New York.

Bridgwater, A. V., Chinthapalli, R. and Smith, P. W. (2010). Identification and market analysis of most promising added-value products to be co-produced with the fuels, Bioref-Integ, Deliverable 2 Total.

Bruhns, G., Riffer, R., van Bekkum, H., Schiweck, H., Heitz, F. and Mauch, W. (1998). Sugar. In *Sugar Technology: Beet and Cane Sugar Manufacture*. P. W. van der Poel, H. Schiweck and T. Schwartz, eds. Verlag Dr Albert Bartens KG: Berlin.

Cox, M., Hogarth, M. and Smith, G. (2000). Cane breeding and improvement. *Manual of Cane Growing*, pp. 91–108. M. Hogarth and P. Allsop. Bureau of Sugar Experiment Stations: Indooroopilly.

Day, D. F., Dequeiroz, G., Chung, C. H. and Kim, M. (2008). By-products from bagasse. *International Sugar Journal* **110**(1309):7–11.

Edye, L. A., Doherty, W. O. S., Blinco, J. A. and Bullock, G. E. (2006). The sugarcane biorefinery: energy crops and processes for the production of liquid fuels and renewable commodity chemicals. *International Sugar Journal* **108**(1285):19–20, 22–27.

Erickson, J. C. (2007). Overview of thermochemical biorefinery technologies. *International Sugar Journal* **109**(1299):163–173.

FAO (2011). *Energy-Smart Food for People and Climate: Issue Paper*. Food and Agriculture Organization of the United Nations: Rome, Italy.

FAO (2015). FAOSTAT database. Retrieved 09-7-2015, from http://faostat.fao.org/site/339/default.aspx.

Garnaut, R. (2008). *The Garnaut Climate Change Review: Final Report*. Garnaut Climate Change Review: Melbourne.

Godshall, M. A. (2005). Enhancing the agro-industrial value of the cellulosic residues of sugarcane. *International Sugar Journal* **107**(1273):53–60.

Holladay, J. E., Bozell, J. J., White, J. F. and Johnson, D. (2007). *Top Value Added Chemicals from Biomass: Results of Screening for Potential Candidates from Biorefinery Lignin*, vol. **II**. Pacific Northwest National Laboratory: Richland, WA.

IEA (International Energy Agency) (2006). *World Energy Outlook 2006*. OECD/IEA: Paris.

IEA (International Energy Agency) (2007). *Contributions of Renewables to Energy Security*. OECD/IEA: Paris.

Intercontinental Exchange Inc. (2012). ICE Futures US Sugar No. 11 and Sugar No. 16. Retrieved 11-9-2013, 2013, from https://www.theice.com/publicdocs/ICE_Sugar_Brochure.pdf.

IPCC (2007). Climate change 2007: synthesis report. Contribution of working groups I, II and III to the fourth assessment report of the Intergovernmental Panel on Climate Change. Intergovernmental Panel on Climate Change. Core Writing Team, R. K. Pachauri and A. Reisinger. Geneva, Switzerland, IPCC.

Kojima, M. and Johnson, T. (2005). *Potential for Biofuels in Transport in Developing Countries*. The International Bank for Reconstruction and Development/The World Bank: Washington, DC.

Martinelli, L. A. and Filoso, S. (2008). Expansion of sugarcane ethanol production in Brazil: environmental and social challenges. *Ecological Applications* **18**(4):885–898.

Martinelli, L. A., Nayloy, R., Vitousek, P. M. and Moutinho, P. (2010). Agriculture in Brazil: impacts, costs and opportunities for a sustainable future. *Current Opinion in Environmental Sustainability* **2**:431–438.

NYSE Euronext (2013). White sugar futures and options summary. Retrieved 11-9-2013, 2013, from https://globalderivatives.nyx.com/sites/globalderivatives.nyx.com/files/white_sugar_eng_130207.pdf.

O'Hara, I. M., Zhang, Z., Rackemann, D. W., Dunn, K. G., Hobson, P. A. and Doherty, W. O. S. (2013). *Prospects for the Development of Sugarcane Biorefineries*. International Society of Sugar Cane Technologists: Sao Paulo, Brazil.

Peterson, J. B. D. (2006). Ethanol production from agricultural residues. *International Sugar Journal* **108**(1287):177–180.

Pye, E. K. (2005). Biorefining; a major opportunity for the sugar cane industry. *International Sugar Journal* **107**(1276):222–253.

Rao, P. J. M. (1997). *Industrial Utilization of Sugarcane and Its Co-products*. ISPCK Publishers and Distributors: New Delhi.

Sparovek, G., Barretto, A., Berndes, G., Martins, S. and Maule, R. (2009). Environmental, land use and economic implications of Brazilian sugarcane expansion 1996–2006. *Mitigation and Adaptation Strategies for Global Change* **14**:285–298.

Spath, P. L. and Dayton, D. C. (2003). Preliminary screening - Technical and economic assessment of synthesis gas to fuels and chemicals with emphasis on the potential for biomass-derived syngas. Report prepared for U.S. Department of Energy, National Renewable Energy Laboratory. Report number NREL/TP-510-510-34929: http://www.nrel.gov/docs/fy04osti/34929.pdf.

Stephen, J. D., Mabee, W. E. and Saddler, J. N. (2012). Will second generation ethanol be able to compete with first-generation ethanol? Opportunities for cost reduction. *Biofuels, Bioproducts and Biorefining* **6**:159–176.

Stern, N. (2006). *The Economics of Climate Change: The Stern Review*. Cambridge University Press: Cambridge, UK.

Taupier, L. O. G. and Bugallo, S. R., eds. (2000). *Handbook of Sugarcane Derivatives*. Cuban Research Institute of Sugar Cane By-Products: Havana, Cuba.

The Royal Society (2008). *Sustainable Biofuels: Prospects and Challenges*. The Royal Society: London, UK.

Thorburn, P. J., Archer, A. A., Hobson, P. A., Higgins, A. J., Sandel, G. R., Prestwidge, D. B., Andrew, B., Antony, G., McDonald, L. M., Downs, P. and Juffs, R. (2006). Value chain analyses of whole crop harvesting to maximise co-generation. *Proceedings of the Australian Society of Sugar Cane Technologists* **28**:37–48.

Werpy, T., Petersen, G., Aden, A., Bozell, J., Holladay, J., White, J. and Manheim, A. (2004). *Top Value Added Chemicals from Biomass: Results of Screening for Potential Candidates from Sugars and Synthesis Gas*, vol. **1**. Pacific Northwest National Laboratory: Richland, WA.

CHAPTER 2

Sugarcane biotechnology: tapping unlimited potential

Sudipta S. Das Bhowmik, Anthony K. Brinin, Brett Williams and Sagadevan G. Mundree
Centre for Tropical Crops and Biocommodities, Queensland University of Technology (QUT), Brisbane, Australia

2.1 Introduction

Saccharum officinarum (sugarcane) is a monocotyledonous crop of the Poaceae family, which is cultivated in tropical and subtropical regions of the world primarily for its ability to store high concentrations of sucrose in the stem. Around 70% of worldwide raw table sugar production is obtained from sugarcane, with the remaining production coming from sugar beet (Contreras *et al.* 2009).

Sugarcane is a C4 grass that vigorously accumulates biomass and sugar under tropical and subtropical climatic conditions (Figure 2.1).

In 2013, worldwide sugarcane production was approximately 1.88 billion tons over an area of 26.0 million hectares (FAO (Food, Agriculture Organization of the United Nations) 2015). Brazil is the largest sugarcane producer, contributing 40% of world production (700 Mt in 2009), followed by India (285 Mt), China (114 Mt), Thailand (67 Mt), Pakistan (50 Mt), Colombia (38.5 Mt), Australia (31 Mt), Argentina (30 Mt), United States (27.5 Mt), Indonesia (26.5 Mt), and the Philippines (23 Mt) (http://faostat.fao.org).

Sugarcane is the world's largest biomass-producing crop. It is a semi-perennial cash crop that matures 12–18 months after planting. Initially, sugarcane "seed" is planted. The seed (set) consists of three to five nodes and internodes that have been taken from a mature plant. Depending on the variety, up to eight ratoon crops can be vegetatively grown from the original set before there is a need to replant due to reduced sucrose yield. More commonly, three to five ratoon crops can be grown from the initial set without a significant decrease in sucrose yield.

Almost every country around the world has realized the benefits of growing sugarcane. It is a rich source of food (sucrose, jaggery, molasses, and syrup), fibers (cellulose), fodder (green leaves and tops of sugarcane plants), fuels and chemicals (bagasse and alcohols), and fertilizers (Solomon 2011). Demand for sugarcane and sugarcane-derived products is set to increase as the world population increases and as new technology extends the diversity of sugarcane-derived

Sugarcane-Based Biofuels and Bioproducts, First Edition. Edited by Ian M. O'Hara and Sagadevan G. Mundree.

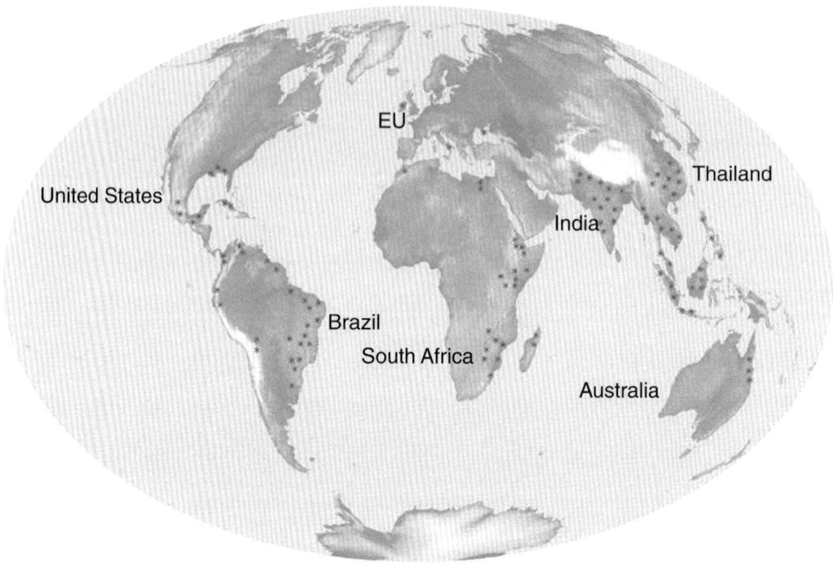

Figure 2.1 Global distribution of sugarcane. (Redrawn on Google map: www.earthobservatory.nasa.gov.)

products. Climate change offers additional challenges that the sugarcane industry needs to consider to maintain a sustainable industry. The sucrose yield from sugarcane has remained relatively unchanged for decades (Wu and Birch 2007). Therefore, the improvement of sugarcane through biotechnology is essential to improve the value and sustainability of the sugarcane industry.

This chapter describes the sugarcane crop, its history and genetics, conventional breeding versus GM (genetic modification) improvement, advancement through biotechnology and its limitations, GM approaches for improving biotic and abiotic stress tolerance, industrial applications of sugarcane biotechnology, and the effect of climate change on sugarcane.

2.2 History of sugarcane, sugarcane genetics, wild varieties

The genus name *Saccharum* is derived from the Sanskrit word "Shakkara" and is adduced as linguistic evidence of Indian origin on the basis of the meaning "a new crop from the east." Sugarcane is believed to have been cultivated in India over 5000 years ago (Daniels and Daniels 1975, Daniels and Roach 1987).

The genus *Saccharum* includes six species, namely *S. spontaneum*, *S. robustum*, *S. officinarum*, *S. barberi*, *S. sinense*, *and S. edule*. The two wild species, *S. spontaneum* and *S. robustum*, have the basic chromosome number $n=8$ and $n=10$, respectively (D'Hont *et al.* 1998, Ha *et al.* 1999), and are thought to be the founding species of this genus (Daniels and Roach 1987). India is the origin and centre of diversity of *S. spontaneum* L., which is considered to be the most primitive sugarcane species that originated in the cold regions of subtropical India, whereas *S. officinarum* and *S. robustum* originated in Papua New Guinea (Daniels and Roach 1987, Daniels *et al.* 1975). The remaining three species are hybrids. *S. sinense* is likely to have originated in China and *S. barberi* is likely to have originated in India (D'Hont *et al.* 2002, Daniels and Roach 1987). Sugarcane is an ancient crop plant with a long history of cultivation.

The commercial hybrid cultivar of sugarcane is achieved by interspecific hybridization between *S. officinarum* and *S. spontaneum*. The early hybridization of these two cultivars lead to the release of an elite cultivar in 1921 called POJ2878, commonly known as Java Wondercane (Jesweit 1929, Bull and Glasziou 1979). Repeated back crosses (BC) of this hybrid to *S. officinarum* resulted in cultivars with increased sugar content, improved ratooning ability and improved disease resistance (Cox *et al.* 2000, Lakshmanan *et al.* 2005). The modern cultivars developed from these hybrids are aneuploid, with chromosomes ranging from $2n = 100$ to $2n = 130$, of which 70–80% of the genome consists of *S. officinarum*, 10–20% *S. spontaneum*, and 10% recombinant chromosomes from the two species (D'Hont *et al.* 1996). Sugarcane exhibits the most complex genome of any hybridized crop with varying genome sizes and ploidy levels in hybrids and wild cultivars. The genome size of *S. officinarum* ($2n = 8x = 80$) is about 7.5 Gbp and that of *S. spontaneum* ($2n = 8x = 64$) is about 6.7 Gbp (Zhang *et al.* 2012). The genome size of more recent hybrid cultivar R570 is about 10 Gbp with ploidy level about 12x (D'Hont 2005).

2.3 Uses of sugarcane

Sugarcane contributes significantly to the economies of many countries where it serves as an important food and bioenergy crop. In addition to the use of sugarcane for food, almost every part of sugarcane can be utilized for producing a range of valuable products.

2.3.1 Food and beverages

Sugar, mainly sucrose (a disaccharide of glucose and fructose), is an important primary source of energy used in many foods and drinks. One of the best known by-products of sugarcane is molasses, which is known for its unusual flavor and sweetness. Molasses is the residual syrup from which no further crystalline sucrose can be obtained by simple means. A large number of food products can be derived from molasses such as jaggery (solidified molasses), rum, beer, dark rye bread, flavoring agent, citric acid, and animal feed (Troiani 2009). In addition to its use as food, sugarcane and its by-products (bagasse, filter mud, and molasses) are utilized in many other industrial applications.

2.3.2 Biofuels and bioenergy

Bioethanol is widely generated by the fermentation of sugarcane molasses to ethanol (Sreenivasan *et al.* 1987). Most of the ethanol is sold as either ethanol fuel or blended with gasoline (IBGE 2010). All gasoline sold in Brazil is a blend of 18–25% ethanol (Meyer *et al.* 2013). A significant number of sugarcane factories in Brazil and around the world coproduce sugar and bioethanol in integrated factories (Szwarc 2009).

Bioenergy (mainly electricity and biogas) is also generated from sugarcane bagasse. Bagasse is the fibrous residue of the sugarcane stalk, which remains after crushing and extraction of the juice. It consists of fiber, water, and relatively small quantities of soluble solids, mostly sugar (Murugan *et al.* 2013). Bagasse is burnt to generate the electricity that powers the sugar mill. Cogeneration of electricity through the burning of sugarcane bagasse can yield up to 130 kWh/t bagasse (Seabra and Macedo 2011). In addition, cellulose rich sugarcane bagasse may be digested by bacteria to produce methane, a principle product of biogas for domestic and industrial use.

2.3.3 Fibers and textiles

Bagasse is also used to produce building products such as particle board and is also a component of some good-quality wrapping and magazine paper (Murugan *et al.* 2013). Other products derived from sugarcane bagasse are furfural, which is a valuable platform chemical that is used as a selective solvent for the refining of lubricating oils, nylon 6.6, and resins used for molding powders (Murugan *et al.* 2013).

2.3.4 Value-added products

Sugarcane juice contains impurities that are removed by filtration. The filtrates form a cake of varying moisture content called filter mud. Filter mud contains around 1% w/w phosphate (P_2O_5) and 1% w/w nitrogen and is commonly used as a fertilizer (Paturau 1986). The filter mud also contains a mixture of waxy and fatty lipids which, after extraction and refinements, are used as natural additives for functional foods, medicines, and cosmetics (Chen *et al.* 2005).

2.4 Sugarcane biotechnology

There is increasing global demand for sucrose, fossil fuel alternatives, and green alternatives for fossil fuel–derived products such as plastics. Sugarcane is a candidate crop that can assist in meeting these demands. Sugarcane has a large global growing region, high sucrose content, and it is fast growing, producing massive volumes of biomass. These characteristics make sugarcane an attractive candidate for the application of biotechnology. Biotechnology may assist the crop to meet future world energy demands, thereby improving the value and sustainability of the crop. Improvement of sugar yield through selective breeding may have reached its potential; however, there still exists scope for the application of biotechnology to improve sugarcane by enhancing desirable traits and adding novel traits that may add value and improve the sustainability of this crop (Kinkema *et al.* 2014a).

Biotechnology research on sugarcane began in the 1960s with the discovery of differences between protein and metabolite expression in different sugarcane cultivars. These discoveries and the invention of *in vitro* cultivation techniques of sugarcane led to the notion that some of these proteins and metabolites could be manipulated *in vitro* to improve desirable traits. The application of *in vitro* techniques to sugarcane biotechnology began with research into micropropagation. Micropropagation is an *in vitro* method of cloning large numbers of plants from a small piece of explant. Pioneering biotechnology research demonstrated the large-scale production of sugarcane clones *in vitro* from dedifferentiated sugarcane cells (callus) of elite parent plants (Barba and Nickell 1969, Heinz and Mee 1969, Nickell 1964).

Research around *in vitro* micropropagation of sugarcane also led to the finding that micropropagation could be used to regenerate pathogen-free clones from infected parent plants (Kristini 2004, Leu 1978, Wagih *et al.* 1995) and for germplasm conservation (Reinert and Bajaj 1977). It was found that in contrast to conventional vegetative propagation, tissue culture reduced the time that it took to produce seed sugarcane and boosted the propagation potential of sugarcane by 20–35 times (Snyman *et al.* 2007, Geijskes *et al.* 2003).

Micropropagation in sugarcane has been achieved mainly through axillary shoot formation, adventitious shoot formation, and somatic embryogenesis (Tiwari *et al.* 2011). However, there are reports of variability that arises during tissue culture and that this variability is exacerbated by prolonged *in vitro* culture, culture conditions, and tissue selection. This somaclonal variation is likely due to genetic/epigenetic changes (Larkin and Scowcroft 1983, Silvarolla 1992). Numerous studies into somaclonal variation have demonstrated that axillary shoot formation poses lesser risk of somaclonal variation and is the safest explant to ensure the genetic stability of variety (Vasil 1987, Zucchi *et al.* 2002, Lal *et al.* 2008).

Genetic similarity is highly desirable for maintaining desirable traits in elite germplasm. Polymerase chain reaction (PCR) is a useful molecular technique that can be used to characterize genetic differences in micropropagated plants (Martins *et al.* 2004). Genetic similarity analyses between tissue culture raised sugarcane propagules and the mother plant have been successfully achieved through PCR-based techniques such as simple sequence repeats (SSR) markers (Tiwari *et al.* 2011, Pandey *et al.* 2012), random amplified polymorphic DNA (RAPD) (Saini *et al.* 2004, Devarumath *et al.* 2007, Lal *et al.* 2008), and inter-simple sequence repeats (ISSR) markers (Srivastava and Gupta 2008).

Although genetic similarity is highly desirable for maintaining the required traits in elite germplasm, some genetic variability is also essential to maintain crop health by avoiding disease and pest problems that can catastrophically destroy monoculture crops. Tissue culture-derived somaclonal variation can be useful for generating plants with desirable traits. Selective breeding of plants with desirable traits that are derived from somaclonal variation helps to speed up the breeding efficiency and improves the accessibility of existing and new varieties of germplasm for sugarcane crop improvement. Somaclonal variation arising from clonal or callogenic regeneration can further be enhanced through mutation breeding using radiation (Suprasanna 2010).

Unlike conventional mutation, inducing mutations in tissue-cultured plants is fast and manageable due to lower spatial requirements and the availability of small plantlets. Somaclonal variation in combination with *in vitro* mutagenesis has been successfully used to develop salt- and herbicide-tolerant lines in sugarcane (Patade *et al.* 2008, Koch *et al.* 2012).

Sugarcane biotechnology has evolved through numerous paradigm shifts. The first paradigm shift for sugarcane biotechnology came with the invention of methods to transform heterologous genes (transgenes) into the sugarcane genome. The second paradigm shift was the improvement of the expression of heterologous and autologous genes in sugarcane protoplasts, callus, young plants, and mature plants. The third paradigm shift was the improvement of desirable traits such as disease and pest resistance, sugar content, and drought resistance through the expression of autologous and heterologous genes. The fourth paradigm shift was the ability to improve the versatility, sustainability, and value of the crop by producing novel products in sugarcane such as bioplastics, enzymes, and pharmaceutical compounds. The current paradigm shift in sugarcane biotechnology is taking place *in silica*. The sequencing and annotation of the sugarcane genome will lead to a greater understanding of the morphology and physiology of sugarcane, which in the context of biotechnology will no doubt lead to further improvement of desirable traits and will assist in improving the value and sustainability of the crop.

Sugarcane has numerous properties that make it a promising candidate for the application of biotechnology. They can be listed as follows:

1. Sugarcane is grown in six of the seven continents. A large global growing-region and genetic variability means that the crop is unlikely to suffer significant loss of production on a global scale. In 2013, 1.9 billion tons of sugarcane was harvested globally (FAOstat 2015).
2. Sugarcane has been cultivated globally for hundreds of years. Therefore, a substantial body of research has addressed many knowledge gaps surrounding

sugarcane pests and diseases. The knowledge gained from this research has been implemented to develop strategies that assist in controlling pests and diseases that attack sugarcane (Lakshmanan *et al.* 2005, Rott 2000).

3. Sugarcane has high sucrose content, which means that it has a pool of metabolites and precursor chemicals that can be readily used for the improvement of traits or for production of novel compounds in the crop (Naik *et al.* 2010).
4. The life cycle of sugarcane promotes reduced production costs compared with other crops. Sugarcane is only replanted every three to five seasons because the vegetatively propagated ratoon crops continue to yield high sucrose concentrations (James 2004).
5. Sugarcane produces massive volumes of biomass that can be used to generate energy or that can be used as a feedstock for second generation bioethanol (O'Hara 2011, O'Hara *et al.* 2010, 2013).
6. Sugarcane grown and harvested under normal regimes is a sterile crop, which is vegetatively propagated (James 2004). Crop sterility limits the potential of gene flow from GM crops to related or nearby species (Cheavegatti-Gianotto *et al.* 2011).
7. The yield of sugar extracted from sugarcane decreases with increased time between harvest and processing (O'Hara *et al.* 2013). Therefore, sugarcane processing mills are usually located in proximity to where the sugarcane is grown.
8. Bioethanol produced from sugarcane contributes significantly less carbon dioxide emissions than fossil fuels when burned (Goldemberg *et al.* 2008, Naik *et al.* 2010).
9. Sugarcane can be sustainably cultivated for the production of second-generation biofuels, platform chemicals, and novel products (Goldemberg *et al.* 2008).

2.4.1 Limitations of sugarcane biotechnology

1. Public acceptance and regulatory hurdles can delay and halt the development of GM crops. Socioeconomic issues, environmental effects, and food safety are the major concerns surrounding GM crops (James 2014). GM crops are currently grown on more than 10% of the world's arable land (181 million hectares of 1.5 billion hectares); however, public acceptance of GM food is divided, with the majority of people in European countries taking exception to GM foods (Bonsch *et al.* 2015, James 2014).

2.5 Improvement of sugarcane – breeding versus genetic modification through biotechnology

To capitalize on the increasing global demand for sugar and sugarcane-derived products, thereby optimizing the value and sustainability of the sugarcane industry, it is essential to continuously improve the sugarcane crop. Further development of high sucrose and biomass yielding sugarcane is fundamental for improving the value and sustainability of the sugarcane industry. Meeting the increasing demand for sugarcane-derived products and improving the value of the sugarcane industry depends largely upon the uninterrupted bulk supply of quality sugarcane into the future. Intractable forces that can negatively impact sugarcane production fall into two categories: (i) abiotic stresses, such as water deficit or excess, temperature variability, soil properties, and the amount of solar radiation (Lakshmanan and Robinson 2014) and (ii) biotic stresses of sugarcane, which usually result from fungal, bacterial, viral, or phytoplasma infection (Sengar *et al.* 2011). In addition to these prominent influences, the unknown impact of climate change may also present challenges to the growth and development of sugarcane in the future (De Souza *et al.* 2008).

Although conventional hybridization breeding programs have been used to develop new hybrid varieties yielding high biomass and sugar, there are several limitations to the conventional breeding of sugarcane (Suprasanna 2010). Efficient, conventional, selective breeding of sugarcane is hampered by its narrow gene pool, complex genome, poor fertility, and a long breeding and selection cycle (Manickavasagam *et al.* 2004). Unlike some other plants, hybridization and selection of superior germplasm is very tedious and needs skilled labor over a prolonged time. Because of the complex physiology of sugarcane it can take 10–15 years to develop and release an elite sugarcane variety. Due to slow multiplication and limited availability of seed sugarcane of a new variety at the time of its release, it may take a further 8–10 years to cover an area sufficient for commercial cultivation (Sengar *et al.* 2011). In some instances, biotic and abiotic stresses can further delay the growth of the sugarcane or destroy the seed sugarcane that has been generated for commercial release.

2.6 Genetic modification of sugarcane

The greatest leap forward in sugarcane biotechnology came with the ability to transform foreign genes (transgenes) into sugarcane cells and regenerate sugarcane plants that harbored the metabolically active transgene product. This technology can significantly improve plant characteristics rapidly, and it can reduce the breeding time when compared with conventional breeding techniques (Scortecci *et al.* 2012). Furthermore, in contrast to conventional breeding, gene transfer techniques allow the introduction of desired genes from any origin without species boundary (Qaim 2009).

Numerous methods were initially applied to transfer foreign DNA into sugarcane cells. *Agrobacterium tumefaciens*–mediated transformation (Agro-transformation), electroporation, and polyethylene glycol (PEG) transformations were successfully used to transform DNA into protoplasts and callus. However, the transformation efficiency was unacceptably low and the regeneration of plants from the transformed cells proved too difficult (Chen *et al.* 1987, Franks and Birch 1991). Franks and Birch (1991) successfully adapted a biolistic approach for introducing foreign DNA into sugarcane cells, a technique that had been proven to be successful in other plant species (Franks and Birch 1991, Klein *et al.* 1988). Through the use of biolistics, sugarcane callus and sugarcane cell suspensions were able to be transformed more efficiently (Franks and Birch 1991).

The biolistic approach yielded, on average, one stably transformed transgenic plant per bombardment (Bower and Birch 1992). In addition to generating stably transformed sugarcane plants, biolistics has been used to assess the functionality of genes and gene expression elements in transient assays in sugarcane leaf and callus (Bower and Birch 1992, Franks and Birch 1991, Gallo-Meagher and Irvine 1993). Through biolistics, it was possible to produce relatively large numbers of stably transformed sugarcane plants and to quickly screen genes and gene expression elements for their utility in sugarcane.

Trials on sugarcane transformation using *Agrobacterium* yielded poor results initially; however, the method was eventually optimized (Arencibia *et al.* 1998, Dong *et al.* 2014, Enríquez-Obregón *et al.* 1998). An important breakthrough in sugarcane biotechnology was the development of somatic embryogenesis. Somatic embryos arise from single cells and, therefore, reduce the formation of chimeras. These embryos are, therefore, suitable material for high-efficiency genetic transformation and large-scale clonal propagation in sugarcane (Kamle *et al.* 2011, Ravindra *et al.* 2011, Snyman *et al.* 2006, Lakshmanan *et al.* 2006). Recent research suggests that both biolistics and Agro-transformation in conjunction with somatic embryogenesis can be used to efficiently generate large numbers of low transgene copy number, stably transformed sugarcane plants (Joyce *et al.* 2014).

Using a GM approach to improve the value of sugarcane is showing great promise. GM has been used to improve sugarcane herbicide resistance, resistance to diseases and pests (Falco *et al.* 2000, Joshi 2009, Setamou *et al.* 2002, Weng *et al.* 2011), drought tolerance (Zingaretti *et al.* 2014), and sugar content (Ma *et al.* 2000, Mudge *et al.* 2013, Wu and Birch 2007).

Sugarcane has shown great potential for its use as a biofactory for the production of novel products. The production of bioplastics and biopharmaceuticals has been demonstrated in sugarcane (Brumbley *et al.* 2003, Brumbley *et al.* 2007, McQualter *et al.* 2005, Petrasovits *et al.* 2013, Wang *et al.* 2005). Harnessing the potential of using sugarcane as a biofactory may assist in improving the value and sustainability of the sugarcane industry.

Another valuable GM approach that may be used to improve sugarcane is RNA interference (RNAi). RNAi has been successfully used to control pests, increase native gene expression, decrease native gene expression, inhibit native gene expression, and to decrease transgene expression in plants (Gan *et al.* 2010,

Joyce *et al.* 1998, Jung *et al.* 2012, Viswanathan *et al.* 2014, Zhuo *et al.* 2014). The mechanism of RNAi relies on the plant perceiving double-stranded RNA as foreign RNA. The foreign double-stranded RNA is cleaved into smaller pieces that are used for future sequence recognition or they are degraded (Baulcombe 2004). Current research into RNAi technology in sugarcane is focused on the downregulation of enzymes involved in the lignin biosynthesis pathway and also in pathogen resistance (Jung *et al.* 2012). Downregulation of these enzymatic *in planta* would assist in making sugarcane bagasse more amenable to enzyme hydrolysis, which would, therefore, decrease the cost of producing second-generation biofuel.

2.7 Paucity of high-quality promoters

A major opportunity for sugarcane biotechnology is to achieve reliable, high-level constitutive, tissue-specific, and inducible transgene expression in sugarcane (Lakshmanan *et al.* 2005). Historically, the gold standard for constitutive, high-level transgene expression in sugarcane has been the maize ubiquitin 1 promoter (*Zm-Ubi1*) (Kinkema *et al.* 2014b). Recently, additional promoters that have demonstrated potential, or have demonstrated constitutive, tissue-specific, or inducible transgene expression in sugarcane have been identified (Table 2.1).

Although there has been a substantial research effort to identify a number of constitutive and tissue-specific promoters in sugarcane, the number of promoters available is still relatively few compared with other plant systems that are regularly transformed (Kinkema *et al.* 2014b). In addition to research on gene promoters in sugarcane, molecular techniques exploiting enhancers of gene expression (Kinkema *et al.* 2014b, Rathus *et al.* 1993), codon optimization of transgenes (Kinkema *et al.*, 2014b), transgene silencing suppression (Gao *et al.* 2013), and RNAi interference (Gan *et al.* 2010, Gao *et al.* 2013, Jung *et al.* 2012) are additional valuable tools that may assist in improving transgene expression in sugarcane.

2.8 Opportunities for GM-improved sugarcane

Linkage maps are valuable tools that help to elucidate genetic and genomic organization, particularly in sugarcane which is polyaneuploid. However, high inbreeding depression caused by endogamy limits the production of experimental mapping populations as noted in F2, BC, recombinant inbred lines (RILs), and double haploid (DH) lines (Palhares *et al.* 2012). The *S. spontaneum* "SES 208" ($2n=64$) linkage map was the first map constructed directly from a complex polyploid species based on single-dose markers (SDMs), which

Table 2.1 Potential promoters identified for constitutive, tissue-specific, or inducible transgene expression in sugarcane.

Gene	Origin	Specificity	Levels compared with ZmUbi1	Author
ZmPepC	Maize	Leaf	Up to 10-fold	Harrison *et al.* (2011), Matsuoka *et al.* (1994)
UQ67P	Sugarcane	Stem	No data	Hansom *et al.* (1999)
RUBQ	Rice	Constitutive	+1.6-fold	Liu *et al.* (2003)
Dirigent	Sugarcane	Stem	+34-fold	Damaj *et al.* (2010)
SHOMT	Sugarcane	Stem	+8.4-fold	Damaj *et al.* (2010)
ScLSG5	Sugarcane	Stem	−7-fold	Moyle and Birch (2013)
BSV	Sugarcane bacilliform badnavirus	Banana roots and leaves. Not tested in sugarcane	Equivalent in banana	Schenk *et al.* (1999)
pRD29a	*Arabidopsis thaliana*	Tested in callus only	No data	Wu *et al.* (2008)
Rice Actin 1	Rice	Leaves and roots in rice. No specificity shown in sugarcane	No data in sugarcane	Molinari *et al.* (2007), Su *et al.* (1998)
PScMYBAS1	Sugarcane	Stress-inducible tested transiently in tobacco	No data	Prabu and Prasad (2012)
AtCOR15a	*Arabidopsis thaliana*	Cold stress inducible	No data	Belintani *et al.* (2012)
Scr1MYB1	Sugarcane	Stem and roots	−30-fold	Mudge *et al.* (2013)
PRol	Sugarcane	Constitutive	−25-fold	Mudge *et al.* (2013)
ScCIPK-21	Sugarcane	Stem and roots	−50-fold	Mudge *et al.* (2013)
Improved *alc* gene switch	*Aspergillus nidulans*	Inducible constitutive expression	Equivalent	Kinkema *et al.* (2014a, b)
Enhanced *Zm-Ubi1*	Maize	Constitutive	+5-fold	Kinkema *et al.* (2014a, b)
Enhanced *ZmPepC*	Maize	Constitutive	+4.5-fold	Kinkema *et al.* (2014a, b)

considers the use of simplex (single allele copy from one parent) markers to obtain the genetic map (Wu *et al.* 1992, Al-Janabi *et al.* 1993). More recently, molecular genetic maps, although incomplete, have also been constructed from 10 segregating populations (Zhang *et al.* 2014). Follow-up mapping of quantitative trait loci (QTLs) has been constructed on several populations for sugar content, sugar yield, disease resistance, and other agronomic traits (Zhang *et al.* 2014). In addition, a large collection of expressed sequence tags (ESTs) has been generated and used for mining single-nucleotide polymorphisms (SNPs), gene expression profiling, and gene discovery (Zhang *et al.* 2014). Although these techniques have been useful, an annotated map of the sugarcane genome is not yet available.

Current efforts in genetic mapping of the sugarcane genome involve the use of the sorghum (*Sorghum bicolor* (L.) Moench) genome as a template for mapping and assembly of the sugarcane genomic sequences (Wang *et al.* 2010). The diploid sorghum genome has been mapped and the synteny between the sugarcane genome and the sorghum genome has been validated (D'Hont 2015, Wang *et al.* 2010). Further mapping of the sugarcane genome will assist to improve the opportunities for identifying novel genetic elements that may be manipulated and exploited to improve the crop.

Conventional breeding of sugarcane has primarily focused on achieving high sucrose content (maximum of 20%). Further maximization of sucrose content could potentially be achieved by identifying genes associated with sugar accumulation and transferring the genes into sugarcane or manipulating the expression of native genes using a transgenic approach. In addition to an overall increase in carbohydrate yield (sucrose or cellulose), other traits are important to consider in addressing the major challenges in sugarcane production systems. Potential targets for the application of biotechnology are improved fitness against biotic (pests, diseases, weed competition) and abiotic (drought, salinity, cold, aluminum toxicity, nutrient poor and compacted soils) stress, flowering, plant vigor (fast growth under limiting conditions), and plant architecture including height, stalk number, tillering, leaf angle, root length, and biomass (Carvalho 2010, Creste *et al.* 2010).

Characterization of genes associated with various traits in any crop is fundamental to biotechnology. Sugarcane genes associated with, or conferring salt, drought, and oxidative stress tolerance have been identified from sugarcane genome databases, and their functionality has been established in heterologous systems (Begcy *et al.* 2012).

Numerous studies report on the improvement of drought tolerance in sugarcane through transgenesis with differing rates of success (Wang *et al.* 2005, Zhang *et al.* 2006, Wu *et al.* 2008, Yang *et al.* 2010). The first approval to commercially grow a GM sugarcane crop was given to Indonesia in 2013 (ISAAA, 2015). The approved GM sugarcane crop was transformed with the *EcBetA* gene from *Escherichia coli*. The product of the *EcBetA* gene is the enzyme choline dehydrogenase, which catalyzes the production of the osmoprotectant "glycine betaine" and confers tolerance to drought stress. This crop was scheduled to be planted in 2014 and has the potential to be a valuable contribution to the field of sugarcane biotechnology.

2.9 Improved stress tolerance and disease resistance

2.9.1 Stress tolerance

A stress is defined as an environmental factor that restricts normal function and development to the degree that it may even kill the plant. Sugarcane productivity is greatly affected by biotic and abiotic stress. Globally, sugarcane production is challenged by abiotic stress, mainly water deficit, salinity, and drought. Furthermore, biotic stresses such as devastating fungi such as mildew, eye spot, smut, rust, and red rot also contribute to significant crop or yield losses.

2.9.2 Drought

On semi-arid continents such as Australia, approximately 15% of sugarcane production is lost annually due to water deficit (Inman-Bamber *et al.* 2012). Increasing the water use efficiency (WUE) of sugarcane by modulating key regulatory factors through a transgenic approach may assist in maintaining or improving a sustainable sugarcane industry (Inman-Bamber *et al.* 2012). Through the use of biotechnology, it may be possible to engineer plants with altered root physiology or architecture as well as improved hydraulic efficiency in the aerial organs of the plants. These improvements may assist the plants to withstand extended periods of water deficit (Lakshmanan and Robinson 2014).

2.9.3 Salinity

Globally, about 22% of agricultural land is saline (FAO (Food, Agriculture Organization of the United Nations) 2004). Being a glycophyte, the growth and sucrose content of sugarcane is severely affected by drought and salinity stresses. Salt stress severely impedes the photosynthetic activity of sugarcane, which in turn results in a decrease in sucrose concentration and quality. Biotechnology could offer a plausible and rapid solution to mitigate these losses. Recent research has identified the *Arabidopsis AVP1* gene that encodes the vacuolar pyrophosphatase protein (AVP1). AVP1 confers resistance against high concentrations of NaCl and also against water deprivation. Overexpression of AVP1 improves vegetative development of roots, which in turn provides increased water absorption and water retention capacity (Kumar *et al.* 2013). Similar transgenic approaches with other membrane transporters such as the Na^+/H^+ antiporter gene and osmoprotectants such as proline and glycine betaine appear to be promising techniques for conferring resistance to salinity. These approaches may allow the cultivation of sugarcane using brackish water.

In the future, expansion of the sugarcane growing regions will be essential to fulfill the increasing demand for sugar and fiber. Therefore, growing sugarcane in cold areas would be beneficial. Significant progress has been made toward improving cold tolerance and adaptation to freezing in sugarcane germplasm, at the Subtropical Sugarcane Breeding Program of the US Department of Agriculture (Lakshmanan and Robinson 2014).

To optimize the potential of a sustainable crop yield, it is necessary to design and develop better crop varieties that can tolerate the adverse effects of constantly changing environmental factors. One of the most useful approaches to develop more resilient and higher sucrose yielding sugarcane varieties is by identifying novel stress tolerance–associated genes from a range of related resilient plant species. The recent advent of next generation sequencing revealed approximately 1670 genes that are differentially expressed in sugarcane plants under water deficit conditions (Rodrigues *et al.* 2011). The identification of these differentially expressed genes may provide targets for manipulating the transcriptome of sugarcane to improve desirable traits.

2.10 Naturally resilient plants as a novel genetic source for stress tolerance

A small group of plants including *Tripogon lolliformis, Sporobolus stapfianus, Xerophyta viscosa,* and so on are termed "resurrection plants." Resurrection plants are naturally resilient for various stresses and can withstand prolonged adverse conditions. These plants wither but can withstand dehydration to water contents of <5% while retaining the ability to resurrect to full metabolic activity within 48–72 h of watering (Gaff and Latz 1978, Gaff *et al.* 2009, Peters *et al.* 2007). Resurrection plants are primed for desiccation tolerance even in the hydrated state. Furthermore, studies have indicated that resurrection plants express genes in vegetative tissues that are normally found only in the seeds of the majority of the land plants. It has been suggested that the expression of these genes in resurrection plants confers a seed-like dormant state when the plants are desiccated (Farrant *et al.* 2007). Since stress tolerance is genetically encoded, naturally resilient species such as resurrection plants may serve as genetic resources for the improvement of crop tolerance.

Sugarcane belongs to the Poaceae family and, therefore, has a close physiological and genetic similarity to the drought-tolerant sorghum. Similarly, the wild sugarcane species *S. spontaneum* is naturally resilient and can grow in drought-affected areas by producing deep, highly branched rope roots (Moore 1987). Therefore, these groups of drought-tolerant plants probably represent a novel source of abiotic stress tolerance genes and could provide opportunities to develop resilient sugarcane varieties to overcome abiotic and biotic stresses in future. Transcriptome analysis through next generation sequencing technology of expressed genes under different stress conditions may identify these novel genes. Functional characterization and overexpression of these genes in sugarcane through transgenesis may lead to the development of stress-tolerant sugarcane crops.

2.11 Disease resistance

Sugarcane is adversely affected by a range of pests and pathogens causing more than 200 types of diseases that contribute to significant loss in yield globally (Chakraborty *et al.* 1998, Gonclaves *et al.* 2012). These pathogens can be viruses such as sugarcane mosaic virus (SCMV), Fiji disease virus (FMV), sorghum mosaic potyvirus (SrMV), sugarcane streak virus (SSV), and sugarcane yellow leaf virus (SCYLV) (Lakshmanan *et al.* 2005); bacteria that cause gumming disease, leaf scald, mottled stripe, and red stripe (Cottrell-Dormer 1932, 1936); fungi such as red rot and root rot (Ricaud *et al.* 2012); phytoplasmas that cause sugarcane grassy shoot and yellow leaf syndrome (Cronje *et al.* 1998, Viswanathan 1997); insects, which are mainly stem borers; invertebrates (cane grubs); or vertebrates (rats). Furthermore, there are many diseases with unknown etiology (Rott 2000), most recently, yellow canopy syndrome in Australia (Botha 2015).

Viral pathogens of greatest concern in Australia include SCMV, SrMV, SSV, and SCYLV (Lakshmanan *et al.* 2005), whereas Brazil is affected only by SCMV and SCYLV (Gonclaves *et al.* 2012). Indian sugarcane crops, in addition to SCMV and SrMV, are also affected by maize dwarf mosaic virus, Johnson grass mosaic virus, Zea mosaic virus, cocksfoot streak virus, and Pennisetum mosaic virus (Shukla *et al.* 1989, Mckern *et al.* 1991, Seifers *et al.* 2000, Gotz and Maiss 2002, Fan *et al.* 2003).

Pathogen-derived resistance (PDR) (Sanford and Johnston 1985) is a technique that has been successfully used to impart resistance to some plant viral pathogens. PDR may be conferred through the expression of RNA, DNA, or proteins (Baulcombe 1996). PDR functions through the expression of viral wild-type or mutated nonfunctional transgenic RNA or DNA in the plant. The transgenic RNA or DNA acts as a competitive decoy to the viral native RNA or DNA, interfering with the native viral proliferation or being incorporated into the viral proliferative process as a nonfunctional component thus rendering the virus ineffective (Baulcombe 1996). An RNAi-PDR strategy has been shown to be successful in inhibiting SCMV infection (Gan *et al.* 2010).

PDR was first successfully demonstrated in tobacco to confer resistance to tobacco mosaic virus (TMV) (Abel *et al.* 1986) and has been successfully demonstrated in sugarcane to assist in controlling the viral pathogens, namely

FDV, SCMV, and SrMV (Ingelbrecht *et al.* 1999, Joyce *et al.* 1998, McQualter *et al.* 2004).

The nature of the mechanism of PDR and the improved ability to sequence the genome of pathogens suggests that the PDR approach will be a valuable tool for improving pathogen resistance in sugarcane in the future against a wider range of sugarcane pathogens (Lakshmanan *et al.* 2005).

Sugarcane does possess some natural resistance to microbial and fungal colonization; however, it is still susceptible to some bacterial and fungal diseases (Lakshmanan *et al.* 2005). Common fungal pathogens of sugarcane include *Colletotrichum falcatum* Went (red rot) and *Pythium graminicola* Subr (root rot), and it is considered to be a major research opportunity to develop a resistant variety to combat their infection.

Opportunity exists for research into the development of fungal resistance in sugarcane. Fungal pathogens can elicit the production of pathogenesis-related (PR) proteins from the plant that assist the plants to withstand the pathogen attack. One particular PR protein of interest is the tobacco osmotin protein (TBOSM). The overexpression of TBOSM has been shown to confer resistance to osmotic stress (Barthakur *et al.* 2001), salt tolerance, drought stress (Goel *et al.* 2010), and fungal development in plants (Li *et al.* 1998). TBOSM has been successfully expressed in solanaceous plants (Barthakur *et al.* 2001, Goel *et al.* 2010, Li *et al.* 1998, Sharma *et al.* 2015); however, there are few if any reports of transgenic expression of osmotin in monocotyledonous plants. The expression of TBOSM or an osmotin-like PR protein in sugarcane may assist to improve the fungal tolerance of the crop. The generation of *Xanthamonas albilineans*-resistant sugarcane has been successfully achieved by the transformation of sugarcane with a gene that inactivates *X. albilineans* pathogenesis factors (Zhang *et al.* 1999).

Insect damage accounts for approximately 10% of sugarcane crop loss worldwide (Weng *et al.* 2011). In addition to direct damage, insects can damage the crop indirectly acting as vectors for the transmission of other pathogens. Stem borers are the most important insect pest of sugarcane (Weng *et al.* 2011). A successful strategy for dealing with insect pathogens is engineering sugarcane plants with proteins that are toxic to insects. One example of this strategy is the transformation and expression of *cry* gene in the leaves and stem of the plant. *The cry* gene is derived from the bacterium *Bacillus thuringiensis* that encodes for a protein toxic to insects such as moth, butterflies, flies, and beetles upon ingestion (Weng *et al.* 2011). An additional example of similar strategies includes the *Nicotiana alata* proteinase inhibitor, which is active against sugarcane grubs (Atkinson *et al.* 1993). The expression of the *cry1* gene has been successfully used for the control of pests in cotton. However, due to the fact that the *cry1* gene encodes a toxin, it is unlikely that such a strategy could be reasonably implemented in a food crop.

Another promising approach to control insect infestation is through the engineering of plants with nontoxic α-amylase inhibitors. Plants including sugarcane produce α-amylase inhibitors as insect feeding deterrents. α-Amylase inhibitors disrupt α-amylases and proteinases in insect guts. Overexpression of α-amylase inhibitors has been shown to halt the development of pea weevil larvae (Schroeder *et al.* 1995). α-Amylase inhibitors are likely to inhibit human α-amylases; however, this is unlikely to occur in the food product sugar due to denaturation of α-amylase inhibitors as a result of substantial processing of the sugar. The overexpression of α-amylase inhibitors in sugarcane, therefore, may assist in controlling the infestation and destruction of sugarcane by insects.

2.12 Industrial application of sugarcane

Sugarcane has wide range of industrial applications; the first being its use for biofuels. First-generation bioethanol is produced from sucrose in sugarcane. Although sucrose is readily fermentable to ethanol, this can lead to a diversion of sucrose from use as a food and hence result in an impact on food price and availability. An alternative to first-generation bioethanol is second-generation biofuel production. The production of second-generation biofuels utilizes lignocellulosic plant material.

Sugarcane bagasse and sugarcane leaves/trash are lignocellulosic plant materials that are left over after sugarcane harvesting and crushing. Sugarcane bagasse, leaves, and trash are mostly made up of cellulose and hemicellulose. Cellulose and hemicellulose are long-chain sugars that are not easily fermentable and, therefore, methods employing high temperature, high pressure, strong acids, and enzymes are used to break down the long-chain sugars to more easily fermentable sugars such as glucose, xylose, arabinose, mannose, and galactose (Singh *et al.* 2008, Krishnan *et al.* 2010). Such sugar streams obtained from sugarcane bagasse and leaves can be converted into bioethanol and value-added products of commercial significance (Pandey *et al.* 2000, da Costa Sousa *et al.* 2009). Recent studies suggest that it is possible to engineer sugarcane to produce cellulases that are capable of hydrolyzing the bagasse to fermentable sugars (Harrison *et al.* 2014). Commercial production of sugarcane that produces *in vivo* cellulases would reduce the cost of generating second-generation bioethanol, and it would help to improve the value and sustainability of sugarcane.

Apart from ethanol, sugarcane is also used for producing D-xylitol, a five carbon sugar alcohol from microbial fermentation (Carvalho *et al.* 2008, Prakasham *et al.* 2009). D-Xylitol exhibits anticarcinogenicity and tooth rehardening and remineralization properties. D-Xylitol has been widely used in the pharmaceutical industry, odontological formulations, and in food industries (Prakasham *et al.* 2009).

Sugarcane bagasse has been also used for the production of industrial enzymes such as xylanase, cellulase, amylase, and laccase by certain bacteria and fungi, employing solid-state fermentation (SSF) or submerged fermentation (SmF) systems (Chandel *et al.* 2012). Bagasse has also been explored for use in the production of various organic acids such as citric acid, lactic acid, and gluconic acid using different cultivation techniques and with a variety of microorganisms such as *Actinobacillus succinogenes* (Borges and Pereira 2010, Chandel *et al.* 2012). Sugarcane bagasse has been employed to make other value-added products of commercial interest such as antibiotics, animal feed, biohydrogen, alkaloids, pigments, and alternative packing material for biofiltration of benzene-polluted gaseous streams (Pandey *et al.* 2000, Nampoothiri and Pandey 1996, Nigam 2000). The growth regulator, gibberellic acid (GA), was produced under SSF conditions by *Gibberella fujikuroi*, using bagasse (Nampoothiri and Pandey 1996).

In addition to the numerous, above-mentioned applications, sugarcane has many other nonbiotechnological applications in steam generation, electricity generation, pulp and paper production industries, and removal of heavy metal ions from waste water (Wan Ngah and Hanafiah 2008, Inyang et al. 2010). Sugarcane and sugarcane bagasse are valuable and versatile, cheap feedstocks for industrial applications. Biotechnology has the potential to improve sugarcane by improving abiotic and biotic stress tolerance, the generation of novel products in sugarcane, and by increasing the available feedstock biomass.

2.13 How will climate change and expanded growing-region affect vulnerability to pathogens?

Sugarcane is cultivated in every continent between the Tropic of Cancer and the Tropic of Capricorn. Biotechnology has the demonstrated potential to improve drought tolerance of sugarcane. The ability of sugarcane to withstand drought and thus thrive in more arid regions could potentially expand the growing regions for sugarcane around the world and assist the crop to tolerate climate change effects. If the growing region of the crop is expanded or the climate in established growing regions is altered, sugarcane may face new challenges from pest and diseases.

Studies have suggested that the impact of known sugarcane pathogens on sugarcane cultivation in new regions depends upon the dispersal methods of the pathogen, the suitability of the environment for dispersal, the seasonal survival rate of the pathogen, the suitability of the physiology of the pathogen to the new environment, and the mode of pathogen fitness into the native ecology of the new geographical location (Chakraborty *et al.* 1998). These aggregate factors may combine to make a particular pathogen more aggressive or a futile threat to sugarcane (Ghini *et al.* 2011). Improved drought tolerance of sugarcane would also assist in maintaining healthy plants in current growing regions. Healthy plants would therefore be less susceptible to pathogen attack (Chakraborty *et al.* 1998). The use of biotechnology to improve the biotic and abiotic stress tolerance of sugarcane would assist the crop to tolerate climatic changes and may improve the value and sustainability of the sugarcane industry.

2.14 Conclusion and perspectives

Sugarcane is an excellent candidate crop for biotechnological improvement. Few disadvantages in terms of genome complexity, however, can be viewed as untapped opportunities for future work. As compared with conventional breeding, the GM approach for sugarcane improvement shows great potential for overcoming the future challenges. The availability of efficient transformation systems, the availability of novel genetic resources from naturally resilient plants, and high-throughput technological advancement together provide a substantial arsenal for the "future proofing" of sugarcane for sustainable production in future climates.

Public acceptance on GM sugarcane is blurred and contrasting at present; however, public perception of GM crops may be positively influenced through refined and rational education as well as through improvements to GM techniques. Transformation systems that do not incorporate any transgenic DNA into the plant, and utilize nonantibiotic, plant gene–based selection system strategies would help in overcoming regulatory and negative public perception. Innovative research into the potential of engineering native genes without significant genetic rearrangements is essential.

Sugarcane grows in relatively stable climatic conditions; however, there is potential to increase the growing region of the crop through GM approaches by improving the biotic and abiotic stress tolerance of the crop. Biotechnology may improve the potential for sugarcane to meet or exceed the future demands for green alternatives to fossil fuels and fossil fuel derived products. Sugarcane is a globally important crop that is essential for sustaining food, fuel, and industrial resources. The improvement of sugarcane through the use of biotechnology will assist to improve the value and sustainability of the crop in the future.

References

Abel, P. P., Nelson, R. S., De, B., Hoffmann, N., Rogers, S. G., Fraley, R. T. and Beachy, R. N. (1986). Delay of disease development in transgenic plants that express the tobacco mosaic virus coat protein gene. *Science* **232**(4751):738–743.

Al-Janabi, S. M., Honeycutt, R. J., McClelland, M. and Sobral, B. W. S. (1993). A genetic linkage map of Saccharum spontaneum L. 'SES 208'. *Genetics* **134**:1249–1260.

Arencibia, A. D., Carmona, E. R., Tellez, P., Chan, M. T., Yu, S. M., Trujillo, L. E. and Oramas, P. (1998). An efficient protocol for sugarcane (Saccharum spp. L.) transformation mediated by Agrobacterium tumefaciens. *Transgenic Research* **7**(3):213–222.

Atkinson, A. H., Heath, R. L., Simpson, R. J., Clarke, A. E. and Anderson, M. A. (1993). Proteinase inhibitors in Nicotiana alata stigmas are derived from a precursor protein which is processed into five homologous inhibitors. *The Plant Cell* **5**(2):203–213.

Barba, R. and Nickell, L. (1969). Nutrition and organ differentiation in tissue cultures of sugarcane, a monocotyledon. *Planta* **89**(3):299–302.

Barthakur, S., Babu, V. and Bansa, K. (2001). Over-expression of osmotin induces proline accumulation and confers tolerance to osmotic stress in transgenic tobacco. *Journal of Plant Biochemistry and Biotechnology* **10**(1):31–37.

Baulcombe, D. C. (1996). Mechanisms of pathogen-derived resistance to viruses in transgenic plants. *The Plant Cell* **8**(10):1833.

Baulcombe, D. (2004). RNA silencing in plants. *Nature* **431**(7006):356–363.

Begcy, K., Eduardo, D., Gentile, M. A., Lembke, C. G., Zingaretti, S. M., Glaucia, M. and Menossi, S. M. (2012). A novel stress-induced sugarcane gene confers tolerance to drought, salt and oxidative stress in transgenic tobacco plants. *PLoS One* **7**(9):1–14.

Belintani, N., Guerzoni, J., Moreira, R. and Vieira, L. (2012). Improving low-temperature tolerance in sugarcane by expressing the ipt gene under a cold inducible promoter. *Biologia Plantarum* **56**(1):71–77.

Bonsch, M., Popp, A., Biewald, A., Rolinski, S., Schmitz, C., Weindl, I., Stevanovic, M., Högner, K., Heinke, J. and Ostberg, S. (2015). Environmental flow provision: Implications for agricultural water and land-use at the global scale. *Global Environmental Change* **30**:113–132.

Borges, E. R. and Pereira, N. J. (2010). Succinic acid production from sugarcane bagasse hemicellulose hydrolysate by Actinobacillus succinogenes. *Journal of Industrial Microbiology and Biotechnology* **38**:1001–1011.

Botha, F. C. (2015). Metabolic Changes Associated with the Development of Yellow Canopy Syndrome in Sugarcane. Paper presented at the Plant and Animal Genome XXIII Conference.

Bower, R. and Birch, R. G. (1992). Transgenic sugarcane plants via microprojectile bombardment. *Plant Journal* **2**(3):409–416.

Brumbley, S. M., Petrasovits, L. A., Bonaventura, P. A., O'Shea, M. J., Purnell, M. P. and Nielsen, L. K. (2003). Production of polyhydroxyalkanoates in sugarcane. Paper presented at the 4th ISSCT Molecular Biology Workshop, Montpelier, France, 7–14 April, 2003.

Brumbley, S. M., Purnell, M. P., Petrasovits, L. A., Nielsen, L. K. and Twine, P. H. (2007). Developing the sugarcane biofactory for high-value biomaterials. *International Sugar Journal* **1297**:5–15.

Bull, T. A. and Glasziou, K. T. (1979). Sugarcane. Chapter 4. In *Australian Field Crops Volume 2. Tropical Cereals, Oilseed, Grain Legumes and Other Crops*, pp. 95–113. J. V Lovett and A. Lazenby, eds. Angus and Robertson publishers: London.

Carvalho, L. C. C. (2010). Evolution of sugarcane industry in the state of São Paulo. In *Sugar Cane Bioethanol: R&D for Productivity and Sustainability*, pp. 3–16. L. A. B. Cortez, ed. Blucher: São Paulo. ISBN: 9788521205302.

Carvalho, W., Canilha, L. and Silva, S. S. (2008). Semi-continuous xylose-to-xylitol bioconversion by Ca-alginate entrapped yeast cells in a stirred tank reactor. *Bioprocess and Biosystems Engineering* **31**:493–498.

Chakraborty, S., Murray, G., Magarey, P., Yonow, T., O'Brien, R., Croft, B., Barbetti, M., Sivasithamparam, K., Old, K. and Dudzinski, M. (1998). Potential impact of climate change on plant diseases of economic significance to Australia. *Australasian Plant Pathology* **27**(1):15–35.

Chandel, A. K., Silva, S. S. D., Carvalhoa, W. and Singh, O. V. (2012). Sugarcane bagasse and leaves: foreseeable biomass of biofuel and bio-products. *Journal of Chemical Technology and Biotechnology* **87**:11–20.

Cheavegatti-Gianotto, A., de Abreu, H. M. C., Arruda, P., Bespalhok Filho, J. C., Burnquist, W. L., Creste, S., Di Ciero, L., Ferro, J. A., de Oliveira Figueira, A. V. and de Sousa Filgueiras, T. (2011). Sugarcane (Saccharum X officinarum): a reference study for the regulation of genetically modified cultivars in Brazil. *Tropical Plant Biology* **4**(1):62–89.

Chen, W., Gartland, K., Davey, M., Sotak, R., Gartland, J., Mulligan, B., Power, J. and Cocking, E. (1987). Transformation of sugarcane protoplasts by direct uptake of a selectable chimeric gene. *Plant Cell Reports* **6**(4):297–301.

Chen, F., Cai, T. Y., Zhao, G. H., Liao, X. J., Guo, L. Y. and Hu, X. S. (2005). Optimizing conditions for the purification of crude octacosanol extract from rice bran wax by molecular distillation analyzed using response surface methodology. *Journal of Food Engineering* **70**:47–53.

Contreras, A. M., Rosa, E., Perez, M., Langenhove, H. V. and Dewulf, J. (2009). Comparative Life Cycle Assessment of four alternatives for using by-products of cane sugar production. *Journal of Cleaner Production* **17**:772–779.

Cottrell-Dormer, W. (1932). Redstripe disease of sugar-cane in Queensland. *Bulletin, Bureau of Sugar Experiment Stations, Queensland, Division of Pathology* **3**:25–59.

Cottrell-Dormer, W. (1936). The variability of plant pathogens. Paper presented at the Proceedings of the International Society of Sugarcane Technologists; 5th Congress, Brisbane, 1935.

Cox, M., Hogarth, M. and Smith, G. (2000). Cane breeding and improvement. Chapter 5. In *Manual of Cane Growing*, pp. 263–289. M. Hogarth and P. Allsopp, eds. Bureau of Sugar Experimental Stations: Indooroopilly, Australia.

Creste, S., Pinto, L. R., Xavier, M. A. and Landell, M. G. A. (2010). Sugarcane breeding method and genetic mapping. In *Sugar Cane Bioethanol: R&D for Productivity and Sustainability*, pp. 353–357. L. A. B. Cortez, ed. Blucher: São Paulo. ISBN: 9788521205302.

Cronje, C., Tymon, A., Jones, P. and Bailey, R. (1998). Association of a phytoplasma with a yellow leaf syndrome of sugarcane in Africa. *Annals of Applied Biology* **133**(2):177–186.

D'Hont, A. (2005). Unraveling the genome structure of polyploids using FISH and GISH; examples of sugarcane and banana. *Cytogenetic and Genome Research* **109**:27–33.

D'Hont, A., Grivet, L., Feldmann, P., Rao, P. S., Berding, N. and Glauszmann, J. C. (1996). Characterisation of the double genome structure of modern sugarcane cultivars (Saccharum spp.) by molecular cytogenetics. *Molecular Genetics and Genomics* **250**:405–413.

D'Hont, A., Ison, D., Alix, K., Roux, C. and Glaszmann, J. C. (1998). Determination of basic chromosome numbers in the genus Saccharum by physical mapping of ribosomal RNA genes. *Genome* **41**:221–225.

D'Hont, A., Paulet, F. and Glaszmann, J. C. (2002). Oligoclonal interspecific origin of "North Indian" and "Chinese" sugarcanes. *Chromosome Research* **10**:253–262.

Da Costa Sousa, L., Chundawat, S. P., Balan, V. and Dale, B. E. (2009). 'Cradle-to-grave' assessment of existing lignocellulose pretreatment technologies. *Current Opinion in Biotechnology* **20**(3):339–347.

Damaj, M. B., Kumpatla, S. P., Emani, C., Beremand, P. D., Reddy, A. S., Rathore, K. S., Buenrostro-Nava, M. T., Curtis, I. S., Thomas, T. L. and Mirkov, T. E. (2010). Sugarcane DIRIGENT and O-METHYLTRANSFERASE promoters confer stem-regulated gene expression in diverse monocots. *Planta* **231**(6):1439–1458.

Daniels, J. and Daniels, C. A. (1975). Geographical, Historical and cultural aspects of the origin of the Indian and Chinese sugarcanes S. barberi and S. sinense. *Sugarcane Breeding Newsletter* **36**:4–23.

Daniels, J. and Roach, B. T. (1987). Taxonomy and evolution in sugarcane. In *Sugarcane Improvement Through Breeding*, pp. 7–84. D. J. Heinz, ed. Elsevier Press: Amsterdam.

Daniels, J., Smith, P., Paton and N. (1975). The origin of sugarcane and centres of genetic diversity in Saccharum; an enquiry to determine the most profitable areas to explore for sugarcane germplasm. *Sugarcane Breeding Newsletter* **35**:4–18.

De Souza, A. P, Gaspar, M., Da Silva, E. A., Ulian, E. C and Waclawovsky, A. J. (2008). Elevated CO_2 increases photosynthesis, biomass and productivity, and modifies gene expression in sugarcane. *Plant, Cell and Environment* **31**:1116–1127.

Devarumath, R. M., Doule, R. B., Kawar, P. G., Naikebawane, S. B. and Nerkar, Y. S. (2007). Field performance and RAPD analysis to evaluate genetic fidelity of tissue culture raised plants vis-à-vis conventional setts derived plants of sugarcane. *Sugar Techology* **9**(1):17–22.

D'Hont, A. (2015). Toward a reference sequence of the gene-rich part of the highly polyploid sugarcane genome. Paper presented at the Plant and Animal Genome XXIII Conference.

Dong, S., Delucca, P., Geijskes, R. J., Ke, J., Mayo, K., Mai, P., Sainz, M., Caffall, K., Moser, T. and Yarnall, M. (2014). Advances in Agrobacterium-mediated sugarcane transformation and stable transgene expression. *Sugar Technology* **16**(4):366–371.

Enríquez-Obregón, G. A., Vázquez-Padrón, R. I., Prieto-Samsonov, D. L., De la Riva, G. A. and Selman-Housein, G. (1998). Herbicide-resistant sugarcane (Saccharum officinarum L.) plants by Agrobacterium-mediated transformation. *Planta* **206**(1):20–27.

Falco, M., Neto, A. T. and Ulian, E. (2000). Transformation and expression of a gene for herbicide resistance in a Brazilian sugarcane. *Plant Cell Reports* **19**(12):1188–1194.

Fan, Z., Chen, H., Cai, S., Deng, C., Wang, W., Liang, X and Li, H. (2003). Molecular characterization of distinct potyvirus from white grass in China. *Archives of Virology* **148**:1219–1224.

FAO (Food, Agriculture Organization of the United Nations) (2004). *FAO Production Year Book.* FAO: Rome.

FAO (Food, Agriculture Organization of the United Nations) (2015). *Statistical Pocketbook World Food and Agriculture*, pp 1–28. FAO: Rome.

FAOstat (2015). Food and agriculture organization of the United Nations. Retrieved January 1, 2015, from http://faostat3.fao.org/browse/Q/*/E.

Farrant, J. M., Brandt, W. and Lindsey G. G. (2007). An overview of mechanisms of desiccation tolerance in selected angiosperm resurrection plants. *Plant Stress* **1**(1):72–84, Global Science Books.

Franks, T. and Birch, R. G. (1991). Gene transfer into intact sugarcane cells using microprojectile bombardment. *Australian Journal of Plant Physiology* **18**:471–480.

Gaff, D. F. and Latz, P. K. (1978). The occurrence of resurrection plants in the Australian Flora. *Australian Journal of Botany* **26**:485–492.

Gaff, G. F., Blomstedt, C. K., Neale, A. D., Le, T. N., Hamill, J. D. and Ghasempour, H. R. (2009). Sporobolus stapfianus, a model desiccation-tolerant grass. *Functional Plant Biology* **36**:589–599.

Gallo-Meagher, M. and Irvine, J. E. (1993). Effects of tissue type and promoter strength on transient GUS expression in sugarcane following particle bombardment. *Plant Cell Reports* **12**(12):666–670.

Gan, D., Zhang, J., Jiang, H., Jiang, T., Zhu, S. and Cheng, B. (2010). Bacterially expressed dsRNA protects maize against SCMV infection. *Plant Cell Reports* **29**(11):1261–1268.

Gao, S., Damaj, M. B., Park, J.-W., Beyene, G., Buenrostro-Nava, M. T., Molina, J., Wang, X., Ciomperlik, J. J., Manabayeva, S. A. and Alvarado, V. Y. (2013). Enhanced transgene expression in sugarcane by co-expression of virus-encoded RNA silencing suppressors. *PLoS One* **8**(6):e66046.

Geijskes, R. J., Wang, L., Lakshmanan, P., McKeon, M. G., Berding, N., Swain, R. S., Elliott, A. R., Grof, C. P. L., Jackson, J. A. and Smith, G. R. (2003). Smartsett^{TM} seedlings: tissue cultured seed plants for the Australian sugar industry. *Sugarcane International* **25**:13–17.

Ghini, R., Bettiol, W. and Hamada, E. (2011). Diseases in tropical and plantation crops as affected by climate changes: current knowledge and perspectives. *Plant Pathology* **60**(1):122–132.

Goel, D., Singh, A., Yadav, V., Babbar, S. and Bansal, K. (2010). Overexpression of osmotin gene confers tolerance to salt and drought stresses in transgenic tomato (Solanum lycopersicum L.). *Protoplasma* **245**(1–4):133–141.

Goldemberg, J., Coelho, S. T. and Guardabassi, P. (2008). The sustainability of ethanol production from sugarcane. *Energy Policy* **36**(6):2086–2097.

Gonclaves, M. C., Pinto, L. R., Souza, S. C. and Landell, M. G. A. (2012).Virus diseases of sugarcane. A constant challenge to sugar cane breeding in Brazil. Functional Plant Science and Biotechnology. *Global Science Books* **6**(2):108–116.

Gotz, R. and Maiss, E. (2002). The complete sequence of the genome of Cocksfoot streak virus (CSV): a grass infecting Potyvirus. *Archives of Virology* **147**:1573–1583.

Ha, S., Moore, P., Heinz, D., Kato, S., Ohmido, N. and Fukui, K. (1999). Quantitative chromosome map of the polyploid Saccharum spontaneum by multicolor fluorescent in situ hybridisation and imaging methods. *Plant Molecular Biology* **39**(6):1165–1173.

Hansom, S. P., Bower, R. S., Zhang, L., Potier, B. A., Elliott, A. R., Basnayake, S. W. V., Codeiro, G. M., Hogarth, D. M., Cox, M. and Berding, N. (1999). Regulation of transgene expression in sugarcane. Paper presented at the The International Society of Sugarcane Technologists XXIII Congress, New Delhi, India.

Harrison, M. D., Geijskes, J., Coleman, H. D., Shand, K., Kinkema, M., Palupe, A., Hassall, R., Sainz, M., Lloyd, R. and Miles, S. (2011). Accumulation of recombinant cellobiohydrolase and endoglucanase in the leaves of mature transgenic sugar cane. *Plant Biotechnology Journal* **9**(8):884–896.

Harrison, M. D., Geijskes, R. J., Lloyd, R., Miles, S., Palupe, A., Sainz, M. B. and Dale, J. L. (2014). Recombinant cellulase accumulation in the leaves of mature, vegetatively propagated transgenic sugarcane, *Molecular Biotechnology* **56**(9):795–802.

Heinz, D. J. and Mee, G. W. P. (1969). Plant differentiation from callus tissue of Saccharum species. *Crop Science* **9**(3):346–348.

IBGE (2010). Pesquisa Industrial Anual. Instituto Brasileiro de GeografiaeEstatistica. http://www.sidra.ibge.gov.br/bda/tabela/listabl.asp?z=t&c=3463.

Ingelbrecht, I. L., Irvine, J. E. and Mirkov, T. E. (1999). Posttranscriptional gene silencing in transgenic sugarcane. Dissection of homology-dependent virus resistance in a monocot that has a complex polyploid genome. *Plant Physiology* **119**(4):1187–1198.

Inman-Bamber, N. G., Lakshmanan, P. and Park, S. (2012). Sugarcane for water-limited environments: theoretical assessment of suitable traits. Sugarcane for water-limited environments: theoretical assessment of suitable traits. *Field Crops Research* **134**:95–104.

Inyang, M., Gao B., Pullammanappallil, P., Ding, W. and Zimmerman, A. R. (2010). Biochar from anaerobically digested sugarcane bagasse. *Bioresource Technology* **101**:8868–8872.

ISAAA (2015). International service for the acquisition of agri-biotech applications. GM approval data base. Retrieved 01-04-2015, 2015, from http://www.isaaa.org/gmapprovaldatabase/event/default.asp?EventID=340&Event=NXI-1T.

James, G. L. (2004). An Introduction to sugarcane. In *Sugarcane*, 2nd edition. G. L. James, ed. Blackwell Publishing Ltd: Oxford, UK.

James, C. (2014). Global Status of Commercialized Biotech/GM Crops: 2014. Retrieved February 1, 2015, from http://www.isaaa.org/resources/publications/briefs/46/executivesummary/.

Jesweit, J. (1929). The development of selection and breeding of the sugar cane in Java. Proceedings of the Third Congress of the International Society of Sugar Cane Technologists. The Executive Committee Soerabaia, pp. 44–57.

Joshi, S. (2009). Molecular improvement for fall armyworm resistance in sugarcane, TifEagle and Tifton 85. Doctor of Philosophy Dissertation, University of Florida.

Joyce, P., McQualter, R., Handley, J., Dale, J., Harding, R. and Smith, G. (1998). Transgenic sugarcane resistant to sugarcane mosaic virus. Paper presented at the Proceedings-Australian Society of Sugar Cane Technologists.

Joyce, P., Hermann, S., O'Connell, A., Dinh, Q., Shumbe, L. and Lakshmanan, P. (2014). Field performance of transgenic sugarcane produced using Agrobacterium and biolistics methods. *Plant Biotechnology Journal* **12**(4):411–424.

Jung, J. H., Fouad, W. M., Vermerris, W., Gallo, M. and Altpeter, F. (2012). RNAi suppression of lignin biosynthesis in sugarcane reduces recalcitrance for biofuel production from lignocellulosic biomass. *Plant Biotechnology Journal* **10**(9):1067–1076.

Kamle, M., Bajpai, A., Chandra, R., Kalim, S. and Kumar, R. (2011). Somatic embryogenesis for crop improvement. *GERF Bulletin of Bioscience* **2**(1):54–59.

Kinkema, M., Geijskes, J., Palupe, A., Shand, K., Coleman, H. D., Brinin, A., Williams, B., Sainz, M. and Dale, J. L. (2014a). Improved molecular tools for sugar cane biotechnology. *Plant Molecular Biology* **84**(4–5):497–508.

Kinkema, M., Geijskes, R. J., Shand, K., Coleman, H. D., De Lucca, P. C., Palupe, A., Harrison, M. D., Jepson, I., Dale, J. L. and Sainz, M. B. (2014b). An improved chemically inducible gene switch that functions in the monocotyledonous plant sugar cane. *Plant Molecular Biology* **84**(4–5):443–454.

Klein, T. M., Gradziel, T., Fromm, M. E. and Sanford, J. C. (1988). Factors influencing gene delivery into Zea mays cells by high–velocity microprojectiles. *Nature Biotechnology* **6**(5):559–563.

Koch, A. C., Ramgareeb, S., Rutherford, R. S., Snyman, S. J. and Watt, M. P. (2012). An in vitro mutagenesis protocol for the production of sugarcane tolerant to the herbicide imazapyr. *In Vitro Cellular and Developmental Biology—Plant* **48**:417–427.

Krishnan, C., Sousa, L. C., Jin, M., Chang, L., Dale, B. E. and Balan, V. (2010). Alkali-based AFEX pretreatment for the conversion of sugarcane bagasse and cane leaf residues to ethanol. *Biotechnology and Bioengineering* **107**:441–450.

Kristini, A. (2004). The use of tissue culture to eliminate some important diseases in sugarcane. M.Sc. thesis, University of Queensland, Australia.

Kumar, T., Uzma, T., Khan, M. R., Abbas, Z. and Ali, G. M. (2013). Genetic improvement of sugarcane for drought and salinity stress tolerance using Arabidopsis vacuolar pyrophosphatase (AVP1) gene. *Molecular Biotechnology* **56**:199–209.

Lakshmanan, P. and Robinson, N. (2014). Stress physiology: abiotic stresses. In *Sugarcane Physiology, Biochemistry and Functional Biology*. P. H Moore and F. C. Botha, eds. Wiley & Sons publication, Inc.: Ames, IA.

Lakshmanan, P., Geijskes, R. J., Aitken, K. S., Grof, C. L., Bonnett, G. D. and Smith, G. R. (2005). Sugarcane biotechnology: the challenges and opportunities. *In Vitro Cellular and Developmental Biology—Plant* **41**(4):345–363.

Lakshmanan, P., Geijskes, R. J., Wang, L., Elliott, A., Grof, C. P. L., Berding, N., and Smith, G. R. (2006). Developmental and hormonal regulation of direct shoot organogenesis and somatic embryogenesis in sugarcane (Saccharum spp. interspecific hybrids) leaf culture. *Plant Cell Reports* **25**(10):1007–1015.

Lal, M., Singh, R. K., Srivastava, S., Singh, N, Singh, S. P. and Sharma, M. L. (2008). RAPD based analysis of micropropagated plantlets of sugarcane for early assessment of genetic fidelity. *Sugar Technology* **10**:99–103.

Larkin, P. J. and Scowcroft, W. R. (1983). Somaclonal variation and eyespot toxin tolerance in sugarcane. *Plant Cell, Tissue and Organ Culture* **2**:111–121.

Leu, L. S. (1978). Apical meristem culture and redifferentiation of callus masses to free some sugarcane systemic-diseases. *Plant Protection Bulletin* **20**(1):77–82.

Li, R., Wu, N., Fan, Y. and Song, B. (1998). Transgenic potato plants expressing osmotin gene inhibits fungal development in inoculated leaves. *Chinese Journal of Biotechnology* **15**(2):71–75.

Liu, D., Oard, S. V. and Oard, J. H. (2003). High transgene expression levels in sugarcane (Saccharum officinarum L.) driven by the rice ubiquitin promoter RUBQ 2. *Plant Science (Limerick)* **165**(4):743–750.

Ma, H., Albert, H. H., Paull, R. and Moore, P. H. (2000). Metabolic engineering of invertase activities in different subcellular compartments affects sucrose accumulation in sugarcane cells. *Australian Journal of Plant Physiology* **27**(11):1021–1030.

Manickavasagam, M., Ganapathi, A., Anbazhagan, V. R., Sudhakar, B., Selvaraj, N., Vasudevan, A. and Kasthurirengan, S. (2004). Agrobacterium-mediated genetic transformation and development of herbicide-resistant sugarcane (Saccharum species hybrids) using axillary buds. *Plant Cell Reports* **23**:134–143.

Martins, M., Sarmento, D. and Oliveira, M. M. (2004). Genetic stability of micropropagated almond plantlets as assessed by RAPD and ISSR markers. *Plant Cell Reports* **23**:492–496.

Matsuoka, M., Kyozuka, J., Shimamoto, K. and Kano-Murakami, Y. (1994). The promoters of two carboxylases in a C4 plant (maize) direct cell-specific, light-regulated expression in a C3 plant (rice). *The Plant Journal* **6**(3):311–319.

Mckern, N. M., Shukla, D. D., Toler, R. D., Jensen, S. G., Tosic, M., Ford, R. E., Leon, O. and Ward, C. W. (1991). Confirmation that the sugarcane mosaic virus subgroup consist of four distinct potyvirus by using peptide profiles of coat proteins. *Phytopathology* **81**:1025–1029.

McQualter, R. B., Dale, J. L., Harding, R. M., McMahon, J. A. and Smith, G. R. (2004). Production and evaluation of transgenic sugarcane containing a Fiji disease virus (FDV) genome segment S9-derived synthetic resistance gene. *Australian Journal of Agricultural Research* **55**(2):139–146.

McQualter, R. B., Chong, B. F., Meyer, K., Van Dyk, D. E., O'Shea, M. G., Walton, N. J., Viitanen, P. V. and Brumbley, S. M. (2005). Initial evaluation of sugarcane as a production platform for p-hydroxybenzoic acid. *Plant Biotechnology Journal* **3**(1):29–41.

Meyer, P. M., Rodrigues, P. H. M. and Millen, D. D. (2013). Impact of biofuel production in Brazil on the economy, agriculture, and the environment. *Animal Frontiers* **3**(2):29.

Molinari, H. B. C., Marur, C. J., Daros, E., De Campos, M. K. F., De Carvalho, J. F. R. P., Pereira, L. F. P. and Vieira, L. G. E. (2007). Evaluation of the stress-inducible production of proline in transgenic sugarcane (Saccharum spp.): osmotic adjustment, chlorophyll fluorescence and oxidative stress. *Physiologia Plantarum* **130**(2):218–229.

Moore, P. H. (1987). Anatomy and morphology. In *Sugarcane Improvement Through Breeding*, pp. 85–142. D. Heinz, ed. Elsevier: Amsterdam.

Moyle, R. L. and Birch, R. G. (2013). Sugarcane Loading Stem Gene promoters drive transgene expression preferentially in the stem. *Plant Molecular Biology* **82**(1–2):51–58.

Mudge, S. R., Basnayake, S. W., Moyle, R. L., Osabe, K., Graham, M. W., Morgan, T. E. and Birch, R. G. (2013). Mature-stem expression of a silencing-resistant sucrose isomerase gene drives isomaltulose accumulation to high levels in sugarcane. *Plant Biotechnology Journal* **11**(4):502–509.

Murugan, A., Singh, M. and Ranjit, A. J. A. (2013). Sugarcane. In *Valorization of Food Processing By-Products*, vol. **15**, pp. 416–449. M. Chandrasekaran, ed. CRC Press: Boca Raton.

Naik, S., Goud, V. V., Rout, P. K. and Dalai, A. K. (2010). Production of first and second generation biofuels: a comprehensive review. *Renewable and Sustainable Energy Reviews* **14**(2):578–597.

Nampoothiri, K. M. and Pandey, A. (1996). Solid state fermentation for L-glutamic acid production using Brevi bacterium sp. *Biotechnology Letters* **18**:199–204.

Nickell, L. (1964). Tissue and cell cultures of sugarcane: another research tool. *Hawaiian Planters' Record*, **57**:223–229.

Nigam, J. N. (2000). Cultivation of Candida langeronii in sugarcane bagasse hemicellulosic hydrolysate for the production of single cell protein. *World Journal of Microbiology and Biotechnology* **16**:367–372.

O'Hara, I. M. (2011). Cellulosic ethanol from sugarcane bagasse in Australia: exploring industry feasibility through systems analysis, techno-economic assessment and pilot plant development. Doctor of Philosophy Dissertation, Queensland University of Technology, Brisbane, Queensland. Australia.

O'Hara, I. M., Edye, L. A., Doherty, W. O. and Kent, G. (2010). Demonstration of cellulosic ethanol production from sugarcane bagasse in Australia: the Mackay Renewable Biocommodities Pilot Plant. Paper presented at the International Society of Sugar Cane Technologists, Veracruz, Mexico.

O'Hara, I. M., Zhang, Z., Rackemann, D. W., Dunn, K. G., Hobson, P. A. and Doherty, W. O. (2013). Prospects for the development of sugarcane biorefineries. Paper presented at the Proceedings of the 28th International Society of Sugar Cane Technologists Conference.

Palhares, A. C., Taislene, B., Rodrigues, M., Marie-Anne, V. S., Douglas, S. D., Walter, M. Jr., Hamilton, J. Jr., Anete, P S., Thiago, G. M., Mollinari, M., Gazaffi, R., Garcia, A. A. F. and Vieira, M. L. C. (2012). A novel linkage map of sugarcane with evidence for clustering of retrotransposon-based markers. *BMC Genetics* **13**:51.

Pandey, A., Soccol, C. R., Nigam, P. and Soccol, V. T. (2000). Biotechnological potential of agro-industrial residues. I: sugarcane bagasse. *Bioresource Technology* **74**:69–80.

Pandey, R. N., Singh, S. P., Rastogi, J., Sharma, M. L. and Singh, R. K. (2012). Early assessment of genetic fidelity in sugarcane (Saccharum officinarum) plantlets regenerated through direct organogenesis with RAPD and SSR markers. *Australian Journal of Crop Science* **6**(4):618–624.

Patade, V. Y., Suprasanna P. and Bapat, V. A. (2008). Effects of salt stress in relation, to osmotic adjustment on sugarcane (Saccharum officinarum L.) callus cultures. *Plant Growth Regulation* **55**:169–173.

Paturau, J. M. (1986). Alternative uses of sugarcane and its by-products in agro industries. In *Sugarcane as Feed*, pp. 37–64. R. Sansoucy, G. Aarts and T. R. Preston, eds. Proceedings of an FAO Expert Consultation, Santo Domingo, Dominican Republic.

Peters, S., Mundree, S. G., Thomson, J. A., Farrant, J. M. and Felix, K. (2007). Protection mechanisms in the resurrection plant Xerophyta viscosa (Baker): both sucrose and raffinose family oligosaccharides (RFOs) accumulate in leaves in response to water deficit. *Journal of Experimental Botany* **58**(8):1947–1956.

Petrasovits, L. A., McQualter, R. B., Gebbie, L. K., Blackman, D. M., Nielsen, L. K. and Brumbley, S. M. (2013). Chemical inhibition of acetyl coenzyme A carboxylase as a strategy to increase polyhydroxybutyrate yields in transgenic sugarcane. *Plant Biotechnology Journal* **11**(9):1146–1151.

Prabu, G. and Prasad, D. T. (2012). Functional characterization of sugarcane MYB transcription factor gene promoter (PScMYBAS1) in response to abiotic stresses and hormones. *Plant Cell Reports* **31**(4):661–669.

Prakasham, R. S., Rao, R. S. and Hobbs, P. J. (2009). Current trends in biotechnological production of xylitol and future prospects. *Current Trends in Biotechnological and Pharmacology* **3**:8–36.

Qaim, M. (2009). The economics of genetically modified crops. *Annual Review of Resource Economics* **1**:665–693.

Rathus, C., Bower, R. and Birch, R. G. (1993). Effects of promoter, intron and enhancer elements on transient gene expression in sugar-cane and carrot protoplasts. *Plant Molecular Biology* **23**(3):613–618.

Ravindra, B., Malabadi, G. S., Mulgund, K., Nataraja, S. and Kumar, V. (2011). Induction of somatic embryogenesis in different varieties of sugarcane (Saccharum officinarum L.). *Research in Plant Biology* **1**(4):39–48.

Reinert, J. and Bajaj, Y. P. S. (1977). Anther culture: haploid production and its significance. In *Applied and Fundamental Aspects of Plant Cell, Tissue, and Organ Culture*, pp. 251–267. J. Reinert and Y. P. S. Bajaj, eds. Springer-Verlag: Berlin, Heidelberg.

Ricaud, C., Egan, B., Gillaspie, A. and Hughes, C. (2012). *Diseases of Sugarcane: Major Diseases*. Elsevier: Amsterdam, The Netherlands.

Rodrigues, F. A., Da Graca, J. P., De Laia, M. L., Nhani, A. and Galbiati, J. A. (2011). Sugarcane genes differentially expressed during water deficit. *Biologia Plantarum* **55**:43–53.

Rott, P. (2000). *A Guide to Sugarcane Diseases*. CIRAD: Montpellier, France.

Saini, N., Saini, M. L. and Jain, R. K. (2004). Large scale production, field performance and RAPD analysis of micropropagated sugarcane plants. *Indian Journal of Genetics and Plant Breeding* **64**:102–107.

Sanford, J. and Johnston, S. (1985). The concept of parasite-derived resistance—deriving resistance genes from the parasite's own genome. *Journal of Theoretical Biology* **113**(2):395–405.

Schenk, P. M., Sagi, L., Remans, T., Dietzgen, R. G., Bernard, M. J., Graham, M. W. and Manners, J. M. (1999). A promoter from sugarcane bacilliform badnavirus drives transgene expression in banana and other monocot and dicot plants. *Plant Molecular Biology* **39**(6):1221–1230.

Schroeder, H. E., Gollasch, S., Moore, A., Tabe, L. M., Craig, S., Hardie, D. C., Chrispeels, M. J., Spencer, D. and Higgins, T. J. (1995). Bean [alpha]-amylase inhibitor confers resistance to the pea weevil (Bruchus pisorum) in transgenic peas (Pisum sativum L.). *Plant Physiology* **107**(4):1233–1239.

Scortecci, K. C., Creste, S., Calsa, T Jr., Xavier, M. A., Landell, M. G. A., Figueira, A. and Benedito, V. A. (2012). Challenges, opportunities and recent advances in sugarcane breeding. In *Plant Breeding*. I. Abdurakhmonov, ed. InTech: London. DOI: 10.5772/28606, ISBN: 978-953-307-932-5.

Seabra, J. E. and Macedo, I. C. (2011). Comparative analysis for power generation and ethanol production from sugarcane residual biomass in Brazil. *Energy Policy* **39**(1):421–428.

Seifers, D. L., Salomon, R., Marie-Jeanne, V., Alliot, B., Signoret, P., Harber, S., Laboda, A., Ens, W., She, Y. M. and Standing, K. G. (2000). Characterization of novel potyvirus isolated from maize in Israel. *Phytopathology* **90**:505–513.

Sengar, R. S., Sengar, K. and Garg, S. K. (2011). Biotechnological approaches for high sugarcane yield. *Plant Sciences Feed* **1** (7):101–111.

Setamou, M., Bernal, J., Legaspi, J., Mirkov, T. and Legaspi Jr., B. (2002). Evaluation of lectin-expressing transgenic sugarcane against stalk borers (Lepidoptera: Pyralidae): effects on life history parameters. *Journal of Economic Entomology* **95**(2):469–477.

Sharma, P., Gaur, S. N., Goel, N. and Arora, N. (2015). Engineered hypoallergenic variants of osmotin demonstrate hypoallergenicity with in vitro and in vivo methods. *Molecular Immunology* **64**(1):46–54.

Shukla, D. D., Tosic, M., Ford, R. E., Taker, R. W. and Langham, M. A. C. (1989). Taxonomy of Potyvirus infecting maize, sorghum and sugarcane in Australia and United State as determined by reactivities of polyclonal antibodies directed towards virus specific N-termini of coat proteins. *Phytopathology* **79**:223–229.

Silvarolla, M. B. (1992). Plant genomic alterations due to tissue culture. *Journal of Brazilian Association of Advance Science* **44**:329–335.

Singh, P., Suman, A., Tiwari, P., Arya, N., Gaur, A. and Shrivastava, A. K. (2008). Biological pretreatment of sugarcane trash for its conversion to fermentable sugars. *World Journal of Microbiology and Biotechnology* **24**:667–673.

Snyman, S. J., Meyer, G. M., Richards, JM., Haricharan, N., Ramgareeb, S. and Huckett, B. I. (2006). Refining the application of direct embryogenesis in sugarcane: effect of the developmental phase of leaf disc explants and the timing of DNA transfer on transformation efficiency. *Plant Cell Reports* **25**:1016–1023.

Snyman, S. J., Van A. T., Ramdeen, V., Meyer, G. M., Richards, J. M. and Rutherford, R. S. (2007). Micropropagation by direct somatic embryogenesis: is disease elimination also a possibility? *Proceedings of the International Society of Sugar Cane Technologists* **26**:943–946.

Solomon, S. (2011). The Indian Sugar Industry: an overview. *Sugar Technology* **13**(4):255–265.

Sreenivasan, J., Sreenivasan, T. V. and Alexander, K. C. (1987). Somaclonal variation for rust resistance in sugarcane. *Indian Journal of Genetics and Plant Breeding* **47**:109–114.

Srivastava, S. and Gupta, P. S. (2008). Inter simple sequence repeat profile as a genetic marker system in sugar. *Sugar Technology* **10**(1):48–52.

Su, J., Shen, Q., Ho, T.-H. D. and Wu, R. (1998). Dehydration-stress-regulated transgene expression in stably transformed rice plants. *Plant Physiology* **117**(3):913–922.

Suprasanna, P. (2010). Biotechnological interventions in sugarcane improvement: strategies, methods and progress. *BARC Newsletter* **316**:47–53.

Szwarc, A. (2009). Brazilian Sugar Mill Types. Gopal, A. R., personal communication, Berkeley, CA.

Tiwari, A. K, Mishra, N., Tipasthi, S., Lal, M., Singh, R. K. and Sharma, M. L. (2011). Assessment of genetic stability in micro propagated population of sugarcane variety CoS 07250 through SSR markers. *International Journal of Plant Research* **24**(2):75–81.

Troiani, E. (2009). Molasses Sugar Content. Gopal, A. R., Personal communication, Berkeley, CA.

Vasil, I. K. (1987). Developing cell and tissue culture systems for improvement of cereal and grass crops. *Journal of Plant Physiology* **128**:193–218.

Viswanathan, R. (1997). Detection of phytoplasmas associated with grassy shoot disease of sugarcane by ELISA techniques. *Journal of plant diseases and protection* **104**(1):9–16.

Viswanathan, C., Anburaj, J. and Prabu, G. (2014). Identification and validation of sugarcane streak mosaic virus-encoded microRNAs and their targets in sugarcane. *Plant Cell Reports* **33**(2):265–276.

Wagih, M., Gordon, G., Ryan, C. and Adkins, S. (1995). Development of an axillary bud culture technique for Fiji disease virus elimination in sugarcane. *Australian Journal of Botany* **43**(1):135–143.

Wan Ngah, W. S. and Hanafiah, M. A. (2008). Removal of heavy metal ions from waste water by chemically modified plant wastes as adsorbents: a review. *Bioresource Technology* **99**:3935–3948.

Wang, M. L., Goldstein, C., Su, W., Moore, P. H. and Albert, H. H. (2005). Production of biologically active GM-CSF in sugarcane: a secure biofactory. *Transgenic Research* **14**(2):167–178.

Wang, J., Roe, B., Macmil, S. L., Yu, Q., Murray, J. E. Tang, G., Chen, C., Najar, F., Wiley, G., Bowers, J., Vaan Sluys, M.-A.., Rokhsar, D. S., Hudson, M. E., Moose, S. P., Peterson, A. H. and Ming, R. (2010). Microcollinearity between autopolyploid sugarcane and diploid sorghum genomes. *BMC Genomics* **11**:261.

Weng, L.-X., Deng, H.-H., Xu, J.-L., Li, Q., Zhang, Y.-Q., Jiang, Z.-D., Li, Q.-W., Chen, J.-W. and Zhang, L.-H. (2011). Transgenic sugarcane plants expressing high levels of modified cry1Ac provide effective control against stem borers in field trials. *Transgenic Research* **20**(4):759–772.

Wu, L. G. and Birch, R. G. (2007). Doubled sugar content in sugarcane plants modified to produce a sucrose isomer. *Plant Biotechnology Journal* **5**(1):109–117. DOI: 10.1111/j.1467-7652.2006.00224.x.

Wu, K. K., Burnquist, W., Sorrells, M. E., Tew, T. L., Moore, P. H. and Tanksley, S. D. (1992). The detection and estimation of linkage in polyploids using single-dose restriction fragments. *Theoretical and Applied Genetics* **83**:294–300.

Wu, Y., Zhou, H., Que, Y.-X., Chen, R.-K. and Zhang, M.-Q. (2008). Cloning and identification of promoter Prd29A and its application in sugarcane drought resistance. *Sugar Technology* **10**(1):36–41.

Yang, S., Vanderbeld, B. and Wan, J. (2010). Narrowing down the targets towards successful genetic engineering of drought tolerant crops. *Molecular Plant* **3**:469–490.

Zhang, L., Xu, J. and Birch, R. G. (1999). Engineered detoxification confers resistance against a pathogenic bacterium. *Nature Biotechnology* **17**(10):1021–1024.

Zhang, S. Z., Yang, B. P. and Feng, C. L. (2006). Expression of the Grifola frondasa trehalose synthase gene and improvement of drought tolerance in sugarcane (S. officinarum L.). *Journal of Integrative Plant Biology* **48**:453–459.

Zhang, J., Nagai, C., Yu, Q., Pan, Y., Ayala-Silva, T., Schnell R. J., Comstock, J. C., Arumuganathan, A. K. and Ming, R. (2012). Genome size variation in three Saccharum species. *Euphytica* **185**:511–519.

Zhang, J., Zhou, M., Walsh, J., Zhu, L., Chen, Y. and Ming, R. (2014). Sugarcane genetics and genomics. In *Sugarcane: Physiology, Biochemistry and Functional Biology*, pp. 623–643. H. P. Moore and F. C., Botha, eds. John Wiley & Sons, Inc.: Ames, IA.

Zhuo, Z., PingHua, C., LiPing, X., ZhongWei, C., GuiJiao, S., HengBo, W., LianSai, T., Wei, X., SanJi, G. and Di, L. (2014). Construction of tetravalent expression vector carrying Cry1Ac, Bar and RNAi fragments resistant to virus diseases and genetic transformation in sugarcane. *Genomics and Applied Biology* **33**(3):661–673.

Zingaretti, S., Demore, P., Morceli, T. and Mantovanini, L. (2014). Glycine betaine biosynthesis genes differentially expressed in sugarcane under water stress. Paper presented at the BMC Proceedings.

Zucchi, M. I., Arizono, H., Morais, V. A., Fungaro, M. H. P., Lúcia, M. and Vieira, C. (2002). Genetic instability of sugarcane plants derived from meristem cultures. *Genetics and Molecular Biology* **25**(1):91–96.

PART II
Biofuels and bioproducts

CHAPTER 3

Fermentation of sugarcane juice and molasses for ethanol production

Cecília Laluce[1], Guilherme R. Leite[1], Bruna Z. Zavitoski[1], Thamires T. Zamai[1] and Ricardo Ventura[2]

[1] *Biochemistry and Chemical Technology Department, Institute of Chemistry, Univ Estadual Paulista (UNESP), São Paulo, Brazil*

[2] *Integra Consultoria Química LTDA, Ribeirão Preto, Brazil*

3.1 Introduction

Sugarcane is the most important feedstock for bioethanol production in tropical and subtropical countries. The composition of sugarcane juice varies with sugarcane variety over the harvesting season and with cultivation conditions. The quality of both sugarcane juice and molasses is affected by the nutrients and topology of the soil, variety of sugarcane, climate (temperature, humidity), and harvesting. One of the main advantages of sugarcane ethanol for biofuels production is its low cost per hectare compared with other crops (Rudorff *et al.* 2010). Another significant advantage is the capability to carry out repetitive sugarcane harvests from the same crop area, because ratoon sugarcane crops are able to regrow for up to three or four cycles before replanting (Rudorff *et al.* 2010).

The ethanol yield from sugarcane depends on the yeast strains, process design, and operational conditions. In 2012, the United States produced 13.3 billion gallons of ethanol, while Brazil produced 7.0 billion gallons (RFA, Renewable fuels association 2013). Nevertheless, Brazil is typically the biggest ethanol exporter to the world market (Balat and Balat 2009, Mussatto *et al.* 2010).

The microbial organism most commonly used to ferment sugarcane juice and molasses is *Saccharomyces cerevisiae*. This organism is widely used in other traditional fermentation processes, such as bread making and beverage production, and it is a model organism to study diseases such as human aging (Bitterman *et al.* 2003) and cancer therapies (Matuo *et al.* 2012). In addition to diversified industrial applications, *S. cerevisiae* is characterized by fast growth, easy genetic manipulation, a detailed understanding of the genome (Liang *et al.* 2012), and has been widely researched from both a basic and applied perspective (Zhou *et al.* 2004). This yeast is highly tolerant to ethanol and able to convert sugars

Sugarcane-Based Biofuels and Bioproducts, First Edition. Edited by Ian M. O'Hara and Sagadevan G. Mundree.

into ethanol at high rates, particularly in high-cell-density cultures initiated with inoculum of high cell quantities (Basso *et al.* 2011). This chapter describes the natural microbial ecology of the ethanol production process, methods for yeast identification, bioreactors and fermentation processes, cleaning and sanitization, and process control for industrial ethanol production from sugarcane juice and molasses.

3.2 Natural microbial ecology

Ethanol production is an open process that is constantly exchanging materials and energy with the environment. A diversity of strains from the genera *Saccharomyces*, along with bacteria and wild yeasts, are constantly isolated and identified from the industrial niches of ethanol production. Such organisms compete with each other for nutrients and produce antimicrobials generating a dynamic microbial population. This dynamics is due to a constant evolutionary process that results in a succession of sub-microbial populations within each bioreactor during fermentation (Replansky *et al.* 2008). Stress conditions in the bioreactors stimulate the presence of dynamic conditions.

3.2.1 Saccharomyces yeasts

S. cerevisiae is a nonfilamentous fungus that reproduces by vegetative growth or sexual reproduction (Hartwell 1974, Kurtzman *et al.* 2011), as illustrated in Figure 3.1. Industrial performance of a yeast strain depends on its ancestor's genetic background and chromosomal DNA recombination. During vegetative growth, new cells are formed from emergent buds attached to the mother cell that die after generating 30–40 new cells. In sexual reproduction (meiotic process), two to four small spores are formed per ascus of diploid cells that comprise two sets of complementary chromosomal DNA. Spores germinate forming haploid cells having just one set of chromosomal DNA. Hybridization or chromosomal recombination between haploid cells from distinct strains gives rise to a new diploid with improved industrial characteristics (Wongkhalaung *et al.* 2004, Zheng *et al.* 2013). In a nutritionally deficient medium, new budding cells remain attached to each other showing a typical filamentous growth (Gimeno *et al.* 1992).

Yeasts from the genera *Saccharomyces* and non-*Saccharomyces* or contaminants can be found in industrial ethanol bioreactors. Based on evolutionary theory (Replansky *et al.* 2008), yeast strains evolve in natural niches giving rise to taxonomically closer species belonging to the *sensu stricto* group; the species within the genus *Saccharomyces* are as follows (Replansky *et al.* 2008): *S. cerevisiae, S. paradoxus, S. cariocanus, S. bayanus, S. mikatae, S. kudriavzevii*, and *S. pastorianus*. *S. cerevisiae* is the preferred species to produce ethanol due to its ability to grow

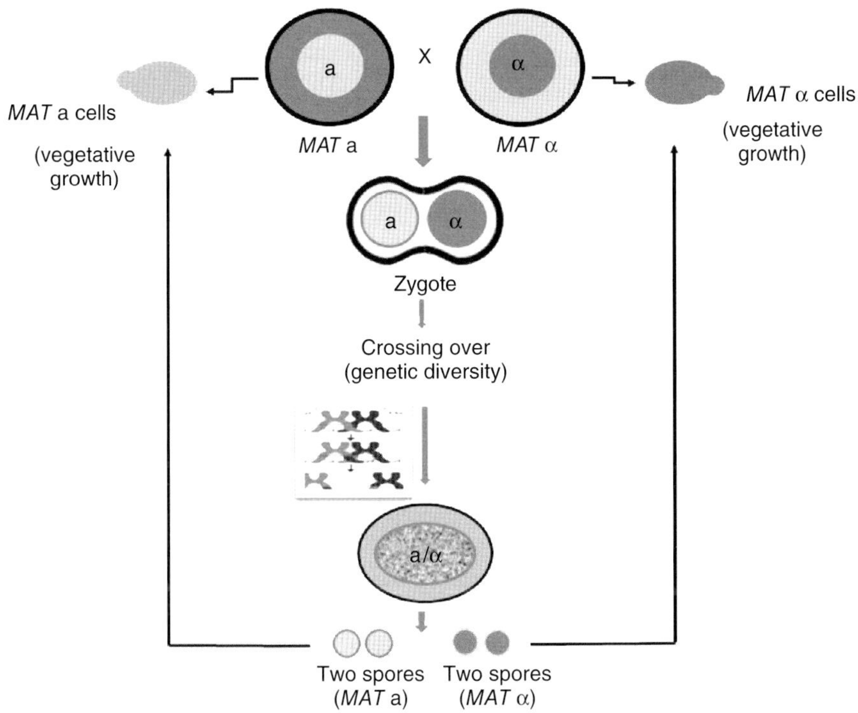

Figure 3.1 Cell cycle of *S. cerevisiae* illustrating vegetative and sexual reproduction and chromosomal recombination between homologous chromosomes in a cell of *Saccharomyces cerevisiae* during meiosis.

and ferment at temperatures ranging from 15 to 35°C, as well as its capacity to supplant contaminant yeasts and bacteria during fermentation (G-Alegria *et al.* 2004, Abdel-Banat *et al.* 2010). Yeast strains tolerant to high temperatures are becoming increasingly important due to climatic changes and an increasing water shortage in the world.

Yeasts isolated from industrial processes are species of *S. cerevisiae* selected based on fermentation performance at laboratory and industrial scale (Basso *et al.* 2008). According to the authors, the most widely adopted industrial strain is PE-2, which is a naturally heterothallic diploid available on the market in active dry form (ADY, active dry yeast) and used in amounts ranging from 10 to 300 kg per distillery.

As organic acids can act as antibacterial agents, the search for new yeast strains, which are able to produce increased levels of organic acids, is important. The adaptation of yeast cells to organic acids leads to growth at low pH values. Increases in the resistance of yeast cells to organic acids has been described (Piper *et al.* 2001), while the inhibition of the bacterial growth by succinic acid has been observed during the ethanol production process (Basso *et al.* 1997).

Antibacterial activity, generated by the synergistic interaction between lactic acid and ethanol, occurs as a way to control bacterial contaminants in favor of ethanol production at laboratory or industrial scale (Dorta *et al.* 2006). In addition, yeast cells get adapted when exposed to organic acids at low pH (Jarboe *et al.* 2013, Piper *et al.* 2001, Basso *et al.* 1997)

3.2.2 Wild yeasts

There is a great variety of conventional yeasts (bread, beer, and wine) and promising nonconventional yeasts within industrial microbial populations (Spencer *et al.* 2002). Wild yeasts exhibit characteristics that are different from those exhibited by domesticated or laboratory yeasts, since phenotypes can be altered by process conditions and climate. Nevertheless, the following yeasts were able to remain physiologically active up to the end of repetitive fermentation runs in cold climates, as shown by PCR-RFLP assays (Díaz *et al.* 2013): *Metschnikowia pulcherrima, Rhodotorula mucilaginosa, Pichia kluyveri, Pichia membranifaciens, and S. cerevisiae.* In five fuel-ethanol distilleries, located in Pernambuco and Paraıba in the northeast of Brazil, the following wild yeasts were identified as causes of acute episodes of infections (Basílio *et al.* 2008): *Dekkera bruxellensis, Pichia galeiformis,* and *Candida tropicalis.* On the other hand, in a cachaça distillery (a Brazilian sugarcane spirit) located in the northeast region of Brazil, the following non-*Saccharomyces* species were identified (Gomes *et al.* 2007): *Torulaspora mucinaginosa, P. membranifaciens, Torulaspora delbrueckii, Zygosaccharomyces bailii, Candida intermedia, Candida sake,* and *Candida zeylanoides.*

Wild yeasts can be identified by characteristics such as the ability to flocculate or sediment poorly, resistance to high temperatures, and growth on plates containing high concentrations (>0.2 ppm) of cycloheximide (Fall 1994). Wild yeasts also show abnormal cell and colony morphologies with irregular borders. In addition, the preference for growth instead of fermentation was observed with a new strain of *Issatchenkia orientalis* (Gallardo *et al.* 2011). This strain formed greater colonies on YPD medium compared with other yeasts.

3.2.3 Bacterial contaminants

The incidence of bacterial contaminants, particularly *Lactobacillus,* is a continuing problem in industrial fermentations, where bacterial proliferation takes place despite the competition for nutrients against other microbial cells and the production of acetic acid and lactic acid, which inhibit the yeast activity at low pH (Skinner and Leathers 2004).

Species of *Lactobacillus* grow well during the ethanol production processes at pH < 5.0 due to fast growth rate and high tolerance to ethanol. Strains of *Lactobacillus fermentum* and *Lactobacillus brevis* isolated from bioreactors of an ethanol factory, during an acute infection, were identified as the cause of inhibition of yeast fermentation (Skinner and Leathers 2004). Ethanol production by yeasts is predominately inhibited by *L. fermentum* and *Lactobacillus plantarum* due to the excretion of organic acids (Muthaiyan *et al.* 2010). In addition, both lactic and acetic acids excreted in sugarcane must by *L. fermentum* are able to inhibit fermentation processes after a few recycles (Oliva-Neto and Yokoya 1994).

Lactic acid bacteria can act as homofermentative (producing only lactic acid) or heterofermentative (producing ethanol, lactic acid, acetic acid, and CO_2) organisms during the production of ethanol from glucose (Hofvendahl and Hahn-Hägerdal 2000). Both lactic and acetic acids are strong inhibitors of fermentation activity at low pH, while lactic acid is generally the most abundant weak acid in industrial fermentations (Narendranath *et al.* 2001). However, acetic acid is the stronger inhibitor of fermentation activity at low pH. High ethanol concentrations may cause leakage of essential metabolites in the cytoplasmic membrane (Ingram 1986).

Regarding temperature, *Lactobacillus* sp. isolated from sugarcane juice are resistant to low pH and able to survive for short periods at temperatures above 40–45°C (Shafiee *et al.* 2010). Furthermore, *L. plantarum and Oenococcus oeni* isolated from wine fermentation grew well at pH 3.2, exhibiting resistance to lyophilization and proliferation in the presence of up to 13% v/v ethanol (G-Alegria *et al.* 2004). Lastly, an ordinary fermentation can be converted into a stuck fermentation depending on the extracellular and intracellular amounts of weak acids (acetic and lactic acids) and other stress factors aggravated by the process conditions (O'Connor-Cox and Ingledew 1989), and this condition requires a costly shutdown of facilities for cleaning and sterilization.

A successful fermentation depends on the proper use of bacteria identification. Gram staining is a simple and quick assay to distinguish morphologies physiological properties (e.g., resistance to antibiotics) between different bacterial species. During this assay, cells of Gram-positive bacteria become dark blue or violet blue, while cells of Gram-negative bacteria do not retain the crystal violet stain, showing red or pink colors that allow a better cell count under the microscope. Gram-positive bacteria have thick cell walls, are more resistant to antibiotics, drying, and physical disruption than Gram-negative bacteria, which have thinner cell walls and exhibit higher sensitivity to penicillin (Danielsen and Wind 2003, McCarter 2005). Based on cell morphology and physiological properties, Gram-positive and Gram-negative bacteria can be divided into subgroups (Brenner et al. 2005, Narendranath and Brey 2009), as follows:

- Gram-negative aerobic bacteria that do not produce acetic acid (*Zymomonas*);
- Gram-negative aerobic bacteria that do not oxidize acetic acid (*Gluconobacter*) or oxidize it slowly (*Acetobacter*);
- elongated cells of wild yeasts (*Dekkera*) that are catalase-positive and produce acetic acid;

- Gram-positive cocci that are catalase-negative and produce lactic acid (*Pediococcus*);
- Gram-positive cocci that are catalase-negative and produce lactic acid and CO_2 (*Leuconostoc*);
- Gram-positive rods that are catalase-negative and produce both lactic and acetic acids (*Lactobacillus*);
- Gram-positive cocci not typically found in fermenters as they require oxygen (*Micrococcus*).

3.3 Yeast identification

The inoculum used to start a fermentation process may contain more than one yeast strain of the same species, while complex consortia of yeast cells arise during fermentation with cell recycling. Yeast strains showing genetic traits different from those shown by ancestors were identified during industrial fermentations (Basso *et al.* 2008). Nevertheless, the main challenge is the identification and/or characterization of the contribution of each strain or similar groups of strains to the industrial processes, particularly when cell recycling is used (Antonangelo *et al.* 2013).

3.3.1 Identification of genetic and physiological phenotypes

Several methods are described in the literature to identify phenotypic variants in yeast populations; however, most of the procedures are very time-consuming (2–7 days to grow). The literature describes media containing dyes and other procedures to identify the incidence of mutant phenotypes in a yeast population, which can be described as follows:

- detection of nutritional deficiencies, sporulation frequency, and mating-types in culture medium (Amberg *et al.* 2005);
- WLN medium (Wallerstein Nutrient Broth) to isolate and identify contaminant yeast and mutants (Powel and Diacetis 2007);
- magdala red (Horn and Wilkie 1996) or mixtures of eosin and trypan blue (Nagai 1963) to identify the respiratory-deficient mutants and auxotrophic diploids;
- suppression of growth of super-secreting lysine mutants by *S*-2-aminoethyl-L-cysteine added at increasing concentrations to a minimal medium (Gasent-Ramírez and Benítez 1997).

Taxonomic keys based on physiological and morphological assays can be discriminatory and convenient to identify a high number of yeasts (Middelhoven 2002). Rapid test kits (API Kits from BioMerieux, France) allow easy and preliminary classification of yeasts and bacteria.

3.3.2 Molecular identification methods

Sophisticated molecular techniques are required to predict and characterize the dynamic behavior of yeasts during fermentation. Currently, developments in the genetic analysis are leading to more discriminatory techniques, which allow a better understanding of population biodiversity. The modern techniques are based on the extraction of the total DNA (direct methods) or DNA extracted from isolated colonies (indirect methods). An overview of most relevant techniques can be summarized as described in the following text (Ivey and Phister 2011).

PFGE (pulsed-field gel electrophoresis) is a technique, based on the determination of numbers and sizes of chromosomal DNA on agarose gel (Bidenne *et al.* 1992). Data obtained are reproducible, easy to interpret, and discriminatory. However, this is also a laborious and expensive method that requires a skilled operator.

RAPD (randomly amplified polymorphic DNA) is a PCR-based fingerprinting method that utilizes arbitrary primers or short-length polynucleotides (5–15 nucleotides), annealing at low temperature to loci distributed at random throughout the genome. As a result of amplification, an increasing number of DNA copies are visualized by gel electrophoresis (Quesada and Cenis 1995).

RFLP (restriction fragment length polymorphism) is another PCR-based fingerprinting method that uses short sequences annealing to complementary regions of ribosomal rDNA for amplification and visualization by gel electrophoresis (Barros Lopes *et al.* 1998).

Real-time quantitative polymerase chain reaction (QPCR) is similar to traditional PCR, except that a fluorescent dye is incorporated to the procedure (Ivey and Phister 2011). This technique remains the most widely used method for evaluation of wild yeast dynamics during fermentation. The major disadvantage of QPCR (Hierro *et al.* 2006, Martorell *et al.* 2005), apart from the required cost and skill, is the lack of its ability to differentiate between viable and nonviable cells. This method is quick and no more than 5 h is required to analyze a sample (Martorell *et al.* 2005).

Multilocus sequence typing (MLST) gives the numbers and sizes of repetitive and hypervariable sequences (genetic markers) distributed in loci along the mitochondrial DNA (Ayoub *et al.* 2006).

SSRs sequences (simple sequence repeats) consist of repeated sequences usually having 10 bp (base pairs), which originate from DNA-replication errors (Strand *et al.* 1993).

Select SSR genetic markers are able to detect the high frequencies of different alleles at each locus located on different chromosomes of *S. cerevisiae.* This repeated sequence of *S. cerevisiae* allows detecting remarkable differences between genotypes and heterozygosity levels (Pérez *et al.* 2001). This sequence is a stable and powerfully discriminatory indicator of evolutionary relationships between strains and ancestors that are extensively used for paternity tests, and yeast evolution studies (Ezov *et al.* 2006). Databases containing microsatellites and SNP (single nucleotide polymorphism) sequences of the *S. cerevisiae* genome are available (www.pasteur.edu.uy/yeast) to evaluate the evolutionary distance between strains of closely related species (Ayoub *et al.* 2006).

S. cerevisiae genome contains repetitive DNA sequences, rRNA and mitDNA sequences, that allow sequential fingerprinting of strains. A survey on molecular methods for typing yeast strains (Schuller *et al.* 2004) suggests that interdelta typing with an optimized primer pair has almost the same level of discrimination as PFGE. The authors also concluded that PCR amplification of delta sequences is a very convenient method, since it is rapid, reproducible, and very sensitive. Two new primers for interdelta typing (delta21 and delta2) were assayed, and both were unequivocally able to differentiate between laboratory strains from wild-type yeasts giving results equivalent to PFGE (Legras and Karst 2003). In addition, microsatellite sequencing does not require a high investment in equipment and highly skilled human resources (Schuller *et al.* 2004).

3.4 Cell surface and cell–cell interactions

Yeast cells possess a remarkable capacity to adhere to abiotic surfaces, cells, and tissues (Verstrepen and Klis 2006). Film formation, flotation, and flocculation are fungal adhesion phenomena that exemplify the impressive phenotypic plasticity of yeasts, allowing cells to adapt quickly and survive in stressful environments. However, surface interactions may be a problem to industrial ethanol production processes.

3.4.1 Dissolved air flotation

Dissolved air flotation (DAF) includes bubble formation and size, bubble–particle interactions, and measures of supplied air (Edzwald 1995). Flotation assays can be performed at the laboratory scale using small columns (Sousa *et al.* 2001). After inducing flotation in liquid medium (Palmieri *et al.* 1996), the foam enriched with cells of *S. cerevisiae* was dried at room temperature (Figure 3.2a), while dense layers of cells adsorbed at surfaces of air bubbles were able to preserve the integrity of the bubbles during drying as illustrated in Figure 3.2b. In addition, flotation systems have been designed to separate cells, other particles, and solutes from liquid media (Hossain and Saleh 2001, Viehweg and Schügerl 1983). Generation of favorable hydrophobic regions, where the differences in hydrophobicity between the two surfaces (cell wall and bubble surfaces) are low, is required for flotation (Sousa *et al.* 2006). Flotation is also associated with foam formation, which is dependent on surface characteristics, particularly concerning the hydrophobic and hydrophilic balance established at cell surfaces (Sousa *et al.* 2003). Foam formation depends (Miyazu and Yano 1974) on cell age, cultivation conditions, presence of ions, cell wall characteristics, and

Figure 3.2 Foam dried at room temperature (a, magnified at 40x) obtained by flotation of cells of *Saccharomyces cerevisiae* suspended in synthetic medium and the micrograph (b, magnified at 750x), showing dense layers of dry cells adsorbed onto surfaces of three neighboring air bubbles in the dry foam.

surface tension of the medium. However, the main disadvantage of foaming is the sequestering of cells from the medium that leads to decreases in ethanol production. In order to minimize cell sequestering, antifoam agents are added to fermenters.

Defoamers or antifoaming agents are chemical additives that reduce and hinder the formation of foam in industrial process. A variety of antifoam agents are available on the market. In addition, acetic, propionic, and succinic acids are also good inducers of flotation at low pH (Sosa and Farías 2012). If an antifoam agent is not added to the fermentation, the flotation phenomenon may cause problems, such as decreases in the rate of ethanol formation.

Strains (strains FLT) isolated from wort sample of sugarcane molasses came into flotation in laboratory only in the presence of air bubbles (Palmieri *et al.* 1996). Using such a type of yeast strains, the flotation was induced at the end of fermentation by injection of air to the bottom of the fermentation vessel and this resulted in foam enriched with cells that was directly collected into sampling flasks. Due to a high flotation yield, the medium remaining in the fermentation vessel became apparently free of cells. Collapse of this foam can be induced by adding drops of isoamyl alcohol.

3.4.2 Flocculation

Flocculation is a homotypic phenomenon of aggregation involving the same type of cells. The process is dependent on the strain and pH of medium, being defined as a nonsexual aggregation that leads to separation of the cells from the medium as illustrated in Figure 3.3. It is a process inhibited by EDTA (ethylenediaminetetraacetic acid) or sugars, but induced by Ca^{2+} ions (Stratford 1992, Stewart 2009). Calcium ions are able to establish bridges between two carboxylic groups of lectins (surface proteins) located in different cells and this is required for flocculation (Mill 1964, Stratford 1992). The dependence of flocculation on pH observed during microbial cultivations confirms that carboxyl groups are probably the combining groups located at cellular adhesion sites. Calcium removal is required to avoid or minimize the incidence of flocculation in yeast cell suspensions, while restoration of flocculation activity in cultures of nonflocculating cells does not require treatment with solution of chelating agent (EDTA), and washing of the cells with water (Stewart 2009). Other metals are potentially able to induce flotation in yeast suspensions, but none of them can replace calcium ions. Nevertheless, weakly flocculating yeast cells are able to coflocculate in the presence of Ca^{2+}. Flocculation can also be induced by limiting nutrients, which act as strong signals of induction of flocculation (Smit *et al.* 1992). Indeed, the development of flocculation is affected by the chemical nature of the medium, particularly by salts and sugars, pH and temperature of the medium, aeration, and agitation (Claro *et al.* 2007, Sosa and Farías 2012). As *S. cerevisiae* cells may form aggregates in different ways, flocculation should not be confused with other adhesion phenomena such as cell chain formation (Soares 2011), coflocculation, or the aggregation of cells during the sexual reproduction induced by peptide hormones (Lipke and Kujan 1992).

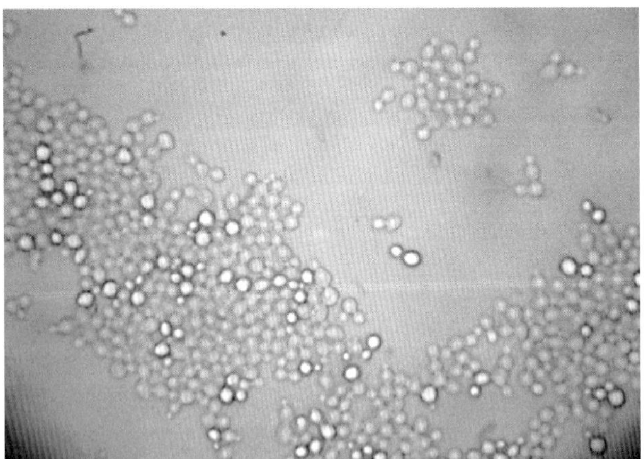

Figure 3.3 Flocculation observed (optical microscope, magnified at 40x) in samples obtained at an industrial alcohol plant during fermentation of sugarcane molasses.

The induction of flocculation is repressed by cycloheximide, due to inhibition of cytoplasmic protein synthesis at ribosomal level, although fermenting yeasts also show flocculation in the presence of high ethanol concentrations (Dengis *et al.* 1995). In beer fermentation, yeast cells showing moderate degree of flocculation also showed increases in hydrophobicity after serial repitching, despite replicative cell age (Powell *et al.* 2003). In addition, the effects of respiratory inhibitors on glucose uptake by yeast cells indicate that the degree of hydrophobicity is influenced by the availability of oxygen (Straver *et al.* 1993). High levels of bacterial contaminants may impair the centrifuging step due to coflocculation with yeast cells (Amorim *et al.* 2009). However, flocculation provides protection against ethanol stress and prevents viability losses. Due to the possibility of facilitating downstream processing, continuous process operating at high cell densities offers the same advantages as immobilized systems (Domingues *et al.* 2000).

3.4.3 Biofilms

Bacteria can form biofilms on surfaces of facilities in alcohol production plants (Skinner-Nemec *et al.* 2007), where cells are embedded in a viscous matrix containing polymeric substances. Maximal adhesion of bacteria to hydrophilic surfaces depends upon the free energy and wettability of surfaces. Within biofilms, bacterial populations are protected against sanitation procedures; consequently, bacterial cells survive, and the likelihood of contamination also increases (Kokare *et al.* 2009). Biofilms possess a diversity of bacterial species that generally reflect the level of contamination of the yeast inoculum (Skinner-Nemec *et al.* 2007). Ethanol production by *S. cerevisiae* in reactors with biofilm has been reported in the literature (Demirci *et al.* 1997).

Xanthan gum is used as a thickening or suspending agent in processed foods. This gum is produced by the Gram-negative bacterium *Xanthomonas campestris* or by other bacteria from the genus *Xanthomonas* during utilization of molasses as carbon source (Moosavi-Nasab *et al.* 2010).

3.5 Sugarcane juice and bagasse

The main steps in the production of anhydrous bioethanol from sugarcane are as follows: reception and cleaning of the sugarcane, extraction of sugars, juice heating, clarification treatments, concentration, fermentation, distillation, and dehydration. Around 75% of Brazilian sugarcane bagasse is burned to produce steam and electricity (Seabra *et al.* 2011).

3.5.1 Harvesting of the sugarcane

Although mechanical harvesting leads to gains in productivity with less environmental risks, sugarcane is still being burnt in tropical countries (Núñez and

Spaans 2008, Aguiar *et al.* 2010). Nevertheless, it is likely that the manual harvest of sugarcane is only adopted today by some Brazilian distilleries located in the northeast regions of the country. It is likely that all mechanical harvesting in Brazil will be of unburned green sugarcane in the near future.

3.5.2 Reception and cleaning of sugarcane

Sugarcane containing impurities (minerals, salts, acids, and fibers) is unloaded on feed tables, where sugarcane is washed with water or dry-cleaned to remove impurities. Then, rotary knives cut sugarcane into pieces before shredding. The opening of the sugarcane structure during defibrillation facilitates juice extraction (Steindl 2010).

3.5.3 Juice extraction

The composition of the sugarcane bagasse (Olivério *et al.* 2010, Canilha *et al.* 2012) varies with culture conditions and the following average values (% w/w, dry basis) are obtained: 50% water, 38.4–45.5% cellulose, 22.7–27.0% hemicellulose, 19.1–32.4% lignin, 1.0–2.8% ashes, and 4.6–9.1% extractives. The primary juice containing around 15% sucrose (16.2°Brix) is extracted by a series of rollers in sugar mill. Then, concentrated sugarcane juice is obtained in concentration facilities, while bagasse is burnt in boilers to cogenerate heat and power (Dias *et al.* 2011).

3.5.4 Juice clarification

The clarification of the juice is a recommended practice to produce ethanol from sugarcane. In addition, the concentrated must, resulting from the evaporation of the clarified juice, shows a darker color due to the formation of products (e.g., aldehydes) toxic to the yeast cells (Prati and Moretti 2010). Hot defecation by liming, sulphitation, carbonation, and double carbonation were the procedures used in clarifications (Bellissimi and Ingledew 2005, Zossi *et al.* 2012). Nevertheless, discoloration of the juice is routinely achieved by defecation (Eggleston 2000), while turbidity is minimal within the pH range of 6.5–7.6. Minimal degradation of sugars by organic acids occurs at pH below 6.5 (Steindl 2010). In conclusion, clarification of sugarcane juice comprises coagulation, flocculation, and the precipitation of the complex colloids having pigments, followed by further elimination by decantation and filtration.

3.5.5 Juice concentration

To reduce energy consumption during distillation, sugarcane juice containing 15% sucrose is concentrated into syrups in MEE (multiple effect evaporator) systems to reach around 60–70°Brix (Dias *et al.* 2009). The concentrated juice is mixed with the remaining clarified fraction to produce the final juice containing 22.5% (w/v), which is cooled before the feeding of fermenters.

3.5.6 Quality of clarified juice

The optimal conditions shown by a high-quality juice are as follows (Steindl 2010, Godoy *et al.* 2008): pH of the juice adjusted to 5.8–6.0, clarified juice adjusted to pH 5.05, calcium <800 mg/kg, manganese and zinc <2 mg/kg, phosphorous 50–65 mg/kg, potassium 700–1300 mg/l, and nitrogen 60–70 mg/kg. Nevertheless, turbidity of the sugarcane juice varies from one alcohol plant to another in the same country. Organic acid concentrations should stay below 5000 ppm to avoid degradation of sugars. As an essential pre-requisite required by a standard clarification process, formation of precipitates of calcium phosphate is due the addition of lime that reacts with the phosphate occurring naturally in the sugarcane juice (Steindl 2010).

3.6 Fermentation of juice and molasses

3.6.1 Starters yeasts

The preparation of ADY begins with the spread of yeast cells. The classical propagation is carried out aerobically at temperatures within the range of 29–32°C and pH adjusted to 4.5–5.0 (Maxon and Johnson 1953). A stock culture of selected yeast is exponentially grown in substrates of low sugar concentrations, so that increasing growth rates and decreasing fermentation rates (<10%) take palace, during sequential batch cultivations of increasing volumes. Then, the resulting yeast mass is washed, concentrated, and dried to obtain the ADY. Hydration is usually made as recommended by manufacturers. In order to obtain better results during fermentations, the resulting ADY must be metabolically acclimatized (Bellissimi and Ingledew 2005) before use to reduce the lag phase and overall fermentation times. When the growth ceases, large-volume cells (maturated cells) can be observed in the culture.

Recently, multistage-based fermentation systems have been designed for the propagation of yeast cells, using supplemented molasses due to increasing demand for starter yeast by food and bioethanol industries (Gómez-Pastor *et al.* 2011).

3.6.2 Raw materials used in fermentation

Sugarcane juice, syrups, and molasses resulting from the sugarcane processing in sugar mills are used in fermentations. In the sugarcane mills, the sugarcane juice is concentrated to syrup in MEE (60–70°Brix). Molasses "B" is an enriched sugar substrate, since it contains 50–65% sugars, 18–33% solids, vitamins, and minerals (Albers *et al.* 1996, Fernández-López *et al.* 2012). Molasses "C" (blackstrap molasses) is the preferred substrate to produce ethanol (Converti *et al.* 2003). However, the three types of molasses (A, B, or C) can be utilized as carbon source to produce ethanol.

3.6.3 The fermentation

Fermentations are usually carried out using the Melle–Boinot process, which is based on the cell recycling procedure. The cell reuse reduces fermentation time and increases ethanol yields, improving the adaptation of the yeast cells to higher sugar and ethanol concentrations (Basso *et al.* 2011, Oliva-Neto and Yokoya 1994). However, glucose is not totally consumed (0.2–0.4% residual sugar) during fermentation. At the end of discontinuous fermentation, yeast cells present in the wine (fermented broth) are separated in centrifuges (Dias *et al.* 2009) to form the yeast cream (cell suspension, not yet treated with sulfuric acid).

Despite increases in ethanol yield, decreases in viability and biomass can be observed during fermentation (Souza *et al.* 2007). It is usually accepted that temperatures above 30–32°C should be avoided in industrial bioreactors to obtain optimal levels of ethanol production (Aldiguier *et al.* 2004). Nevertheless, it is difficult to maintain temperatures within this range in tropical countries. However, maintenance of temperature of the broth during fermentation is critical to maintain the viability.

3.7 Cogeneration of energy from bagasse

The cogeneration of energy in most power plants around the world is based on the Rankine cycle principle. This thermodynamic cycle is a commercially mature technology, which is important to the continuous generation of heat, steam, and condensates in industrial plants.

The ethanol production from sugarcane depends on the energy (heat and steam) generated from bagasse. In alcohol plants, cogeneration systems are basically constituted of boilers, backpressure turbines, deaerators, and condensers. Approximately, 250 kg bagasse with 50% humidity is produced per ton of sugarcane in an autonomous distillery with a few kilograms of bagasse surplus/sugarcane ton (Ensinas *et al.* 2007), which is equivalent to about 120 kWh/t of bagasse (Dias *et al.* 2011, Basso *et al.* 2011). The steam produced in boilers is used to obtain electricity in steam turbines and represents thermal energy required to run the process (Dias *et al.* 2010). Both the sugarcane preparation and juice extraction also depend on steam, while the energy surplus is sold as electricity.

Absorption chillers or cooling towers installed in cooling system are used to improve the refrigeration of the fermentation media. Due to the capacity of the lithium bromide solutions to absorb steam, absorption chillers are of good help to maintain fermentation temperatures within an optimal range (Olivério *et al.* 2010). In addition, the constant circulation of water (pretreated and refrigerated in cooling towers) from lakes or rivers allows maintaining temperatures in heat exchangers. Steam at low pressures (20 bar and 350°C) and back pressure steam turbines result in a low energetic efficiency, high bagasse consumption, and low bagasse surplus (Dias *et al.* 2011). Gains in cogenerated energy could be obtained by increasing boiler efficiency and decreasing losses by diffusion and exhaustion of the steam energy (Ensinas *et al.* 2007). Concerning fermentation, when temperatures are within the range of 28–32°C, cell viability and the yields of ethanol are higher, while energies required for distillation are significantly reduced (Magazoni *et al.* 2010).

3.8 Bioreactors and processes

After harvest, the sugarcane is processed to obtain sugars in sugar mills, ethanol in the distilleries and both the sugar and ethanol in integrated alcohol industries (Cardona *et al.* 2010). Around 75% of the sugarcane plants in Brazil are integrated alcohol industries (Andrietta *et al.* 2011, Basso *et al.* 2011), while other countries also produce sugar and ethanol in integrated factories, though not on the scale of the Brazilian industry (Magazoni *et al.* 2010).

Figure 3.4 shows a block diagram of integrated alcohol industries to describe the production of sugar and ethanol. This figure summarizes the clarification

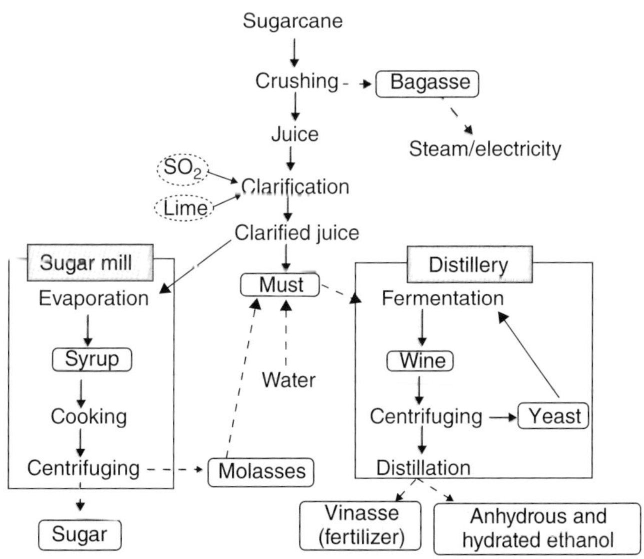

Figure 3.4 Diagram of an annexed distillery illustrating the interactions between sugar mill and distillery in an integrated alcohol factory.

of the juice, and formation of syrup and molasses used to produce crystallized sugar and ethanol. Molasses (76–80°Brix) is diluted to around 25°Brix before being used as substrate to produce ethanol.

Fermentation is the bioconversion process that converts monosaccharides (mainly the glucose) into alcohol and CO_2 via a sequence of reactions known as glycolysis. However, fermentation of molasses and juice is affected by the composition, mainly of organic acids and levels of sugar and nitrogen that are required to synthesize amino acids and proteins (Yoshimi and Masazumi 1981).

Batch, fed-batch, and continuous operation are the three main types of processes used in fermentations. Fed-batch process allows obtaining average yields of 91–92% in Brazilian alcohol industries (Della-Bianca *et al.* 2013).

3.8.1 Batch fermentation

In this type of process, the production of the metabolites as ethanol is usually associated with growth. Nutritional supplements, inoculum added initially, and agitation at controlled aeration rates are the factors that lead to growth and simultaneous fermentation. Chemical and antibiotics are added to the medium to control pH and contamination, but the withdrawal of samples cannot cause significant variations in the workload volumes of the bioreactors (Maxon and Johnson 1953, Zhang 2009).

Modern yeast propagation systems are equipped with more than one fermenter operating in series to meet the high demand of the food and bioethanol industries to display fermentations (Gómez-Pastor *et al.* 2011). Assessment of growth performance with varying nutrients can be undertaken in small volume Erlenmeyer flask trials. A new strategy based on the use of a feeding containing limiting glucose was recently developed to produce large amounts of yeast biomass (Vieira *et al.* 2013).

3.8.2 Fed-batch fermentation

Unlike the batch fermentation systems, the volume of the medium is not constant in the fed-batch fermentation systems. Fed-batch processes with increasing working volumes are operated in order to obtain higher product yields and reduced inhibition by the sugar concentration (Hong 1986).

Figure 3.5 describes a fed-batch process designed to operate with four independent reactors in parallel. Before the feeding with must of sugarcane, about 30% of the working volume of the bioreactors is filled with yeast milk, a suspension containing around 30% cells (wet weight) and previously treated with sulfuric acid. In each fermenter, the fermentation time usually takes about 6–8 h to complete, including the feeding (2–4 h) and fermentation phases. During fermentation, part of the sugarcane juice is continuously cooled by passing through heat exchangers. At the end of the fermentation, the fermented broth is sent to centrifuges where yeast cells are separated from the wine

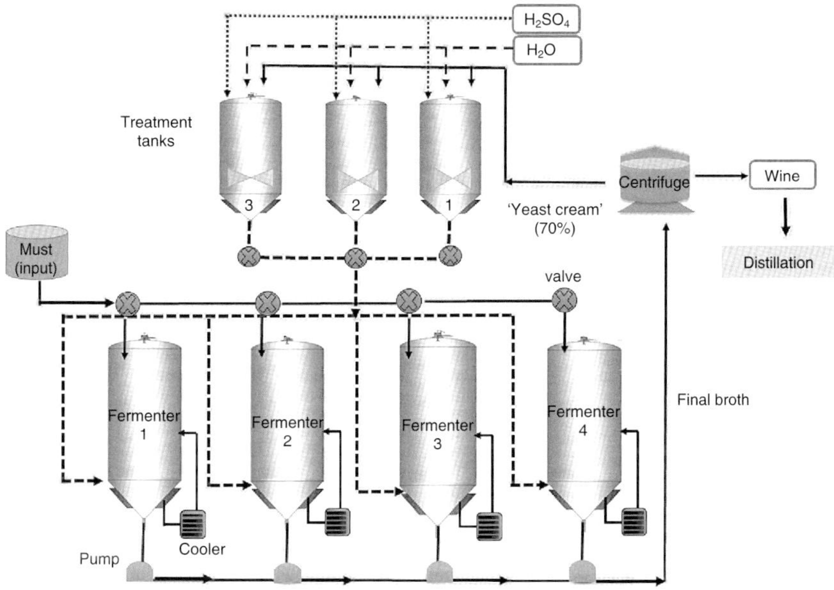

Figure 3.5 Fed-batch fermentation with cell recycling.

(fermented broth). The yeast cream (70% cells) is diluted to 30% cells (wet weight basis) with water and then treated with sulfuric acid. As bacteria are less resistant to low pH than the yeast cells, acidification reduces the levels of contaminants, which increases ethanol yields. However, acid treatment can lead to significant drops in the yeast viability, depending on conditions and adaptation of the yeast population.

Due to the simultaneous and concomitant operation of the independent bioreactors in the system described earlier (Figure 3.5), the following is observed from time to time: reactor-1 is fed by yeast suspension and must of sugarcane, while reactor-2 remains completely filled by the medium under full fermentation. Meantime, fermenting broth is pumped from reactor-3 to the centrifuges, where cells are separated from the wine. Reactor 4 remains empty for cleaning-in-place (CIP) and sanitization for a period of usually 2 h using flegmass or steam as described in the literature (Troller 2012) before restarting a new fermentation cycle. The broth enriched with ethanol is continuously sent to centrifuges, while the yeast cream is pumped back to the treatment tanks.

The fed-batch system differs from the batch process as follows (Stanbury *et al.* 2003): high yields of ethanol and productivities with less residual sugar and lower risk of inhibition by substrates and products. Other advantages are decreased fermentation time, less toxic effects of the medium components on the yeast cells, and maintenance and control of the physiological state of the cells at desired levels. However, when inoculum density is high, lower cells viabilities are observed

(Nagodawithana *et al.* 1974, Laluce *et al.* 2009), as confirmed by central composite design and response surface methodology (RSM). Thus, when the cell mass (>15%, wet weight) returns to the bioreactors after sulfuric acid treatment, significant drops in viability are observed due to rapid exhaustion of nutrients from the sugarcane must. Optimizations of state variables (cell mass, concentration of substrates, and ethanol) are required to reach the maximum process performance (Carvalho *et al.* 1990, Converti *et al.* 2003, Rivera *et al.* 2013). In another study (Cheng *et al.* 2009), the performance of a batch process was compared with a fed-batch processes, in a defined medium under aeration and containing yeast extract. Under such conditions, the following results were observed in fed-batch process: decreases in the biomass yields ($Y_{P/X,}$ g ethanol/g sugar) followed by simultaneous increases in specific rate (h^1) of sugar consumption, and formation of ethanol.

Compared with the simple batch process, the fed-batch fermentation system shows the following characteristics (Sheppard 2009):

- maintenance of substrate at low levels;
- reduced risks of substrate or product inhibition;
- maintenance of the physiological state of the cells at desired levels by manipulating the composition of the feeding medium;
- an extended stationary phase by adding nutrients to the feed and/or changing the substrate concentration;
- larger amounts of products resulted from increases in medium volume;
- lower levels of inhibitors that shorten the fermentation times.

3.8.3 Multistage Stage Continuous Fermentation (MSCF) system

Figure 3.6 describes an MSCF (multistage stage continuous fermentation) system designed to operate continuously with four reactors linked in series, similar to others reported in the literature (Bayrock and Ingledew 2001). In such a system, the maximum fermentation rate is reached in the first reactor, which is simultaneously fed by must of sugarcane and consequently requires more cooling to maintain the broth temperature within the range of 32–35°C. A broad stationary phase can last for months in this kind of system, depending on operational and bacterial controls. Fermented broth is pumped between reactors linked in series to each other, while the decreases in sugars in the bioreactors are observed with the resultant increases in ethanol concentration. The final broth containing low levels of sugar and high amounts of ethanol is pumped from the last reactor to centrifuges. Theoretically, this type of system is designed to be continuously fed by the must of sugarcane juice in order to ensure a continuous flow of wine toward distillation columns. The advantage of the use of this system is that maximum levels of ethanol are obtained without the need for cleaning and sanitizing

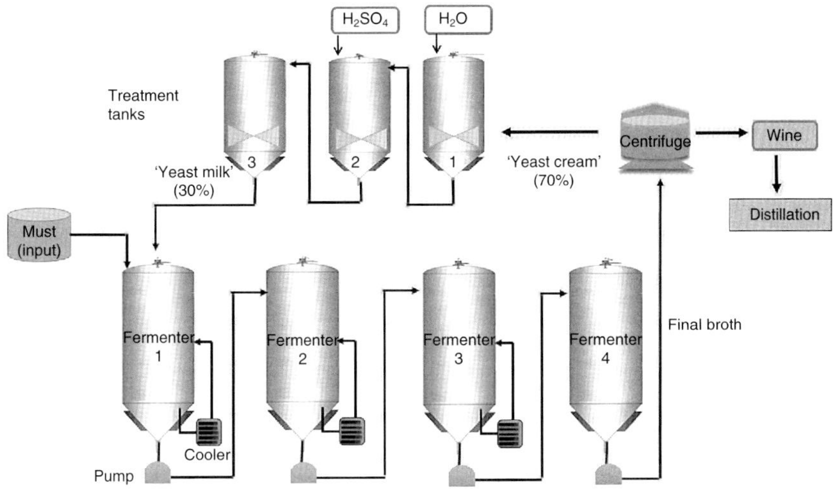

Figure 3.6 Multistage continuous system with cell recycling.

of the bioreactors, over months of the operation at a quite constant steady state (Deindoerfer and Humphrey 1959).

Other advantages of operating continuous stirred-tank reactors (CSTR) are the higher volumetric productivities and lower capital required for facilities and operational control (Deindoerfer and Humphrey 1959). However, in a continuous process with yeast cell recycling, the bacterial contaminations are greater than in fed-batch processes (Domingues *et al.* 2000). Thus, increases in the amounts of antibiotics and/or sulfuric acid are required. Less than 10% of the Brazilian distilleries are currently operating a continuous process (Abreu-Cavalheiro and Monteiro 2013).

3.9 Control of microbial infections

Bacterial contamination in industrial fermentations is a common and persistent problem. However, the levels of bacterial contaminants are significantly reduced with the use of effective cleaning procedures, sanitizing of facilities, pasteurizing of substrates, and addition of antibiotics. Despite prevention procedures, propagation of contaminants above acceptable levels is a persistent risk in the industrial ethanol production (Narendranath and Brey 2009). Due to this, a great diversity of antimicrobials is commercially available. As dust, air, water, and the raw materials are sources of contaminants, the ethanol producers should maintain bacterial contaminants at acceptable levels ($\leq 10^6$ CFU/mL) to avoid outbreaks of infections.

The choice of an antimicrobial product is determined by its efficacy at low concentrations, low costs, and ecological safety. Several chemical antiseptics (hydrogen peroxide and potassium metabisulfite) and antibiotics such as penicillin, streptomycin, and tetracycline are able to limit growth of contaminants (Narendranath *et al.* 2001). Minor *Lactobacillus* growth can be contained with the addition of virginiamycin (innocuous to yeasts) to the fermentation process (Bischoff *et al.* 2007). Natural or synthetic organics substances antimicrobials can also be used in sanitizing procedures of facilities, as well as to reduce contaminants in the fermentation broth (McDonnell and Russell 1999). Hop acids appeared to be efficient during fermentation, when used to replace antibiotics of chemical origin (Rückle and Senn 2006).

Although not preferred in the ethanol industry, the hops used in beer fermentation shows antimicrobial activity, similar to Kamoran (sodium monensin, antibiotic produced by *Streptomyces cinnamonensis*), during the fermentation of sugarcane juices (Leite *et al.* 2013). However, monensin is the most used antibacterial used in Brazilian distilleries (Stroppa et al. 2000). Chlorine dioxide is widely used as a powerful sanitizing agent that has a broad germicidal activity against Gram-positive and Gram-negative bacteria and yeasts (Gómez-López *et al.* 2009, Zhu *et al.* 2013). During alcohol fermentation, growth of bacteria (*Bacillus subtilis, L. plantarum, L. fermentum, and Leuconostoc mesenteroides*) and yeasts is efficiently inhibited by chlorine dioxide at concentration between 10 ppm to 125 ppm (Meneghin *et al.* 2008). Since chlorine dioxide is easily decomposed, without leaving residues, risks of generating mutants resistant to antibiotics are significantly reduced. Although used in juice clarification for color removal, sulfur dioxide (SO_2) can be another problem to the ethanol industry (Oliva-Neto *et al.* 2013). The high values of minimum inhibitory concentration (MIC) determined for SO_2 (ranging from 10 to 40 mg/l for lactic acid bacteria and 5000 mg/l for *S. cerevisiae*) indicate that the use of sulfuric acid to control bacterial infections may be not as effective as expected. In this case, use of yeast strains resistant to the sulfite is recommended (Stannard 1997).

Lastly, the consumption of fermentation products by animals, such as dry yeast including the antibiotics, are not allowed in some countries due to the possibility of developing resistance to antibiotics in animals.

3.10 Monitoring and controlling processes

The monitoring of the fermentation process is mainly based on variations in fermentation parameters, as yields and flow rates of materials in the input and output of the systems. State variables or response variables (Farinas *et al.* 2014) are dependent on the disorder variables (environmental variables) and control variables, which are dependent on the valve operations. The control variables are useful to elaborate the complex and dynamic mathematical models to optimize the operation of the system. Nevertheless, the precise monitoring of the industrial fermentations is not easy, since bioreactors are open systems that operate within a narrow range of optimal conditions.

Stress conditions can be enumerated as follows (Ingledew 2009, Nofemele *et al.* 2012): nutritional starvation, physical stress (e.g., temperature), and chemical stress (acidity, ethanol concentration), as well as the effects of organic acids on the yeast cells. Nevertheless, stress can be minimized by manipulating variables (Liu *et al.* 2012) such as temperature (optimum at 34–35°C), osmotic factors (≤18–20°Brix), and pH (2.0–2.5) obtained during the sulfuric acid treatment. Other inhibitory factors are by-products, such as ethanol concentration (10–13%, v/v) and the presence of toxic ions as sulfides (100 mg/ml).

Temperature is an important key factor in the control of the process performance (Torija *et al.* 2003, Dias *et al.* 2009, 2011). Due to this, the variables energy demands, yeast viability, alcohol levels in the wine (fermented broth), growth rates, length of lag phases, specific rate of ethanol formation, and cellular enzyme activities depend on temperature. However, it is generally accepted that temperatures below 30–32°C should be maintained in the fermenting broth in order to obtain high levels of ethanol. Alterations in membrane functions (e.g., permeability to weak acids), and glycerol and trehalose accumulation are also other responses to stresses (Basso *et al.* 2011).

Concerning chemical analysis, the key factors to control fermentation processes are based on the determination of three different classes of analytes, which are carbohydrates, organic acids, and alcohols. To perform assays of analytes, there is a great variety of instrumental and analytical tools, as follows:

- high-performance liquid chromatography (HPLC);
- near-infrared spectroscopy (NIRS);
- enzymatic instruments (YSI biochemical analyzers, Yellow Springs Instruments, USA) to measure sugars, ethanol, glycerol, lactic acid, and other by-products;
- sophisticated instruments, such as probes for the determination of soluble solids (°Brix);
- optical fibers to assay sugars and ethanol (Fujiwara *et al.* 2012).

Electronic sensors to determine pH, dissolved oxygen, and CO_2 are the most commonly used instruments in bioprocess monitoring (Harms *et al.* 2002). The use of inline biosensors in bioreactors is limited by the difficulties in preventing unwanted effects of the uncontrolled variables, and those related to climate.

Compared with biosensors, HPLC has the advantage in allowing more detailed assessment and simultaneously analyzing a variety of compounds in the broth. However, chemical sensors for individual assays in complex media show limited sensitivity, but this can be solved using other systems such as electronic tongues (Rudnitskaya and Legin 2008).

Spectroscopy techniques allow simultaneous evaluation of different components of fermentation broths of complex composition. Spectroscopic methods are fast and nondestructive techniques, which require little or no previous preparations of samples, as reported by the FTIR (Fourier transform infrared spectroscopy) or infrared spectroscopy (Veale *et al.* 2007).

Viable cells of contaminants (bacteria, yeasts, and molds) are easily enumerated in cultures by optical microscopy. In addition, Petri films (3M Corporation, USA) are available in the market for fast identification and enumeration of microbial cells. Depending on nutritional deficiencies, unusual cellular (dimorphic states) and colony morphologies can be observed in cultures of *S. cerevisiae* (Gimeno *et al.* 1992).

Control panels (Figure 3.7) of fermentation rooms in alcohol plants give an integrated view of the entire process, and variations in ethanol production due to variations in physicochemical parameters of the raw materials can be corrected in a real time with slight interventions.

3.11 Concluding remarks and perspective

The interest in the production of ethanol and energy from sugarcane was stimulated in 1975 with the implementation of the National Alcohol Program in Brazil. This program was launched in response to rising oil prices and the crisis in the international sugar market. Due to the current economic crisis and severe droughts, reduced numbers of sugar mills are expected to be in operation during the harvesting of the 2015/2016 sugarcane crop (UNICA 2015).

Figure 3.7 Control room of a commercial ethanol plant.

Cell reuse during fermentation leads to significant changes in the microbial population of the process in any alcohol plant of tropical countries. Stress conditions, mainly related to high temperatures and ethanol concentrations, induce the formation of subpopulations of yeast strains. During these dynamic and evolutionary processes, new daughter cells different from the mother cells arise in the yeast population (Silva-Filho *et al.* 2005).

The benefits of cell reuse during crop seasons are due to the development of resistant and stable yeast strains, mainly related to the capacity to ferment sucrose. Thus, finding new strains and improving the management of the process control would be great contributions to industrial processing. Development of more efficient controls to determine contaminants would allow a better cleaning of facilities and the obtainment of a must of sugarcane showing better quality for the fermentation. Strains resistant to high levels of sugars, ethanol, and temperatures would allow performing VHG (very high gravity) fermentations to obtain higher yields of ethanol and less vinasse.

Monitoring and control of the evolutionary process that takes place in an alcohol plant during the entire crop season is not an easy task. Nevertheless, the major challenge is to reduce the costs associated with raw materials and required operations to maintain Brazil in its position as the most important ethanol supplier to world markets and supply the increasing domestic demands. Lastly, the main scientific challenges, in regard to the production of the first- (1G ethanol) and second-generation ethanol (2G ethanol), are the development of technologies to use new feedstocks that will lead to more positive energy balances, meeting environmental sustainability criteria (Amorim *et al.* 2011).

Acknowledgments

We are deeply grateful to Dr. Ian O'Hara for inviting us to write this chapter. We are also grateful to FAPESP (Fundação de Amparo à Pesquisa do Estado de São Paulo) for financial aids granted to our research over the years, as well as the scholarships awarded by CNPq (Conselho Nacional de Desenvolvimento Científico e Tecnológico, Brasília), and CAPES (Coordenação de Aperfeiçoamento de Pessoal de Nível Superior) to the students.

References

Abdel-Banat, B. M. A., Hoshida, H., Ano, A., Nonklang, S., and Akada, R. (2010). High-temperature fermentation: how can processes for ethanol production at high temperatures become superior to the traditional process using mesophilic yeast? *Applied Microbiology and Biotechnology* **85**(4):861–867.

Abreu-Cavalheiro, A. and Monteiro, G. (2013). Solving ethanol production problems with genetically modified yeasts. *Brazilian Journal of Microbiology* **44**(3):665–671.

Aguiar, D. A., Silva W. F., Rudorff, B. F. T. and Adami, M. (2010). Canasat Project: Monitoring the Sugarcane Harvest Type in the State of São Paulo, Brazil. In ISPRS TC VII Symposium-100 Years ISPRS, Advancing Remote Sensing Science, July 5–7, Vienna, Austria.

Albers, E., Larsson, C., Lidén, G., Niklasson C. and Gustafsson, L. (1996). Influence of the nitrogen source on Saccharomyces cerevisiae anaerobic growth and product formation. *Applied and Environmental Microbiology* **62**(9):3187–3195.

Aldiguier, A. S., Alfenore, S., Cameleyre, X., Goma, G., Uribelarrea, J. L., Guillouet, S. E. and Molina-Jouve, C. (2004). Synergistic temperature and ethanol effects on Saccharomyces cerevisiae dynamic behavior in ethanol bio-fuel production. *Bioprocess and Biosystems Engineering* **26**(4):217–222.

Allen, F. (1994). The microbrewery laboratory manual - Part III: wild yeast detection and remediation. *Brewing Techniques* **2**(6).

Amberg, D. C., Burke, D. J. and Stranthern, J. N. (2005). In *Experiment in yeast genetics: Cold Spring Harbor Laboratory Course Manual*, pp. 1–95. CSHL: Cold Spring Harbor Press Publishing: Cold Spring Harbour, New York.

Amorim, H. V., Basso, L. C. and Lopes, M. L. (2009). Sugarcane juice and molasses, beet molasses and sweet sorghum: composition and usage. In *The Alcohol Text Book*, pp. 39–46. W. M. Ingledew, D. R. Kelsall, G. D. Austin and C. Kluhspies, eds. Nottingham University Press Publishing: Nottingham.

Amorim, H. V., Lopes, M. L., Oliveira, J. V. C., Buckeridge, M. S. and Goldman, G. H. (2011). Scientific challenges of bioethanol production in Brazil. *Applied Microbiology and Biotechnology* **91**(5):1267–1275.

Andrietta, M. G. S., Andrietta, S. R. and Stupiello, E. (2011). Bioethanol - What has Brazil learned about yeasts inhabiting the ethanol production processes from sugar cane? In *Biofuel Production-Recent Developments and Prospects*, pp. 67–85. M. A. dos Santos Bernardes, ed. InTech Publishing: Rijeke (Croatia).

Antonangelo, A. T. B. F., Alonso, D. P., Ribolla, P. E. M. and Colombi, D. (2013). Microsatellite marker-based assessment of the biodiversity of native bioethanol yeast strains. *Yeast* **30**(8):307–317.

Ayoub, M. C., Legras, J. L., Saliba, R. and Gaillardin, C. (2006). Application of multi locus sequence typing to the analysis of the biodiversity of indigenous Saccharomyces cerevisiae wine yeasts from Lebanon. *Journal of Applied Microbiology* **100**(4):699–711.

Balat, M. and Balat, H. (2009). Recent trends in global production and utilization of bio-ethanol fuel. *Applied Energy* **86**(11):2273–2282.

Barros Lopes, M., Sonden, A., Martens, A. L., Henschke, P. A. and Langridge, P. (1998). Differentiation and species identification of yeasts using PCR. *International Journal of Systematic Bacteriology* **48**(1):279–286.

Basílio, A. C. M., Araujo, P. R. L., Morais, J. O. F., Silva-Filho, E. A., Morais Jr., M. A. and Simões, D. A. (2008). Detection and identification of wild yeast contaminants of the industrial fuel ethanol fermentation process. *Journal of Applied Microbiology* **10**(4):2438–2447.

Basso, L. C., Amorim, H. V. and de Oliveira, A. J. (1997). The antibacterial action of succinic acid formation by yeast during fermentation. *Revista de Microbiologia (São Paulo, Brazil)* **28**(1):77–82.

Basso, L. C., Amorim, H. V., Oliveira A. J. and Lopes, M. L. (2008). Yeast selection for fuel ethanol production in Brazil. *FEMS Yeast Research* **8**(7):1155–1163.

Basso, L. C., Basso, T. O. and Rocha, S. N. (2011). Ethanol production in Brazil: the industrial process and its impact on yeast fermentation. In *Biofuel Production-Recent Developments and Prospects*, pp. 55–100. M. A. dos Santos Bernardes, ed. InTech Publishing: Vienna (Austria).

Bayrock D. P. and Ingledew, W. M. (2001). Application of multistage continuous fermentation for production of fuel alcohol by very-high-gravity fermentation technology. *Journal of Industrial Microbiology and Biotechnology* **27**(2):87–93.

Bellissimi, E. and Ingledew, W. M. (2005). Metabolic acclimatization: preparing active dry yeast for fuel ethanol production. *Process Biochemistry* **40**(6):2205–2213.

Bidenne, C., Blondin, B., Dequin, S. and Vezinhet, F. (1992). Analysis of the chromosomal DNA polymorphism of wine strains of Saccharomyces cerevisiae. *Current Genetics* **22**(1):1–7.

Bischoff, K. M., Skinner-Nemec, K. A. and Leathers, T. D. (2007). Antimicrobial susceptibility of Lactobacillus species isolated from commercial ethanol plants. *Journal of Industrial Microbiology and Biotechnology* **34**(1):739–744.

Bitterman, K. J., Medvedik, O. and Sinclair, D. A. (2003). Longevity regulation in Saccharomyces cerevisiae: linking metabolism, genome stability, and heterochromatin. *Microbiology and Molecular Biology Reviews* **67**(3):376–399.

Brenner, D. J., Staley, J. T. and Krieg, N. R. (2005). Classification of prokaryotic organisms and concepts of bacterial speciation. In *Bergey's Manual of Systematic Bacteriology*, vol. **2**, pp. 27–32. G. Garrity, ed. Lippincott Williams & Wilkins Publishing: Baltimore.

Canilha, L., Chandel, A. K., Milessi, T. S. S., Antunes, F. A. F., Freitas, W. L. C., Felipe, M. G. A. and Silva, S. S. (2012). Bioconversion of sugar cane biomass into ethanol: an overview about composition, pretreatment methods, detoxification of hydrolysates, enzymatic saccharification and ethanol fermentation. *Journal of Biomedicine and Biotechnology* **2012**:1–15.

Cardona, C. A., Sanches, O. J. and Gutierrez, L. F. (2010). Sucrose containing materials used for ethanol production. In *Process Synthesis for Fuel Ethanol Production*, pp. 50–52. Taylor and Francis Group (CRC Press) Publishing: New York.

Carvalho, J. C. M., Aquarone, A., Sato S., Brazzach, M. L., Almeida, K. A. and Borzani, W. (1990). Influence of exponentially decreasing feeding rates on fed-batch ethanol fermentation of sugar cane blackstrap molasses. *Biotechnology Letters* **12**(10):777–782.

Cheng, N. G., Hasan, M., Kumoro, A. C., Ling, F. C. and Tham, M. (2009). Production of ethanol by fed-batch fermentation. *Pertanika Journal of Science and Technology (Malaysia)* **17**(2):399–408.

Claro, F. B., Rijsbrack, K. and Soares, E. V. (2007). Flocculation onset in Saccharomyces cerevisiae: effect of ethanol, heat and osmotic stress. *Journal of Applied Microbiology* **102**(3):693–700.

Converti, A., Arni, S., Sato, S., Carvalho, J. C. M. C. and Aquarone, E. (2003). Simplified modeling of fed-batch alcoholic fermentation of sugarcane blackstrap molasses. *Biotechnology and Bioengineering* **84**(1):88–95.

Danielsen, M. and Wind, A. (2003). Susceptibility of Lactobacillus spp. to antimicrobial agents. *International Journal of Food Microbiology* **82**(1):1–11.

Deindoerfer, F. H. and Humphrey, A. E. (1959). Design of multistage systems for simple fermentation processes. *Industrial and Engineering Chemistry* **51**(7):809–812.

Della-Bianca, B. E., Basso T. O., Stambuk, B. U., Basso, L. C. and Gombert, A. K. (2013). What do we know about the yeast strains from the Brazilian fuel ethanol industry? *Applied Microbiology and Biotechnology* **97**(3):979–991.

Demirci, A., Pometto III, H. and Kal G. (1997). Ethanol production by Saccharomyces cerevisiae in biofilm reactors. *Journal of Industrial Microbiology and Biotechnology* **19**(4):299–304.

Dengis, P. B., Nelissen, L. R. and Rouxhet, P. G. (1995). Mechanisms of yeast flocculation: comparison of top- and bottom-fermenting strains. *Applied and Environmental Microbiology* **61**(2):718–728.

Dias, M. O. S., Ensinas, A. V., Nebra, S. A., Filho, R. M., Rossell, C. E. V. and Maciel, M. R. W. (2009). Production of bioethanol and other bio-based materials from sugarcane bagasse: integration to conventional bioethanol production process. *Chemical Engineering Research and Design* **87**(9):1206–1216.

Dias, M. O. S., Cunha, M. P., Jesus, C. D. F., Scandiffio, M. I. G., Rossell, C. E. V., Filho, R. M. and Bonomi, A. (2010). Simulation of ethanol production from sugarcane in Brazil: economic study of an autonomous distillery. *Computer Aided Chemical Engineering* **218**:733–736.

Dias, M. O. S., Cunha, M. P., Filho, R. M., Bonomi, A., Jesus, C. D. F. and Rossell, C. E. V. 2011. Simulation of integrated first and second generation bioethanol production from sugarcane: comparison between different biomass pretreatment methods. *Journal of Industrial Microbiology and Biotechnology* **38**(8):955–966.

Díaz, C., Molina, A. M., Nähring, J. and Fischer, R. (2013). Characterization and dynamic behavior of wild yeast during spontaneous wine fermentation in steel tanks and amphorae. *BioMed Research International* **2013**:1–13.

Domingues, L., Lima, N. and Teixeira, J. A. (2000). Contamination of a high-cell-density continuous bioreactor. *Biotechnology and Bioengineering* **68**(5):584–587.

Dorta, A. C., Oliva-Neto, P., Abreu-Neto, M. S., Nicolau-Junior, N. and Nagashima, A. I. (2006). Synergism among lactic acid, sulfite, pH and ethanol in alcoholic fermentation of Saccharomyces cerevisiae (PE-2 and M-26). *World Journal of Microbiology and Biotechnology* **22**(2):177–182.

Edzwald, J. K. (1995). Principles and applications of dissolved air flotation. *Water Science and Technology* **31**(3–4):1–23.

Eggleston, G. (2000). Hot and cold lime clarification in raw sugar manufactory I: juice quality differences. *International Sugar Journal* **102**(1220):406–416.

Ensinas, A. V., Nebra, S. A., Lozano, M. A. and Serra, L. M. (2007). Analysis of process steam demand reduction and electricity generation in sugar and ethanol production from sugarcane. *Energy Conversion and Management* **48**(11):2978–2987.

Ezov, T. K., Boger-Nadjar, E., Frenkel, Z., Katsperovski, I., Kemeny, S., Nevo, E., Korol, A. and Kashi, Y. (2006). Molecular-genetic biodiversity in a natural population of the yeast Saccharomyces cerevisiae from "Evolution Canyon": microsatellite polymorphism, ploidy and controversial sexual status. *Genetics* **174**(3):1455–1468.

Farinas, C. S., Ladislau, M. N. and Giordano, R. C. (2014). Instrumentation and automation in the sugarcane ethanol agroindustry. In *Sugarcane Bioethanol – R. D. for Productivity and Sustainability*, pp. 601–618. L. A. B. Cortez, ed. Edgard Blücher Publishing: São Paulo.

Fernández-López, C. L., Torrestiana-Sánchez, B., Salgado-Cervantes, M. A., Mendoza García, P. G. and Aguilar-Uscanga, M. G. (2012). Use of sugarcane molasses "B" as an alternative for ethanol production with wild-type yeast Saccharomyces cerevisiae ITV-01 at high sugar concentrations. *Bioprocess and Biosystems Engineering* **35**(4):605–614.

Fujiwara, E., Ono, E., Manfrim, T. P., Santos, J. S. and Suzuki C. K. (2012). Measurement of sucrose and ethanol concentrations in process streams and effluents of sugarcane bioethanol industry by optical fiber sensor. *IEEE Sensors Journal* **12**(9):2839–2843.

G-Alegria, E., López, I., Ruiz, J. I., Sáenz, J., Fernández, E., Zarazaga, M., Dizy, M., Torres, C. and Ruiz-Larrea, F. (2004). High tolerance of wild Lactobacillus plantarum and Oenococcus oeni strains to lyophilisation and stress environmental conditions of acid pH and ethanol. *FEMS Microbiology Letters* **230**(1):53–61.

Gallardo, J. C. M., Souza, C. S., Cicarelli, R. M. B., Oliveira, K. F., Morais, M. R. and Laluce, C. (2011). Enrichment of a continuous culture of Saccharomyces cerevisiae with the yeast Issatchenkia orientalis in the production of ethanol at increasing temperatures. *Journal of Industrial Microbiology and Biotechnology* **38**(3):405–414.

Gasent-Ramírez, J. M. and Benítez, T. (1997). Lysine-overproducing mutants of Saccharomyces cerevisiae baker's yeast isolated in continuous culture. *Applied and Environmental Microbiology* **63**(12):4800–4806.

Gimeno, C. J., Ljungdahl, P. O., Styles, C. A. and Fink, G. R. (1992). Unipolar cell divisions in the yeast S. cerevisiae lead to filamentous growth: regulation by starvation and RAS. *Cell* **68**(6):1077–1090.

Godoy, A., Amorim, H. V., Lopes, M. L. and Oliveira, A. J. (2008). Continuous and batch fermentation processes: advantages and disadvantages of these processes in the Brazilian ethanol production. *International Sugar Journal* **110**(1311):175–181.

Gomes, F. C. O., Silva C. L. C., Marini, M. M., Oliveira, E. S. and Rosa, C. A. (2007). Use of selected indigenous Saccharomyces cerevisiae strains for the production of the traditional cachaça in Brazil. *Journal of Applied Microbiology and Biotechnology* **103**(6):2438–2447.

Gómez-López, V. M., Rajkovic, A., Ragaert, P., Smigic, N. and Devlieghere, F. (2009). Chlorine dioxide for minimally processed produce preservation: a review. *Trends in Food Science & Technology* **20**(1):17–26.

Gómez-Pastor, R. M., Pérez-Torrado, R., Garre, E. and Matallana, E. (2011). Recent advances in yeast biomass production. In *Biomass - Detection, Production and Usage*, pp. 201–222. D. Matovic, ed. InTech Publishing: Rijeka (Croatia).

Harms, P., Yordan, K. and Govind, R. (2002). Bioprocess monitoring. *Current Opinion in Biotechnology* **13**(2):124–127.

Hartwell, L. H. (1974). Saccharomyces cerevisiae cell cycle. *Bacteriological Reviews* **38**(2):164–198.

Hierro, N., Esteve-Zarzoso, B., González, A., Mas, A. and Guillamón, J. M. (2006). Real-time quantitative PCR (QPCR) and reverse transcription-QPCR for detection and enumeration of total yeasts in wine. *Applied and Environmental Microbiology* **72**(11):7148–7155.

Hofvendahl, K. and Hahn-Hägerdal, B. (2000). Factors affecting the fermentative lactic acid production from renewable resources. *Enzyme and Microbial Technology* **26**(2–4):87–107.

Hong, J. S. (1986). Optimal substrate feeding policy for a fed-batch fermentation with substrate and product inhibition kinetics. *Biotechnology and Bioengineering* **28**(9):1421–1431.

Horn, P. and Wilkie, D. (1996). Use of magdala red for the detection of auxotrophic mutants of Saccharomyces cerevisiae. *Journal of Bacteriology* **91**(3):1388–2226.

Hossain, M. M. and Saleh, Z. (2001). Enrichment of α-lactalbumin in whey permeate solution by a semi-batch foaming process. *Australasian Biotechnology* **11**(2):29–31.

Ingledew, W. M. (2009). Yeast stress in fermentation process. In *The Alcohol Textbook*, pp. 115–126. W. M. Ingledew, D. R. Kelsall, G. D. Austin and C. Kluhspies, eds. Nottingham University Press Publishing: Nottingham.

Ingram, L. O. (1986). Microbiology tolerance to alcohols: role of the cell membrane. *Trends in Biotechnology* **4**(2):40–44.

Ivey, M. L. and Phister, T. G. (2011). Detection and identification of microorganisms in wine: a review of molecular techniques. *Journal of Industrial Microbiology and Biotechnology* **38**(10):1619–1634.

Jarboe, L. R., Royce, L. A. and Liu, P. (2013). Understanding biocatalyst inhibition by carboxylic acids. *Frontiers in Microbiology* **4**(1):1–8.

Kokare, C. K., Chakraborty, S., Khopade, A. N. and Mahadik, K. R. (2009). Biofilm: importance and applications. *Indian Journal of Biotechnology* **8**(1):159–168.

Kurtzman, C., Fell, J. W. and Boekhout, T. (2011). Ascomycetes and yeast like taxa. In *The Yeasts: A Taxonomic Study*, pp. 179–195. C. P. Kurtzman, J. W. Fell and T. Boekhout, eds. Elsevier Publishing: Amsterdam.

Laluce, C., Tognolli, J. O., Oliveira, K. F., Souza, C. S. and Morais, M. R. (2009). Optimization of temperature, sugar concentration and inoculum size to maximize ethanol production without significant decrease in yeast cell viability. *Applied Microbiology and Biotechnology* **83**(4):627–637.

Legras, J. L. and Karst, F. (2003). Optimisation of interdelta analysis for Saccharomyces cerevisiae strain characterisation. *FEMS Microbiology Letters* **221**(2):249–255.

Leite, I. R., Faria, J. R., Marquez, L. D. S., Reis M. H. M., Resende M. M., Ribeiro, E. J. and Cardoso, V. L. (2013). Evaluation of hop extract as a natural antibacterial agent in contaminated fuel ethanol fermentations. *Fuel Processing Technology* **106**(2013):611–618.

Liang, L., Xu, R., Li, Q., Huang, X., An, Y., Zhang, Y. and Guo, Y. (2012). Simultaneous production of sugar and ethanol from sugarcane in China, the development, research and prospect aspects. In *Bioethanol*, pp. 75–92. M. A. P. Lima and A. P. P. Natalense, eds. InTech Publishing: Shanghai (China).

Lipke, P. N. and Kujan, J. (1992). Sexual agglutination in budding yeasts: structure, functions and regulation of adhesion glycoproteins. *Microbiological Reviews* **56**(1):180–194.

Liu, C. G., Wang, N., Lin, Y. H. and Bai, F. W. (2012). Very high gravity ethanol fermentation by flocculating yeast under redox potential-controlled conditions. *Biotechnology for Biofuels* **5**(1):61–68.

Magazoni, F. C., Monteiro, J. B., Cardemil, J. M. and Colle, S. (2010). Cooling of ethanol fermentation process using absorption chillers. *International Journal of Thermodynamics* **13**(3):111–118.

Martorell, P. A., Querol, A. and Fernandez-Espinar, M. T. (2005). Rapid identification and enumeration of Saccharomyces cerevisiae cells in wine by real-time PCR. *Applied and Environmental Microbiology* **71**(11):6823–6830.

Matuo, R., Sousa, F. G., Soares, D. G., Bonatto, D., Saffi, J., Escargueil, A. E., Larsen, A. K. and Henriques, J. A. P. (2012). Saccharomyces cerevisiae as a model system to study the response to anticancer agents. *Cancer Chemotherapy and Pharmacology* **70**(4):491–502.

Maxon, W. D. and Johnson, M. J. (1953). Aeration studies on propagation of baker's yeasts. *Industrial and Engineering Chemistry* **45**(11):2554–2560.

McCarter, Y. S. (2005). Antimicrobial modes of action. In *Manual of Antimicrobial Susceptibility Testing*, pp. 3–12. M. B. Coyle, ed. University of Washington, American Society for Microbiology publishing: Washington, DC.

Mcdonnell, G. and Russell, D. A. (1999). Antiseptics and disinfectants: activity, action and resistance. *Clinical Microbiology Reviews* **12**(1):147–179.

Meneghin, S. P., Reis, F. C., Almeida, P. G. and Ceccato-Antonini, S. R. (2008). Chlorine dioxide against bacteria and yeasts from the alcoholic fermentation. *Brazilian Journal of Microbiology* **39**(2):337–343.

Middelhoven, W. J. (2002). Identification of yeasts presents in sour fermented foods and fodders. *Molecular Biotechnology* **21**(3):279–292.

Mill, P. J. (1964). The nature of the interaction between flocculent cells in the flocculation of Saccharomyces cerevisiae. *Journal of General Microbiology* **35**(1):61–68.

Miyazu, Y. and Yano, T. (1974). Surface character and absorbability to air hydrocarbon-grown yeasts. *Agricultural and Biological Chemistry* **38**(1):183–188.

Moosavi-Nasab, M. and Pashangeh, S. and Rafsanjani, M. (2010). Effect of time on xanthan gum production from sugar beet molasses. *World Academy of Science, Engineering and Technology* **44**:1244–1247.

Mussatto, S. I., Dragone, G. D., Guimarães, P. M., Silva, J. P. A., Carneiro, L. M., Roberto, I. C., Vicente, A., Domingues, L. and Teixeira, J. A. (2010). Technological trends, global market and challenges of bio-ethanol production. *Biotechnology Advances* **28**(6):817–830.

Muthaiyan, A., Limayem, A. and Ricke, S. C. (2010). Antimicrobial strategies for limiting bacterial contaminants in fuel bioethanol fermentations. *Progress in Energy and Combustion Science* **37**(3):351–370.

Nagai, S. (1963). Diagnostic color differentiation plates for hereditary respiration deficiency in yeast. *Journal of Bacteriology* **86**(2):299–302.

Nagodawithana, T. W., Castellano, C. and Steinkraus, K. H. (1974). Effect of dissolved oxygen, temperature, initial cell count and sugar concentration on the viability of Saccharomyces cerevisiae in rapid fermentations. *Applied Microbiology* **28**(3):383–391.

Narendranath, N. V. and Brey, S. (2009). Bacterial contamination and control. In *The Alcohol Textbook*, pp. 337–356. W. M. Ingledew, D. R. Kelsall, G. D. Austin and C. Kluhspies, eds. Nottingham University Press Publishing: Nottingham.

Narendranath, N. V., Thomas, K. C. and Ingledew, W. M. (2001). Effects of acetic acid and lactic acid on the growth of Saccharomyces cerevisiae in a minimal medium. *Journal of Industrial Microbiology and Biotechnology* **26**(3):171–177.

Nofemele, Z., Shukla, P., Trussler, A., Permaul K. and Singh, S. (2012). Improvement of ethanol production from sugarcane molasses through enhanced nutrient supplementation using Saccharomyces cerevisiae. *Journal of Brewing and Distilling* **3**(2):29–35.

Núñez, O. and Spaans, E. (2008). Evaluation of green-cane harvesting and crop management with a trash-blanket. *Sugar Technology* **10**(1):29–35.

O'Connor-Cox, E. S. C. and Ingledew, W. M. (1989). Wort nitrogenous sources-their use by brewing yeasts: a review. *Journal of the American Society of Brewing Chemists* **47**(4):102–108.

Oliva-Neto, P. and Yokoya, F. (1994). Evaluation of bacterial contamination in a fed-batch alcoholic fermentation process. *World Journal of Microbiology and Biotechnology* **10**(6):697–699.

Oliva-Neto, P., Dorta, C., Carvalho, A. F. A., Lima, V. M. G. and Silva, D. F. (2013). The Brazilian technology of fuel ethanol fermentation – yeast inhibition factors and new perspectives to improve the technology. In *Materials and Processes for Energy: Communicating Current Research and Technological Developments*, pp. 371–379. A. Mendez-Vilas, ed. FORMATEX: Verlag Wiley-VCH Publishing: Badajoz, Spain.

Olivério, J. L., Barreira, S. T., Boscariol, E. D., César, A. R. P. and Yamakawa, C. K. (2010). Alcoholic fermentation with temperature controlled by ecological absorption Chiller—EcoChill. *Proceeding of the International Society of Sugar Cane Technologists* **27**:1–9.

Palmieri, M. C., Greenhalf, W. and Laluce, C. (1996). Efficient flotation of yeast cells grown in batch culture. *Biotechnology and Bioengineering* **50**(3):248–256.

Pérez, M. A., Gallego, F. J., Martinez, I. and Hidalgo, P. (2001). Detection, distribution and selection of microsatellites (SSRs) in the genome of the yeast Saccharomyces cerevisiae as molecular markers. *Letters in Applied Microbiology* **33**(6):461–466.

Piper, P., Claderon, C. O., Hatzixanthis, K. and Mallapour M. (2001). Weak acid adaptation: the stress response that confers yeasts with resistance to organic acid food preservatives. *Microbiology* **147**(10):2635–2642.

Powel, C. D. and Diacetis, A. N. (2007). Long serial repitching and generic and phenotypic stability of brewer's yeast. *Journal of the Institute of Brewing* **113**(1):67–74.

Powell, C. D., Quain, D. E. and Smart, K. A. (2003). The impact of brewing yeast cell age on fermentation performance, attenuation and flocculation. *FEMS Yeast Research* **3** (2003): 149–157

Prati, P. and Moretti, R. H. (2010). Study of clarification process of sugar cane juice for consumption. *Food Science and Technology (Campinas)* **30**(3):776–783.

Quesada, M. P. and Cenis, J. L. (1995). Use of random amplified polymorphic DNA (RAPD-PCR) in the characterization of wine yeasts. *American Journal of Enology and Viticulture* **46**(2):204–208.

Replansky, T., Koufopanou, V., Greig, D. and Bell, G. (2008). Saccharomyces sensu stricto as a model system for evolution and ecology. *Trends in Ecology and Evolution* **23**(9):494–501.

RFA, Renewable Fuels Association (2013). World Fuel Ethanol Production, United State Department of Agriculture.

Rivera, E. C., Yamakawa, C. K., Garcia, M. H., Geraldo, V. C., Rossell, C. E. V., Maciel-Filho, R. and Bonomi, A. (2013). A procedure for estimation of fermentation kinetic parameters in fed-batch bioethanol production process with cell recycle. *Chemical Engineering Transactions* **32**(12):1369–1374.

Rückle, L. and Senn, T. (2006). Hop acids can efficiently replace antibiotics. *International Sugar Journal* **108**(1287):139–147.

Rudnitskaya, A. and Legin, A. (2008). Sensor systems, electronic tongues and electronic noses for the monitoring of biotechnological processes. *Journal of Industrial Microbiology and Biotechnology* **35**(5):443–451.

Rudorff, B. F. T., Aguiar, D. A., Silva, W. F., Sugawara, L. M., Adami, M. and Moreira, M. A. (2010). Studies on the rapid expansion of sugarcane for ethanol production in São Paulo State (Brazil) using Landsat data. *Remote Sensing* **2**(4):1057–1076.

Schuller, D., Valero, E., Dequin, S. and Casal, M. (2004). Survey of molecular methods for the typing of wine yeast strains. *FEMS Microbiology Letters* **231**(1):19–26.

Seabra, J. E. A., Macedo, I. C., Chum, H. L., Faroni, C. E. and Sarto, C. A. (2011). Life cycle assessment of Brazilian sugarcane products: GHG emissions and energy use. *Biofuels, Bioproducts and Biorefining* **5**(5):519–532.

Shafiee, G., Mortazavian, A. M., Mohammadifar, M. A., Koushki, M. R., Mohammadi, A. and Mohammadi, R. (2010). Combined effects of dry matter content, incubation temperature and final pH of fermentation on biochemical and microbiological characteristics of probiotic-fermented milk. *African Journal of Microbiology Research* **4**(12):1265–1274.

Sheppard, J. D. (2009). Continuous ethanol fermentation. In *The Alcohol Textbook*, pp. 259–274. W. M. Ingledew, D. R. Kelsall, G. D. Austin and C. Kluhspies, eds. Nottingham University Press Publishing: Nottingham.

Silva-Filho, E. A., Santos, S. K. B., Resende, A. M., Morais, J. O. F., Morais Jr., M. A. and Simões, D. A. (2005). Yeast population dynamics of industrial fuel-ethanol fermentation process assessed by PCR-fingerprinting. *Antonie Van Leeuwenhoek* **88**(1):13–23.

Skinner, K. A. and Leathers, T. D. (2004). Bacterial contaminants of fuel ethanol production. *Journal of Industrial Microbiology and Biotechnology* **31**(90):401–408.

Skinner-Nemec, K. A., Nichols, N. N. and Leathers, T. D. (2007). Biofilm formation by bacterial contaminants of fuel ethanol production. *Biotechnology Letters* **29**(3):379–383.

Smit, G., Straver, M. H., Lugtenberg, B. J. and Kijne, J. W. (1992). Flocculence of Saccharomyces cerevisiae cells is induced by nutrient limitation, with cell surface hydrophobicity as a major determinant. *Applied and Environmental Microbiology* **58**(11):3709–3714.

Soares, E. V. (2011). Flocculation in Saccharomyces cerevisiae: a review. *Journal of Applied Microbiology* **110**(1):1–18.

Sosa, O. and Farías, M. A. (2012). Influence of culture and nutritional conditions on self-flocculation of a Kloeckera apiculata wine strain. *Food Technology and Biotechnology* **50**(1):66–72.

Sousa, S. R., Perez, M. F. S. and Laluce, C. (2001). Flotation assays in small volumes of yeast cultures. In *Food Microbiology Protocols*, pp. 349–353. J. F. T. Spencer and A. L. Ragout de Spencer, eds. Humana Press Inc. Publishing: Totowa (New Jersey).

Sousa, S. R., Oliveira, K. F., Souza, C. S., Kilikian, B. V. and Laluce, C. (2003). Yeast flotation viewed as the result of the interplay of supernatant composition and cell-wall hydrophobicity. *Colloids and Surfaces B: Biointerfaces* **29**(4):309–319.

Sousa, S. R., Laluce, C. and Jafellici Jr., M. (2006). Effects of organic and inorganic additives on flotation recovery of washed cells of Saccharomyces cerevisiae resuspended in water. *Colloids and Surfaces B: Biointerfaces* **48**(1):77–83.

Souza, C. S., Oliveira, K. F., Trevisan, H. C. and Laluce, C. (2007). A strategy to compare yeast strains and improve cell survival in ethanol production process above 30°C. In *Communicating current Research and Education Topics and Trends in Applied Microbiology*, pp. 410–417. A. Mendez-Vilas, ed. FORMATEX, Verlag Wiley-VBC Publishing: Badajoz, Spain.

Spencer, J. F. T., Spencer, A. L. R. and Laluce, C. (2002). Non-conventional yeasts. *Applied Microbiology and Biotechnology* **58**(2):147–156.

Stanbury, P. F., Whitaker, A., and Hall, S. J. (2003). Media for industrial fermentations. In *Principles of Fermentation Technology*, pp. 93–122. Butterworth-Heinemann Elsevier Publishing: Oxford.

Stannard, C. (1997). Criteria for foods. *Food Science and Technology Today* **11**(3):137–177.

Steindl, R. J. (2010). Clarification of cane juice for fermentation. *Proceedings of the International Society of Sugarcane Technologists* **27**(1):1–10.

Stewart, G. G. (2009). The Horace Brown Medal lecture: forty years of brewing research. *Journal of the Institute of Brewing* **115**(1):3–29.

Strand, M., Prolla, T. A., Liskay, M. R. and Petes, T. D. (1993). Destabilization of tracts of simple repetitive DNA in yeast by mutations affecting DNA mismatch repair. *Nature* **365**(6443):274–276.

Stratford, M. (1992). Lectin-mediated aggregation of yeasts - yeast flocculation. *Biotechnology and Genetic Engineering Reviews* **10**(1):283–342.

Straver, M. H., Van der Aar, P. C., Smit, G. and Kijne, J. W. (1993). Determinants of flocculence of brewer's yeast during fermentation in wort. *Yeast* **9**(5):527–532.

Stroppa, C. T., Andrietta, M. G. S., Andrietta, S. R., Steckelberg, C. and Serra, G. E. (2000). Use of penicillin and monensin to control bacterial contamination of Brazilian alcohol fermentations. *International Sugar Journal* **102**(1):78–82.

Torija, M. J., Rozes, N., Poblet, M., Guillamon, J. M. and Mas, A. (2003). Effects of fermentation temperature on the strain population of Saccharomyces cerevisiae. *International Journal of Food Microbiology* **80**(1):47–53

Troller, J. A. (2012). Sanitation in food storage and facilities. In *Sanitation in Food Processing*, pp. 353–375. Academic Press Inc. Publishing: San Diego (California).

UNICA (2015). Brazilian Sugarcane Industry Association, Drought, lower yields reduce cane supply for 2014/2015 sugarcane season in South-Central Brazil: www.unica.com.br.

Veale, E. L., Irudayaraj, J. and Demirci, A. (2007). An on-line approach to monitor ethanol fermentation using FTIR spectroscopy. *Biotechnology Progress* **23**(2):494–500.

Verstrepen, K. J. and Klis, F. M. (2006). Flocculation, adhesion and biofilm formation in yeasts. *Molecular Microbiology* **60**(1):5–15.

Viehweg, H. and Schügerl, K. (1983). Cell recovery by continuous flotation. *European Journal of Applied Microbiology and Biotechnology* **17**(2):96–102.

Vieira, E. D., Andrietta, M. G. S. and Andrietta, S. R. (2013). Yeast biomass production: a new approach in glucose liming feeding strategy. *Brazilian Journal of Microbiology* **44**(2):551–558.

Wongkhalaung, C., Nakatomi, Y. and Hiroyuki, T. (2004). Hybridization and selection of Saccharomyces cerevisiae strains from industrial baker's yeasts. *Kasetsart Journal (Natural Science)* **38**(2):255–266.

Yoshimi, S. and Masazumi, W. (1981). Effects of yeast strains and environmental conditions on formation of organic acids in must during fermentation: studies on organic acids in grape must and wine (III). *Journal of Fermentation Technology* **59**(1):27–32.

Zhang, Z. (2009). *Batch Fermentation and Fermenter Design*, pp. 257–299. Nottingham University Press Publishing: Nottingham.

Zheng, D. Q., Liu, T. Z., Chen, J., Zhang, K., Li, O., Zhu, L., Zhao, Y. H., Wu, X. C. and Wang, P. M. (2013) Comparative functional genomics to reveal the molecular basis of phenotypic diversities and guide the genetic breeding of industrial yeast strains. *Applied Microbiology and Biotechnology* **97**(5):2067–2076.

Zhou, J., Thompson, D. K. and Tiedje, J. M. (2004). Genomics: toward a genome-level understanding of the structure, functions, and evolution of biological systems. In *Microbial Functional Genomics*, pp. 1–20. J. Zhou, D. K. Thompson, Y. Xu and J. M. Tiedje, eds. Wiley-Blackwell Publishing: Hoboken (New Jersey).

Zhu, C., Chen, Z. and Yu, G. (2013). Fungicidal mechanism of chlorine dioxide in Saccharomyces cerevisiae. *Annals of Microbiology* **66**(3):495–502.

Zossi, B. S., Cárdenas, G. J., Sorol, N. and Sastre, M. M. (2012). Análisis del proceso de sulfitación en la remoción de compuestos no azúcares en jugos de variedades de caña de Tucumán. Revista. *Industrial y Agrícola de Tucumán (Argentina)* **89**(2):9–24.

CHAPTER 4

Production of fermentable sugars from sugarcane bagasse

Zhanying Zhang, Mark D. Harrison and Ian M. O'Hara
Centre for Tropical Crops and Biocommodities, Queensland University of Technology (QUT), Brisbane, Australia

4.1 Introduction

Sugarcane residues including bagasse, tops, and leaves are one of the most abundant agricultural residues with an annual global production estimated at 279 million metric tons (Chandel *et al.* 2012). These residues, similar to other agricultural residues, forestry and municipal wastes, and dedicated energy crops, are primarily composed of plant cell walls. Plant cell walls are a resilient, intractable barrier composed of complex polysaccharides (cellulose and hemicellulose), lignin, and proteins. Bagasse typically consists of 35–45% cellulose, 25–36% hemicellulose, 11–27% lignin, 3–17% ash, and other components (including protein), with variation arising from the sugarcane variety, climate, location, plant age, and soil conditions (Zhao *et al.* 2009a, Canilha *et al.* 2011, Diedericks *et al.* 2012, Zhang *et al.* 2013a).

Cellulose is an unbranched, homopolysaccharide consisting of D-anhydroglucose repeating units (i.e., C6 sugars) joined by 1,4-β-D-glycosidic linkages. Cellulose in the plant cell wall is ordered into fibrils, which are chemically bound to lignin and hemicellulose. Most cellulose in plant biomass is crystalline, while the remainder is amorphous (Klemm 1998).

Hemicellulose is a heterogeneous, branched heteropolysaccharide consisting of C5 sugars (xylose, arabinose), C6 sugars (mannose, glucose, and galactose), and uronic acids. Hemicellulose provides the connection between lignin and the cellulose fibers and gives the polymer network rigidity (Laureano-Perez *et al.* 2005).

Lignin is a complex molecule containing cross-linked polymers of phenolic monomers with both aliphatic and aromatic constituents (John and Thomas 2008). The lignin polymer is amorphous and hydrophobic (Hendriks and Zeeman 2009).

Bagasse has long been used as a fuel to generate steam and electricity for sugar mill operations (O'Hara 2011, Macrelli *et al.* 2012). Bagasse is rich in carbohydrates and lignin, and therefore can also be used as substrate for the production of

Sugarcane-Based Biofuels and Bioproducts, First Edition. Edited by Ian M. O'Hara and Sagadevan G. Mundree.

fermentable sugars and other bioproducts. However, accessing the fermentable sugars (monosaccharides) in sugarcane residues is challenging because of the structure of the sugarcane cell wall polymers and the connections between them. Tops and leaves of sugarcane, often called trash, also consist of lignocellulose. In traditional sugarcane-harvesting processes, trash is removed from the sugarcane stalks and is left on the fields to maintain the nutritional properties of the soil, retain soil moisture, and/or control pests and weeds (O'Hara 2011, Macrelli *et al.* 2012).

However, it has been estimated that up to 66% of sugarcane trash can be used for industrial processes without significantly affecting sugarcane production (Macrelli *et al.* 2012). The recovery and utilization of sugarcane trash will significantly increase the total amount of available biomass for producing bioproducts. Although this chapter focuses on bagasse, the technologies described can also be readily adapted for utilization of sugarcane trash.

In contrast to other agricultural residues such as wheat straw and rice straw, bagasse is already collected as part of the sugarcane-harvesting system and transported to industrial sites (sugar mills), which significantly reduces costs related to biomass collection and transportation. However, high processing cost is the key barrier to commercialization of biofuels and biochemicals from bagasse. Current commercial use of bagasse in sugar mills is often limited to inefficient combustion to supply steam and electrical power for sugar mills and to generate minor amounts of electricity for export to the local electricity network (O'Hara 2011, Macrelli *et al.* 2012).

4.2 Bioethanol from bagasse

Bagasse has been widely studied as a feedstock for bioethanol production. In some sugar mills, existing facilities for the production of bioethanol from sugarcane juice (a first-generation biofuel) can also be used to produce bioethanol from the fermentable sugars from sugarcane residues. Although the production of first-generation bioethanol from sugarcane juice has been commercialized in countries such as Brazil, the high cost of production of fermentable sugars from bagasse has limited commercial opportunities to produce second-generation bioethanol from this abundant feedstock.

Direct acid hydrolysis of lignocellulosic biomass including bagasse to fermentable sugars has been studied since the early 20th century (O'Hara *et al.* 2011). However, acid hydrolysis processes have inherent challenges including low glucose yields when dilute or solid acids are used (Lenihan *et al.* 2010, O'Hara *et al.* 2011, Huang and Fu 2013), cost-effective acid recovery when concentrated acids are employed (Camacho *et al.* 1996, Iranmahboob *et al.* 2002, Liao *et al.* 2006, O'Hara *et al.* 2011), and the formation of compounds that inhibit fermentation of the resulting glucose. Alternatively, fermentable sugars

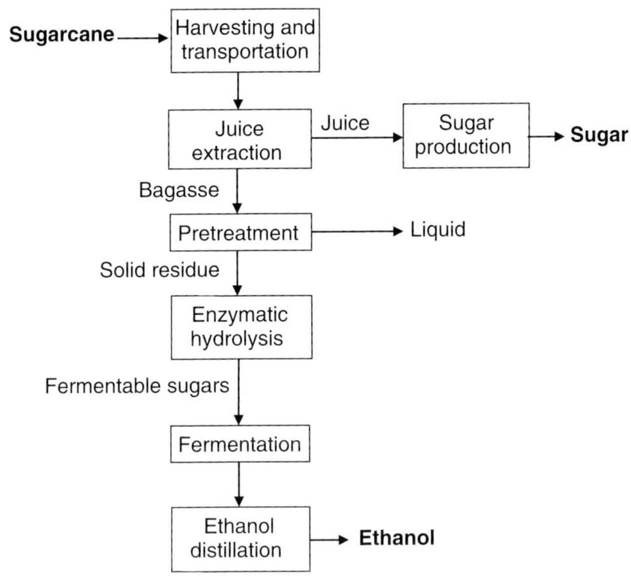

Figure 4.1 Typical ethanol production process from bagasse.

can be produced enzymatically from biomass in a biochemical process. As is the case with lignocellulosic biomass, however, cellulose is embedded in a matrix of hemicellulose and lignin, and biomass pretreatment is essential in order to make cellulose more accessible for efficient enzymatic hydrolysis to glucose (Yang and Wyman 2008).

Biochemical processes have significant advantages over direct acid hydrolysis processes, including high sugar yields and a reduction in the requirement for acid. In some cases, the need for acid is eliminated. A typical biochemical process for second-generation bioethanol production from bagasse is shown in Figure 4.1.

Ongoing technology improvements have substantially reduced the cost of biomass pretreatment and enzymatic hydrolysis to fermentable sugars for bioethanol production; however, recent estimates indicate that these processes still account for ~30% of the total process costs (Macrelli *et al.* 2012). The cost of enzymatic hydrolysis depends largely upon enzyme dosage and the hydrolysis rate, which in turn depends upon enzyme accessibility to cellulose. As a result, the cost of enzymatic hydrolysis is pretreatment dependent. Studies have shown that the amount of enzymes used for hydrolysis can be reduced by >80% (Del Rio *et al.* 2011, Sathitsuksanoh *et al.* 2011), and the time for completion of hydrolysis can be shortened by >50% (Zhu *et al.* 2009) if effective pretreatment processes are used. Despite these advances, however, the cost of bioethanol production using biochemical processes remains greater than that of petrochemical processes for producing fuels.

In this chapter, technologies for producing ethanol from fermentable sugars in bagasse are reviewed with an emphasis on pretreatment technologies. In addition, the technologies for enzymatic hydrolysis and fermentation are also briefly reviewed and discussed.

4.3 Overview of pretreatment technologies

The recalcitrance of lignocellulosic biomass (including bagasse) can be reduced by disruption of the crystalline structure of cellulose, removal of hemicellulose, or removal or modification of the lignin (Zheng *et al.* 1998, Mosier *et al.* 2005). In the past two decades, significant efforts have been made to improve sugar yield from lignocellulosic biomass, using biological, physical, or chemical pretreatments (Alvira *et al.* 2010). These pretreatment methods are often combined to improve pretreatment effectiveness. A number of recent reviews have summarized the principles of these pretreatment processes (Yang and Wyman 2008, Hendriks and Zeeman 2009, Zhao *et al.* 2009b, 2012, Alvira *et al.* 2010, Maki-Arvela *et al.* 2010, Agbor *et al.* 2011, O'Hara *et al.* 2011, Menon and Rao 2012, Brandt *et al.* 2013).

Biological pretreatments employ fungi, typically white-rot fungi, to degrade lignin in biomass, thereby increasing carbohydrase accessibility to their substrates (Hendriks and Zeeman 2009, Alvira *et al.* 2010, Agbor *et al.* 2011, O'Hara *et al.* 2011).

Physical processes include steam explosion, mechanical comminution, extrusion, liquid hot water (LHW), ultrasound, and irradiation by microwave, gamma ray, and electron beams (Hendriks and Zeeman 2009, Alvira *et al.* 2010, Agbor *et al.* 2011, O'Hara *et al.* 2011). Physical pretreatments are often combined with a chemical catalyst or solvent to improve pretreatment efficiency. These physicochemical pretreatments include acid-steam explosion, SO_2-steam explosion, CO_2-explosion, ammonia fiber explosion (AFEX), and acid-catalyzed microwave pretreatment (Hendriks and Zeeman 2009, Alvira *et al.* 2010, Agbor *et al.* 2011, O'Hara *et al.* 2011).

Chemical pretreatments are carried out in the presence of catalysts and/or nonaqueous solvents. Chemical pretreatments include the use of dilute acids (e.g., H_2SO_4 and HCl), dilute alkalis (e.g., NaOH, ammonia), oxidizing agents (e.g., ozone, peroxyacetic acid and hydrogen peroxide), organosolv processes (e.g., ethanol, glycerol and ethylene glycol (EG)), organic acids (e.g., acetic acid and formic acid), ionic liquids, and inorganic salts (e.g., $FeCl_3$ and $CaCl_2$) (Hendriks and Zeeman 2009, Alvira *et al.* 2010, Agbor *et al.* 2011, O'Hara *et al.* 2011). In order to improve pretreatment effectiveness, combinations of chemical catalysts and solvents are often used, including combinations of oxidative chemicals (e.g., H_2O_2 and O_2) with alkaline pretreatments and combinations of dilute acid or alkaline pretreatments with polyols, such as glycerol and EG (Lee *et al.* 2010, Martin *et al.* 2011, Zhang *et al.* 2013a), or ionic liquids (Nguyen *et al.* 2010, Brandt *et al.* 2011, Zhang *et al.* 2012, 2013b).

4.4 Pretreatment of bagasse

Chemical and physicochemical pretreatments are more promising than biological and pure physical processes as the latter require long pretreatment times and high energy inputs respectively, making them unacceptable at industrial scale. In this section, pretreatment of bagasse by chemical and physicochemical processes are reviewed and discussed.

4.4.1 Dilute acid pretreatment

Efficient dilute acid pretreatments hydrolyze the majority of hemicellulose and dissolve a small proportion of lignin in biomass, thereby making cellulose more accessible to carbohydrases (Alvira *et al.* 2010, O'Hara *et al.* 2011). Dilute acid pretreatment has been the subject of extensive research (Alvira *et al.* 2010) but pretreatment conditions that maximize cellulose digestibility also result in degradation of significant quantities of xylose to furfural. While furfural is an important feedstock chemical, it is also an inhibitor of fermentation and so is undesirable in the present context.

Therefore, a number of studies have been undertaken to examine the solubilization of hemicellulose and improve the recovery of xylose during dilute acid (e.g., H_2SO_4, H_3PO_4, HCl) pretreatment of bagasse (Lavarack *et al.* 2000, Silva *et al.* 2005, Martin *et al.* 2007a,b, Sumphanwanich *et al.* 2008, Betancur and Pereira 2010, Geddes *et al.* 2010, Canilha *et al.* 2011, Candido *et al.* 2012). These studies demonstrated that bagasse was more susceptible to hemicellulose hydrolysis than many other herbaceous materials, such as rice hulls and cassava stalks (Martin *et al.* 2007a,b).

The combination of dilute acid pretreatment with steam explosion for pretreatment of sugarcane bagasse has also been reported (Morjanoff and Gray 1987, Fox *et al.* 1989). A total sugar yield of 65.1 g/100 g bagasse (equivalent to 83% of the anhydroglucan and 84% of the anhydroxylan content of untreated bagasse) was reported under optimized conditions (1 g H_2SO_4/100 g dry bagasse; water/solid = 2; 220 °C for 30 s; 2.22 MPa) (Morjanoff and Gray 1987). Interestingly, the use of mixtures of dilute acids in a similar system (190 °C for 10 min with 1% w/v H_2SO_4 and 1% w/v acetic acid) improved hemicellulose solubilization and cellulose digestibility to 75% (compared with 7% of untreated bagasse) (Rocha *et al.* 2011).

Our research group has undertaken a series of experiments to assess dilute acid pretreatment of bagasse at the pilot-scale using a 150 l horizontal reactor (Andritz, Austria; Figure 4.2). Dilute acid pretreatment was also assessed in combination with physical pretreatment using a second-stage, vertical reactor (69 l) capable of steam explosion. Cellulose digestibility was improved to 87% by the addition of steam explosion at 185 °C for 5 min and 2 MPa (Harrison *et al.* 2013).

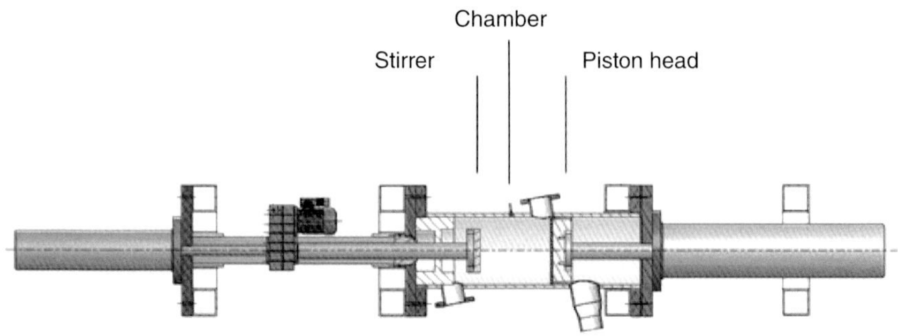

Figure 4.2 Schematic of pilot-scale horizontal reactor (Zhang *et al.* 2013a).

4.4.2 Alkaline pretreatment

Efficient alkaline pretreatments solubilize the majority of the lignin and partially solubilize hemicellulose in lignocellulosic biomass, thereby improving carbohydrase accessibility to cellulose (O'Hara *et al.* 2011). Alkaline pretreatments of bagasse are often combined with other pretreatment methods to improve the sugar yield. For example, pretreatment of bagasse using dilute NaOH at 90 °C for 1.5 h, followed by pretreatment with peracetic acid at 75 °C for 2.5 h, resulted in a total reducing sugar yield (i.e., glucose and xylose) of 92% after 120 h enzymatic hydrolysis (Zhao *et al.* 2009a).

$Ca(OH)_2$ (lime) is a low-cost chemical commonly used in industrial processes. A recent study showed that the optimal condition for lime pretreatment of sugarcane bagasse was 90 °C for 90 h at a dosage of 0.47 g lime/g bagasse (Rabelo *et al.* 2013). This pretreatment resulted in a total glucose yield of 86–91% and a total xylose yield of 94–98% (including the sugars in the hydrolysate) with the production of low concentrations of fermentation inhibitors (Rabelo *et al.* 2013). In another study, lime pretreatment of screened bagasse was shown to be more effective than alkaline-H_2O_2 pretreatment at pH 11.5 (adjusted using NaOH) (Rabelo *et al.* 2008). However, there was no difference between lime and alkaline-H_2O_2 pretreatments when unscreened bagasse was used as the substrate (Rabelo *et al.* 2008). Lime has been reported to be a less expensive alkali than NaOH and $NH_3 \cdot H_2O$ for biomass pretreatment (Playne 1984).

Ammonia is another well-studied alkaline chemical used for biomass pretreatment. Dilute ammonia, aqueous ammonia, and ammonia fiber expansion (AFEX) have been used to pretreat bagasse (Kurakake *et al.* 2001, Krishnan *et al.* 2010, Cao and Aita 2013). Furthermore, the combination of surfactants (PEG 4000 or Tween 80) with dilute ammonia pretreatment at 160 °C for 1 h has been shown to improve cellulose digestibility in bagasse by 24% (Cao and Aita 2013). The enhancement in cellulose digestibility was attributed to the ability of the nonionic surfactant to improve delignification and reduce cellulose crystallinity (Cao and Aita 2013). AFEX pretreatment of bagasse has been shown to result in a maximum cellulose digestibility of ~85%, while supplementation with hemicellulases during enzymatic hydrolysis improved the xylan digestibility to 95–98% (Krishnan *et al.* 2010).

Wet oxidation with dilute Na_2CO_3-O_2 has been shown to be more effective for pretreatment of bagasse than either dilute H_2SO_4-O_2 or water-O_2 under the same pretreatment conditions (195 °C for 15 min) (Martín *et al.* 2006). In another study, a glucose yield of 97% was achieved after 72 h enzymatic hydrolysis of bagasse pretreated using Na_2CO_3-O_2 at 120 °C for 20 h (Kallioinen *et al.* 2013). Oxidative agents target lignin and hemicellulose structures, thereby improving the accessibility of carbohydrases to cellulose (O'Hara *et al.* 2011). High concentrations of oxidized compounds, such as carboxylic acids, aldehydes, and ketones, are generally observed in wet oxidation hydrolysates (Martin *et al.* 2007a,b).

Combinations of alkaline pretreatments with steam explosion have also been applied to bagasse. Steam explosion of bagasse (200 °C for 7 min) followed by alkaline delignification (1% (w/v) NaOH at 100 °C for 1 h) significantly improved glucose yield from ~15 to ~40 g/l after enzymatic hydrolysis, compared with steam explosion alone (Wanderley *et al.* 2013). In another study, bagasse was first pretreated by NaOH steaming (200 °C for 5 min, 1.72 mPa) followed by steam explosion, which resulted in the solubilization of 40–45% of the bagasse fibers and final glucose yield of 80% after 8 h of enzymatic hydrolysis, which was significantly higher than that observed for NaOH steaming alone (58% for glucose and 52% for xylose) (Puri and Pearce 1986). Our research group has demonstrated that pilot-scale, alkaline pretreatment (170 °C for 30 min; 15.5 w% NaOH on dry fiber) of bagasse followed by steam explosion (185 °C for 5 min; 2 MPa) results in a cellulose digestibility of 86% (Harrison *et al.* 2013).

4.4.3 Liquid hot water pretreatment

LHW pretreatment does not generally require added catalysts. Pressure is applied to maintain water in the liquid state at temperatures above 100 °C and to induce alterations in biomass structure, while the acidity of water and released organic acids from biomass at elevated temperatures (160–240 °C) solubilizes hemicellulose and partially solubilizes lignin (Alvira *et al.* 2010, Agbor *et al.* 2011, O'Hara *et al.* 2011). Most of the C5 sugars from hemicellulose degradation can be recovered after LHW pretreatment and the process produces only low concentrations of degradation products. As a result, there is only minimal inhibition to subsequent enzymatic hydrolysis and fermentation (Alvira *et al.* 2010, Agbor *et al.* 2011, O'Hara *et al.* 2011), which eliminates the need for a final wash step after pretreatment or neutralization (Agbor *et al.* 2011).

A limited number of studies describing LWH pretreatment of bagasse (either alone or in combination with other methods) have been reported (VanWalsum *et al.* 1996, Bigelow and Wyman 2002, Laser *et al.* 2002, Gao *et al.* 2013, Yu *et al.* 2013a, b). Under optimal conditions, LHW pretreatment achieved cellulose digestibilities in the solid residues comparable to those from conventional dilute acid pretreatment (Laser *et al.* 2002). LHW pretreatment also resulted in a higher ethanol yield after simultaneous saccharification and fermentation (SSF) than steam explosion (Laser *et al.* 2002).

A recent study showed that, although LHW pretreatments produced low levels of inhibitors (i.e., <1.2 g/l furfural) after pretreatment at 180 °C for 60 min, total sugar yield (71.6%) was lower than that obtained from pretreatments with dilute HCl (76.6%) and NaOH (77.3%) (Yu *et al.* 2013a). Furthermore, the same research group showed that the combination of NaOH pretreatment with LHW could improve pretreatment efficiency (NaOH-LHW > NaOH > LHW-NaOH > NaOH) (Gao *et al.* 2013) and that bagasse pretreatment with LHW-ammonia resulted in high recovery of hemicellulose-derived (75.5%) and cellulose-derived (87%) sugars (Yu *et al.* 2013b).

4.4.4 Organosolv pretreatment

Organosolv pretreatments are very promising processes because of the high enzymatic digestibility of the resulting solid residue. Organosolv pretreatments primarily dissolve lignin in the presence or absence of alkaline catalysts but dissolve both lignin and hemicellulose in the presence of acid catalysts (Zhao *et al.* 2009b).

Ethanol and glycerol, both of which are low cost, low toxicity, renewable, and widely used industrial solvents, have been used to pretreat bagasse in a number of research laboratories (Rossell *et al.* 2006, Area *et al.* 2009, Teramoto *et al.* 2009, Mesa *et al.* 2010, 2011). The combination of dilute acid pretreatment (120 °C for 40 min with 0.2 M H_2SO_4) with ethanol pretreatment in the presence of NaOH (3% w/w on fiber) has been reported to improve glucose yield (Mesa *et al.* 2011). In this study, optimum bagasse pretreatment was achieved using 30% (v/v) ethanol at 195 °C for 60 min (Mesa *et al.* 2011). The same research group also optimized pretreatment of bagasse in aqueous ethanol (50% v/v) in the presence of either H_2SO_4 or NaOH and demonstrated that optimal pretreatment was achieved at 175 °C for 60 min in 1.25% H_2SO_4, with a glucose yield of 25.1 g/g dry bagasse (Mesa *et al.* 2010). Replacing H_2SO_4 with acetic acid reduced the degradation of monosugars during bagasse pretreatment (Teramoto *et al.* 2009).

Although recovery of low boiling point solvents, such as ethanol, can be achieved by distillation with low energy consumption, pretreatments using low boiling point solvents must be performed under conditions that prevent the loss of volatile solvents into the atmosphere (Zhao *et al.* 2009b).

Glycerol has a significantly higher boiling point (290 °C) than solvents such as ethanol and methanol. In addition, large quantities of glycerol are being produced annually as a by-product of the biodiesel industry (Zhang *et al.* 2013a). Recent studies have shown that pretreatment of bagasse with 80 wt% glycerol and 0.94 wt% H_2SO_4 at 190 °C for 60 min resulted in a glucan digestibility of >90% (Martin *et al.* 2011).

Our research group has evaluated the pretreatment of bagasse at mild temperatures (≤130 °C) using acidified aqueous glycerol at both the laboratory and pilot scales (Zhang *et al.* 2013a). Laboratory-scale pretreatment of bagasse (4.00 g dry mass in 40.00 g liquid) with glycerol solutions containing ≤20 wt% water and 1.2 wt% HCl at 130 °C for 60 min resulted in biomass with glucan digestibility of ≥88%. Despite a fourfold reduction in reaction time (15 min) and threefold reduction in acid catalyst concentration (0.4% wt HCl), comparable glucan digestibility (90%) was observed after pilot-scale (10 kg dry mass in 60 kg liquid) bagasse pretreatment using glycerol containing 17 wt% water at 130 °C. We attributed the increased efficiency of pretreatment at the pilot scale to improved mixing and heat transfer in the horizontal reactor.

Compared with pretreatment in dilute acid (HCl) alone, pretreatment in acidified glycerol solutions improved cellulose digestibility by 25–30% (Figure 4.3) (Zhang *et al.* 2013a).

We have also assessed the effectiveness of bagasse pretreatment using three acidified aqueous polyol solutions (Zhang *et al.* 2013c). Our results have demonstrated that, under the same pretreatment conditions, EG and propylene glycol (PG) were more effective at solubilizing lignin than was glycerol, with a corresponding improvement in glucan digestibility (Table 4.1). Despite this, pretreatment of bagasse using glycerol is preferred because of the relatively low cost of the solvent. We have also demonstrated that glycol-glycosides are produced in significant concentrations during acid-catalyzed polyol pretreatment of bagasse (Zhang *et al.* 2013b, c).

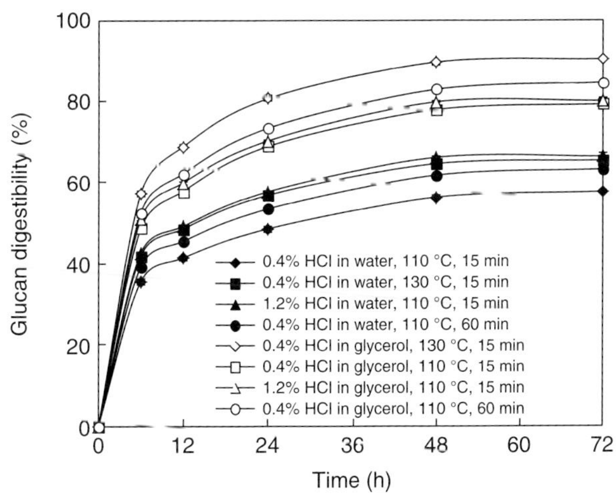

Figure 4.3 Kinetics of enzymatic hydrolysis of bagasse pretreated at different acid contents, temperatures, and times in a pilot scale reactor (Zhang *et al.* 2013a).

Table 4.1 Comparison of pretreatment effectiveness by glycerol, PG and EG solutions in the presence of 1.2% H_2SO_4 and 10% water at 130 °C for 30 min (Zhang *et al.* 2013c).

Polyol	Polymer content, %			Polymer recovery, %			Glucan digestibility, %	Glucose yield, %
	Glucan	Xylan	Lignin	Glucan	Xylan	Lignin		
Glycerol	69.9	7.6	19.6	96.7	22.8	43.2	76.8	74.3
PG	81.9	10.1	5.8	97.3	26.0	11.0	94.8	92.2
EG	81.6	9.7	5.8	96.8	25.0	11.0	94.7	91.7
Untreated bagasse	43.8	20.2	27.5	100.0	100.0	100.0	6.0	6.0

Pretreatment of lignocellulosic biomass at relatively low temperatures (≤100 °C) can result in energy savings of up to 50% compared with pretreatments (e.g., dilute acid pretreatments with water as solvent) operated at temperatures of 160–180 °C (Zhu *et al.* 2010). In order to reduce the energy consumption during pretreatment and the capital cost of constructing pretreatment reactors, we developed a low-temperature pretreatment process operating at atmospheric pressure using acid-catalyzed mixtures of alkylene carbonate and alkylene glycol (Zhang *et al.* 2013d, e). It was found that pretreatment of bagasse with mixtures of ethylene carbonate (EC) and EG was more effective than the individual components alone (Zhang *et al.* 2013d).

Cellulose digestibility of ≥87% could be achieved with a wide range of EC/EG ratios. Pretreatment of bagasse using EC/EG in a ratio of 4:1 in the presence of 1.2% H_2SO_4 at 90 °C for 30 min resulted in the highest glucan digestibility of >90%. The high glucan digestibility in the solid residues was due to (i) the ability of EC to significantly reduce biomass particle size, (ii) the ability of EG to defibrillate biomass, (iii) the ability of EC and EG to remove xylan and lignin, and (iv) the combination of these in the mixtures of EC/EG.

Furthermore, it was found that pretreatment by glycerol carbonate (GC) alone also led to a cellulose digestibility of 90% (Zhang *et al.* 2013e), which was significantly higher than that (16%) by EC pretreatment (Table 4.2). The higher digestibility obtained with GC-pretreated bagasse sample possibly was due to the presence of a hydroxyl group in GC's molecular structure, resulting in more significant biomass delignification and defibrillation though both solvent pretreatments reduced bagasse particles to a similar size range. Under optimal conditions for alkylene carbonate/alkylene glycol pretreatments, 5-hydroxymethyl furfural (a glucose degradation product) was not detected and furfural (a xylose degradation product) was only detected at very low concentrations (≤0.3% of initial mole xylose) in the hydrolysates (Zhang *et al.* 2013d, e).

Peracetic acid is a strong oxidant that can remove lignin from biomass under relatively mild conditions and has been used as an organic solvent for

Table 4.2 Results from pretreatment by AC/AG solvents at 90 °C for 30 min in the presence of 1.2% H_2SO_4 (Zhang *et al.* 2013e).

Solvent type	AC: AG	Content in solid residue, % Glucan	Xylan	Lignin	Component recovery, % Biomass	Glucan	Xylan	Lignin	Glucan digestibility, %	Total glucose yield, %
GC:glycerol	1:0	75.6	7.6	8.6	51.9	89.6	19.5	16.3	89.9	80.5
	9:1	74.8	9.9	9.7	53.9	92.0	26.3	19.0	90.2	83.0
	4:1	74.9	10.2	9.1	55.6	95.1	28.0	18.4	87.1	82.9
	2:1	72.0	11.9	11.3	57.6	94.7	34.0	23.7	86.9	82.3
	1:1	68.7	12.3	12.0	59.5	93.3	36.3	25.9	87.6	81.7
	0:1	56.6	13.8	23.4	73.6	95.1	50.3	62.5	61.3	58.3
EC:EG	1:0	64.2	7.0	20.4	61.8	90.6	21.3	45.8	16.3	14.8
	4:1	76.7	10.1	7.5	53.1	93.0	26.5	14.4	97.1	90.3
	0:1	67.2	14.1	13.4	62.4	95.7	43.6	30.6	74.7	71.5
Untreated bagasse		43.8	20.2	27.5	100.0	100.0	100.0	100.0	12.0	12.0

pretreatment of bagasse. Peracetic acid solutions of different concentrations have been assessed for pretreatment of bagasse at ambient temperature (20 °C) for 7 days (Teixeira *et al.* 1999a, b). Pretreatment of bagasse at liquid/solid ratio of 6:1 with peracetic acid (50% v/v) containing 1.4% H_2SO_4 (w/w) resulted in the effective removal of hemicellulose at 80 °C for 2 h, which resulted in a cellulose digestibility of 80% (Zhao *et al.* 2007).

4.4.5 Ionic liquid pretreatment

Ionic liquids are salts that exist as liquids at relatively low temperature (<100 °C) and are typically composed of large organic cations and small inorganic anions (Maki-Arvela *et al.* 2010, Brandt *et al.* 2013). Some ionic liquids are chemically and thermally stable, nonflammable, and have immeasurably low vapor pressure, all of which are desirable properties in biomass solvents (Maki-Arvela *et al.* 2010, Brandt *et al.* 2013). Ionic liquids are typically used to dissolve biomass with regeneration of the dissolved biomass being achieved via the addition of antisolvents, such as water (Maki-Arvela *et al.* 2010, Brandt *et al.* 2013). Biomass dissolution occurs because ionic liquids disrupt the hydrogen bonds between cellulose chains and form new hydrogen bonds between the cellulose chains and ionic liquid anion (Maki-Arvela *et al.* 2010, Brandt *et al.* 2013). The addition of antisolvent destroys the newly formed hydrogen bonds and so dissolved cellulose is precipitated in its amorphous form, which is significantly more susceptible to enzymatic hydrolysis than crystalline cellulose (Maki-Arvela *et al.* 2010, Brandt *et al.* 2013).

In the last decade, the numbers of studies describing pretreatment of bagasse using ionic liquids has increased significantly (Kimon *et al.* 2011, Lan *et al.*

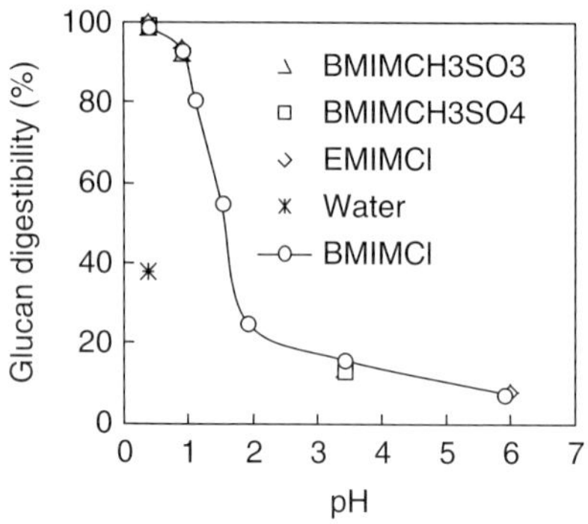

Figure 4.4 Relationships between glucan digestibility and solution pH.

2011, Wu *et al.* 2011, Yoon *et al.* 2011, Karatzos *et al.* 2012, Zhang *et al.* 2012, 2013b, da Silva *et al.* 2013, Hou *et al.* 2013, Qiu and Aita 2013). Our research group has demonstrated that complete dissolution of bagasse is not necessary to achieve high cellulose digestibility (Kimon *et al.* 2011). Recently, new strategies have been developed in order to decrease the cost associated with ionic liquid pretreatment of biomass. Continuous pretreatment of bagasse at 140 °C for 8 min with a relatively high biomass loading (25%) in 1-ethyl-3-methylimidazolium acetate produced residues with cellulose digestibility of >90% after 24 h of enzymatic hydrolysis (da Silva *et al.* 2013). Biomass characterization suggested that the use of an extruder as a pretreatment reactor significantly decreased the crystallinity of cellulose in the bagasse and increased the specific surface area by more than 100-fold (da Silva *et al.* 2013).

Our studies have demonstrated that it is possible to replace expensive ionic liquids with significant amounts of water (up to 30%) without negatively affecting pretreatment effectiveness of bagasse (Zhang *et al.* 2012). This ionic liquid pretreatment delignifies and removes hemicellulose from bagasse, rather than dissolving the biomass as is the case in a conventional ionic liquid pretreatment (Zhang *et al.* 2012). After pretreatment at 130 °C for 30 min in the presence of 30% water, cellulose digestibility of bagasse was >90% (Zhang *et al.* 2012). Furthermore, our study showed that aqueous ionic liquid pretreatment was more significantly impacted by pH than the nature of the ionic liquid anion (Zhang *et al.* 2013b) (Figure 4.4).

Recently, the use of renewable cholinium (cation)-amino acid (anion) ionic liquid for pretreatment of bagasse was reported (Hou *et al.* 2013). Sugar yields of 80% for glucose and 84% for xylose were obtained without bagasse size reduction after pretreatment at a biomass loading of 5 wt % in a 50% cholinium lysine–water mixture at 90 °C after 6 h (Hou *et al.* 2013).

4.4.6 SO_2- and CO_2-associated pretreatments

SO_2 pretreatment of biomass solubilizes lignin and hydrolyses hemicellulose. Martin *et al.* (2002a) have reported that the total sugar yield from SO_2-impregnated and exploded bagasse was higher than that from H_2SO_4-impregnated and exploded bagasse under the same pretreatment conditions. In another study, SO_2-catalyzed steam pretreatment of bagasse at 190 °C for 5 min resulted in only minimal sugar degradation and a sugar yield of 87% (Carrasco *et al.* 2010).

Supercritical CO_2 has been utilized for the pretreatment of bagasse. CO_2 forms carbonic acid when it dissolves in water. Even though it is a relatively weak acid it is expected to partially hydrolyze hemicellulose, as well as cellulose. Upon explosive release of pressure, the resulting disruption to the biomass is expected to increase the accessible surface area of the substrate for enzymatic hydrolysis (Zheng *et al.* 1998). However, low-temperature pretreatment of bagasse using supercritical CO_2 (80 °C at 3000 psi or 20.7 MPa) and subsequent explosive decompression only resulted in cellulose digestibility of <45% (Zheng *et al.* 1998). A recent study at higher temperature has demonstrated pretreatment of bagasse using supercritical CO_2 (5 MPa CO_2, 100 min, and 160 °C) with a total sugar yield of 76% (i.e., the combined amount of xylose and glucose in the pretreatment hydrolysate and enzymatic hydrolysate) (Zhang and Wu 2013).

Furthermore, the optimal conditions for CO_2-impregnated steam explosion (205 °C for 15 min) have been demonstrated to result in a comparable glucose yield (86.6% of theoretical) to that observed for SO_2-impreganated steam explosion at 190 °C for 5 min (Ferreira-Leitao *et al.* 2010). CO_2 is considered the more attractive pretreatment agent of the two because of its high availability, low cost, low toxicity, low corrosivity, and low occupational risk (Ferreira-Leitao *et al.* 2010).

4.5 Enzymatic hydrolysis

The discovery of *Trichoderma reesei* and its natural ability to secrete high titers of complex mixtures of biomass-degrading enzymes (Reese 1976) enabled enzymatic hydrolysis of biomass to replace chemical hydrolysis using concentrated acid, a process with typically poor yields and which degrades sugars (Saeman 1945). Biomass-degrading enzymes were subsequently identified in a wide range of bacteria and fungi (Sun and Cheng 2002). Fungi and some bacteria secrete biomass-degrading enzymes to function individually, while some bacteria link their biomass-degrading enzymes into large complexes called cellulosomes (Bayer *et al.* 2004).

Commercial cellulase mixtures are primarily derived from fungal cultures and typically contain a complex mixture of enzymes that degrade cellulose (endoglucanase, exoglucanase, β-glucosidase (βG), and copper-dependent oxygenase) and hemicellulose (xylanase, β-xylosidase, acetyl xylan esterase, arabinofuranosidase, β-mannanase, β-galactosidase, β-glucuronidase, and β-galactosidase), as well as proteins that enhance the performance of the other components (expansin-like proteins) (Foreman *et al.* 2003).

The enzymes required for efficient and complete degradation of cellulose to glucose are derived from three functional classes; (i) endoglucanases (endo-1,4-β-galactosidase-D-glucanases, EG) that attack regions of low crystallinity (i.e., amorphous regions), creating free chain ends; (ii) exoglucanases or cellobiohydrolases (exo-1,4-β-D-glucanases, CBH), that cleave cellobiose units from either the reducing (CBH I) or nonreducing (CBH II) free chain-ends, and (iii) β-βGs that hydrolyze cellobiose to glucose (Coughlan and Ljungdahl 1988). EG and CBH enzymes act synergistically to effect the extensive hydrolysis of crystalline cellulose (Wood, 1969, 1969) and the optimum EG/CBH ratio is strongly influenced by the pretreatment method used to generate the cellulosic substrate (Banerjee *et al.* 2010a).

There are numerous studies describing the hydrolysis of pretreated biomass by mixtures of purified cellulases, including barley straw (Meyer *et al.* 2009), wheat straw (Szijarto *et al.* 2011, Billard *et al.* 2012), corn stover (Zhou *et al.* 2009, Banerjee *et al.* 2010b, Gao *et al.* 2010), switchgrass (Banerjee *et al.* 2010b, Barr *et al.* 2012), *Miscanthus* (Banerjee *et al.* 2010b), willow (Karlsson *et al.* 1999), poplar (Banerjee *et al.* 2010b, Barr *et al.* 2012), and Douglas fir wood (Gusakov *et al.* 2007). These studies clearly demonstrated that the choice of pretreatment process has a profound effect on the overall susceptibility of a given biomass to hydrolysis by commercial cellulase mixtures or mixtures of purified cellulases.

Despite the importance of sugarcane as a biomass crop in the tropics and subtropics, similar studies are lacking for bagasse. We recently reported that glucan conversions obtained with mixtures of CBH I and *β*G on bagasse pretreated at pilot scale with acidified glycerol were 10–30% higher than those obtained with NaOH or H_2SO_4 pretreatments (Harrison *et al.* 2013). Furthermore, we demonstrated that the susceptibility of bagasse pretreated using acidified glycerol, NaOH, or H_2SO_4 to enzymatic hydrolysis by a commercial cellulase mixture was not indicative of susceptibility to enzymatic hydrolysis by a simple mixture of purified cellulase enzymes (Harrison *et al.* 2013).

4.6 Fermentation

Ethanol fermentation is an anaerobic process that has long been applied in human history for the production of alcohol drinks. Ethanol can be readily produced from starchy materials (wheat, corn, cassava, etc.) and plant juices (sugarcane juice, agave juice, etc.). However, cellulosic ethanol production can be problematic unless the sugars in lignocellulosic biomass are released after pretreatment and enzymatic hydrolysis.

Conventionally, ethanol is produced by yeasts such as *Saccharomyces, Pichia, Candida, Filobasidium, Lipomyces,* and *Saccharomycopsis* while *Saccharomyces cerevisiae* is the most commonly used and is generally recognized as safe (Lin and Tanaka 2006). Other microorganisms such as some bacteria of the genus *Clostridium* and *Zymomonas,* and some filamentous fungi of the genus *Aspergillus, Monilia, Mucor, Neurospora, Paecilomyces, Rhizopus, Trichoderma,* and *Zygosaccharomyces* also have the ability to produce ethanol but their production is limited to laboratory scale because of the low productivity and/or yield (Lin and Tanaka 2006, Abedinifar *et al.* 2009). With genetically engineered yeasts (e.g., *S. cerevisiae*) and bacteria (e.g., *Escherichia coli, Zymomonas mobilis, Klebsiella oxytoca*), sugars such as xylose and cellobiose can be utilized together with glucose (Doran *et al.* 1994, Golias *et al.* 2002, Martin *et al.* 2002b, Dien *et al.* 2003, Lin and Tanaka 2006, Rudolf *et al.* 2008).

In general, fermentation processes can be classified as solid-state fermentation (SoSF) and submerged fermentation (SMF) (Couto and Sanroman 2006). The latter, which consists of separate hydrolysis and fermentation (SHF), SSF processes, and consolidated bioprocessing (CBP) (Olofsson *et al.* 2008), has wider applications in industry because it is more amenable to industrial process control, design, and scale-up to commercial scale.

CBP integrates all the single steps (production of saccharolytic enzymes, hydrolysis of carbohydrate components in pretreated biomass to sugars, fermentation of both hexose and pentose sugars) in one process. It is distinguished from other less highly integrated configurations in that it does not involve a dedicated process step for cellulase production (Lynd *et al.* 2005). Although CBP has attracted great attention since the concept was proposed, its industrial application is still in development.

Currently, the most promising and reliable process for industrial ethanol production using lignocellulosic biomass is SSF, in which enzymes and ethanol-producing microorganisms are mixed with pretreated biomass in an integrated process. Compared with the SHF, the principle benefits of the SSF are reduced end-product inhibition of the enzymatic hydrolysis, and the reduced investment costs (Olofsson *et al.* 2008).

Ethanol fermentation can be impaired by inhibitors (e.g., aromatic compounds, aliphatic acids, furan aldehydes) generated from biomass during pretreatment (Jonsson *et al.* 2013). Detoxification is often used to improve the growth of microorganism and ethanol production. A recent review has summarized the methods used for detoxification of hydrolysates after pretreatment (Jonsson *et al.* 2013). Common methods include post-pretreatment alkali cooking, addition of oxidase enzymes or co-cultivation with oxidase-producing enzymes, addition of lime and activated carbon, and treatment with ion exchange resins (Jonsson *et al.* 2013).

It has been reported that the toxicity of dilute phosphoric acid pretreated bagasse was reduced by adding a reducing agent, sodium metabisulfite (0.5 mM), which led to ethanol fermentation by a recombinant xylose-utilizing strain *E. coli* MM160 without solid–liquid separation or cleanup of sugars (Nieves *et al.* 2011). Studies have shown that recombinant *S. cerevisiae* strains can utilize both xylose and glucose from pretreated sugarcane bagasse in SHF and SSF processes (Martin *et al.* 2002b, Rudolf *et al.* 2008). A recombinant xylose-utilizing yeast strain of *S. cerevisiae* TMB3400 consumed all glucose and large fraction of the xylose in SSF of steam-pretreated sugarcane bagasse (Rudolf *et al.* 2008). This strain also produced more ethanol than a natural xylose-utilization strain (*Pichia stipitis*) with a higher productivity (Rudolf *et al.* 2008). Compared with SHF by another recombinant xylose-utilization yeast strain (*S. cerevisiae* TMB3001) (Martin *et al.* 2002b), SSF resulted in much higher ethanol concentration (26.7 vs. 9.9 g/l) but lower ethanol yield (0.37 vs 0.30 g ethanol/g sugar) and productivity (0.27 vs 0.41 g/l/h). It is worth noting in these studies that, although detoxification was not used for the steam-exploded bagasse samples, ethanol fermentation performed compared with those studies with detoxification by activated carbon, ion-exchange resin, and overliming (Martin *et al.* 2002b, Rudolf *et al.* 2008).

4.7 Conclusions and future perspectives

Although significant efforts are being made to commercialize second-generation bioethanol from bagasse, the costs associated with pretreatment and enzymatic hydrolysis are the major obstacles to price parity with the production of first-generation ethanol from sugarcane juice. While dilute acid pretreatment has received significant attention, the high inhibitor levels resulting from this process and the low cellulose digestibility of the resulting residue continue to pose significant challenges. Alkaline pretreatment can achieve high sugar recovery with low levels of inhibitor production. However, this process is not economically viable without technology for the efficient recycling of the alkaline catalysts (such as NaOH and ammonia). Although LHW and steam explosion are more environmentally acceptable processes due to the lack of catalysts required, they generally result in lower sugar yields as a result.

The ideal bagasse pretreatment for second-generation bioethanol production would be a single-step process that operates at low temperature and results in high cellulose digestibility and high recovery of xylose with the production of only low levels of fermentation inhibitors. This can be achieved with the use of solvents such as polyols and ionic liquids. Although significant progress has been made in the last decade, the high cost of ionic liquids necessitates highly efficient solvent recovery. As a result, low-cost polyol solvents, such as glycerol, may provide the most promising pretreatment process for the future economic production of second-generation bioethanol from bagasse.

References

Abedinifar, S., Karimi, K., Khanahmadi, M. and Taherzadeh, M. J. (2009). Ethanol production by Mucor indicus and Rhizopus oryzae from rice straw by separate hydrolysis and fermentation. *Biomass & Bioenergy* **33**(5):828–833.

Agbor, V. B., Cicek, N., Sparling, R., Berlin, A. and Levin, D. B. (2011). Biomass pretreatment: fundamentals toward application. *Biotechnology Advances* **29**(6):675–685.

Alvira, P., Tomas-Pejo, E., Ballesteros, M. and Negro, M. J. (2010). Pretreatment technologies for an efficient bioethanol production process based on enzymatic hydrolysis: a review. *Bioresource Technology* **101**(13):4851–4861.

Area, C. M., Felissia, F. E. and Vallejos, M. E. (2009). Ethanol-water fractionation of sugar cane bagasse catalyzed with acids. *Cellulose Chemistry and Technology* **43**(7–8):271–279.

Banerjee, G., Car, S., Scott-Craig, J. S., Borrusch, M. S. and Walton, J. D. (2010a). Rapid optimization of enzyme mixtures for deconstruction of diverse pretreatment/biomass feedstock combinations. *Biotechnology for Biofuels* **12**(3):22.

Banerjee, G., Car, S., Scott-Craig, J. S., Borrusch, M. S., Bongers, M. and Walton, J. D. (2010b). Synthetic multi-component enzyme mixtures for deconstruction of lignocellulosic biomass. *Bioresource Technology* **101**(23):9097–9105.

Barr, C. J., Mertens, J. A. and Schall, C. A. (2012). Critical cellulase and hemicellulase activities for hydrolysis of ionic liquid pretreated biomass. *Bioresource Technology* **104**:480–485.

Bayer, E. A., Belaich, J. P., Shoham, Y. and Lamed, R. (2004). The cellulosomes: multienzyme machines for degradation of plant cell wall polysaccharides. *Annual Review of Microbiology* **58**:521–554.

Betancur, G. J. and Pereira, N. (2010). Sugar cane bagasse as feedstock for second generation ethanol production. Part I: diluted acid pretreatment optimization. *Electronic Journal of Biotechnology* **13**(3):1–9.

Bigelow, M. and Wyman, C. (2002). Cellulase production on bagasse pretreated with hot water. *Applied Biochemistry and Biotechnology* **98–100**(1):921–934.

Billard, H., Faraj, A., Lopes Ferreira, N., Menir, S. and Heiss-Blanquet, S. (2012). Optimization of a synthetic mixture composed of major *Trichoderma reesei* enzymes for the hydrolysis of steam-exploded wheat straw. *Biotechnology for Biofuels* **5**:9.

Brandt, A., Ray, M. J., To, T. Q., Leak, D. J., Murphy, R. J. and Welton, T. (2011). Ionic liquid pretreatment of lignocellulosic biomass with ionic liquid-water mixtures. *Green Chemistry* **13**(9):2489–2499.

Brandt, A., Grasvik, J., Hallett, J. P. and Welton, T. (2013). Deconstruction of lignocellulosic biomass with ionic liquids. *Green Chemistry* **15**(3):550–583.

Camacho, F., GonzalezTello, P., Jurado, E. and Robles, A. (1996). Microcrystalline-cellulose hydrolysis with concentrated sulphuric acid. *Journal of Chemical Technology and Biotechnology* **67**(4):350–356.

Candido, R. G., Godoy, G. G. and Gonçalves, A. R. (2012). Study of sugarcane bagasse pretreatment with sulfuric acid as a step of cellulose obtaining. *World Academy of Science, Engineering and Technology* **61**:101–105.

Canilha, L., Santos, V. T. O., Rocha, G. J. M., Silva, J. B. A. E., Giulietti, M., Silva, S. S., Felipe, M. G. A., Ferraz, A., Milagres, A. M. F. and Carvalho, W. (2011). A study on the pretreatment of a sugarcane bagasse sample with dilute sulfuric acid. *Journal of Industrial Microbiology & Biotechnology* **38**(9):1467–1475.

Cao, S. and Aita, G. M. (2013). Enzymatic hydrolysis and ethanol yields of combined surfactant and dilute ammonia treated sugarcane bagasse. *Bioresource Technology* **131**:357–364.

Carrasco, C., Baudel, H. M., Sendelius, J., Modig, T., Roslander, C., Galbe, M., Hahn-Hagerdal, B., Zacchi, G. and Liden, G. (2010). SO_2-catalyzed steam pretreatment and fermentation of enzymatically hydrolyzed sugarcane bagasse. *Enzyme and Microbial Technology* **46**(2):64–73.

Chandel, A. K., da Silva, S. S., Carvalho, W. and Singh, O. V. (2012). Sugarcane bagasse and leaves: foreseeable biomass of biofuel and bio-products. *Journal of Chemical Technology and Biotechnology* **87**(1):11–20.

Coughlan, M. P. and Ljungdahl, L. G. (1988). Comparative biochemistry of fungal and bacterial cellulolytic systems. In *Biochemistry and Genetics of Cellulose Degradation*, pp. 11–30. J.-P. Aubert, P. Béguin and J. Millet, eds. Academic Press: London.

Couto, S. R. and Sanroman, M. A. (2006). Application of solid-state fermentation to food industry—A review. *Journal of Food Engineering* **76**(3):291–302.

Del Rio, L. F., Chandra, R. P. and Saddler, J. N. (2011). The effects of increasing swelling and anionic charges on the enzymatic hydrolysis of organosolv-pretreated softwoods at low enzyme loadings. *Biotechnology and Bioengineering* **108**(7):1549–1558.

Diedericks, D., van Rensburg, E., Garcia-Aparicio, M. D. and Gorgens, J. F. (2012). Enhancing the enzymatic digestibility of sugarcane bagasse through the application of an ionic liquid in combination with an acid catalyst. *Biotechnology Progress* **28**(1):76–84.

Dien, B. S., Cotta, M. A. and Jeffries, T. W. (2003). Bacteria engineered for fuel ethanol production: current status. *Applied Microbiology and Biotechnology* **63**(3):258–266.

Doran, J. B., Aldrich, H. C. and Ingram, L. O. (1994). Saccharification and fermentation of Sugar-Cane bagasse by *Klebsiella-oxytoca* P2 containing chromosomally integrated genes encoding the *Zymomonas-mobilis* ethanol pathway. *Biotechnology and Bioengineering* **44**(2):240–247.

Ferreira-Leitao, V., Perrone, C. C., Rodrigues, J., Franke, A. P. M., Macrelli, S. and Zacchi, G. (2010). An approach to the utilisation of CO2 as impregnating agent in steam pretreatment of sugar cane bagasse and leaves for ethanol production. *Biotechnology for Biofuels* **3**:7.

Foreman, P. K., Brown, D., Dankmeyer, L., Dean, R., Diener, S., Dunn-Coleman, N. S., Goedegebuur, F., Houfek, T. D., England, G. J., Kelley, A. S., Meerman, H. J., Mitchell, T., Mitchinson, C., Olivares, H. A., Teunissen, P. J. M., Yao, J. and Ward, M. (2003). Transcriptional regulation of biomass-degrading enzymes in the filamentous fungus *Trichoderma reesei*. *Journal of Biological Chemistry* **278**(34):31988–31997.

Fox, D. J., Gray, P. P., Dunn, N. W. and Marsden, W. L. (1989). Comparison of alkali and steam (acid) pretreatments of lignocellulosic materials to increase enzymic susceptibility Evaluation under optimised pretreatment conditions. *Journal of Chemical Technology and Biotechnology* **44**(2):135–146.

Gao, D. H., Chundawat, S. P. S., Krishnan, C., Balan, V. and Dale, B. E. (2010). Mixture optimization of six core glycosyl hydrolases for maximizing saccharification of ammonia fiber expansion (AFEX) pretreated corn stover. *Bioresource Technology* **101**(8):2770–2781.

Gao, Y., Xu, J., Zhang, Y., Yu, Q., Yuan, Z. and Liu, Y. (2013). Effects of different pretreatment methods on chemical composition of sugarcane bagasse and enzymatic hydrolysis. *Bioresource Technology* **144**:396–400.

Geddes, C. C., Peterson, J. J., Roslander, C., Zacchi, G., Mullinnix, M. T., Shanmugam, K. T. and Ingram, L. O. (2010). Optimizing the saccharification of sugar cane bagasse using dilute phosphoric acid followed by fungal cellulases. *Bioresource Technology* **101**:1851–1857.

Golias, H., Dumsday, G. J., Stanley, G. A. and Pamment, N. B. (2002). Evaluation of a recombinant *Klebsiella oxytoca* strain for ethanol production from cellulose by simultaneous saccharification and fermentation: comparison with native cellobiose-utilising yeast strains and performance in co-culture with thermotolerant yeast and *Zymomonas mobilis*. *Journal of Biotechnology* **96**(2):155–168.

Gusakov, A. V., Salanovich, T. N., Antonov, A. I., Ustinov, B. B., Okunev, O. N., Burlingame, R., Emalfarb, M., Baez, M. and Sinitsyn, A. P. (2007). Design of highly efficient cellulase mixtures for enzymatic hydrolysis of cellulose. *Biotechnology and Bioengineering* **97**(5):1028–1038.

Harrison, M. D., Zhang, Z., Shand, K., O'Hara, I. M., Doherty, W. O. S. and Dale, J. L. (2013). Effect of pretreatment on saccharification of sugar cane bagasse by complex and simple enzyme mixtures. *Bioresource Technology* **148**:105–113.

Hendriks, A. and Zeeman, G. (2009). Pretreatments to enhance the digestibility of lignocellulosic biomass. *Bioresource Technology* **100**(1):10–18.

Hou, X. D., Li, N. and Zong, M. H. (2013). Facile and simple pretreatment of sugar cane bagasse without size reduction using renewable ionic liquids-water mixtures. *ACS Sustainable Chemistry & Engineering* **1**(5):519–526.

Huang, Y. B. and Fu, Y. (2013). Hydrolysis of cellulose to glucose by solid acid catalysts. *Green Chemistry* **15**(5):1095–1111.

Iranmahboob, J., Nadim, F. and Monemi, S. (2002). Optimizing acid-hydrolysis: a critical step for production of ethanol from mixed wood chips. *Biomass and Bioenergy* **22**:401–404.

John, M. J. and Thomas, S. (2008). Biofibres and biocomposites. *Carbohydrate Polymers* **71**(3):343–364.

Jonsson, L. J., Alriksson, B. and Nilvebrant, N. O. (2013). Bioconversion of lignocellulose: inhibitors and detoxification. *Biotechnology for Biofuels* **6**:16.

Kallioinen, A., Hakola, M., Riekkola, T., Repo, T., Leskela, M., von Weymarn, N. and Siika-aho, M. (2013). A novel alkaline oxidation pretreatment for spruce, birch and sugar cane bagasse. *Bioresource Technology* **140**:414–420.

Karatzos, S. K., Edye, L. A. and Doherty, W. O. S. (2012). Sugarcane bagasse pretreatment using three imidazolium-based ionic liquids; mass balances and enzyme kinetics. *Biotechnology for Biofuels* **5**:62e73.

Karlsson, J., Medve, J. and Tjerneld, F. (1999). Hydrolysis of steam-pretreated lignocellulose—Synergism and adsorption for cellobiohydrolase I and endoglucanase II of *Trichoderma reesei*. *Applied Biochemistry and Biotechnology* **82**(3):243–258.

Kimon, K. S., Alan, E. L. and Sinclair, D. W. O. (2011). Enhanced saccharification kinetics of sugarcane bagasse pretreated in 1-butyl-3-methylimidazolium chloride at high temperature and without complete dissolution. *Bioresource Technology* **102**(19):9325–9329.

Klemm, D. (1998). *Comprehensive Cellulose Chemistry*. Wiley-VCH:Weinheim, New York.

Krishnan, C., Sousa, L. D., Jin, M. J., Chang, L. P., Dale, B. E. and Balan, V. (2010). Alkali-Based AFEX Pretreatment for the Conversion of Sugarcane Bagasse and Cane Leaf residues to Ethanol. *Biotechnology and Bioengineering* **107**(3):441–450.

Kurakake, M., Kisaka, W., Ouchi, K. and Komaki, T. (2001). Pretreatment with ammonia water for enzymatic hydrolysis of corn husk, bagasse, and switchgrass. *Applied Biochemistry and Biotechnology* **90**(3):251–259.

Lan, W., Liu, C. F. and Sun, R. G. (2011). Fractionation of bagasse into cellulose, hemicelluloses, and lignin with ionic liquid treatment followed by alkaline extraction. *Journal of Agricultural and Food Chemistry* **59**(16):8691–8701.

Laser, M., Schulman, D., Allen, S. G., Lichwa, J., Antal, M. J. and Lynd, L. R. (2002). A comparison of liquid hot water and steam pretreatments of sugar cane bagasse for bioconversion to ethanol. *Bioresource Technology* **81**(1):33–44.

Laureano-Perez, L., Teymouri, F., Alizadeh, H. and Dale, B. (2005). Understanding factors that limit enzymatic hydrolysis of biomass. *Applied Biochemistry and Biotechnology* **124**(1):1081–1099.

Lavarack, B. P., Griffin, G. J. and Rodman, D. (2000). Measured kinetics of the acid-catalysed hydrolysis of sugar cane bagasse to produce xylose. *Catalysis Today* **63**(2–4):257–265.

Lee, D. H., Cho, E. Y., Kim, C. J. and Kim, S. B. (2010). Pretreatment of waste newspaper using ethylene glycol for bioethanol production. *Biotechnology and Bioprocess Engineering* **15**(6):1094–1101.

Lenihan, P., Orozco, A., O'Neill, E., Ahmad, M. N. M., Rooney, D. W. and Walker, G. M. (2010). Dilute acid hydrolysis of lignocellulosic biomass. *Chemical Engineering Journal* **156**:395–403.

Liao, W., Liu, Y., Liu, C., Wen, Z. and Chen, S. (2006). Acid hydrolysis of Wbers from dairy manure. *Bioresource Technology* **97**:1687–1695.

Lin, Y. and Tanaka, S. (2006). Ethanol fermentation from biomass resources: current state and prospects. *Applied Microbiology and Biotechnology* **69**(6):627–642.

Lynd, L. R., van Zyl, W. H., McBride, J. E. and Laser, M. (2005). Consolidated bioprocessing of cellulosic biomass: an update. *Current Opinion in Biotechnology* **16**(5):577–583.

Macrelli, S., Mogensen, J. and Zacchi, G. (2012). Techno-economic evaluation of 2nd generation bioethanol production from sugar cane bagasse and leaves integrated with the sugar-based ethanol process. *Biotechnology for Biofuels* **5**:22.

Maki-Arvela, P., Anugwom, I., Virtanen, P., Sjoholm, R. and Mikkola, J. P. (2010). Dissolution of lignocellulosic materials and its constituents using ionic liquids—A review. *Industrial Crops and Products* **32**(3):175–201.

Martin, C., Galbe, M., Nilvebrant, N. O. and Jonsson, L. J. (2002a). Comparison of the fermentability of enzymatic hydrolyzates of sugarcane bagasse pretreated by steam explosion using different impregnating agents. *Applied Biochemistry and Biotechnology* **98**:699–716.

Martin, C., Galbe, M., Wahlbom, C. F., Hahn-Hagerdal, B. and Jonsson, L. J. (2002b). Ethanol production from enzymatic hydrolysates of sugarcane bagasse using recombinant xylose-utilising *Saccharomyces cerevisiae*. *Enzyme and Microbial Technology* **31**(3):274–282.

Martín, C., González, Y., Fernández, T. and Thomsen, A. B. (2006). Investigation of cellulose convertibility and ethanolic fermentation of sugarcane bagasse pretreated by wet oxidation and steam explosion. *Journal of Chemical Technology & Biotechnology* **81**(10):1669–1677.

Martin, C., Alriksson, B., Sjöde, A., Nilvebrant, N.-O. and Jönsson, L. (2007a). Dilute sulfuric acid pretreatment of agricultural and agro-industrial residues for ethanol production. *Applied Biochemistry and Biotechnology* **137–140**(1):339–352.

Martin, C., Klinke, H. B., Marcet, M., Garcia, L., Hernandez, E. and Thomsen, A. B. (2007b). Study of the phenolic compounds formed during pretreatment of sugarcane bagasse by wet oxidation and steam explosion. *Holzforschung* **61**(5):483–487.

Martin, C., Puls, J., Saake, B. and Schreiber, A. (2011). Effect of Glycerol Pretreatment on Component Recovery and Enzymatic Hydrolysis of Sugarcane Bagasse. *Cellulose Chemistry and Technology* **45**(7–8):487–494.

Menon, V. and Rao, M. (2012). Trends in bioconversion of lignocellulose: biofuels, platform chemicals & biorefinery concept. *Progress in Energy and Combustion Science* **38**(4):522–550.

Mesa, L., Gonzalez, E., Cara, C., Ruiz, E., Castro, E. and Mussatto, S. I. (2010). An approach to optimization of enzymatic hydrolysis from sugarcane bagasse based on organosolv pretreatment. *Journal of Chemical Technology and Biotechnology* **85**(8):1092–1098.

Mesa, L., Gonzalez, E., Cara, C., Gonzalez, M., Castro, E. and Mussatto, S. I. (2011). The effect of organosolv pretreatment variables on enzymatic hydrolysis of sugarcane bagasse. *Chemical Engineering Journal* **168**(3):1157–1162.

Meyer, A. S., Rosgaard, L. and Sorensen, H. R. (2009). The minimal enzyme cocktail concept for biomass processing. *Journal of Cereal Science* **50**(3):337–344.

Morjanoff, P. J. and Gray, P. P. (1987). Optimization of steam explosion as a method for increasing susceptibility of sugar cane bagasse to enzymatic saccharification. *Biotechnology and Bioengineering* **29**(6):733–741.

Mosier, N., Wyman, C., Dale, B., Elander, R., Lee, Y. Y., Holtzapple, M. and Ladisch, M. (2005). Features of promising technologies for pretreatment of lignocellulosic biomass. *Bioresource Technology* **96**(6):673–686.

Nguyen, T. A. D., Kim, K. R., Han, S. J., Cho, H. Y., Kim, J. W., Park, S. M., Park, J. C. and Sim, S. J. (2010). Pretreatment of rice straw with ammonia and ionic liquid for lignocellulose conversion to fermentable sugars. *Bioresource Technology* **101**(19):7432–7438.

Nieves, I. U., Geddes, C. C., Miller, E. N., Mullinnix, M. T., Hoffman, R. W., Fu, Z., Tong, Z. and Ingram, L. O. (2011). Effect of reduced sulfur compounds on the fermentation of phosphoric acid pretreated sugarcane bagasse by ethanologenic *Escherichia coli*. *Bioresource Technology* **102**(8):5145–5152.

O'Hara, I. M. (2011). Cellulosic ethanol from sugarcane bagasse in Australia: exploring industry feasibility through systems analysis, techno-economic assessment and pilot plant development. Doctor of Philosophy, Queensland University of Technology.

O'Hara, I. M., Zhang, Z., Doherty, W. O. S. and Fellows, C. M. (2011). Lignocellulosics as a renewable feedstock for chemical industry: chemical hydrolysis and pretreatment processes. In *Green Chemistry for Environmental Remediation*. R. Sanghi and V. Singh, eds. Scrivener Publishing LLC: Salem.

Olofsson, K., Bertilsson, M. and Liden, G. (2008). A short review on SSF—an interesting process option for ethanol production from lignocellulosic feedstocks. *Biotechnology for Biofuels* **1**:7.

Playne, M. J. (1984). Increased digestibility of bagasses by pretreatment with alkalis and steam explosion. *Biotechnology and Bioengineering* **26**(5):426–433.

Puri, V. P. and Pearce, G. R. (1986). Alkali-explosion pretreatment of straw and bagasse for enzymic hydrolysis. *Biotechnology and Bioengineering* **28**(4):480–485.

Qiu, Z. and Aita, G. M. (2013). Pretreatment of energy cane bagasse with recycled ionic liquid for enzymatic hydrolysis. *Bioresource Technology* **129**:532–537.

Rabelo, S., Filho, R. and Costa, A. (2008). A comparison between lime and alkaline hydrogen peroxide pretreatments of sugarcane bagasse for ethanol production. *Applied Biochemistry and Biotechnology* **144**(1):87–100.

Rabelo, S. C., Maciel, R. and Costa, A. C. (2013). Lime pretreatment and fermentation of enzymatically hydrolyzed sugarcane bagasse. *Applied Biochemistry and Biotechnology* **169**(5):1696–1712.

Reese, E. T. (1976). History of the cellulase program at the U.S. Natick Development Center. *Biotechnology and Bioengineering Symposium* **6**:9–20.

Rocha, G. J. D., Martin, C., Soares, I. B., Maior, A. M. S., Baudel, H. M. and de Abreu, C. A. M. (2011). Dilute mixed-acid pretreatment of sugarcane bagasse for ethanol production. *Biomass and Bioenergy* **35**(1):663–670.

Rossell, C. E. V., Lahr, D., Hilst, A. G. P. and Leal, M. (2006). Saccharification of sugarcane bagasse for ethanol production using the Organosolv process. *Zuckerindustrie* **131**(2):105–109.

Rudolf, A., Baudel, H., Zacchi, G., Hahn-Hagerdal, B. and Liden, G. (2008). Simultaneous saccharification and fermentation of steam-pretreated bagasse using *Saccharomyces cerevisiae* TMB3400 and *Pichia stipitis* CBS6054. *Biotechnology and Bioengineering* **99**:783–790.

Saeman, J. F. (1945). Kinetics of wood saccharification—Hydrolysis of cellulose and decomposition of sugars in dilute acid at high temperatures. *Industrial and Engineering Chemistry* **37**:43–52.

Sathitsuksanoh, N., Zhu, Z. G., Wi, S. and Zhang, Y. H. P. (2011). Cellulose solvent-based biomass pretreatment breaks highly ordered hydrogen bonds in cellulose fibers of switchgrass. *Biotechnology and Bioengineering* **108**(3):521–529.

Silva, S. S., Matos, Z. R. and Carvalho, W. (2005). Effects of sulfuric acid loading and residence time on the composition of sugarcane bagasse hydrolysate and its use as a source of xylose for xylitol bioproduction. *Biotechnology Progress* **21**(5):1449–1452.

da Silva, A. S., Teixeira, R. S. S., Endo, T., Bon, E. P. S. and Lee, S. H. (2013). Continuous pretreatment of sugarcane bagasse at high loading in an ionic liquid using a twin-screw extruder. *Green Chemistry* **15**(7):1991–2001.

Sumphanwanich, J., Leepipatpiboon, N., Srinorakutara, T. and Akaracharanya, A. (2008). Evaluation of dilute-acid pretreated bagasse, corn cob and rice straw for ethanol fermentation by *Saccharomyces cerevisiae*. *Annals of Microbiology* **58**(2):219–225.

Sun, Y. and Cheng, J. Y. (2002). Hydrolysis of lignocellulosic materials for ethanol production: a review. *Bioresource Technology* **83**(1):1–11.

Szijarto, N., Horan, E., Zhang, J. H., Puranen, T., Siika-aho, M. and Viikari, L. (2011). Thermostable endoglucanases in the liquefaction of hydrothermally pretreated wheat straw. *Biotechnology for Biofuels* **4**:2.

Teixeira, L. C., Linden, J. C. and Schroeder, H. A. (1999a). Optimizing peracetic acid pretreatment conditions for improved simultaneous saccharification and co-fermentation (SSCF) of sugar cane bagasse to ethanol fuel. *Renewable Energy* **16**(1–4):1070–1073.

Teixeira, L. C., Linden, J. C. and Schroeder, H. A. (1999b). Alkaline and peracetic acid pretreatments of biomass for ethanol production. *Applied Biochemistry and Biotechnology* **77-9**:19–34.

Teramoto, Y., Lee, S. H. and Endo, T. (2009). Cost reduction and feedstock diversity for sulfuric acid-free ethanol cooking of lignocellulosic biomass as a pretreatment to enzymatic saccharification. *Bioresource Technology* **100**(20):4783–4789.

VanWalsum, G. P., Allen, S. G., Spencer, M. J., Laser, M. S., Antal, M. J. and Lynd, L. R. (1996). Conversion of lignocellulosics pretreated with liquid hot water to ethanol. *Applied Biochemistry and Biotechnology* **57-8**:157–170.

Wanderley, M. C. D., Martin, C., Rocha, G. J. D. and Gouveia, E. R. (2013). Increase in ethanol production from sugarcane bagasse based on combined pretreatments and fed-batch enzymatic hydrolysis. *Bioresource Technology* **128**:448–453.

Wood, T. M. (1969). The cellulase of Fusarium solani. Resolution of the enzyme complex. *Biochemical Journal* **115**:457–464.

Wu, H., Mora-Pale, M., Miao, J. J., Doherty, T. V., Linhardt, R. J. and Dordick, J. S. (2011). Facile pretreatment of lignocellulosic biomass at high loadings in room temperature ionic liquids. *Biotechnology and Bioengineering* **108**(12):2865–2875.

Yang, B. and Wyman, C. E. (2008). Pretreatment: the key to unlocking low-cost cellulosic ethanol. *Biofuels, Bioproducts and Biorefining* **2**(1):26–40.

Yoon, L. W., Ngoh, G. C., Chua, A. S. M. and Hashim, M. A. (2011). Comparison of ionic liquid, acid and alkali pretreatments for sugarcane bagasse enzymatic saccharification. *Journal of Chemical Technology and Biotechnology* **86**(10):1342–1348.

Yu, Q., Zhuang, X., Yuan, Z., Qi, W., Wang, W., Wang, Q. and Tan, X. (2013a). Pretreatment of sugarcane bagasse with liquid hot water and aqueous ammonia. *Bioresource Technology* **144**:210–215.

Yu, Q., Zhuang, X., Lv, S., He, M., Zhang, Y., Yuan, Z., Qi, W., Wang, Q., Wang, W. and Tan, X. (2013b). Liquid hot water pretreatment of sugarcane bagasse and its comparison with chemical pretreatment methods for the sugar recovery and structural changes. *Bioresource Technology* **129**:592–598.

Zhang, H. and Wu, S. (2013). Subcritical CO2 pretreatment of sugarcane bagasse and its enzymatic hydrolysis for sugar production. *Bioresource Technology* **149**:546–550.

Zhang, Z. Y., O'Hara, I. M. and Doherty, W. U. S. (2012). Pretreatment of sugarcane bagasse by acid-catalysed process in aqueous ionic liquid solutions. *Bioresource Technology* **120**:149–156.

Zhang, Z., Wong, H. H., Albertson, P. L., Doherty, W. O. S. and O'Hara, I. M. (2013a). Laboratory and pilot scale pretreatment of sugarcane bagasse by acidified aqueous glycerol solutions. *Bioresource Technology* **138**:14–21.

Zhang, Z., O'Hara, I. M. and Doherty, W. O. S. (2013b). Effects of pH on pretreatment of sugarcane bagasse using aqueous imidazolium ionic liquids. *Green Chemistry* **15**:431–438.

Zhang, Z., Doherty, W. O. S. and O'Hara, I. M. (2013c). Pretreatment of sugarcane bagasse by acidified aqueous polyol solutions. *Cellulose* **20**(6):3179–3190.

Zhang, Z., O'Hara, I. M., Rackemann, D. W. and Doherty, W. O. S. (2013d). Low temperature pretreatment of sugarcane bagasse at atmospheric pressure using mixtures of ethylene carbonate and ethylene glycol. *Green Chemistry* **15**:255–264.

Zhang, Z., Rackemann, D. W., Doherty, W. O. S. and O'Hara, I. M. (2013e). Glycerol carbonate as green solvent for pretreatment of sugarcane bagasse. *Biotechnology for Biofuels* **6**(1):153.

Zhao, X.-B., Wang, L. and Liu, D.-H. (2007). Effect of several factors on peracetic acid pretreatment of sugarcane bagasse for enzymatic hydrolysis. *Journal of Chemical Technology and Biotechnology* **82**(12):1115–1121.

Zhao, X. B., Peng, F., Cheng, K. and Liu, D. H. (2009a). Enhancement of the enzymatic digestibility of sugarcane bagasse by alkali-peracetic acid pretreatment. *Enzyme and Microbial Technology* **44**(1):17–23.

Zhao, X. B., Cheng, K. K. and Liu, D. H. (2009b). Organosolv pretreatment of lignocellulosic biomass for enzymatic hydrolysis. *Applied Microbiology and Biotechnology* **82**(5):815–827.

Zhao, X. B., Zhang, L. H. and Liu, D. H. (2012). Biomass recalcitrance. Part II: fundamentals of different pre-treatments to increase the enzymatic digestibility of lignocellulose. *Biofuels, Bioproducts and Biorefining* **6**(5):561–579.

Zheng, Y. Z., Lin, H. M. and Tsao, G. T. (1998). Pretreatment for cellulose hydrolysis by carbon dioxide explosion. *Biotechnology Progress* **14**(6):890–896.

Zhou, J., Wang, Y. H., Chu, J., Luo, L. Z., Zhuang, Y. P. and Zhang, S. L. (2009). Optimization of cellulase mixture for efficient hydrolysis of steam-exploded corn stover by statistically designed experiments. *Bioresource Technology* **100**(2):819–825.

Zhu, Z. G., Sathitsuksanoh, N., Vinzant, T., Schell, D. J., McMillan, J. D. and Zhang, Y. H. P. (2009). Comparative study of corn stover pretreated by dilute acid and cellulose solvent-based lignocellulose fractionation: enzymatic hydrolysis, supramolecular structure, and substrate accessibility. *Biotechnology and Bioengineering* **103**(4):715–724.

Zhu, J. Y., Pan, X. J. and Zalesny, R. S. (2010). Pretreatment of woody biomass for biofuel production: energy efficiency, technologies, and recalcitrance. *Applied Microbiology and Biotechnology* **87**(3):847–857.

CHAPTER 5
Chemicals manufacture from fermentation of sugarcane products

Karen T. Robins[1,2] and Robert E. Speight[2]
[1] *Sustain Biotech Pty Ltd, Sydney, Australia*
[2] *School of Chemistry, Physics and Mechanical Engineering, Science and Engineering Faculty, Queensland University of Technology (QUT), Brisbane, Australia*

5.1 Introduction

The industrial biotechnology industry offers significant promise for the manufacture of a range of important and useful chemicals from biobased feedstocks. The industry is at a critical and exciting stage with a number of new large-scale commercial projects up and running or coming on line. These new processes typically use engineered microorganisms in optimized fermentations to produce either new molecules on the market or to replace molecules currently derived from petrochemicals. In addition to these new and emerging processes, industrial biotechnology has been used for many years (millennia in the case of brewing for ethanol) for the production of a range of molecules, such as amino acids and organic acids. Both new and old commercial projects vary significantly in the nature and value of the chemical product as well as the microbial strains and fermentation processes that are used to produce them. What is common between the projects is that they all require a source of carbon and that the carbon source is a major cost driver for the overall process (often more than 50% for a commodity product (Chotani *et al.* 2014)). Typically, the carbon in these processes is derived from agricultural biomass with exceptions including gas fermenting microbial processes utilizing methane, such as the production of single-cell protein by Dansk BioProtein in the 1980s, or carbon monoxide in syngas (Kopke *et al.* 2010), an approach that is being commercialized by the company LanzaTech (2015).

Costs and risks of generating fermentable carbon sources from biomass arise from the climate dependence (yield variations) of biomass growth and required inputs (e.g., fertilizer) coupled to the costs of harvesting, transportation, and processing of the biomass to fermentable sugars. During fermentation, the carbon source is also used by the microorganism for cell growth and other biological processes in addition to producing the desired chemical, affecting the overall yield

Sugarcane-Based Biofuels and Bioproducts, First Edition. Edited by Ian M. O'Hara and Sagadevan G. Mundree.

(g product/g carbon). This maximum theoretical yield on carbon varies depending on the production organism and the specific metabolic pathways that generate the target molecule, but due to the significant cost contribution of the carbon source, product yield is always a key consideration. Furthermore, fermentations must be optimized to obtain as close as possible to the maximum theoretical yield, and this is a major objective of the significant strain and bioprocess engineering work that goes into developing and scaling up industrial biotechnological processes.

Apart from feedstock cost, obtaining sufficient quantities of plant biomass is crucial to the commercial success of a project. Biomass must usually be grown in relatively close proximity to the production facility given the significance of transportation costs and the low density of biomass compared with other industrial feedstocks such as petrochemicals. As such many industrial biotechnology commercial projects have been sited close to areas compatible with high biomass growth (areas with large amounts of available growing land, sunshine, rainfall, agricultural expertise, and transportation infrastructure) that are also close to markets or existing export routes. Developing and commercializing new industrial bioprocesses carries a high degree of technological and economic risk and adding further risks associated with developing biomass production, processing, and transportation logistics is undesirable. For these reasons, the majority of commercial industrial biotechnology chemical manufacturing projects have been located in areas where biomass production is already established and strong agricultural industries and expertise are in place for specific biomass crops that are proven to be high yielding and compatible with the geographical location. These areas include the United States for corn, Europe for sugar beet and wheat, and Brazil for sugarcane.

Different biomass types generate different fermentable sugars. For example, sucrose is derived from sugarcane and sugar beet; glucose from the starch in corn and wheat; glucose from cellulose; and xylose and other C5 sugars from hemicellulose (Figure 5.1). As such the production microorganism and fermentation process may also need to be specifically tailored for the particular biomass type and, therefore, the intended geographical location of production.

The focus in this chapter is on the production of chemicals (other than ethanol, which is discussed elsewhere in the book) from sugarcane-derived feedstocks but for comparison some examples of processes that use other biomass sources are detailed in the following:

- DuPont Tate & Lyle Bio Products in Loudon Tennessee, USA using corn-derived glucose to produce 1,3-propanediol (DuPont 2015) and 1,4-butanediol in partnership with Genomatica (2015);
- NatureWorks in Blair Nebraska, USA using corn-derived glucose to produce polylactic acid (NatureWorks 2015);

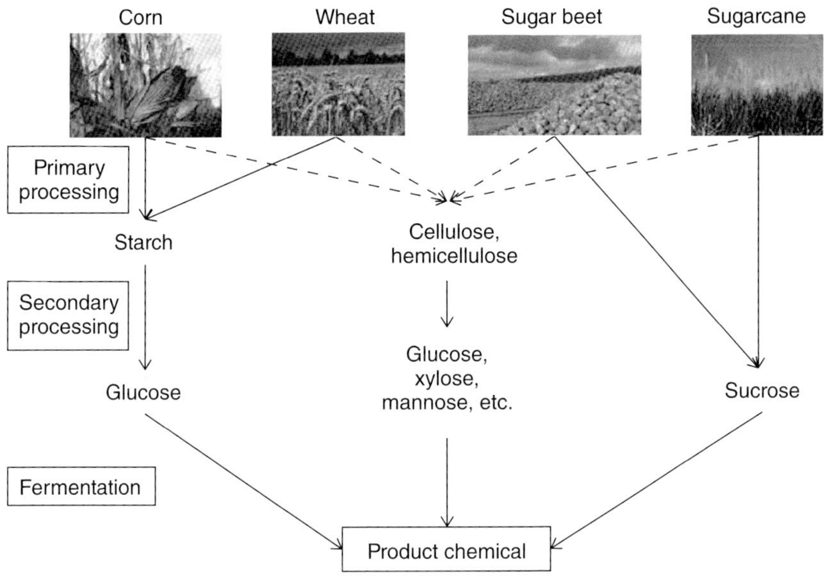

Figure 5.1 Processing steps from each biomass feedstock. Primary processing is required to liberate the carbohydrate polymers starch, cellulose, hemicellulose, and the disaccharide sucrose. Sucrose can be used directly in fermentations, whereas the polymers require secondary processing to release fermentable sugars.

- Corbion Purac looking toward locations including Brazil, United States, Europe, and Thailand to use a variety of sugar sources to produce polylactic acid (Corbion 2015);
- Reverdia (a joint development between Royal DSM and Roquette Frères) in Cassano Spinola, Italy using starch feedstocks (hydrolyzed to glucose) to produce succinic acid (Reverdia 2015);
- BioAmber in Sarnia, Ontario, Canada using corn-derived glucose to produce succinic acid (BioAmber 2015).

In addition to using starch- and sucrose-derived sugars, many companies such as Cobion Purac and NatureWorks are actively researching new processes using lignocellulosic feedstocks in an attempt to further reduce costs and remove competition with food biomass. These lignocellulosic feedstocks are often derived from processing the biomass for the primary application; examples include corn stover (leaves and stalks) and sugarcane bagasse (residual fiber from crushed sugarcane).

The sugar industry already operates a biorefinery model where multiple products are generated from the single biomass crop. These products include crystal sugar, molasses for animal feed, electricity via combustion of bagasse, and ethanol from the fermentation of molasses. It is possible, and has been demonstrated in Brazil, to vary the product mix (such as crystal sugar and ethanol) depending on market prices and conditions to maximize revenue and profit. Adding more products into the biorefinery based on advanced fermentations to supply additional markets such as the fine and commodity chemicals industries offers greater commercial flexibility and the opportunity to generate more value and revenue from the same biomass as well as new uses for current waste streams.

Despite the numerous examples of new processes that use sugars from corn and wheat, sugarcane remains a very important crop and biomass source for industrial biotechnology processes both now and in the future. Again, commercial projects are focused close to the source of biomass, primarily in Brazil with India, China, and Thailand also harvesting and processing significant volumes. Australia is the world's third largest sugar exporter (Canegrowers 2015) due to a significant sugarcane harvest with a relatively small domestic market and as such offers opportunities for adding value to the industry and generating additional uses for the sugar.

Sugarcane processing generates a number of intermediates and side products on the way to producing white sugar that can be fermented to produce bioproducts (Figure 5.2). As the sucrose from sugarcane becomes more refined, the purity increases along with the value. The uses of the crystal sugar and intermediates such as sugarcane juice and molasses as well as lignocellulose from sugarcane bagasse in fermentative processes are discussed in the remainder of the chapter. Commercial examples also include processes that use sucrose from other sources such as sugar beet and sweet sorghum due to the potential to translate these processes to sugarcane-derived sucrose.

5.2 The suitability of sugarcane-derived feedstocks in industrial fermentation processes

Sugarcane delivers a number of potential industrial biotechnology feedstocks such as purified sucrose, molasses, and bagasse. Each of these feedstocks has a different cost and purity profile that affects how it can integrate with a bioprocess

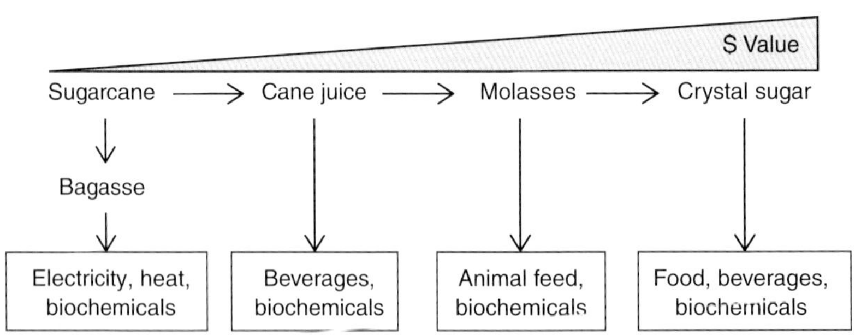

Figure 5.2 Processing streams from harvested sugarcane to crystal sugar, which increases in value as the sucrose becomes more refined. Examples of products that may be derived from each stream with chemical products from fermentation possible at each stage.

and what current uses must be displaced. The production of ethanol from sugarcane in Brazil is well established but there are a number of considerations for translating this success to other regions and to other bioproducts.

5.2.1 Competing current applications of sugarcane products

The various intermediates and side products generated during sugarcane processing have a variety of characteristics and alternative uses depending on their cost and market forces. The main application of sugar is human consumption in food and beverages and it is a globally traded commodity that has experienced a high degree of price volatility, varying between around 9 and almost 30 US cents per pound since the start of 2005 (IndexMundi 2015). As such, the price of sucrose has a very significant effect on the economic viability of an industrial biotechnology process that uses it. When the sugar price is high it is more difficult to economically deliver chemical manufacture by fermentation given that the sale price of the product chemical is determined more by market factors, such as customer demand and the price of competing products derived from petrochemicals, rather than the manufacturing or raw material cost. Furthermore, the volume of chemical manufacture must match available biomass supply volumes as well as the market demand for the product. This requirement means that it is not possible to use very large amounts of sucrose from sugarcane to produce very large volumes of high value chemicals without quickly saturating the market and driving down product prices. Similarly, it is challenging to achieve significant market penetration for very high volume products such as fuels when biomass availability would become limiting. Overall, this means that when the sugar price is high, making high-value products is not an option that could utilize significant amounts of biomass, and a low sugarcane price is desirable for large-scale deployment of industrial biotechnology projects.

When the sugar price is low it is clearly more attractive for the sugar industry to seek alternative uses for the sugar that would generate more profit by converting it into chemical products of higher value, after considering conversion costs. Since 2011, the sugar price has decreased from almost 30 cents per pound to just under 15 cents per pound in 2015 and if these lower prices continue for extended periods then alternative uses for the sugar will become increasingly desirable.

The sugar price has a knock-on effect to the value of intermediates such as molasses as well as the economic viability of extracting additional purified sucrose from these streams. Molasses has traditionally been used to generate ethanol for either beverages such as rum or for biofuel and industrial uses. Sugarcane juice may also be used for beverage production, for example, in making the popular Brazilian cachaça spirit. Molasses is produced in different grades and purities (i.e., A, B, and C molasses) and may also be used for human consumption when produced as treacle or golden syrup. Molasses streams also find use as a livestock feed supplement, especially when the quantity and quality of the usual feed are low (DAF 2015).

Traditionally, sugarcane bagasse has been used to generate heat, steam, and/or electricity to power the sugar mill. With modern cogeneration plants at sugar mills, the burning of bagasse also supplies electricity to the local area, and this is common practice in many operating mills around the world. Mackay Sugar Ltd.'s Racecourse Mill in Australia operates a 38 MW cogeneration plant supplying renewable electricity to around 30% of the town of Mackay as well as powering the sugar mill (Mackay Sugar 2015). Significantly the burning of bagasse displaces the burning of coal at the mill and as such has important environmental benefits. Were the bagasse to be used to generate fermentable sugars and chemical products through fermentation, then mills would likely need to revert to other energy feedstocks such as coal and forego the revenue from supplying electricity to the local community. Conversely, when molasses or purified sucrose is used as the feedstock for fermentation, then the ability to power the process using green electricity and heat from bagasse combustion can have a significant overall positive effect on carbon footprint (Staples *et al.* 2014).

An example of where significant market penetration with an industrial biotechnology process using sugarcane feedstock has been achieved is bioethanol in Brazil. In the United States, similar success has been achieved with ethanol derived from corn. In Brazil, 597 million tons of sugarcane was produced in 2013–2014 that generated 38 million tons of sugar and 27.5 billion liters of ethanol (UNICA 2015). In the United States, in 2014, 13.5 billion gallons (51.1 billion liters) of fuel ethanol was used (EIA 2015a) compared with 136.7 billion gallons (517 billion liters) of motor gasoline, giving a market penetration of 10% (EIA 2015b). In both cases, government support and incentives have been instrumental in achieving such high market penetration. For example, the US Energy Independence and Security Act of 2007 set yearly Renewable Fuel Standard volumes, peaking at 36 billion gallons (136 billion liters)in 2022 with the volume of conventional ethanol (from starch) capped at 15 billion gallons. In both Brazil and the United States, a substantial proportion of the yearly crop of sugarcane or corn is used to deliver such market penetration. In Australia, 30–35 million tons of sugarcane is harvested per year and an equivalent product distribution to Brazil would achieve a similar percentage market penetration to the United States with 1.2–1.4 billion liters of ethanol produced compared with the 18.7 billion liters of gasoline sold in 2012–2013 (Australian Government 2014). To achieve this level of production and use in Australia, however, would require significant infrastructure investment and incentives to increase the installed capacity (2013 data) from around 440 million liters (from both sucrose (60 million liters capacity) and starch (380 million liters capacity) feed stocks. Considering these figures for the examples of Brazil, the United States, and Australia, the development of new and advanced fuel and other chemical manufacturing processes that can achieve meaningful market penetration is a challenge requiring significant investment and government policy support.

5.2.2 Use of sugarcane products in fermentations

When technical and economic criteria are met and chemical manufacture by fermentation is desired, the whole variety of intermediate and crystal sugar (high-purity sucrose) products from the sugarcane milling and refining process can be used. The degree of processing of each intermediate is reflected in the price, impurity profile, and sucrose content of the feedstock (Table 5.1). Bagasse is the lowest cost intermediate but the fermentable sugars must typically be released from lignocellulose by pretreatment and enzymatic saccharification meaning that the cost of fermentable sugar (glucose or various pentoses) can also be comparable to or higher than other sucrose sources depending on the efficiency and costs of processing. Furthermore, it may not be possible to ferment both the hexose and pentose sugars depending on the production strain used, leading to process complexities.

Molasses

Molasses is the most used fermentation feedstock from the sugar industry. It is one of the cheapest sources of fermentable sugars. It has a pH of about 5 and is composed of carbon (64%), nitrogen (6%), inorganic salts such as phosphorous (0.3%), sodium (0.33%), potassium (5.5%), calcium (about 0.7%), and trace elements such as copper (2.2 ppm), zinc (3.91 ppm), manganese (4.74 ppm), iron (78.37 ppm), and magnesium (1370 ppm), and vitamins. The carbon-containing compounds mainly comprise sugars including sucrose (68.4%), glucose (18.5%),

Table 5.1 Sugar composition and relative price of various sugarcane processing intermediates and products

Sugar product	Fermentable sugar, %	Price, US$/kg hexose	Comment
Sugarcane	7–20	–	–
Bagasse	0[a]	–	Sugars polymerized as cellulose and hemicellulose
Raw juice	10–15	–	Seasonal, can be a problem of microbial degradation on storage
Thick syrup	60–70%	–	–
Raw sugar syrup	93–96	–	–
Molasses	40–60	0.15–0.25	Contains sucrose, glucose and fructose
Crystal sugar	>99	0.20–0.40	Nearly pure sucrose

[a]See Section 5.3.2 for the types of sugars and concentrations.
Source: Peters 2006.

and fructose (13.1%) and about 2–8% of nonsugar compounds including citric acid and oxalic acid (El-Gendy *et al.* 2013). The nitrogen-containing compounds mainly comprise a mixture of betaine and amino acids (Olbrich 1963). Both the concentrations and impurity profiles can vary between seasons, location, climate, variety, maturity of the crop, and processing conditions. Molasses is usually used as a carbon feedstock at a concentration of 10–20% in batch or fed-batch processes for the production of yeast (Brewer's and Baker's), ethanol, organic acids (citric acid and itaconic acid), and amino acids (monosodium glutamate and lysine). Some of the compounds other than sugar contained in sugarcane molasses can be stimulatory or deleterious depending on the production strain used in the fermentation.

Relatively high concentrations of biotin are contained in sugarcane molasses and in the case of the production of *Saccharomyces cerevisiae* (Brewer's and Baker's) these levels of biotin stimulate the growth of the yeast, whereas biotin must be added to the medium when beet molasses is used as a carbon source in these processes (Bekatorou *et al.* 2006). Alcohol production with sugarcane molasses is also preferred to molasses from sugar beet as sugarcane molasses contains sufficient biotin and B group vitamins, which are required by the strain for growth and the production of alcohol. In addition to the vitamin content, sugarcane molasses has a much lower buffering capacity than sugar beet molasses, which means that significantly smaller quantities of acid (usually sulfuric) are required to control the pH during fermentation (Olbrich 1963). Conversely, in the case of *Corynebacterium glutamicum*, the biotin is deleterious to the production of monosodium glutamate. This production strain is auxotrophic for biotin, and it was found that when the medium was limiting in biotin the strain secreted large amounts of glutamate into the medium. Various mechanisms have been suggested but currently the evidence supports the theory that a change in membrane tension is caused by the biotin limitation, which triggers the opening of the glutamate export protein responsible for the excretion of glutamate into the medium (Sano 2009). Initial attempts to use the cheaper carbon feedstock, molasses, resulted in much lower glutamate production. It was found that the concentration of biotin in sugarcane molasses inhibited the secretion of L-glutamate. This problem was solved by the inclusion of fatty acid derivatives such as oleic acid in the medium, which restored the secretion of glutamate into the medium despite the presence of biotin in molasses.

Aspergillus niger is used almost exclusively for the commercial production of citric acid, and high concentrations of sucrose (60–240 g/l) are preferred for this microorganism. One of the critical parameters associated with the accumulation of citric acid is that the enzyme in the tricarboxylic acid pathway (TCA) responsible for further metabolism of citric acid, aconitase, is inactivated. This inactivation is achieved by removing iron, an activator of this enzyme, from the medium. Before molasses can be used as a cheap source of sucrose the high levels of iron must be removed (Yalcin *et al.* 2010). Pretreatment of molasses with sulfuric acid

(Liu *et al.* 2008), precipitation with ferrocyanide, and use of ion exchange resins or hydrogen peroxide (Olbrich 1963) remove trace elements including iron and suspended colloids. High concentrations of other trace elements such as zinc, copper, and manganese, which are also found in blackstrap molasses, inhibit the growth of *A. niger* and inactivate enzymes associated with citric acid metabolism in the TCA cycle (Yalcin *et al.* 2010). These trace elements are also removed in the pretreatment process.

As can be seen with the examples mentioned earlier, it is difficult to predict the effect of molasses on the fermentation. Therefore, it is important to have carried out initial development work with pure sugars so that the effect of molasses can be detected. If a problem arises with the change to this cheaper feedstock, steps can then be taken to rectify the problem.

Molasses is composed of about 40% dry weight of nonfermentable compounds. These residues eventually pose a waste problem that leads to increased costs due to the need for a waste management solution, which includes using the residue as animal feed or fertilizer (Bekatorou *et al.* 2006). Costs can be reduced if these residues can be reduced and the yield of the fermentation increased at the same time. The enzyme production company Novozymes (2015) has developed an enzyme preparation called Sanferm Yield® that begins to solve this problem. In the case of alcohol production by fermentation, this preparation has been proven to improve yields by 5–10 l/t molasses when used on molasses wash and juice before fermentation. The mixture of amylases, proteases, glucanases, and xylanases hydrolyses residual starch, dextrans, and proteins into more accessible sugars and amino acids for fermentation.

Crystal sugar

Raw and refined sugar can be used in a defined medium for process development or in large-scale fermentations where product purification costs are very high or reproducibility is extremely important. The use of a defined medium allows results to be more simply interpreted due to the lack of interference on the strain metabolism by impurities contained in less-processed intermediates such as molasses.

The choice of specific sugar stream is also determined by the level of commercial development of an industrial process. One area where reproducibility is extremely important is in the initial stages of strain and process development in the laboratory and as such defined media tends to be used. Given the significant differences between defined media and media based on molasses or sugarcane juice, it is necessary to start using the intended source of sugars before and during pilot studies unless it is envisaged that a process can be economical using purified sucrose in defined media or if downstream processing issues prohibit the use of molasses. The same is true when the use of sugars derived from the lignocellulose in bagasse is intended at large scale. As with molasses, the impurity profiles of these lignocellulosic sugar streams means that they must also be tested alongside refined sucrose during process development. Due to the challenges and costs involved in generating fermentable sugars from this carbon source, processes using bagasse are still mainly at the pilot and early development stage and are likely to follow commercial development using defined media or molasses.

Bagasse

As already stated, bagasse has the potential in the future to provide a cheap carbon feedstock but only if two criteria can be met. The first criterion is that the processing of the bagasse to produce the C5 and C6 sugar monomers must be cheap and efficient and the second is that the production microorganisms must be able to utilize these sugars either naturally or by strain engineering to efficiently use the mixture of pentoses and hexoses provided.

At the moment, pretreatment of bagasse (comprising cellulose, hemicellulose, and lignin) is usually carried out at moderate temperatures (160–180 °C) and pressures (about 6–20 bar) in the presence of mild acid or alkali and/or steam explosion. Pretreatment removes the lignin and renders the fibers more amenable to enzymatic hydrolysis of the cellulose (e.g., by cellobiohydrolase, endoglucanase, and β-glucosidase) and hemicellulose (e.g., by endoxylanase, β-xylosidase, and endomannanase) (Van Dyk and Pletschke 2012). The pretreatment process produces inhibitory compounds such as phenolics, heavy metals, furfurals, and weak acids. These hydrolysates generally have to be detoxified before the monosaccharides released after saccharification can be used for fermentation (Parisutham *et al.* 2014).

In a laboratory-scale experiment, bagasse was pretreated with aqueous ammonia at 65 °C (Montanti *et al.* 2011). After this pretreatment, ammonia was evaporated and the bagasse washed before hydrolyzing the bagasse with cellulase for 90 h at 50 °C. The enzymatic hydrolysis released the sugar monomers, glucose, xylose, and arabinose from the bagasse. These sugar monomers were used as the carbon feedstock in the fermentation of *Phaffia rhodozyma* for the production of astaxanthin. The growth rate on the bagasse hydrolysate was higher than on an artificial mix of these sugars (same ratio as the hydrolyzed bagasse feedstock) but the yield of astaxanthin was lower. This result was probably due to the change in carbon to nitrogen ratio due to the presence of the enzyme in the sugar mix from bagasse, which can be used as a carbon and nitrogen source.

Efforts are being directed at combining the expression of the enzymes required for the hydrolysis of cellulose and hemicellulose with the production of high titers of the desired chemical product in the same microorganism (Parisutham *et al.* 2014). This strategy is called consolidated bioprocessing (CBP). The simultaneous saccharification and production by fermentation can be achieved by engineering a microorganism that already produces the necessary cellulytic enzymes to produce high titers of the target compound or alternatively by using a production microorganism and introducing the cellulytic enzymes for the hydrolysis of the lignocellulose (Hasunuma *et al.* 2013). An example of engineering a production strain to simultaneously carry out saccharification and production of a target compound is described by Zheng *et al* (2012). This group created an engineered *Escherichia coli* Z6373 capable of producing succinate from beechwood xylan. Three hemicellulases (endoxylanase, xylosidase, α-arabinofuranosidase) were introduced into the strain as fusion proteins with protein Y, which is an osmotically induced excretion protein in *E. coli*. The strain was also engineered to overexpress pyruvate decarboxylase. In addition to these changes, lactate dehydrogenase and pyruvate formate lyase were knocked out to avoid side-product formation. The strain produced 0.37 g succinate/g xylan, which corresponds to a 76% yield compared with pretreatment by acid hydrolysis of xylan. The lower yield was in part due to the incomplete xylan hydrolysis by the hemicellulases (Zheng *et al.* 2012).

The bigger challenge of enzymatic lignocellulose degradation, which will ensure an efficient and cost-effective hydrolysis that does not require pretreatment or detoxification, will ideally also combine lignin-degrading enzymes such as laccases and other proteins such as expansin, swollenin, and loosinin in the production strain as well. The latter proteins aid the accessibility of cellulases to cellulose and its subsequent hydrolysis by expanding cell wall structure, promoting breakdown of the cellulose, and cleaving the hemicellulose–lignin linkages (Parisutham *et al.* 2014).

Even with significant improvements in the rates and efficiencies of enzymatic hydrolysis of bagasse it is unlikely that simultaneous saccharification will approach the rates desirable for the cycle times required for economic fermentation processes. If the saccharification and fermentation phases are uncoupled then cellulolytic enzyme loadings can be optimized without compromising chemical production by fermentation and different optimal organisms can be used for both cellulolytic enzyme production and chemical biosynthesis. Cellulosic ethanol production reached commercial scale in 2014, albeit at a relatively smaller scale than corn starch–derived ethanol, with more advanced fermentation products using engineered microorganisms likely to be in the next wave of processes. CBP is a step further and is likely to be some way off from commercial reality.

5.3 Metabolism and industrial host strains

5.3.1 Metabolism of sucrose

Sucrose (from sugarcane juice, molasses, and crystal sugar) is a disaccharide formed by the condensation of α-D-glucose and β-D-fructose. For a microorganism to metabolize sucrose as a carbon source, it must possess the necessary enzymes. Hydrolysis of the disaccharide to form the monosaccharides, glucose and fructose, is the first crucial step. Hydrolysis occurs in some cases extracellularly and in others intracellularly. In the case of intracellular hydrolysis, a sucrose transport system is also necessary for its transport into the cell. In most cases, glucose and fructose are phosphorylated and then fed directly into the central metabolism of the microorganism (Figure 5.3). Not all microorganisms can utilize sucrose as a carbon source, which restricts the use of sugarcane feedstocks in fermentation. An understanding of the different sucrose metabolizing systems is important so that sucrose utilization can be improved and non-sucrose-utilizing production strains can be modified to enable them to utilize these low-cost sugarcane feedstocks.

Most industrially used strains of *E. coli* cannot utilize sucrose. There are some exceptions, however, where sucrose is processed via the sucrose phosphotransferase system (PTS) or the chromosomally carried sucrose catabolism (*csc*) gene cluster (Sabri *et al.* 2013). The strain *E. coli* W metabolizes sucrose via the *csc*

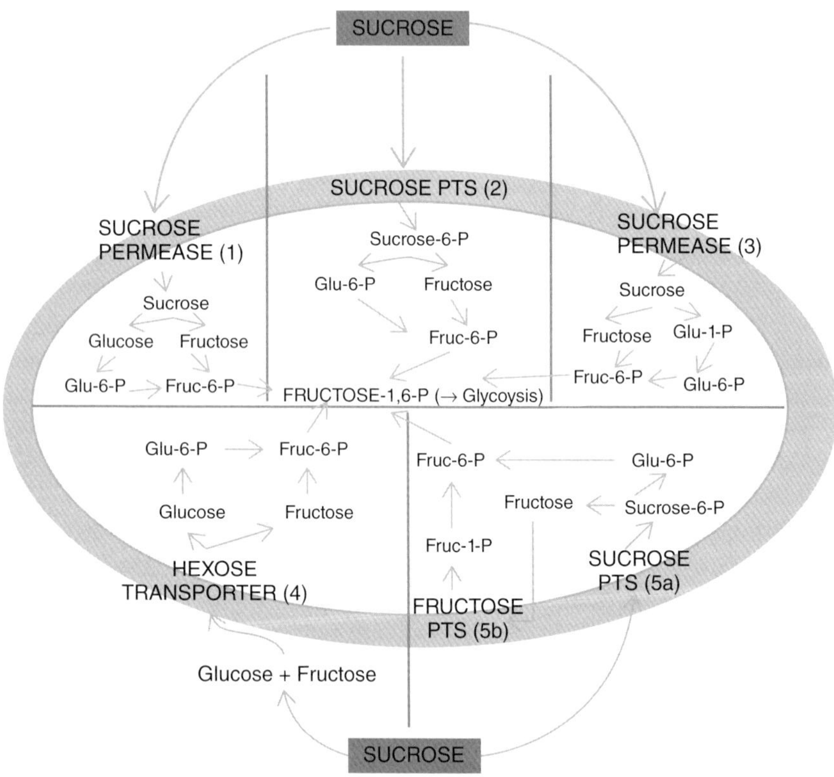

Figure 5.3 Sucrose metabolism. Sucrose permease (1), sucrose PTS (2), and sucrose permease (3) are prokaryotic systems; hexose transporter (4) is the *Saccharomyces cerevisiae* system; sucrose and fructose PTS (5a and 5b) is the *Mannheimia succiniciproducens* system; periplasmic space is shaded grey and the cytoplasm is the inner area shaded white.

gene cluster, which encodes for a transcriptional repressor (CscR), an invertase (CscA), a sucrose permease (CscB), and a fructokinase (CscK). The expression of these genes is controlled by the repressor protein, which represses the expression of the sucrose-catabolizing enzymes in the absence of sucrose or when sucrose is present only in low concentrations. In the presence of >2 g/l sucrose, these enzymes are active and the sucrose is transported into the cell and hydrolyzed. The glucose and fructose are phosphorylated in the 6-position by glucokinase and fructokinase, respectively, and metabolized via the glycolytic pathway. Modification of other industrial strains of *E. coli* by chromosomal integration (or plasmid-encoded) of the *csc* gene cluster will enable other strains to utilize sucrose as a carbon source. For example, the production of the short-peptide surfactant DAMP4 in *E. coli* BL21(DE3) was achieved at over 6.5 g/l from both glucose and sucrose (Bruschi *et al.* 2014). The BL21(DE3) strain naturally utilizes glucose, and sucrose metabolism was enabled using a plasmid encoding the *cscAKB* operon from *E. coli* W.

The PTS, which is found in gram-negative and gram-positive bacteria, includes a sucrose-dependent PTS uptake system, a sucrose 6-phosphate hydrolase, and a fructokinase. Usually, a regulator protein is also expressed (Lee *et al.* 2010). In the case of the PTS system of the succinic acid producer, *Mannheimia succiniciproducens*, the sucrose represser and fructokinase are absent. After hydrolysis of sucrose 6-phosphate the fructose is secreted and then taken up again by a fructose PTS. The absence of the regulator protein also means that no diauxie is detected with sucrose, glucose, and fructose being utilized simultaneously.

S. cerevisiae is an important industrial strain capable of producing a diverse range of products by fermentation. Currently, the major commercial products generated by this microorganism are baker's yeast and ethanol with the potential for a wide variety of products with engineered strains (Borodina and Nielsen 2014). Metabolism of sucrose in *S. cerevisiae* begins with extracellular hydrolysis of sucrose by the enzyme invertase to produce glucose and fructose. These sugars are then transported into the cell by hexose transporters and metabolized further via the glycolytic pathway. A minor pathway for sucrose metabolism is also present. In this case, sucrose is actively transported through the plasma membrane and hydrolyzed within the cell (Batista *et al.* 2004) (Figure 5.3).

Some other examples of microbial strains that have been used to make chemical bioproducts from sucrose include both *A. niger* (Papagianni 2007) and *Yarrowia lipolytica* (Forster *et al.* 2007) for citric acid manufacture. The *A. niger* process is well established with an industrial process in manufacture as early as 1919 in Belgium and Pfizer implementing a process in 1923 that used molasses (Mattey and Kristiansen 1999). Other early examples of fermentations based on molasses include the production of acetone and butanol from *Clostridium* species, for example, by Commercial Solvents Corporation in 1920 (Licht 2014).

5.3.2 Metabolism of lignocellulose-derived sugars

The main sugars released from the enzymatic hydrolysis of bagasse are glucose (43%), xylose (21%), and arabinose (1.9%) and smaller amounts of galactose (0.5%), and mannose (0.3%). Lignin (23.9%) and ash (2.8%) are also present (Parisutham *et al.* 2014). The proportions can vary between different varieties, location, seasons, and processing to release the sugar monomers. Although many production organisms can utilize glucose, not all possess the capability of metabolizing the pentose sugars xylose and arabinose. D-Xylose and L-arabinose are metabolized via the pentose phosphate pathway (PPP). This pathway is present in most organisms. Both D-xylose and L-arabinose enter the PPP via the intermediate D-xylulose-5-phosphate. In the case of yeast, filamentous fungi, and other eukaryotes, D-xylose is converted to this intermediate via a two-step reduction and oxidation carried out by an NADH/NADPH-dependent aldose reductase and an NAD^+-dependent xylitol dehydrogenase followed by an ATP-dependent D-xylulose kinase. In prokaryotes such as *E. coli*, *Bacillus*, and *Lactobacillus* species, a xylose isomerase converts xylose to D-xylulose and a xylulose kinase phosphorylates the D-xylulose in the five positions to produce D-xylulose 5-phosphate (Jeffries 2006, Bettiga *et al.* 2009). The conversion of L-arabinose to the intermediate D-xylulose-5-phosphate in fungi requires multiple steps. The L-arabinose is converted to L-arabitol by an aldose reductase, which also reduces D-xylose. An L-arabitol dehydrogenase converts the arabitol to L-xylulose, which is then reduced to xylitol by a xylulose reductase. At this point, the D-xylose and L-arabinose metabolism in fungi converge. Xylitol is then converted to D-xylulose by xylitol dehydrogenase, which is subsequently phosphorylated by xylulokinase. In bacteria, the conversion of L-arabinose starts with an L-arabinose isomerase producing L-ribulose, which is then phosphorylated by ribulose kinase to L-ribulose-5-phophate. The last step is an epimerization with the enzyme ribulose 5-phophate epimerase to produce D-xylulose-5-phosphate (Bettiga *et al.* 2009) (Figure 5.4).

Two lesser known, closely related, pathways for the degradation of D-xylose and L-arabinose are the Weimberg (1961) and Dahms (1974, Watanabe *et al.* 2006a, b) pathways (Stephens *et al.* 2007). These pathways metabolize D-xylose and L-arabinose via a common intermediate, 3-keto-3-deoxy-pentonate. In the Weimberg pathway, D-xylose is oxidized by xylose dehydrogenase to D-xylono-lactone, which is then hydrolyzed to D-xylonic acid by a lactonase. Water is removed from the D-xylonic acid by xylonite dehydratase to produce 2-keto-3-deoxy xylonate and another dehydratase converts this intermediate to 2-ketoglutarate semialdehyde, which is then oxidized to 2-ketoglutarate. The Dahms pathway is the same as the Weimberg pathway until 2-keto-3-deoxy xylonate is produced. At this point, the Dahms pathway diverges from the Weimberg pathway and 2-keto-3-deoxy xylonite is split by an aldolase to produce glyceraldehyde and pyruvate.

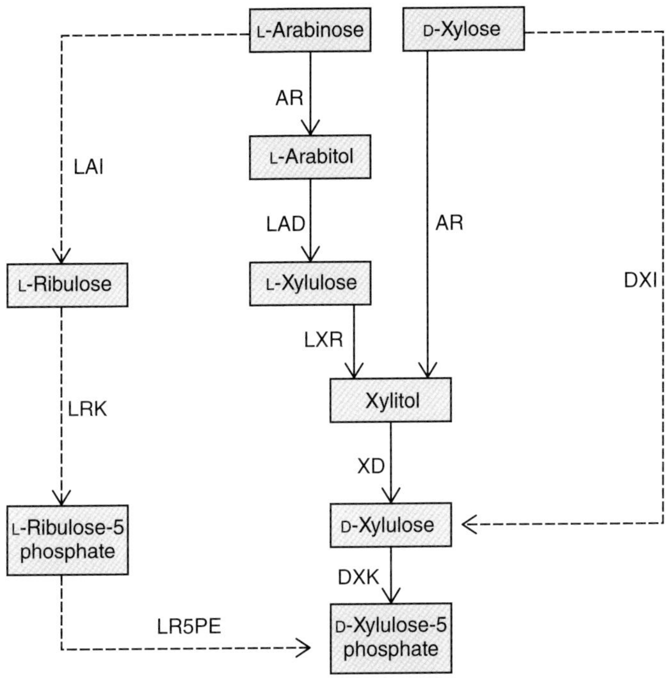

Figure 5.4 Metabolism of D-xylose and L-arabinose. Solid arrows indicate pathways found in yeast and filamentous fungi, and dashed arrows indicate prokaryote pathways. AR, aldose reductase; LAD, L-arabitol dehydrogenase; LXR, L-xylulose reductase; XD, xylitol dehydrogenase; DXR, D-xylulose kinase; LAI, L-arabinose isomerase; LRK, L-ribulose kinase; LR5PE, L-ribulose 5-phosphate epimerase; DXI, D-xylose isomerase.

Attempts to confer the ability to coferment D-xylose and L-arabinose in S. *cerevisiae* have been made, which combine the fungal reduction/oxidation of D-xylose with the bacterial L-arabinose isomerase pathway or the combination of both bacterial isomerization pathways for D-xylose and L-arabinose. These changes led to various problems including low growth rates and low alcohol production caused by cofactor imbalance and, in the case of the fungal reduction step with aldose reductase, the production of the dead-end metabolite L-arabitol, which inhibits the bacterial enzymes (Bettiga *et al.* 2009). A more complex but more successful strategy was achieved by overexpression of the fungal PPP and xylulose kinase in the host *S. cerevisiae* combined with a new fungal pathway, which was reconstructed, and combined NAD^+- and NADH-dependent enzymes responsible for fungal L-arabinose metabolism sourced from different microorganisms (Bettiga *et al.* 2009). The aldose reductase and the xylitol dehydrogenases of the reconstructed pathway also convert D-xylose to the common intermediate, D-xylulose. In addition, the aldose reductase in this construct was engineered for increased preference for NADH. These improvements combined with the improved metabolic flux through the PPP solved the cofactor imbalance problem and led to good growth and ethanol production.

5.3.3 Optimization of strains and metabolism

In processes that were developed before the generation of molecular biology and metabolic engineering tools, it was common practice to identify and use native strains that naturally produce the desired chemical. These strains could be improved through classical chemical or ultraviolet mutagenesis methods with screening for improved traits such as product titer. The fermentations would also be optimized through the use of different carbon sources, nutrients, temperatures, and mixing methods. Since the generation of molecular biology tools for metabolic engineering, it is now common to use a host organism that is amenable to genetic manipulation (such as *S. cerevisiae*) or to develop molecular tools for an organism suited to a particular chemical. For example, plasmids and transformation methods have been developed for *Clostridium ljungdahlii* to allow butanol production from synthesis gas (CO/H_2) or a carbon dioxide/hydrogen mixture (Kopke *et al.* 2010). The organism was previously able to ferment a range of carbon sources including sugars and synthesis gas and produce ethanol but was unable to naturally generate butanol. By incorporating a plasmid containing six genes of the butanol pathway small amounts of butanol could be detected. Since this proof of principle experiment, chromosomal manipulation has been achieved (Leang *et al.* 2013).

The above-mentioned example shows the benefit of taking an organism suited to the use of a particular feedstock (in that case synthesis gas) and then using biotechnology approaches to modify the pathways to particular chemical products. Despite the challenges in developing molecular biology methods for new organisms and introducing new chemical pathways, it is far easier than the wholesale introduction of new carbon metabolism. As such, metabolic engineering for the optimal production of chemicals from sucrose has tended to focus on organisms that already have efficient pathways for sucrose utilization. The most studied is *S. cerevisiae* where metabolic engineering is well-established. Recent developments in the tools available for this organism have been recently reviewed (Borodina and Nielsen 2014) and include genome-scale metabolic models, synthetic biology tools, new genetic elements for the control of gene expression (for example, promoters and terminators), and genome manipulation methods (Jensen *et al.* 2014).

Regardless of the desired end-product chemical, the initial steps for sucrose metabolism described earlier are generally the same and any attempts to improve these initial steps would generally have benefit for any process. For example, the invertase step has been manipulated to increase ethanol yields in anaerobic cultures through engineering cytosolic localization of invertase and intracellular sucrose hydrolysis (Basso *et al.* 2011).

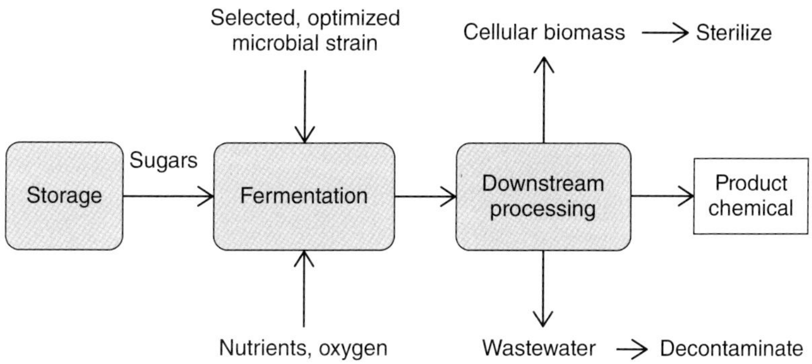

Figure 5.5 Overall bioprocessing steps from stored fermentable sugars to the desired product and waste streams. After fermentation, the cellular biomass is typically separated from the aqueous stream by centrifugation before purification of the product chemical from the aqueous stream. The wastewater may be disposed of or recycled following decontamination, and the cellular biomass may be anaerobically digested to generate heat and power from biogas or used as animal feed.

5.4 Bioprocess considerations

Establishing a successful commercial fermentation process from both an economical and sustainable point-of-view requires a holistic approach, of which strain selection is just the very beginning (Figure 5.5).

Sugarcane juice, molasses, and bagasse (if used in the future) contain other nutrients and possible inhibitors. If pure sucrose is used initially to establish the process, then these data can be used to compare the growth rates, total biomass produced, yields, and changes to the side-product profile when other cheaper sugarcane feedstocks are being assessed in the fermentation. In this manner, the medium can be adjusted and balanced accordingly to compensate for changes that are not beneficial for the process. Once a feedstock has been chosen and the process adjusted accordingly, it is extremely important that the quality of the feedstock is assured. This consistent quality is not always possible, especially with sugarcane juice and molasses, where variations in quality will occur between sources of sugarcane, batches, and seasons. To avoid problems during the production campaign, the equivalence of each new batch of feedstock must be tested. It is often necessary to compare several batches and then reserve the best batch for the campaign so that no batch changes occur during the campaign that would lead to changes in the fermentation profile and yield. In the case of lignocellulose, there will be other contaminants that have been introduced during the treatment of the bagasse to release the sugar monomers. As already mentioned, the enzymes added during the pretreatment may be used as an additional carbon and nitrogen source by the production strain so the carbon-to-nitrogen ratio may need to be readjusted accordingly.

The traditional fermentation process options are batch, fed-batch, and continuous processes. The choice of process will be governed by the production strain and plant. A batch process may be advantageous when a simple process is required due to plant and personnel restraints but the process will be limited by the concentration of carbon feedstock that can be used, which will lead to reduced yields of biomass and product. If the concentration of the feedstock is too high the growth of the microorganism will be inhibited, and the side-product profile will most probably be altered due to several possible mechanisms. The first mechanism is the increased osmotic pressure of the medium, which can inhibit growth. Another mechanism is catabolite repression, which will effect gene expression and subsequent repression of enzyme expression and can lead to low yields of the desired product. In the generation of yeast products, catabolite repression is known as the Crabtree effect, which leads to the repression of respiratory enzymes and can lead to unfavorable side-products and decrease in biomass yields (Bekatorou *et al.* 2006). Overflow metabolisms and fermentative production of mixed acids may also be induced, which handle the excess carbon and its metabolic intermediates.

For example, acetate or mixed acids can be produced in large-scale *E. coli* fermentations in the presence of high sugar concentrations (Xu *et al.* 1999). At the start of the fermentation, the presence of high sugar concentrations and sufficient oxygen can induce the production of acetate by the aerobic pyruvate dehydrogenase, which converts excess pyruvate to acetyl-CoA, a precursor of acetate. This accumulation of acetate can lead to problems such as reduced growth rates. A second effect that can occur later in the fermentation, where mixing problems induce localized oxygen limitations due to high glucose concentrations, is the production of mixed acids under anaerobic conditions. In this case, the initial enzymes of this pathway, pyruvate formate lyase and lactate dehydrogenase, which are otherwise suppressed by oxygen, produce formic acid and lactate, respectively. Other metabolites produced by the same mechanism are succinate, carbon dioxide, hydrogen, and acetate.

These problems can be controlled if mixing in the fermenters is optimized and a fed-batch regime is used after the initial batch phase, where the sugar is kept at acceptable concentrations for the particular microorganism. Typically, a feed rate of about 0.3 h^{-1} is used although this is strain dependent. The fed-batch process allows higher concentrations of biomass to be produced and also higher product concentrations compared with a batch process. A continuous process has the advantage of increased plant output but is often not the process of choice due to the increased complexity of the process, the requirement of maintaining pure culture over long periods of time, and the high impact of plant failure on the process.

Production of commodity chemicals by fermentation needs economy of scale to be financially viable. These processes require extremely large fermentation capacities, including appropriate number and scale of the seed tanks and also laboratories and analytical equipment to monitor the process. The optimization of a process must address problems of adequate oxygen supply and adequate mixing at this scale. The aeration volumes, impeller types and number, stirring speeds, and fermenter geometry are all important aspects of the scale-up and optimization of the process. At this scale, it is important to have multiple sites of addition for acid, base, feeds, antifoam, and air supply, which will improve the mixing profiles in these very large fermenters. Other aspects that should be considered when establishing a new process are

1. sufficient steam production for cleaning and sterilization;
2. sufficient heat removal and cooling capacities;
3. the possibility of biomass inactivation and decontamination of waste streams if necessary.

Inactivation of the biomass is in most cases critical for a variety of reasons. Sometimes, it is important to stop any enzyme activity after the production phase that may lead to the decomposition of the product (e.g., protease degradation is required if enzymes, proteins, or peptides are the product) or to stop further metabolism of a target compound to an unwanted side-product. Trace amounts of product remaining in waste streams after the isolation of the products need to be destroyed if they are bioactive or detrimental to the environment in any way, for example, steroids or antibiotics. The treatment of waste streams contributes significantly to the cost of the process. The waste stream containing the biomass is often too high in organic carbon to be combined with the local waste water. In some cases, the waste stream can be concentrated and as much water as possible is removed before it is burned. If other incineration processes are occurring on site, such as bagasse for heat and power for the sugar mill, it may be possible to co-incinerate to further reduce waste disposal costs and generate additional power.

Depending on the use of native or genetically modified organisms and local regulations, it is sometimes possible to use the waste microbial biomass as animal feed. Overall for large processes, the volume of the remaining waste streams and/or the concentration of organic carbon may be too high for the municipal waste water treatment plants. In this case, treatment of the waste streams on site may be required.

Storage capacity is another important aspect of the plant design. Storage facilities must be large enough for the sugarcane feedstock and other medium components required for a campaign. Sugarcane juice supply is seasonal and would need sterilization before storage as the sugar concentration is too low to be inhibitory to the growth of contaminating microorganisms. Changes in quality due to sterilization and storage should also be monitored. Pasteurization is adequate for molasses during storage as the sugar concentration is much higher (about 60%). At this concentration, the sugar concentration should be bacteriostatic for most microorganisms. Pure sucrose requires dry storage, and measures to avoid dust issues during handling need to be implemented. Fermentation plants also have a very large requirement for water. The water is used for steam production, medium preparation, biomass separation by filtration or centrifugation, product recovery, and also during the cleaning steps.

Product recovery can be very costly and, therefore, should be simple with minimal unit operations. In most cases, simple and cheap product-recovery steps can be implemented if the side-products and impurities have been minimized by strain choice and pathway engineering, the use of higher-quality feedstocks, carbon and nitrogen sources, and process design. There is often a trade-off between purity of medium components and the cost of isolating product in line with specifications. This trade-off must be analyzed carefully when process economics are considered. In a lot of cases, traditional methods of product isolation such as precipitation, solvent extraction, distillation, or crystallization are used.

There are, however, a number of more environmentally friendly technologies that are applicable for fermentation and avoid the excess use of solvents and reduce the volume of the filtrate substantially by removing large amounts of water from the dilute product streams. This water removal will minimize solvent and energy inputs in later steps depending on the technology used for product purification. Biomass can be removed from the filtrate by centrifugation and/or flocculation but alternatively ultrafiltration and microfiltration can be used. In some cases, by using different molecular weight cutoff membranes, a degree of product isolation and concentration can also be achieved with theses filtration methods in the first steps of product recovery.

Reverse osmosis is an excellent method of water removal and concentrates the product further. The use of activated charcoal and nanofiltration can be used to remove cellular impurities such as polysaccharides and nucleosides before the final step of product isolation such as spray drying if the product stream is pure enough or conversely crystallization, solvent extraction, or distillation before drying.

Electrodialysis is an excellent method of isolating charged compounds such as organic acids and avoids the use of solvents. If solvent extraction is used in the process, the possibility of recycling the solvent should be investigated. In fact, other waste streams such as process water and heat (if generated) should also be recycled if possible.

5.5 Sugarcane-derived chemical products

The preceding sections discuss various factors that must be considered when choosing sugarcane as the feedstock for an industrial biotechnology process. Location of the manufacturing plant close to the available feedstock and being in a position to take advantage of existing agricultural infrastructure has meant most commercial sugarcane-based projects to date have been located in Brazil. The Brazilian government, industry, and investors have supported projects, encouraged by the experience of ethanol production and numerous multinational companies have looked toward Brazil for deployment of their technologies or partnering with technology providers. A high-profile example is the focus of biotechnology company Amyris on scale-up of production in Brazil (Amyris 2015) and subsequent significant investment by the oil major Total. Amyris was founded in 2003, and the company has developed strain engineering and synthetic biology approaches to manipulate and optimize the isoprenoid pathways in both *E. coli* and *S. cerevisiae*.

Initial work in this area by Amyris and colleagues was funded by the Bill & Melinda Gates Foundation for the development of the antimalaria drug precursors amorpha-4,11-diene and artemisinic acid. The manufacture of the final active pharmaceutical ingredient based on this biological approach is now performed under license by Sanofi-Aventis. The extensive strain and process engineering work that went into achieving scalable product titers and viable manufacture is described across many journal articles, patents, and review articles (Paddon and Keasling 2014). *S. cerevisiae* was the preferred strain with titers of over 40 g/l of the biochemical intermediate amorpha-4,11-diene achieved (Westfall *et al.* 2012), a significant improvement over initial levels. The conversion of amorpha-4,11-diene to artemisinic acid was also examined using a biological approach involving a cytochrome P450 enzyme but in the end a chemical conversion for this step was found to be superior.

Isoprenoids are a very large class of natural chemicals, including simple molecules such as isoprene and large complex molecules such as the anticancer compound taxol that could be targets for biosynthesis (Kirby and Keasling 2009). The isoprenoid pathway engineering to generate amorphadiene could be adapted and expanded using different synthases to make a variety of useful products, from flavors and fragrances to fuels. Amyris focused on the production of the 15 carbon isoprenoid molecule farnesene that when hydrogenated to farnesane makes a drop-in diesel fuel and can be blended at up to 10% in aviation fuel under the American Society for Testing and Materials (ASTM) standard D7566.

Despite being founded in the Bay Area of San Francisco and performing pilot-scale operations there, Amyris looked toward Brazil for scale-up and access to sugarcane feedstock. Amyris began operation of its industrial biorefinery in 2012 next to the Tonon Paraiso sugar mill in the state of Sao Paulo using sugar feedstocks (sugarcane syrup) to produce farnesene and related products using engineered strains of *S. cerevisiae*.

Amyris have built numerous partnerships with one of the most significant being with Total for the use of farnesene and the derived farnesane in fuel applications. Total and Amyris have a joint research and development team as well as a commitment to a 50/50 joint venture to produce and commercialize biofuels. Following ASTM certification, the first commercial flight using a blend of Amyris aviation fuel was completed by the Brazilian airline GOL in July 2014 between Orland, Florida and Sao Paulo, Brazil (Amyris 2014).

In addition to the work by Amyris, the group of Amyris's cofounder Professor Jay Keasling has worked extensively across metabolic engineering to produce a wide variety of products. Other isoprenoid pathway engineering approaches using *S. cerevisiae* with the potential to utilize sugarcane feedstocks include casbene (Kirby *et al.* 2010) and the diesel-replacement bisabolene (Peralta-Yahya *et al.* 2011). Other examples of pathways and molecules produced by engineered *S. cerevisiae* include cinnamoyl anthranilates (Eudes *et al.* 2011) and *n*-butanol (Steen *et al.* 2008).

Commercial development of industrial biotechnology processes using sugarcane feedstock in Brazil is dominated by ethanol production as well as the conversion of ethanol to polyethylene. There are several projects in development with the potential to deliver industrial chemical products from Brazilian sugarcane. One example is Genomatica's partnership with Brazilian chemical company Braskem to produce bio-butadiene. Genomatica has achieved rapid strain engineering to generate commercial processes using the principles of systems and synthetic biology. The production of 1,4-butanediol in engineered *E. coli* was achieved at 18 g/l (Yim *et al.* 2011) with the commercial process achieving higher titers and licensed for manufacture by BASF and Novamont. Butadiene processes are currently in development with patent protection in place (Burk *et al.* 2013).

GranBio is a developer of lignocellulosic ethanol manufacture in Brazil and operates a process for the conversion of sugarcane straw and bagasse in Alagoas. GranBio is partnering with Rhodia for the production of *n*-butanol, also from sugarcane straw and bagasse. The partnership was announced in 2013 with the aim of starting operations in 2015. The relationship builds on each partners' existing collaborations with technology providers such as Cobalt Technologies and American Process Incorporated.

Solazyme produces bio-oils from sugar using algae. This approach is distinct from other algal processes that use photosynthesis and carbon dioxide as the feedstock. In May 2014, Solazyme announced that its joint venture with Bunge Global Innovation LLC began commercial production in Brazil using manufacturing-scale fermenters (625,000 l) with a capacity of 100,000 tons per year (Solazyme 2014).

5.6 Summary

Sugarcane has strong potential to be a feedstock of choice for advanced methods to produce a range of chemicals needed by modern society. The current price and alternative uses for sugarcane products such as refined sucrose will continue to fluctuate in both value and popularity. The sugar industry has an established track record in using sugar mills as biorefineries that generate a range of products from crystal sugar to bioethanol and electricity. The production of biochemicals in advanced fermentations is simply an extension of this existing biorefinery approach with the opportunity to generate additional value and revenue streams from each sugarcane subproduct from pure sucrose to bagasse. In order to generate an economic bioprocess and to diversify the range of chemical products required, it will be necessary to utilize the latest scientific tools such as systems and synthetic biology in parallel with bioprocess engineering. Both companies and academic groups have embraced these new tools with impressive results already delivered and the first of the new wave of processes are now in commercial production. Depending on factors such as the price, availability, and consumer acceptance of competing petrochemical feedstocks and competitive uses of sugarcane products, sugarcane-derived industrial chemicals have the potential to be highly important in generating the chemicals and materials needed by society into the future.

References

Amyris (2014). First International Commercial Flight Completed With Newly Approved Amyris-Total Aviation Biofuel, from https://amyris.com/first-international-commercial-flight-completed-with-newly-approved-amyris-total-aviation-biofuel/.

Amyris (2015). Industrial Production, from https://amyris.com/innovation/industrial-production/.

Australian Government (2014). Australian Petroleum Statistics. Australian Government, Department of Industry and Science, Office of the Chief Economist.

Basso, T. O., de Kok, S., Dario, M., do Espirito-Santo, J. C., Muller, G., Schlolg, P. S., Silva, C. P., Tonso, A., Daran, J. M., Gombert, A. K., van Maris, A. J., Pronk, J. T. and Stambuk, B. U. (2011). Engineering topology and kinetics of sucrose metabolism in Saccharomyces cerevisiae for improved ethanol yield. *Metabolic Engineering* **13**(6):694–703.

Batista, A. S., Miletti, L. C. and Stambuk, B. U. (2004). Sucrose fermentation by Saccharomyces cerevisiae lacking hexose transport. *Journal of Molecular Microbiology and Biotechnology* **8**(1):26–33.

Bekatorou, A., Psarianos, C. and Koutinas, A. A. (2006). Production of food grade yeasts. *Food Technology and Biotechnology* **44**(3):407–415.

Bettiga, M., Bengtsson, O., Hahn-Hagerdal, B. and Gorwa-Grauslund, M. F. (2009). Arabinose and xylose fermentation by recombinant Saccharomyces cerevisiae expressing a fungal pentose utilization pathway. *Microbial Cell Factories* **8**:40.

BioAmber (2015). Manufacturing Facilities, from http://www.bio-amber.com/bioamber/en/company#manufacturing_facilities.

Borodina, I. and Nielsen, J. (2014). Advances in metabolic engineering of yeast Saccharomyces cerevisiae for production of chemicals. *Biotechnology Journal* **9**(5):609–620.

Bruschi, M., Kromer, J. O., Steen, J. A. and Nielsen, L. K. (2014). Production of the short peptide surfactant DAMP4 from glucose or sucrose in high cell density cultures of Escherichia coli BL21(DE3). *Microbial Cell Factories* **13**:99.

Burk, M. J., Burgard, A. P., Sun, J., Osterhout, R. E. and Pharkya, P. (2013). Microorganisms and methods for the biosynthesis of butadiene. USPTO. US, Genomatica, Inc. US8580543 B2.

Canegrowers (2015). About the Industry, from http://www.canegrowers.com.au/page/Industry_Centre/industry-focus/about-the-industry/.

Chotani, G., Peres, C., Schuler, A. and Moslemy, P. (2014). Bioprocessing technologies. In *Bioprocessing of Renewable Resources to Commodity Bioproducts*, pp. 133–166. V. S. Bisaria and A. Kondo, eds. John Wiley & Sons, Inc: Hoboken, NJ.

Corbion (2015). Bioplastics for a new generation.

DAF (2015). Fortified molasses for survival feeding of beef cattle, from https://www.daf.qld.gov.au/environment/drought/managing-drought/drought-strategies/fortified-molasses-for-survival-feeding-of-beef-cattle.

Dahms, A. S. (1974). 3-Deoxy-D-pentulosonic acid aldolase and its role in a new pathway of D-xylose degradation. *Biochemical and Biophysical Research Communications* **60**(4): 1433–1439.

DuPont (2015). DuPont Tate & Lyle Bio Products, from http://www.duponttateandlyle.com/.

EIA (2015a). *April 2015 Monthly Energy Review*. U.S. Energy Information Administration: Washington, DC.

EIA (2015b). *Short-Term Energy and Summer Fuels Outlook*. U.S. Energy Information Administration: Washington, DC.

El-Gendy, N., Madian, H. R. and Amr, S. S. (2013). Design and optimization of a process for sugarcane molasses fermentation by Saccharomyces cerevisiae using response surface methodology. *International Journal of Microbiology* **2013**:815631.

Eudes, A., Baidoo, E. E. K., Yang, F., Burd, H., Hadi, M. Z., Collins, F. W., Keasling, J. D. and Loque, D. (2011). Production of tranilast [N-(3′,4 ′-dimethoxycinnamoyl)-anthranilic acid] and its analogs in yeast Saccharomyces cerevisiae. *Applied Microbiology and Biotechnology* **89**(4):989–1000.

Forster, A., Aurich, A., Mauersberger, S. and Barth, G. (2007). Citric acid production from sucrose using a recombinant strain of the yeast *Yarrowia lipolytica*. *Applied Microbiology and Biotechnology* **75**(6):1409–1417.

Genomatica (2015). Partner in commercial-scale production of BDO, from http://www.genomatica.com/partners/dupont-tate-and-lyle/.

Hasunuma, T., Okazaki, F., Okai, N., Hara, K. Y., Ishii, J. and Kondo, A. (2013). A review of enzymes and microbes for lignocellulosic biorefinery and the possibility of their application to consolidated bioprocessing technology. *Bioresource Technology* **135**:513–522.

IndexMundi (2015). Sugar Futures End of Day Settlement Price. Retrieved 05-02-2015, 2015, from http://www.indexmundi.com/commodities/?commodity=sugar&months=120.

Jeffries, T. W. (2006). Engineering yeasts for xylose metabolism. *Current Opinion in Biotechnology* **17**(3):320–326.

Jensen, N. B., Strucko, T., Kildegaard, K. R., David, F., Maury, J., Mortensen, U. H., Forster, J., Nielsen, J. and Borodina, I. (2014). EasyClone: method for iterative chromosomal integration of multiple genes in Saccharomyces cerevisiae. *FEMS Yeast Research* **14**(2):238–248.

Kirby, J. and Keasling, J. D. (2009). Biosynthesis of plant isoprenoids: perspectives for microbial engineering. *Annual Review of Plant Biology* **60**:335–355.

Kirby, J., Nishimoto, M., Park, J. G., Withers, S. T., Nowroozi, F., Behrendt, D., Rutledge, E. J., Fortman, J. L., Johnson, H. E., Anderson, J. V. and Keasling, J. D. (2010). Cloning of casbene and neocembrene synthases from Euphorbiaceae plants and expression in Saccharomyces cerevisiae. *Phytochemistry* **71**(13):1466–1473.

Kopke, M., Held, C., Hujer, S., Liesegang, H., Wiezer, A., Wollherr, A., Ehrenreich, A., Liebl, W., Gottschalk, G. and Durre, P. (2010). *Clostridium ljungdahlii* represents a microbial production platform based on syngas. *Proceedings of the National Academy of Sciences of the United States of America* **107**(29):13087–13092.

LanzaTech (2015). Home Page, from http://www.lanzatech.com/.

Leang, C., Ueki, T., Nevin, K. P. and Lovley, D. R. (2013). A genetic system for *Clostridium ljungdahlii*: a chassis for autotrophic production of biocommodities and a model homoacetogen. *Applied and Environmental Microbiology* **79**(4):1102–1109.

Lee, J. W., Choi, S., Kim, J. M. and Lee, S. Y. (2010). *Mannheimia succiniciproducens* phosphotransferase system for sucrose utilization. *Applied and Environmental Microbiology* **76**(5):1699–1703.

Licht, S. (2014). Fermentation for biofuels and bio-based chemicals. In *Fermentation and Biochemical Engineering Handbook*, pp. 59–82. C. M. Todaro and H. C. Vogel, eds. Elsevier: Philadelphia, PA.

Liu, Y. P., Zheng, P., Sun, Z. H., Ni, Y., Dong, J. J. and Zhu, L. L. (2008). Economical succinic acid production from cane molasses by Actinobacillus succinogenes. *Bioresource Technology* **99**(6):1736–1742.

Mackay Sugar (2015). Cogeneration, from http://www.mackaysugar.com.au/business/green-projects/Pages/Cogeneration.aspx.

Mattey, M. and Kristiansen, B. (1999). A brief introduction to citric acid biotechnology. In *Citric Acid Biotechnology*, pp. 1–9. B. Kristiansen, M. Mattey and J. Linden, eds. Taylor and Francis: London.

Montanti, J., Nghiem, N. P. and Johnston, D. B. (2011). Production of astaxanthin from cellulosic biomass sugars by mutants of the yeast *Phaffia rhodozyma*. *Applied Biochemistry and Biotechnology* **164**(5):655–665.

NatureWorks (2015). From plants to plastics, from http://www.natureworksllc.com/The-Ingeo-Journey/Eco-Profile-and-LCA/How-Ingeo-is-Made.

Novozymes (2015). Healthier fermentation and healthier profits with Novozymes Sanferm® Yield, 2015, from http://www.novozymes.com/en/solutions/food-and-beverages/case-studies-on-Food-Beverages/Pages/Healthier-fermentation-and-healthier-profits-with-Novozymes-Sanferm®-Yield.aspx.

Olbrich, H. (1963). *The Molasses*. Biotechnologie-Kempe GmbH: Kleinmachnow, Germany.

Paddon, C. J. and Keasling, J. D. (2014). Semi-synthetic artemisinin: a model for the use of synthetic biology in pharmaceutical development. *Nature Reviews. Microbiology* **12**(5):355–367.

Papagianni, M. (2007). Advances in citric acid fermentation by *Aspergillus niger*: biochemical aspects, membrane transport and modeling. *Biotechnology Advances* **25**(3):244–263.

Parisutham, V., Kim, T. H. and Lee, S. K. (2014). Feasibilities of consolidated bioprocessing microbes: from pretreatment to biofuel production. *Bioresource Technology* **161**:431–440.

Peralta-Yahya, P. P., Ouellet, M., Chan, R., Mukhopadhyay, A., Keasling, J. D. and Lee, T. S. (2011). Identification and microbial production of a terpene-based advanced biofuel. *Nature Communications* **2**:483.

Peters, D. (2006). Carbohydrates for fermentation. *Biotechnology Journal* **1**(7–8):806–814.

Reverdia (2015). Biosuccinium™ Sustainable Succinic Acid, from http://www.reverdia.com/products/biosuccinium/.

Sabri, S., Nielsen, L. K. and Vickers, C. E. (2013). Molecular control of sucrose utilization in Escherichia coli W, an efficient sucrose-utilizing strain. *Applied and Environmental Microbiology* **79**(2):478–487.

Sano, C. (2009). History of glutamate production. *American Journal of Clinical Nutrition* **90**(3):728S–732S.

Solazyme (2014). Solazyme Bunge Renewable Oils Plant Begins Commercial Production in Brazil, from http://solazyme.com/blog/2014/05/29/solazyme-bunge-renewable-oils-plant-begins-commercial-production-in-brazil/.

Staples, M. D., Malina, R., Olcay, H., Pearlson, M. N., Hileman, J. I., Boies, A. and Barrett, S. R. H. (2014). Lifecycle greenhouse gas footprint and minimum selling price of renewable diesel and jet fuel from fermentation and advanced fermentation production technologies. *Energy & Environmental Science* **7**(5):1545–1554.

Steen, E. J., Chan, R., Prasad, N., Myers, S., Petzold, C. J., Redding, A., Ouellet, M. and Keasling, J. D. (2008). Metabolic engineering of Saccharomyces cerevisiae for the production of n-butanol. *Microbial Cell Factories* **7**:36.

Stephens, C., Christen, B., Fuchs, T., Sundaram, V., Watanabe, K. and Jenal, U. (2007). Genetic analysis of a novel pathway for D-xylose metabolism in Caulobacter crescentus. *Journal of Bacteriology* **189**(5):2181–2185.

UNICA (2015). UNICADATA, from http://www.unicadata.com.br/historico-de-producao-e-moagem.php?idMn=32&tipoHistorico=4.

Van Dyk, J. S. and Pletschke, B. I. (2012). A review of lignocellulose bioconversion using enzymatic hydrolysis and synergistic cooperation between enzymes – factors affecting enzymes, conversion and synergy. *Biotechnology Advances* **30**(6):1458–1480.

Watanabe, S., Kodaki, T. and Makino, K. (2006a). Cloning, expression, and characterization of bacterial L-arabinose 1-dehydrogenase involved in an alternative pathway of L-arabinose metabolism. *Journal of Biological Chemistry* **281**(5):2612–2623.

Watanabe, S., Kodaki, T. and Makino, K. (2006b). A novel alpha-ketoglutaric semialdehyde dehydrogenase: evolutionary insight into an alternative pathway of bacterial L-arabinose metabolism. *Journal of Biological Chemistry* **281**(39):28876–28888.

Weimberg, R. (1961). Pentose oxidation by Pseudomonas fragi. *Journal of Biological Chemistry* **236**:629–635.

Westfall, P. J., Pitera, D. J., Lenihan, J. R., Eng, D., Woolard, F. X., Regentin, R., Horning, T., Tsuruta, H., Melis, D. J., Owens, A., Fickes, S., Diola, D., Benjamin, K. R., Keasling, J. D., Leavell, M. D., McPhee, D. J., Renninger, N. S., Newman, J. D. and Paddon, C. J. (2012). Production of amorphadiene in yeast, and its conversion to dihydroartemisinic acid, precursor to the antimalarial agent artemisinin. *Proceedings of the National Academy of Sciences of the United States of America* **109**(3):E111–E118.

Xu, B., Jahic, M., Blomsten, G. and Enfors, S. O. (1999). Glucose overflow metabolism and mixed-acid fermentation in aerobic large-scale fed-batch processes with Escherichia coli. *Applied Microbiology and Biotechnology* **51**(5):564–571.

Yalcin, S. K., Bozdemir, M. T. and Ozbas, Z. Y. (2010). Citric acid production by yeasts: fermentation conditions, process optimization and strain improvement. In *Current Research, Technology and Education Topics in Applied Microbiology and Microbial Biotechnology*, vol. **2**, pp. 1374–1382. A. Mendez-Vilas, eds. Formatex: Badajoz, Spain.

Yim, H., Haselbeck, R., Niu, W., Pujol-Baxley, C., Burgard, A., Boldt, J., Khandurina, J., Trawick, J. D., Osterhout, R. E., Stephen, R., Estadilla, J., Teisan, S., Schreyer, H. B., Andrae, S., Yang, T. H., Lee, S. Y., Burk, M. J. and Van Dien, S. (2011). Metabolic engineering of Escherichia coli for direct production of 1,4-butanediol. *Nature Chemical Biology* **7**(7):445–452.

Zheng, Z., Chen, T., Zhao, M., Wang, Z. and Zhao, X. (2012). Engineering Escherichia coli for succinate production from hemicellulose via consolidated bioprocessing. *Microbial Cell Factories* **11**:37.

CHAPTER 6

Mathematical modeling of xylose production from hydrolysis of sugarcane bagasse

Ava Greenwood[1], Troy Farrell[1] and Ian M. O'Hara[2]

[1] *Mathematical Sciences, Queensland University of Technology (QUT), Brisbane, Australia*

[2] *Centre for Tropical Crops and Biocommodities, Queensland University of Technology (QUT), Brisbane, Australia*

6.1 Introduction

Bioethanol is a greener, renewable alternative to traditional petroleum-based transport fuels (Nigam and Singh 2011). Bioethanol can be produced by fermentation of sugars, such as glucose, obtained from plant crops. Although these sugars can be obtained through typical milling processes followed by saccharification and fermentation (Solomon *et al.* 2007), this raises concerns that crops (or croplands) otherwise intended for food production may be redirected for fuel manufacture (Nigam and Singh 2011, Harvey and Pilgrim 2011). Therefore, it is often preferred that these sugars be obtained by breaking down, conventionally via enzymatic hydrolysis, the cellulosic material that remains in plant cell walls after milling has occurred. Lignocellulosic biomass (such as sugarcane bagasse) can be resistant to enzymatic attack due to the complex structure of the cell wall, and thus some form of pretreatment is generally required to improve digestibility (Alvira *et al.* 2010, Chang and Holtzapple 2000). One such pretreatment is acid hydrolysis, which primarily removes the hemicellulosic component of the biomass thus improving the accessibility of the cellulosic substrate to enzymes (Sun and Cheng 2002, O'Hara *et al.* 2011). The hydrolysis of sugarcane bagasse primarily produces the pentosan sugar xylose (Gírio *et al.* 2010). This xylose can be collected and cofermented with glucose to further enhance the yield of bioethanol obtained (Lawford and Rousseau 2002, Mohagheghi *et al.* 2002). Alternatively, xylose can be used to create other commercial products such as xylitol or xylonic acid (Mohamad *et al.* 2015, Chun *et al.* 2006).

Sugarcane-Based Biofuels and Bioproducts, First Edition. Edited by Ian M. O'Hara and Sagadevan G. Mundree.

Xylitol is used in the food and pharmaceutical industries as a diabetic-friendly, low-calorie alternative to sucrose, a food preservative agent, and as an anticariogenic agent in dental hygiene products (Mohamad *et al.* 2015, Rafiqul and Mimi Sakinah 2013, Winkelhausen and Kuzmanova 1998). Since xylitol has such vast utility, it is a rapidly growing commodity and sugarcane bagasse has been studied as one of the principal lignocellulosic sources of xylose for xylitol production due to its considerable hemicellulosic content (Mohamad *et al.* 2015; Rafiqul and Mimi Sakinah, 2013). Currently, xylitol is obtained from the catalytic reduction of xylose, which is expensive and energy intensive. New environmentally friendly and cost-effective xylitol production technologies include the use of microbial reagents (particularly yeasts) to undertake the xylose conversion process (Mohamad *et al.* 2015, Rafiqul and Mimi Sakinah 2013, Winkelhausen and Kuzmanova 1998). There is commercial interest in integrating xylitol and bioethanol production processes to create additional revenue sources and reduce overall production costs (Cheng *et al.* 2010). Furthermore, pathways have been discovered for the simultaneous production of xylitol and xylonic acid (Jokic *et al.* 1991). Consequently, maximizing the yield of xylose obtained from the acid pretreatment of sugarcane bagasse is a priority not only for improving the digestibility of biomass for enzymatic hydrolysis, but also as a means of capitalizing on value-adding commercial xylose products such as xylitol, xylonic acid, or hemicellulosic bioethanol.

A drawback of pretreatment, however, is that it contributes substantially to the overall cost of bioethanol production (Yang and Wyman 2008b, Chandra *et al.* 2007). It is, therefore, important to ensure that the efficiency and performance of pretreatment is optimized. Mathematical modeling can be used to provide a better understanding of the mechanisms of biomass recalcitrance and help to inform experimental conditions that lead to maximum yields of hemicellulose solubilization and xylose production. A review of mathematical models of acid pretreatment is presented here.

In particular, when using models to shed light on the chemical and physical processes that occur during acid hydrolysis, it is important to carefully consider how the model implementation may alter the results obtained, and to ensure that the interpretation and validity of the model parameters is investigated. Accepting the model parameters (especially those obtained from parameter fitting) as representative of the true chemical and/or physical reality, without rigorously testing their credibility, may lead to false conclusions about the accuracy and predictability of the model.

Here, a sensitivity analysis is performed on a model of the dilute-acid hydrolysis of sugarcane bagasse previously developed by the authors (Greenwood *et al.* 2015a). This analysis aims to establish the relationships between model parameters and experimental inputs, and propose a course of laboratory or parameter fitting experiments that could be implemented to reduce any existing uncertainties in the model and its parameters.

6.2 Mathematical models of hemicellulose acid pretreatment

Since pretreatment can be expensive, it is imperative that mathematical models of acid hydrolysis are utilized as tools to help predict optimal reactor conditions, and implemented as a means by which to gain better insight into the chemical and physical mechanisms underlying hydrolysis.

6.2.1 Kinetic models of hemicellulose acid hydrolysis

As with any reactive process, kinetic models are favored when modeling dilute-acid hydrolysis since they aim to represent the actual chemical interactions occurring during the course of reaction. Here, the evolution of kinetic modeling of hemicellulose hydrolysis is discussed.

6.2.2 The Saeman kinetic model

In 1945, Saeman studied the kinetics of high temperature (>170 °C) dilute-acid hydrolysis of cellulose in a variety of wood species, particularly milled Douglas fir sawdust (Saeman 1945). Saeman proposed that saccharification of wood is a consecutive two-step process where the reaction proceeds such that cellulose is first hydrolyzed to produce monomeric reducing sugars (in this case glucose), which are in turn decomposed into degradation products. The model exhibits first-order dependence upon the sugar concentrations. Although originally conceived in the context of cellulose degradation, this kinetic pathway has since been adapted to describe the kinetics of hemicellulose hydrolysis as demonstrated in Scheme 6.1.

The system of differential equations that results from the Saeman kinetic model is given by

$$\frac{dC_H}{dt} = -k_1 C_H$$

$$\frac{dC_X}{dt} = k_1 C_H - k_d C_X$$

$$\frac{dC_D}{dt} = k_d C_X$$

where C_H, C_X, and C_D represent the time (t)-dependent concentrations of hemicellulose, xylose, and degradation products, respectively. The rate of hemicellulose scission is described by k_1, and k_d represents the rate of xylose degradation. An advantage of the simplicity of this model is that the differential equations can be solved analytically to produce a closed-form solution for xylose concentration

$$\text{Hemicellulose} \xrightarrow{k_1} \text{Xylose} \xrightarrow{k_d} \text{Degradation products}$$

Scheme 6.1 The Saeman kinetic model (1945).

of the form

$$C_X = C_H^0 \frac{k_1}{k_d - k_1}(e^{-k_1 t} - e^{-k_d t}) \qquad (6.1)$$

where C_H^0 is the initial concentration of hemicellulose, and the rate constants k_1 and k_d are functions of temperature, T (K), as per Arrhenius kinetics, and the concentration of acid C_{H^+} (%). The existence of such a solution is convenient for making prompt comparisons between the model and experimental results and cursory xylose yield predictions.

However, the elementary nature of Saeman reaction pathway also gives rise to a number of limitations, namely, the model does not account for the formation of oligomers, and it assumes that all hemicellulose reacts at the same rate and hence full solubilization of hemicellulose is possible at all reaction severities given a sufficiently long residence time. As such, the model has been found to overestimate yields and has been deemed ineffective when applied to recalcitrant biomass xylan (Zhao *et al.* 2012).

6.2.3 The biphasic model

In 1956, Kobayashi and Sakai conducted a study into the prehydrolysis of hardwood under dilute-acid conditions (Kobayashi and Sakai 1956). They experimentally observed two distinct rates of hemicellulose hydrolysis, one faster than the other, and concluded the existence of two hemicellulose classes each exhibiting first-order kinetics undergoing simultaneous hydrolysis. This discovery gave rise to the biphasic model of hemicellulose hydrolysis depicted in Scheme 6.2.

It follows that the system of differential equations that describe the biphasic model are given by

$$\frac{dC_F}{dt} = -k_1 C_F$$

$$\frac{dC_S}{dt} = -k_2 C_S$$

$$\frac{dC_X}{dt} = k_1 C_F + k_2 C_S - k_d C_X$$

$$\frac{dC_D}{dt} = k_d C_X$$

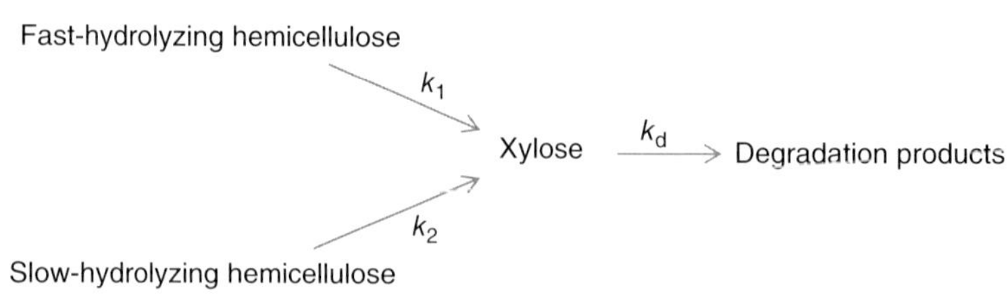

Scheme 6.2 Kobayashi and Sakai's biphasic reaction scheme (1956).

where C_F and C_S represent the concentrations of fast and slow hydrolyzing hemicellulose, respectively. Now, k_1 is the rate of scission of the fast hydrolyzing hemicellulose portion and k_2 is the rate of scission of the slow-to-hydrolyze hemicellulose fraction. Similar to the Saeman model before it, this equation system can be solved analytically to obtain an expression for the xylose concentration, namely,

$$C_X = A^* e^{-k_1 t} + B^* e^{-k_2 t} - (A^* + B^*) e^{-k_d t} \tag{6.2}$$

where

$$A^* = \frac{k_1}{k_d - k_1} C_F^0 \text{ and } B^* = \frac{k_2}{k_d - k_2} C_S^0$$

assuming that the initial concentration of monomers is zero (i.e., $C_X(0) = 0$).

This model has shown significant promise in accurately reproducing xylose yields and consequently has been pervasive in the literature (Jacobsen and Wyman 2000, Zhao *et al.* 2012, Maloney and Chapman 1985, Lavarack *et al.* 2002).

The biphasic model can be simplified by making the assumption that the slow to hydrolyze phase does not undergo hydrolysis, rendering the rate of hydrolysis of the "slow" hemicellulose, k_2, effectively zero. Consequently, $k_2 \to 0$, $C_F^0 = \alpha C_H^0$ and $C_S^0 = (1 - \alpha) C_H^0$, where α represents the portion of hemicellulose susceptible to hydrolysis within the total hemicellulose fraction. These simplifications reduce Equation 6.2 to

$$C_X = \frac{k_1}{k_d - k_1} \alpha C_H^0 (e^{-k_1 t} - e^{-k_d t}) \tag{6.3}$$

which is clearly a modified form of Equation 6.1. Such a simplification may seem unnecessary since the biphasic model is not in itself complex; however, this assumption may prove to be advantageous when a more complicated degradation mechanism is used to describe hemicellulose acid hydrolysis. Equation 6.3 has been used extensively in the literature, although in some instances α is referred to as the potential hydrolysis degree h_d (Zhao *et al.* 2012).

Bustos *et al.* (2003), Aguilar *et al.* (2002), and Zhang *et al.* (2011b) found that Equation 6.3 was able to produce a good fit to experimental xylose yields across a range of conditions, namely, 100–128 °C with 2–6% HCl, 110–128 °C with 2–6% H_2SO_4, and 150–180 °C with 0.2–0.6% HNO_3, respectively. Although Bustos *et al.* showed that this model produced a better fit to the experimental yields than the Saeman model, their results show that this was not by any significant margin, and that the coefficient of determination, r^2 (which determines the goodness of fit), still showed a good correlation between the Saeman model and the data. It is noted that in Bustos *et al.* (2003), the rate parameters for xylose degradation and furfural production, that heretofore have been considered the same parameter, k_d, were calculated separately and were not in agreement. The reasoning behind this may have been to suggest that significant degradation products other than furfural exist. These findings were not consistent with the results of Zhang *et al.* and Aguilar *et al.* who determined that the yields predicted by the Saeman model Equation 6.1 were incongruous with their experimental data. All three studies did, however, find that α is dependent upon the reaction severity, although the functional form of $\alpha(T, C_{H^+})$ has yet to be determined (Zhao *et al.* 2012). Furthermore, Zhao *et al.* (2012) proposed that the relationship between α and the reaction conditions further supports the conclusion that the Saeman model is not sufficiently detailed as to capture the kinetics of biomass acid hydrolysis.

The biphasic model was created as a response to experimental observations. Lavarack *et al.* (2002) claimed that no chemical or physical basis had yet been identified to explain the phenomenon of easy and hard-to-hydrolyze fractions within bagasse. Others have speculated as to the source of this apparent hemicellulose recalcitrance. Perhaps the most popular explanation is that hemicellulose–lignin linkages reduce the accessibility of regions of the hemicellulose substrate (Carrasco and Roy 1992, Conner 1984). Zhao *et al.* (2012) also proposed that lignin and uronic acid could hinder xylan hydrolysis, but further suggested that mass and energy transport throughout the complex layered structure of the plant cell wall could be preventing regions of hemicellulose from being hydrolyzed.

Importantly, the standard Saeman and biphasic models do not consider the formation and subsequent degradation of xylooligomers during dilute-acid hydrolysis. Maloney *et al.* (1986) reported that the rate at which solubilized xylooligomers are hydrolyzed is much faster than the rate of xylan hydrolysis. Consequently, they concluded that xylooligomers could be excluded from the kinetic pathway without adversely affecting the accuracy of the resultant model. Also, when these studies were conducted, it was not possible to compare oligomer profiles to experimental data since xylooligomers of different degrees of polymerization were indistinguishable, making the experimental measurement of xylooligomers in solution infeasible (Bhandari *et al.* 1984). However, with modern technologies, xylooligomer yields are able to be measured in the hydrolysate (Yang and Wyman 2008a, Kumar and Wyman 2008, Samala *et al.* 2012), and the validity of the relatively fast rate of xylooligomer hydrolysis has been questioned (Jacobsen and Wyman 2000).

In order to address these discrepancies, an additional term can be added to the biphasic model such that the fast and slow portions of hemicellulose form an oligomeric intermediate, before the oligomers hydrolyze to produce xylose (Jacobsen and Wyman 2000, Garrote *et al.* 1999, Conner and Lorenz 1986, Zhao *et al.* 2012).

More recently, models have been developed where the kinetics of the hydrolysis of short-chain xylooligomers such as xylose, xylobiose, xylotriose, xylotetraose, and xylopentaose are modeled explicitly (Hosseini and Shah 2009, Lau *et al.* 2014, Kumar and Wyman 2008). As per the Saeman and

Scheme 6.3 Chemical balance of the polymer scission reaction.

$$P_{i\,(s)} + H^{+}_{(aq)} \xrightarrow{k} P_{i-j\,(s)} + P_{j(s)} + H^{+}_{(aq)}$$

biphasic models, these xylooligomer models display a first-order dependence on xylooligomer concentration.

Such models were able to reproduce experimental short-chain oligomer profiles to some extent; however, Lau *et al.* (2014) found that their model underpredicted monomer yield, and Hosseini and Shah (2009) predicted an unrealistically high furfural yield. It must be noted that each of these models require a large number of kinetic parameters. In fact, for each additional oligomer length considered, at least one new rate parameter is added to the system and the overall number of unknown parameters grows considerably. This is problematic, since the rate constants of hemicellulose hydrolysis are not able to be measured and thus must be obtained through parameter fitting. An increase in the number of fitting parameters leads to greater degrees of freedom in the potential parameter space, which may diminish the robustness of the parameters obtained.

6.2.4 The polymer degradation equation

Perhaps a more efficient way to characterize the fragmentation of polymers into their oligomeric constituents due to scission events is to use the polymer degradation equation. The discrete polymer degradation equation was introduced by Simha in 1941 to describe the depolymerization of long-chain polymers. This model characterizes the production and degradation of polymers of every possible chain length; hence, the model encapsulates the cumulative effects of scission events along the chain. The scission reaction can be described by the chemical balance demonstrated in Scheme 6.3 such that polymer chains, P, of length i are scised into two component chains of length i–j and j, respectively, at rate k (m^3/mol/s). Importantly, although k can be a function of the chain length, it is often treated as a constant, thus reducing the number of fitting parameters required compared with the other oligomer models for hemicellulose hydrolysis discussed previously. Acid, $H^+{}_{(aq)}$, appears on both sides of this equation, since it is a catalyst and not a consumable reagent.

A discrete form of the polymer degradation equation can be formulated by summing the probabilities of breakage and formation for the N species being tracked in the reaction sequence, namely,

$$\frac{\mathrm{d}C_i}{\mathrm{d}t} = -C_{\mathrm{H}^+}C_i \sum_{j=1}^{i-1} F(j,i) + \sum_{j=i+1}^{N} (F(i,j) + F(j-i,j))C_{\mathrm{H}^+}C_j,\ (i = 1, 2, \ldots, N) \tag{6.4}$$

where C_i is the concentration of polymer chains of length i, C_{H^+} is the concentration of hydrogen ions in solution, and $F(a, b)$ is the overall rate of formation of a chain of length a due to scission of a chain of length b. Consequently, the first term on the right-hand side represents hemicellulose chains of length i that are lost due to scission, whereas the second term accounts for the accumulation of chains of length i from the scission of longer chains. If it is assumed that there is symmetry to the bond structure of the hemicellulose chain, then the rate of producing a chain of length i from left or right hand scission is the same. Thus, the rate of formation can be simplified to give

$$F(i,j) + F(j-i,j) = 2F_{i,j-i},\ (i = 1, 2, \ldots, N), (j = i+1, \ldots, N)$$

which is the form presented by Ziff and McGrady (1986). The overall rate of scission, $F(a, b)$, can be further broken down into the product of two components $F(a, b) = k\Omega(a, b)$, where k is the standard kinetic rate constant, and $\Omega(a, b)$ is the breakage kernel, which determines how mass is distributed to chains of length a due to the scission of chains of length b, as influenced by the probability of scission. Furthermore, the sum can be removed from the degradation term of Equation 6.4 by noting that the mass lost by chains of length i must be redistributed to chains of length 1 to $i-1$, and hence we observe

$$\sum_{j=1}^{i-1} \Omega(j,i) = 1$$

Consequently, the polymer degradation equation simplifies to

$$\frac{\mathrm{d}C_i}{\mathrm{d}t} = -kC_{\mathrm{H}^+}C_i + 2kC_{\mathrm{H}^+}\sum_{j=i+1}^{N} \Omega_{i,j-i}C_j,\ (i = 1, 2, \ldots, N) \qquad (6.5)$$

In the continuous form of this equation, the chain length is replaced by a continuous independent variable such as molecular weight or length, causing integral terms to replace the summations of the discrete model (Ziff and McGrady 1986). This may not appear to be physically realistic since polymer chains are only comprised of discrete units; however, for very large chain lengths, it is believed to be a reasonable approximation (Kostoglou 2000). When using the continuous polymer degradation equation, difficulties may arise in trying to satisfy conservation of mass for the system when implementing numerical techniques such as the finite volume method. Kumar (2010) found that the moments of the continuous equation are only preserved if the finite volume integrals are thoughtfully applied. Hemicellulose chains in the bagasse cell wall are in the range of 80–200 monomer units long, consequently the small degree of polymerization (DP) suggests that a discrete approach may be more appropriate when considering hemicellulose hydrolysis (O'Hara *et al.* 2011).

The overall rate of degradation, $F(a, b)$, can be formulated in many ways depending upon the mechanism of scission, and the chemical composition of the polymer including the distribution of functional groups within the chain that may influence reactivity. Many such mechanisms have been investigated for both the discrete and continuous cases including random chain scission, preferential end breakage (also often referred to as chain-end scission), scission where the probability of breakage is inversely proportional to the number of bonds in the chain, preferential center breakage, or breakage at a rate proportional to chain length raised to a power (Simha 1941, Ziff and McGrady 1986, Kostoglou 2000, McCoy and Madras 2001). McCoy and Wang (1994) present a generalized form of the breakage kernel for the continuous polymer degradation equation that was shown to be capable of, but not limited to, reproducing random and midpoint scission (Madras and McCoy 1998).

If the breakage kernel, $\Omega(a, b)$, is sufficiently uncomplicated, analytic solutions to the discrete and continuous polymer degradation equation can be found. Hill and Ng (1995) compiled a list of the common solution techniques available in the literature. Under certain circumstances, analytic solutions can also be found for population balance equations that have been modified to include diffusive and advective fluxes. Rayapati *et al.* (2010) solve 2D steady-state discrete population balance equations with diffusion and advection as a representation of the flow of particulates through a pipe. Scanning electron microscopy (SEM) indicates that it may be appropriate to approximate the shape of bagasse fibers with cylindrical geometry and hence their work in cylindrical coordinates may be relevant to modeling the acid hydrolysis of hemicellulose using polymer degradation equations with mass transport considerations in cylindrical bagasse fiber.

In cases where the degradation kinetics are more complex and analytical solutions to the polymer degradation equation are not able to be elucidated, numerical techniques can be implemented for solving the system of differential equations. A limitation of the model is that the maximum DP of the polymer determines the number of equations in the system, and for long polymers this may be computationally inhibitive. However, methods such as the multiphase continuum approach can be implemented to reduce the number of equations solved numerically (Rayapati *et al.* 2010). In this approach, each polymer degradation equation then describes a group of oligomers rather than a single monomer. We note, however, that the multiphase continuum approach is not ideal in the context of hemicellulose hydrolysis, as insight into the mechanism of short-chain oligomer hydrolysis is of interest.

Lloyd and Wyman (2003) applied the polymer degradation equation to the dilute-acid hydrolysis and autohydrolysis of corn cob and corn stover. A version of Simha's random scission model modified to incorporate the degradation of monomer was solved analytically and used to predict the yields of residual xylan, aqueous oligomers (DP 2–8), and monomer content in the hydrolysate as a percentage of the maximum potential xylose conversion. Only two parameters were allowed to vary in the parameter fitting, the rate of hydrolysis, k_h, and the rate of monomer degradation, k_d. Lloyd and Wyman found that for water-only hydrolysis there was some initial agreement between the experiments and model, but that the xylooligomer production was ultimately overestimated and the residual xylan underestimated at long times. For the case of acid hydrolysis, monomer yield appeared to be somewhat accurate, whereas there was little correlation between the modeled and experimental xylooligomer and residual xylan data, even initially. Lloyd and Wyman (2003) postulated that a different formulation of the breakage kinetics or consideration of gelling effects may help to improve this fit. Greenwood *et al.* (2015a) developed a population balance model coupled with simplified biphasic kinetics. The model also incorporated diffusive mass transport and calculations of the temporal and spatial porosity change caused by solubilization of solid xylan chains. This model was shown to be capable of successfully reproducing yield profiles for short-chain xylooligomers across a range of temperatures fitting only four parameters. However, greater inconsistencies between model and experimental yields were observed at longer aqueous chain lengths.

6.2.5 Other mathematical considerations and models of hemicellulose acid hydrolysis

Porosity

The inability of cellulases to access the cellulosic substrate is considered to be a significant contributor to poor conversion of cellulose to glucose (Meng *et al.* 2013), since physical contact between the enzyme and substrate is required to facilitate hydrolysis (Fan *et al.* 1980). Since enzymes are bulky, it may in fact be pore size rather than specific surface area (SSA) that determines the effectiveness of hydrolysis (Tanaka *et al.* 1988, Meng *et al.* 2013). Consequently, the impact of dilute-acid hydrolysis upon the change in porosity of the lignocellulosic material is of interest.

A number of different approaches have been employed to try to quantify some aspect of the porosity and SSA changes induced by pretreatment. Meng *et al.* (2013) reviewed the strengths and weaknesses of Simons' stain, solute exclusion, nitrogen adsorption, and NMR techniques, and the different ways in which these techniques contribute to the understanding of structural changes within the substrate. SEM photo analysis is also a popular qualitative method for determining changes in the material porosity (Zhang *et al.* 2011a, Tsuchida *et al.* 2014, Chen *et al.* 2014). Meng *et al.* (2013) identified that the best measurement techniques could be applied to wet samples, since drying can dramatically alter the internal structure of the sample. Although this is supported by Fan *et al.* (1980), it was determined that some of the structural deformation could be circumvented if a series of organic solvents are applied to the sample before drying.

Zhang *et al.* (2011a) used SEM to observe the formation of microholes in sweet sorghum bagasse after dilute-acid hydrolysis and concluded that steam explosion and dilute acid both increased the accessibility of cellulose to enzymatic attack. The formation of these microholes was also observed by Chen *et al.* upon SEM imaging of dilute acid–hydrolyzed bamboo, concluding that an increase in pore volume resulted. Meng *et al.* (2013) used the Simons' stain technique to elucidate that dilute-acid pretreatment created new pores and increased the size of existing pores in a sample of *Populus*. Their results suggested that dilute-acid pretreatment had a greater impact upon the SSA than steam explosion. Furthermore, cryoporometry indicated an increase in representative pore diameter after dilute-acid pretreatment. Using NMR relaxometry and diffusometry, Meng *et al.* (2013) also found that the increase in cellulose accessibility caused by dilute-acid pretreatment was relatively rapid (occurred primarily within the first 10 min of reaction at 150 °C with 0.15 mol/l H_2SO_4).

Zhou *et al.* (2009a) present a methodology for mathematically modeling the time-evolving substrate (cellulose) morphology in biomass materials during enzymatic hydrolysis. Their model introduces two interesting ideas. The first is that the model geometrically deconstructs the substrate into portions that represent the smallest accessible compartments (SACs) of substrate and smallest accessible voids (SAVs). The SAVs are determined by the size of the enzyme or reagent, and SACs are the minimal volume bounded by reagent accessible pathways that do not contain reagent accessible pathways. Second, Zhou *et al.* describe each SAC as being comprised of several layers, λ_i, corresponding to the fraction of material that would be digested and removed from the SAC surface if all outer bonds are broken. Each layer progressively becomes accessible to the reagent as hydrolysis continues, and the surface-area-to-volume ratio changes. Their layering technique is not limited to a single geometry, and linear, cylindrical, and spherical templates are best suited to materials that are highly crystalline, directionally oriented and amorphous, respectively. Numerical simulations were conducted and the modeling implications of this new formalism were explored by the authors in an accompanying study (Zhou *et al.* 2009b).

However, the porosity change during enzymatic hydrolysis is conceivably more complicated than in the case of acid hydrolysis, since unlike cellulases, the smaller acid reagent is able to interact with both small and large pores across the biomass. Consequently, a simpler model may be appropriate to approximate the time evolution of porosity during acid pretreatment. Given the kinetic models described in Section 6.2.1, it can be seen that the modeling motivation for porosity increase is the solubilization of hemicellulose from solid xylan to aqueous xylose or xylooligomers in which case the solid region is replaced by the acid/oligomer solution. As the porosity increases, more acid is able to interact with the hemicellulose chains and thus the reaction rate is expected to increase. Few models in the literature attempt to describe this phenomenon. Mittal *et al.* (2009) introduce porosity into a model of wood chip hydrolysis where hydrolyzed xylooligomers and degradation products can be transported into the bulk solution via a convective mass transfer coefficient. Mittal *et al.* propose that an advantage of using this approach, over diffusion-based models, is that it is not limited to well-defined geometries; however, they acknowledge that it is a more approximate approach. The overall rate of mass transfer is a function of the surface area of wood chips and their porosity. Porosity is a ratio of the temporally varying pore volume to the constant total volume of wood chips. The pore volume is modeled as a linear function of the mass removed by hydrolysis. This mass transfer coefficient model produced theoretical yields for xylan, xylooligomers, xylose, and furfural that were well correlated to experimental autohydrolysis results for both sugar maple and aspen wood chips. However, a significant number of fitting parameters were present in the model.

Greenwood *et al.* (2015a) present a model for the spatial and temporal evolution of the porous volume fraction within sugarcane bagasse, using model variables and one parameter that can be estimated from mass composition data. This formulation of the porosity equation also has a linear dependence on the time-evolving volume fractions of solid species in the reaction. Although most models do not attempt to calculate the change in porosity during acid hydrolysis, a number of models recognize the influence that porosity has upon the diffusion of species within the biomass.

Mass transport models

Transport limitations have been considered as possible contributors to the rate of hemicellulose hydrolysis (Carrasco and Roy 1992, Garrote *et al.* 1999). Mathematical models of reaction–advection–diffusion have been investigated at multiple size scales including the fiber scale where biomass is hydrolyzed, and solubilized oligomers are transported out of the material into the surrounding hydrolysate (Kostoglou and Karabelas 2004, Rayapati *et al.* 2010); and the reactor scale where the bulk transport of aqueous biomass across the reaction vessel is considered, often coupled with a fiber-scale model (Hosseini and Shah 2009, Maloney *et al.* 1986). In the case of a batch reactor (as investigated here), there is not usually a source of advective flow and hence diffusion is considered as the sole mechanism of mass transport.

Fick's second law states that the temporal rate of change of concentration, $C(x, t)$, depends on diffusion according to the relationship

$$\frac{\partial C}{\partial t} = \nabla \cdot (D \nabla C)$$

where D is the diffusion coefficient. The diffusion coefficient for polymers in solution can be handled in an assortment of ways. Bird *et al.* (2007) present the following equation for diffusion of polymer A in solvent B:

$$D_{\mathrm{A,B}} = \frac{1}{\sqrt{M}} \tag{6.6}$$

where M is the molecular weight of the polymer. This diffusion coefficient is best applied to dilute solutions where the solvent is of low molecular weight.

Alternatively, polymers can coil in solution (Hall 1989) and hence the Stokes–Einstein diffusion coefficient for spherical particles diffusing in dilute solution has been used by Greenwood *et al.* (2015a), as given by

$$D_{\mathrm{A,B}} = \frac{k_{\mathrm{B}} T}{6\pi \eta R_{\mathrm{h}}} \tag{6.7}$$

where k_B is the Boltzmann constant, T is temperature, η represents the dynamic viscosity of the solvent solution, and R_h is the equivalent hydrodynamic radius of the spherical polymers (Cussler 1997). An empirical variation of Equation 6.7, the Wilke–Chang estimation method, has been used by Maloney *et al.* (1986) to depict diffusion of solute A in solvent B for sufficiently dilute solutions. The diffusion coefficient is, therefore,

$$D^0_{A,B} = 7.4 \times 10^{-8} \frac{(\Phi M_B)^{\frac{1}{2}} T}{\eta_B V_A^{0.6}} \tag{6.8}$$

where M_B is the molecular weight of solvent, V_A is the molal volume of solute at normal boiling temperature, and Φ is an "association factor" determined by the type of solvent (purportedly 2.6 for water) (Reid *et al.* 1977).

These diffusion coefficients represent the diffusion of polymer in a bulk solution. However, within the fibrous bagasse material, the lignocellulosic matrix forms a tortuous pathway for the solubilized polymers to navigate, and hence an "effective" diffusion coefficient is required. The effective diffusion coefficient characterizes the influence porosity imparts on the overall diffusion within the material. In their model of short-chain xylooligomer hydrolysis, Hosseini and Shah applied the diffusion coefficient proposed by Weissberg (1963) for porous media, namely,

$$D_{eff} = \frac{\varepsilon D_{bulk}}{1 - 0.5 \ln \varepsilon}$$

where ε is the void fraction. Weissberg's coefficient was developed to describe a bed of spherical particles, a geometry that does not closely align with the internal lignocellulosic structure of bagasse. Alternatively, Lee and Park (2007) apply the random porous cluster model of Zhang and Bishop (1994) to describe mass transport in biofilms, given by

$$D_{eff} = \varepsilon^3 D_{bulk}$$

where ε is the porosity (volume of liquid phase filling the pores per total volume of biofilm). This formulation of the effective diffusion coefficient was used by Greenwood *et al.* (2015a) to describe the tortuous diffusion pathway in a sugarcane bagasse fiber.

Response surface methodology (RSM)

Neureiter *et al.* (2002) used a statistical approach to predict the yield of hemicellulose hydrolysis products (particularly xylose and furfural) as a function of four reaction variables: acid concentration, temperature, residence time, and percentage dry matter. Consequently, such models have the potential to provide insight into the optimal reactor conditions for hemicellulose dilute-acid pretreatment. A "response surface" was constructed from the predicted yields, calculated according to a second-order polynomial regression model given by

$$Y_i = \underbrace{\beta_0}_{\text{intercept}} + \underbrace{\sum_{i=1}^{n} \beta_i x_i}_{\text{linear}} + \underbrace{\sum_{i=1}^{n} \beta_{ii} x_i^2}_{\text{quadratic}} + \underbrace{\sum_{i<j}^{n} \beta_{ij} x_i x_j}_{\text{interaction}} + \underbrace{\xi}_{\text{experimental error}}$$

Here, Y_i is the yield of the hemicellulosic degradation product (such as xylose or furfural), the x_i represent experimental conditions, and the β parameters characterize the regression coefficients for the intercept, linear, quadratic, and interaction effects, respectively. The regression parameters were elucidated by fitting this same model to experimentally determined yield values such that the sum of the least squares residuals was minimized.

The relative influence each experimental condition holds over the xylose or furfural yield was determined through an analysis of the variance (ANOVA). Neureiter *et al.* (2002) found that acid concentration had the greatest impact upon xylose yield, whereas temperature was more influential to the yield of furfural. Diedericks *et al.* (2013) discovered that temperature had the most significance in determining the yields of xylose, glucose, and furfural in the dilute-acid hydrolysis of sugarcane bagasse. Um and Bae (2011) used RSM (response surface methodology) to suggest that the optimal residence time and acid concentration for acid hydrolysis at 170 °C were 15 min and 0.24% (w/v), respectively; however, they did not compare the relative influence of each variable upon the yield obtained.

Unlike the kinetic models described earlier, these statistical response surface methods do not attempt to characterize the underlying chemical interactions of acid-catalyzed hydrolysis. These models also have a high degree of freedom in their parameter fitting due to the significant number of regression coefficients. Consequently, although their utility for identifying which experimental conditions have the greatest impact upon the effectiveness of hemicellulose acid hydrolysis has been demonstrated, statistical models are not revisited in this chapter in favor of the traditional kinetic approach. While kinetic models allow for the investigation of optimal reactor conditions, they additionally afford the opportunity to investigate the chemical (and physical) mechanisms underlying the dilute-acid pretreatment of hemicellulose.

6.3 A mathematical model of sugarcane bagasse dilute-acid hydrolysis

Each of the existing models of dilute-acid hydrolysis has their unique set of advantages and disadvantages. Previously, the authors have chosen to implement a model that combines polymer degradation equations, which allow for chain length–dependent effects to be incorporated into the model, with simplified biphasic kinetics (Greenwood *et al.* 2015a). This model has been shown to capably describe experimentally observed xylooligomer yields. In addition, the time evolution of the bagasse porosity is accounted for by considering the porous volume gained by loss of solid xylan due to solubilization.

The model was formulated to describe the dilute-acid hydrolysis of sugarcane bagasse fibers. The model assumes that the hydrolysis of any individual bagasse fiber within the reaction vessel is independent of any other, and hence the hydrolysis of the convolution of fibers within the experimental reactor can be approximated by the hydrolysis of a single bagasse fiber as depicted in Figure 6.1 for a laboratory-scale reactor. The model domain, therefore, has two subregions, an idealized cylindrical bagasse fiber, surrounded by an associated region of hydrolysate as indicated in Figure 6.1. Mass transport in the longitudinal direction of the cylinder is considered to be slow (since the longitudinal length of the fiber, L, is considerably larger than the fiber radius, R_i) and is consequently ignored. This leaves one radial spatial dimension, r(m).

The fiber region of the domain consists of both fibrous material and acid-filled pores, whereas the hydrolysate region initially consists entirely of acid solution. All mass begins in the fiber region, however xylooligomers of chain length m or smaller are considered to be solubilized and may, therefore, diffuse into the hydrolysate region. All chains of length $m+1$ or greater exist in the solid fraction. The yield of soluble oligomers is calculated from their concentrations in the hydrolysate region of the domain, and the yield of xylan remaining in the solid is calculated from the fiber region.

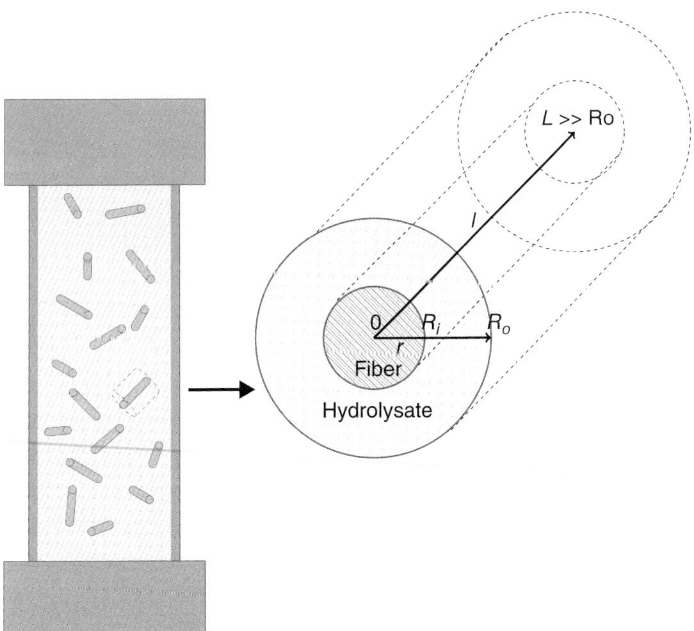

Figure 6.1 Schematic of the laboratory-scale experimental setup and the fiber domain.

Table 6.1 Summary of model equations presented in Greenwood *et al.* (2015a).

Region	Dependent variable	Equation
Fiber region	Effective furfural concentration, ϕ_F	$\frac{\partial \phi_F}{\partial t} = k_d \psi_{H^+} \phi_1 + \frac{1}{r}\frac{\partial}{\partial r}\left(rD^F_{\text{eff}}(\varepsilon_v)\frac{\partial \phi_F}{\partial r}\right)$
	Effective xylose concentration, ϕ_1	$\frac{\partial \phi_1}{\partial t} = -k_d \psi_{H^+} \phi_1 + 2k_a \psi_{H^+} \sum_{j=m+1}^{N} \Omega_{1,j-1}\phi_j + 2k_b \psi_{H^+} \sum_{j=2}^{m} \Omega_{1,j-1}\phi_j + \frac{1}{r}\frac{\partial}{\partial r}\left(rD_{\text{eff}}(\varepsilon_v)\frac{\partial \phi_1}{\partial r}\right)$
	Effective xylooligomer concentration, $\phi_i\,(i = 2, 3, \ldots, m)$	$\frac{\partial \phi_i}{\partial t} = -k_b \psi_{H^+} \phi_i + 2k_a \psi_{H^+} \sum_{j=m+1}^{N} \Omega_{i,j-i}\phi_j + 2k_b \psi_{H^+} \sum_{j=i+1}^{m} \Omega_{i,j-i}\phi_j + \frac{1}{r}\frac{\partial}{\partial r}\left(rD_{\text{eff}}(\varepsilon_v)\frac{\partial \phi_i}{\partial r}\right), (i = 2, 3, \ldots, m-1, m)$
	Effective xylan concentration, $\phi_i\,(i = m+1, m+2, \ldots, N)$	$\frac{\partial \phi_i}{\partial t} = -k_a \psi_{H^+} \phi_i + 2k_a \psi_{H^+} \sum_{j=i+1}^{N} \Omega_{i,j-i}\phi_j, (i = m+1, m+2, \ldots, N-1, N)$
	Porosity, ε_v	$\varepsilon_v = 1 - \left(\hat{F} + \sum_{i=m+1}^{N} \varepsilon_i + \varepsilon_\alpha\right)$
Associated hydrolysate region	Effective furfural concentration, ϕ_F	$\frac{\partial \phi_F}{\partial t} = k_d \psi_{H^+} \phi_1 + \frac{1}{r}\frac{\partial}{\partial r}\left(rD^F_{\infty}\frac{\partial \phi_F}{\partial r}\right)$
	Effective xylose concentration, ϕ_1	$\frac{\partial \phi_1}{\partial t} = -k_d \psi_{H^+} \phi_1 + 2k_b \psi_{H^+} \sum_{j=2}^{m} \Omega_{1,j-1}\phi_j + \frac{1}{r}\frac{\partial}{\partial r}\left(rD_{\infty}(i)\frac{\partial \phi_1}{\partial r}\right)$
	Effective xylooligomer concentration, $\phi_i\,(i = 2, 3, \ldots, m)$	$\frac{\partial \phi_i}{\partial t} = -k_b \psi_{H^+} \phi_i + 2k_b \psi_{H^+} \sum_{j=i+1}^{m} \Omega_{i,j-i}\phi_j + \frac{1}{r}\frac{\partial}{\partial r}\left(rD_{\infty}(i)\frac{\partial \phi_i}{\partial r}\right), (i = 2, 3, \ldots, m-1, m)$
	Porosity, ε_v	$\varepsilon_v = 1$

A summary of the model equations is presented in Table 6.1; however, a full derivation of the model, auxiliary equations, and parameters can be found in Greenwood *et al.* (2015a). The dependent variables ϕ_i, ϕ_F, and ε_v represent the concentration of chains of length i (kg/m^3), the concentration of furfural (kg/m^3), and the porosity of the material (the volume fraction of the porous region of the bagasse), respectively.

This model was found to produce reasonable estimates of short-chain xylooligomer yield profiles obtained from batch dilute-acid pretreatment at a laboratory scale (Greenwood *et al.* 2015a). These promising results were obtained despite fitting only four parameters for the entire system of differential and algebraic equations: k_a, the rate of solid hemicelluloses scission; k_b, the rate of aqueous hemicellulose scission; k_d, the rate of xylose degradation; and $\bar{\alpha}$, the "hard-to-hydrolyze" fraction of hemicelluloses (analogous to $1-\alpha$ discussed earlier). However, the yields predicted by the model have proven inconsistent with experimental results at a reactor scale (hydrolysis conducted on an industrial scale rather than in a benchtop apparatus). This was established based on a fairly limited reactor-scale data set and hence it was indeterminate whether the source of this discrepancy was due to the model formulation and fitting parameters, or fundamental differences in the reaction mechanism between size scales of such disparate magnitudes (Greenwood *et al.* 2015b). Determining the validity of the latter point requires significantly more reactor-scale data than those currently available to the authors. Consequently, the former possibility is considered here, and the influence that certain model parameters have upon the model-predicted yields is investigated. As the relationships between input parameters and the modeling results become better characterized, the model can be used to design new hydrolysis and parameter fitting experiments that better promote the determination of the true mechanism and rate parameters of hemicellulose acid pretreatment.

6.4 Sensitivity analysis

Previously, Greenwood *et al.* (2015a) showed that when fitting model parameters to experimental yield data, the parameter space needs to be adequately constrained in order to elucidate meaningful rate parameters. Consequently, the model was fit to oligomer yields (DP 1–6) in addition to the yield of furfural using the Parameter ESTimation (PEST) software. Despite this, there are many factors that influence the accuracy of the fitting procedure including the precision of the experimental results and the configuration of the stopping criteria provided to PEST. Consequently, there is always potential for some uncertainty in the parameters of best fit. Therefore, it is imperative that the model is robust enough that model outcomes are not completely altered by any such uncertainties. To better characterize the robustness of the model, the sensitivity of the model to small changes in the fitting parameters was examined.

The sensitivity analysis was conducted at 140 °C by changing each of the parameters $k_a = 1.67 \times 10^{-2}$, $k_b = 1.29 \times 10^{-1}$, $k_d = 2.70 \times 10^{-7}$ (m^3/mol/s), and $\bar{\alpha} = 5.21 \times 10^{-2}$ obtained by Greenwood *et al.* (2015a) by ±5% in turn. The resultant percentage change incurred in the maximum xylose yield was recorded for each parameter variation as demonstrated in Figure 6.2. We see that the percentage change incurred in the maximum xylose yield was small for all parameter variations, which suggests that the model is not particularly sensitive to small experimental or parameter fitting errors.

The parameter with the most apparent influence over the maximum xylose yield was $\bar{\alpha}$. This might be expected, since $\bar{\alpha}$ determines the proportion of unhydrolyzable bagasse, and thus directly influences the maximum obtainable yield. However, the maximum obtainable yield can also be influenced by the solid loading of the vessel. This is due to the fact that diffusion smoothes out the concentration gradient, and since k_d is small, the modeled xylose

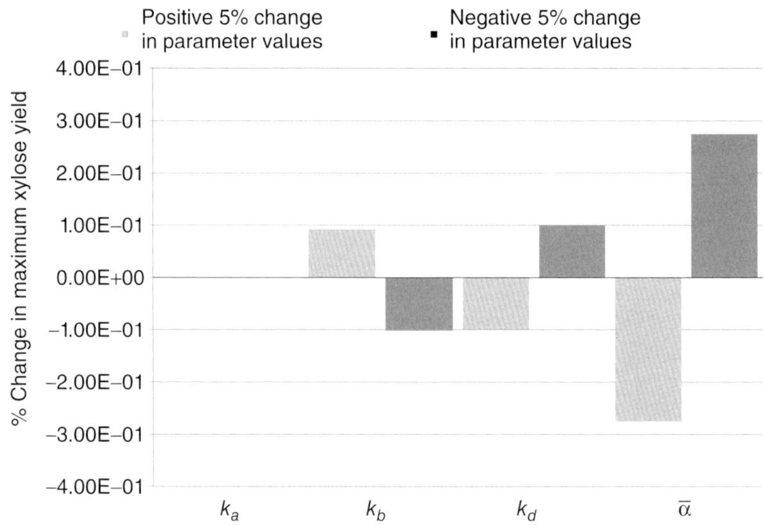

Figure 6.2 Sensitivity analysis examining the change in maximum yield due to small variations in the fitting parameters.

concentration approaches a homogeneous steady state, that is,

$$\nabla\phi_1 \rightarrow 0$$

Therefore, at long times, the monomer concentration is effectively uniformly distributed across the domain (barring small variations due to the degradation reaction). The solids loading determines the volumetric-hydrolysate-to-fiber ratio in the model; hence, it also determines the relative amounts of xylose distributed in the hydrolysate and in the fiber. In the model, sugar yields are calculated based on the concentrations of xylose or xylooligomers in the hydrolysate region and consequently the solids loading can affect the model-predicted yields. Since the hard-to-hydrolyze parameter, $\bar{\alpha}$, is determined by fitting the modeled yields to experimental data, the values of $\bar{\alpha}$ obtained through the fitting process can be influenced by the solids loading. This effect is likely most apparent at high solids loading, when the relative volume of the hydrolysate surrounding the fiber is small and thus a greater proportion of the monomer may be distributed within the fiber pores and thus not counted in the modeled yield. The relationship between $\bar{\alpha}$ and the liquid-to-solid ratio (L:S) is investigated in Section 6.4.1.

The parameter that least significantly influences the maximum xylose yield is k_a, the rate of solid hemicellulose scission. Since k_a is two orders of magnitude larger than the next biggest rate parameter k_b, the rate of aqueous xylooligomer scission, it is unsurprising that it has less impact upon monomer yield, since all solid hemicellulose is hydrolyzed so quickly that small changes to k_a have little impact upon the overall yields of aqueous xylooligomers. The authors' preliminary study of reactor-scale data appears to imply that the fitted k_a may indeed be too fast (Greenwood *et al.* 2015b). Since the sensitivity analysis infers that variations in k_a had little impact upon xylose yield, it may be possible that smaller values of k_a are capable of producing reasonable model outcomes. Two factors that may have influenced the value of k_a obtained through fitting are the input parameters N and m, representing the maximum chain length and solubilization length of hemicellulose, respectively. These two parameters determine how many chains react at rate k_a versus rate k_b. Their influence is further examined in Section 6.4.2.

Consequently, two cases for further investigation have been identified; (i) the relationship between the experimental solids loading and the diffusive transport in the model (and the impact this has upon $\bar{\alpha}$), and (ii) how the fitting procedure determines rate parameters k_a and k_b, given the way that experimentally determined parameters m and N influence the rate of reaction.

6.4.1 Experimental solids loadings and fitting the hard-to-hydrolyze parameter

Often, dilute-acid hydrolysis is not able to experimentally obtain high conversion rates of hemicellulose to monomeric material. As described previously, the "hard-to-hydrolyze" parameter, $\bar{\alpha}$, and the diffusion length, determined by the L:S of biomass in the reaction vessel, effectively present competing mechanisms for modeling the limited yields of xylose observed experimentally. While $\bar{\alpha}$ attributes these limited yields to some aspect of biomass recalcitrance fundamental to the sugarcane bagasse material, high solids loading (i.e., a low L:S ratio) traps a greater proportion of the aqueous oligomers in the pores of the fiber thus limiting the apparent yield measured from the hydrolysate. While it is likely that the combined effect of both phenomena contribute to the modeled oligomer yields, these competing effects must be considered when $\bar{\alpha}$ is obtained from parameter fitting, so that the true value of $\bar{\alpha}$ (independent of the experimental conditions) is obtained in order for the model to be truly predictive.

Therefore, to characterize the change in the modeled yield exclusively due to the liquid-to-solid packing ratio of the reaction, $\bar{\alpha}$ is set to zero, and k_d is similarly set to zero so that the maximum yield is determined by diffusive effects and not degradation reactions. Figure 6.3 demonstrates how the model responds to variations in the solids loading of the reaction vessel. As the volume of liquid in the reactor is reduced, a significant decrease in the maximum obtainable model yield is observed. Furthermore, some simple analysis reveals that the maximum yield is approximately determined by the relationship, max yield $= L/(L+S)$. This confirms two things: (i) the model-predicted yield can be limited by the relative volumes of the fiber and hydrolysate regions of the domain (as influenced by the solids loading in the reaction vessel) due to the concentration smoothing

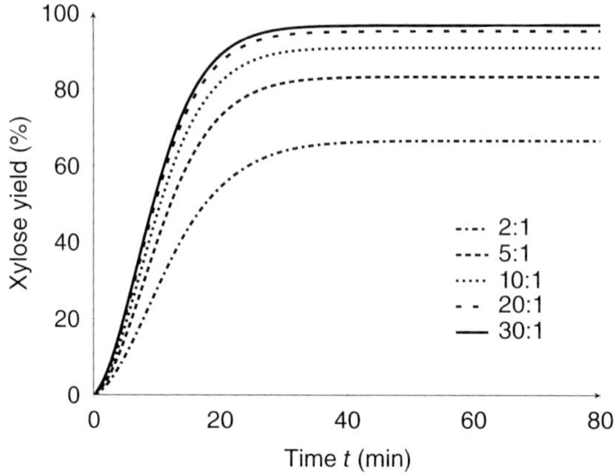

Figure 6.3 Xylose yield curves for different liquid:solid loadings with $\bar{\alpha}=0$, $k_d=0$.

effects of diffusion and (ii) the value of $\bar{\alpha}$ determined from fitting to data may not truly represent the magnitude of the recalcitrant proportion of bagasse due to the reduction in modeled xylose yields that can be attributed to high solids loadings.

It is important to note that this result does not suggest that $\bar{\alpha}$ can be neglected in the model. Experimental results presented in the literature suggest that the maximum obtainable yield is dependent upon the reaction temperature (Zhao *et al.* 2012, Aguilar *et al.* 2002, Zhang *et al.* 2011b, Greenwood *et al.* 2015a). However, if $\bar{\alpha}$ is excluded from the model (as per Figure 6.3) the magnitude of the reduced yields observed due to high solids loading does not change as temperature varies, which contradicts the reality of the experimental results. Consequently, there is certainly still a need for $\bar{\alpha}(T)$ in the model or some other effect that describes the recalcitrance of hemicellulose to dilute-acid hydrolysis as a function of temperature.

The primary concern is that $\bar{\alpha}(T)$ may not represent the true inherent recalcitrance of hemicellulose unless careful consideration has been given to the solids loading used to obtain the experimental data against which the model is fit. Therefore, further work is required across a variety of temperatures for very low solid loadings that approximate (as much as is realistically feasible) the hydrolysis of a bagasse fiber surrounded by an "infinite" amount of hydrolysate. In such a case, limitations imposed upon the yield by diffusion are mitigated, enabling authentic values of the "hard-to-hydrolyze" parameter to be determined.

6.4.2 Hemicellulose chain length characteristics and the parameter fitting of k_a and k_b

The difference $(N-m)$ determines how many hemicellulose chains react with rate k_a, and thus m chains react with rate k_b. In the authors' previous works, the parameters $N=100$ and $m=15$ were chosen to represent the chain length characteristics of hemicellulose. In the literature, O'Hara *et al.* (2011) reported the maximum chain length of hemicelluloses falls in the range $80 \leq N \leq 200$, and Yang and Wyman (2008a) report that chains of up to 30 units long can be measured in solution in various quantities; therefore, $m \leq 30$. Consequently, there is some scope for alternative choices of these parameters.

In Greenwood *et al.* (2015a), values of the parameters k_a, k_b, k_d, and $\bar{\alpha}$ were obtained that produced a good fit between model-predicted and laboratory-scale experimental xylooligomer curves. However, the authors found that this same set of parameters did not enable the model to predict reactor-scale experimental yields of xylan remaining in the fiber (Greenwood *et al.* 2015b). It was hypothesized that one factor that may potentially have contributed to this poor predictive quality was that the rate of solid hemicellulose scission, k_a, was too fast. It was further determined that since $(N-m)$ and m determine how many chains react at rates k_a and k_b, respectively, it may be of interest to investigate how the parameters m and N influence the model results, in particular the time taken to achieve full xylan solubilization and maximum monomer yield.

Figure 6.4 demonstrates how variations to the values of m and N alter the modeled yields of residual xylan remaining in the fiber and monomer yield in the hydrolysate. From Figure 6.4, we observe that neither chain length nor the solubilization length appears to have a significant impact on the time taken for full xylan solubilization (note 100% solubilization is not achieved in these cases because $\bar{\alpha}$ is nonzero). Nor does the maximum chain length, N, appear to have an impact upon xylose yield production. However, the time taken to achieve the maximum xylose yield does appear to be dependent upon the chain length at which hemicellulose solubilizes. The dependence of maximum xylose yield on m and not N may be expected since k_b is two orders of magnitude smaller than k_a, and hence k_b is the limiting rate of xylose production. By increasing the number of chains reacting at the limiting rate, an increase in the timescale of xylose production is intuitive. Furthermore, Figure 6.4 infers that it is specifically the magnitude of m (and not the ratio of m/N) that determines this crucial timescale, since when m is fixed, changing N (and thus the ratio m/N) did not alter the maximum xylose yield.

In our previous work, the parameters k_a, k_b, k_d, and $\bar{\alpha}$ were fit so as to produce a good fit to laboratory-scale data with $N=100$ and $m=15$. Thus, we postulate that if m were to be increased in the model, the rate of aqueous xylooligomer scission, k_b, would need to be faster in order to reduce the timescale of reaction and fit the data. In a scenario where k_b is increased, there may be some latitude for a reduction in k_a since it is the combined influence of these two parameters that ultimately determines the rate at which monomers appear in solution.

Consequently, the parameter fitting was repeated at 110, 125, and 140 °C under the same conditions used in Greenwood *et al.* (2015a) excepting the increase to the value of the solubilization parameter, m, from $m=15$ to $m=30$. The temperature dependence of k_a, k_b, and k_d were determined via an Arrhenius

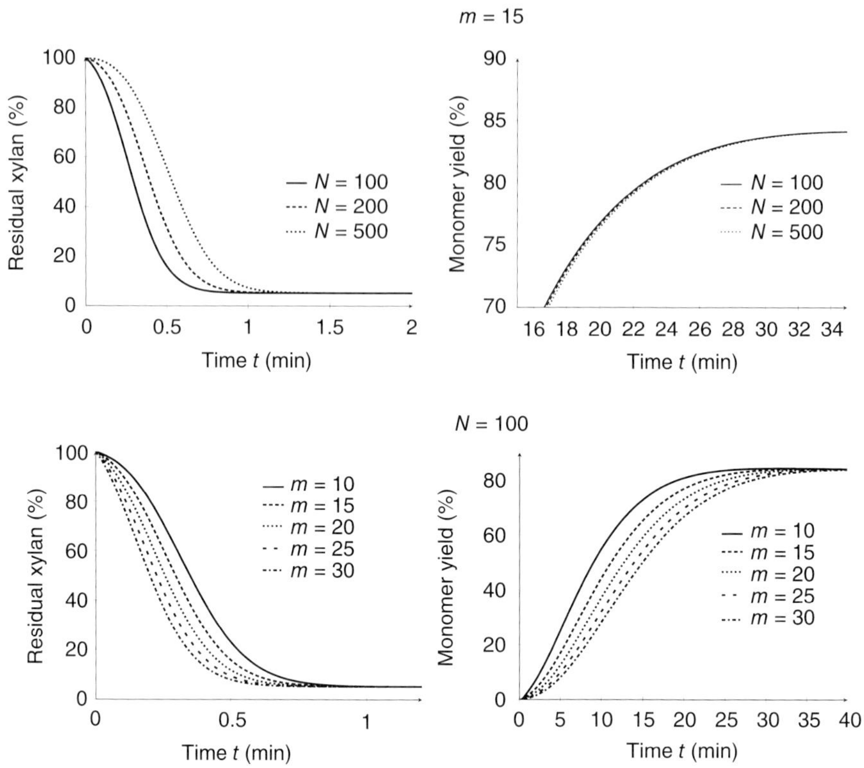

Figure 6.4 Residual xylan and hydrolyzed xylose profiles for variations in the chain length parameter N (fixed m) and solubilization parameter m (fixed N); $k_a = 1.67 \times 10^{-2}$, $k_b = 1.29 \times 10^{-4}$, $k_d = 2.70 \times 10^{-7}$ (m^3/mol/s), and $\bar{\alpha} = 5.21 \times 10^{-2}$.

plot and the "hard-to-hydrolyze" parameter $\bar{\alpha}$ was fit to an exponential curve as per the authors' previous fitting exercise. Figure 6.5 presents a semi-log plot comparing the temperature dependence of the new fitting parameters obtained with $m = 30$ compared with the fitting parameters obtained with $m = 15$ from Greenwood *et al.* (2015a).

Figure 6.5 demonstrates that the refitting increased all of the rate parameters across a majority of the temperature range. The value of k_b was greater with $m = 30$ than $m = 15$ across all temperatures depicted in Figure (6.4); however, the magnitude of this difference was small compared with the change incurred to parameters k_a and k_d. It must be noted that the intersection of the old and new rates of solid hemicellulose scission, k_a, incurred at 140 °C is due to the maximum bound on the parameter space being reached in both fitting exercises. The hard-to-hydrolyze parameter, $\bar{\alpha}$, was largely unaffected by the variation of the solubilization length.

Therefore, although increasing m did to a limited extent result in the expected outcome by marginally increasing k_b, this did not correspond to the desired

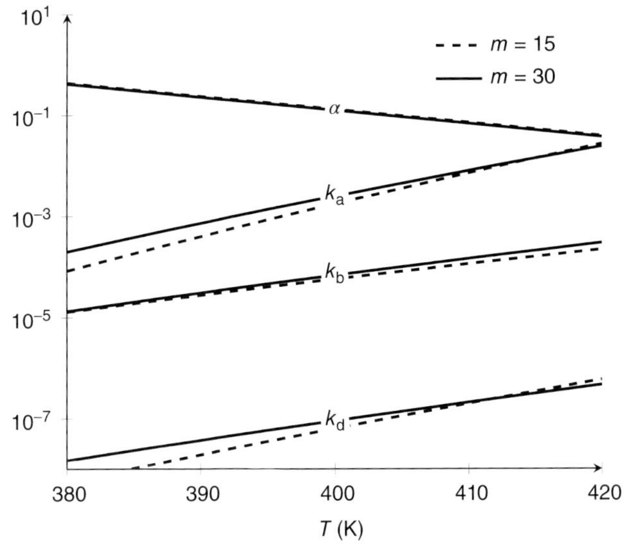

Figure 6.5 Semi-log plot of the temperature dependence of the fitting parameters.

result of simultaneously reducing k_a (and in fact had quite the opposite effect). Consequently, although ascertaining an accurate experimental determination of the solubilization length may contribute to predicting a more accurate determination of k_b, it does not help to address the apparent overprediction of the rate of solid chain scission. Therefore, in order to establish this parameter of concern, k_a, it appears as though the best course of action would be to obtain xylose solubilization data (i.e., experimental data that reflects the reaction rate of solid xylan specifically) and add this to the existing set of experimental xylooligomer yield profiles used to constrain the parameter fitting. This would ensure that both solid-phase and solution-phase kinetics explicitly contribute to determining the parameters of best fit.

6.5 Conclusions

Acid hydrolysis as a pretreatment for bioethanol production has been extensively studied in the literature. A review of the mathematical models and techniques used to characterize either the mechanistic or observed behavior of hemicellulose hydrolysis due to dilute-acid pretreatment has been presented. Although each of these models has their unique advantages and limitations, the authors have championed a polymer degradation model coupled with biphasic kinetics, diffusive mass transport and a consideration of the time evolving porosity of the fibrous material. Such a model has shown significant promise in reproducing and predicting laboratory-scale monomer and short-chain xylooligomer yields.

A sensitivity analysis of this model was conducted, and it was determined that the model does not appear to be sensitive to small random variations in the values of the parameters obtained through data fitting, a first step toward ensuring that the model and its outcomes are reliable and robust. Furthermore, the relationship between the rate and "hard-to-hydrolyze" fitting parameters and pivotal model input parameters was investigated. This study established that the solids loading of bagasse in the reaction vessel, and the experimentally determined solubilization length of hemicellulose, can influence the accuracy and interpretation of the parameters obtained from fitting. Consequently, the sensitivity analysis was used to inform a potential experimental and parameter fitting program that promotes a more robust determination of the fitting parameters by (i) conducting very dilute-acid hydrolysis experiments to isolate a true value of $\bar{\alpha}$ that is not hampered by competing mechanisms that limit xylose yield and (ii) incorporating experimental xylan composition data from the fibrous hydrolysis residue into the data set that constrains the fitting parameter space, so as to better resolve an accurate form of k_a.

Mathematical models of acid pretreatment are almost exclusively used to reproduce the results of a given set of experiments. However, it has been demonstrated here that models can be utilized as a tool for actively directing experimental pathways that are better suited to isolating and characterizing the mechanistic behaviors of the reaction. Consequently, rather than being considered a supplementary tool or accessory for retroactively describing existing experimental results, mathematical modeling has greater utility when implemented in conjunction with the development of experimental programs.

References

Aguilar, R., Ramírez, J., Garrote, G. and Vázquez, M. (2002). Kinetic study of the acid hydrolysis of sugar cane bagasse. *Journal of Food Engineering* **55**:309–318.

Alvira, P., Tomás-Pejó, E., Ballesteros, M. and Negro, M. J. (2010). Pretreatment technologies for an efficient bioethanol production process based on enzymatic hydrolysis: a review. *Bioresource Technology* **101**:4851–4861.

Bhandari, N., Macdonald, D. G. and Bakhshi, N. N. (1984). Kinetic studies of corn stover saccharification using sulfuric acid. *Biotechnology and Bioengineering* **26**:320–327.

Bird, R. B., Stewart, W. E. and Lightfoot, E. N. (2007). *Transport Phenomena*, 2nd edition. Wiley: New York.

Bustos, G., Ramírez, J. A., Garrote, G. and Vázquez, M. (2003). Modeling of the hydrolysis of sugar cane bagasse with hydrochloric acid. *Applied Biochemistry and Biotechnology* **104**:51–68.

Carrasco, F. and Roy, C. (1992). Kinetic study of dilute-acid prehydrolysis of xylan-containing biomass. *Wood Science and Technology* **26**:189–208.

Chandra, R. P., Bura, R., Mabee, W. E., Berlin, A., Pan, X. and Saddler, J. N. (2007). Substrate pretreatment: the key to effective enzymatic hydrolysis of lignocellulosics. *Advances in Biochemical Engineering/Biotechnology* **108**:67–93.

Chang, V. S. and Holtzapple, M. T. (2000). Fundamental factors affecting biomass enzymatic reactivity. *Applied Biochemistry and Biotechnology* **84–86**:5–37.

Cheng, K.-K., Zhang, J.-A., Chavez, E. and Li, J.-P. (2010). Integrated production of xylitol and ethanol using corncob. *Applied Microbiology and Biotechnology* **87**:411–417.

Chen, S., Wang, Z., Fang, K., Fan, G., Peng, W. and Song, J. (2014). Impact of dilute sulfuric acid pretreatment on fermentable sugars and structure of bamboo for bioethanol production. *BioResources* **9**:7636–7652.

Chun, B.-W., Dair, B., Macuch, P. J., Wiebe, D., Porteneuve, C. and Jeknavorian, A. (2006). The development of cement and concrete additive. *Applied Biochemistry and Biotechnology* **129–132**:645–658.

Conner, A. H. (1984). Kinetic modeling of hardwood prehydrolysis. Part I. Xylan removal by water prehydrolysis. *Wood and Fiber Science* **16**:268–277.

Conner, A. H. and Lorenz, L. F. (1986). Kinetic modeling of hardwood prehydrolysis. Part III. Water and dilute acetic acid prehydrolysis of southern red oak. *Wood and Fiber Science* **18**:248–263.

Cussler, E. (1997). *Diffusion: Mass Transfer in Fluid Systems*, 2nd edition. Cambridge University Press: Cambridge.

Diedericks, D., van Rensburg, E. and Görgens, J. F. (2013). Enhancing sugar recovery from sugarcane bagasse by kinetic analysis of a two-step dilute acid pretreatment process. *Biomass and Bioenergy* **57**:149–160.

Fan, L., Lee, Y.-H. and Beardmore, D. H. (1980). Mechanism of the enzymatic hydrolysis of cellulose: effects of major structural features of cellulose on enzymatic hydrolysis. *Biotechnology and Bioengineering* **22**:177–199.

Gírio, F., Fonseca, C., Carvalheiro, F., Duarte, L., Marques, S. and Bogel-Łukasik, R. (2010). Hemicelluloses for fuel ethanol: a review. *Bioresource Technology*, **101**(13):4775–4800.

Garrote, G., Domínguez, H. and Parajó, J. (1999). Hydrothermal processing of lignocellullosic materials. *Holz als Roh- und Werkstoff* **57**:191–202.

Greenwood, A. A., Farrell, T. W., Zhang, Z. and O'Hara, I. M. (2015a). A novel population balance model for the dilute acid hydrolysis of hemicellulose. *Biotechnology for Biofuels* **8**(1):1–13.

Greenwood, A. A., Farrell, T. W., Zhang, Z. and O'Hara, I. M. (2015b). Limitations of a laboratory scale model in predicting optimal pilot scale conditions for dilute acid pretreatment of sugarcane bagasse. In *Proceedings of 37th Conference of the Australian Society of Sugar Cane Technologists*. R. Bruce, ed. Australian Society of Sugar Cane Technologists Ltd: Rockhampton, Queensland.

Hall, C. (1989). *Polymer Materials: An Introduction for Technologists and Scientists*, 2nd edition. Macmillan Education: Basingstoke.

Harvey, M. and Pilgrim, S. (2011). The new competition for land: food, energy, and climate change. *Food Policy* **36**:S40–S51.

Hill, P. J. and Ng, K. M. (1995). New discretization procedure for the breakage equation. *AIChE Journal* **41**:1204–1216.

Hosseini, S. A. and Shah, N. (2009). Multiscale modelling of biomass pretreatment for biofuels production. *Chemical Engineering Research and Design* **87**:1251–1260.

Jacobsen, S. E. and Wyman, C. E. (2000). Cellulose and hemicellulose hydrolysis models for application to current and novel pretreatment processes. *Applied Biochemistry and Biotechnology* **84–86**:81–96.

Jokic, A., Ristic, N., Jaksic, M. M., Spasojevic, M. and Krstajic, N. (1991). Simultaneous electrolytic production of xylitol and xylonic acid from xylose. *Journal of Applied Electrochemistry* **21**:321–326.

Kobayashi, T. and Sakai, Y. (1956). Hydrolysis rate of pentosan of hardwood in dilute sulfuric acid. *Bulletin of the Agricultural Chemical Society of Japan* **20**:1–7.

Kostoglou, M. (2000). Mathematical analysis of polymer degradation with chain-end scission. *Chemical Engineering Science* **55**:2507–2513.

Kostoglou, M. and Karabelas, A. J. (2004). Analytical treatment of fragmentation-diffusion population balance. *AIChE Journal* **50**:1746–1759.

Kumar, R. (2010). Numerical analysis of finite volume schemes for population balance equations. PhD thesis, Otto-von-Guericke University, Magdeburg, Germany.

Kumar, R. and Wyman, C. E. (2008). The impact of dilute sulfuric acid on the selectivity of xylooligomer depolymerization to monomers. *Carbohydrate Research* **343**:290–300.

Lau, C.-S., Thoma, G. J., Clausen, E. C. and Carrier, D. J. (2014). Kinetic modeling of xylose oligomer degradation during pretreatment in dilute acid or in water. *Industrial and Engineering Chemistry Research* **53**:2219–2228.

Lavarack, B., Griffin, G. and Rodman, D. (2002). The acid hydrolysis of sugarcane bagasse hemicellulose to produce xylose, arabinose, glucose and other products, *Biomass and Bioenergy* **23**:367–380.

Lawford, H. G. and Rousseau, J. D. (2002). Performance testing of Zymomonas mobilis metabolically engineered for cofermentation of glucose, xylose and arabinose. *Applied Biochemistry and Biotechnology*: 429–448.

Lee, M. W. and Park, J. M. (2007). One-dimensional mixed-culture biofilm model considering different space occupancies of particulate components. *Water Research* **41**:4317–4328.

Lloyd, T. and Wyman, C. E. (2003). Application of a depolymerization model for predicting thermochemical hydrolysis of hemicellulose. *Applied Biochemistry and Biotechnology* **105–108**:53–67.

Madras, G. and McCoy, B. (1998). Time evolution to similarity solutions for polymer degradation. *AIChE Journal* **44**:647–655.

Maloney, M. T. and Chapman, T. W. (1985). Dilute acid hydrolysis of paper birch: kinetics studies of xylan and acetyl-group hydrolysis. *Biotechnology and Bioengineering* **27**:355–361.

Maloney, M. T., Chapman, T. W. and Baker, A. J. (1986). An engineering analysis of the production of xylose by dilute acid hydrolysis of hardwood hemicellulose. *Biotechnology Progress* **2**:192–202.

McCoy, B. J. and Madras, G. (2001). Discrete and continuous models for polymerization and depolymerization. *Chemical Engineering Science* **56**:2831–2836.

McCoy, B. and Wang, M. (1994). Continuous mixture fragmentation kinetics: particle size reduction and molecular cracking. *Chemical Engineering Science* **34**:3773.

Meng, X., Foston, M., Leisen, J., DeMartini, J., Wyman, C. E. and Ragauskas, A. J. (2013). Determination of porosity of lignocellulosic biomass before and after pretreatment by using Simons' stain and NMR techniques. *Bioresource Technology* **144**:467–476.

Mittal, A., Chatterjee, S. G., Scott, G. M. and Amidon, T. E. (2009). Modeling xylan solubilization during autohydrolysis of sugar maple and aspen wood chips: reaction kinetics and mass transfer. *Chemical Engineering Science* **64**:3031–3041.

Mohagheghi, A., Evans, K., Chou, Y.-C. and Zhang, M. (2002). Cofermentation of glucose, xylose, and arabinose by genomic DNA-integrated xylose/arabinose fermenting strain of Zymomonas mobilis AX101. *Applied Biochemistry and Biotechnology* **98–100**:885–897.

Mohamad, N. L., Mustapa Kamal, S. M. and Mokhtar, M. N. (2015). Xylitol biological production: a review of recent studies. *Food Reviews International* **31**:74–89.

Neureiter, M., Danner, H., Thomasser, C., Saidi, B. and Braun, R. (2002). Dilute-acid hydrolysis of sugarcane bagasse at varying conditions. *Applied Biochemistry and Biotechnology* **98–100**:49–58.

Nigam, P. S. and Singh, A. (2011). Production of liquid biofuels from renewable resources. *Progress in Energy and Combustion Science* **37**:52–68.

O'Hara, I. M., Zhang, Z., Doherty, W. O. and Fellows, C. M. (2011). Lignocellulosics as a renewable feedstock for chemical industry: chemical hydrolysis and pretreatment processes. In: *Green Chemistry for Environmental Remediation*, pp. 505–560. R. Sanghi and V. Singh, eds. John Wiley & Sons: New Jersey.

Rafiqul, I. and Mimi Sakinah, A. (2013). Processes for the production of xylitol - a review. *Food Reviews International* **29**:127–156.

Rayapati, N. P., Bhamidipati, S., Peddieson, J. and Panchagnula, M. V. (2010). Analytical solutions for particulate pipe flows with fragmentation, evaporation, and diffusion. *Mechanics Research Communications* **37**:604–610.

Reid, R. C., Prausnitz, J. M. and Sherwood, T. K. (1977). *The Properties of Gases and Liquids*, 3rd edition. McGraw-Hill: New York.

Saeman, J. F. (1945). Kinetics of wood saccharification, hydrolysis of cellulose and decomposition of sugars in dilute acid at high temperature, *Industrial and Engineering Chemistry* **37**:42–52.

Samala, A., Srinivasan, R., Yadav, M. P., Kim, T.-J. and Prewitt, L. (2012). Xylo-oligosaccharides production by autohydrolysis of corn fiber separated from DDGS. *Bioresources* **7**:3038–3050.

Simha, R. (1941). Kinetics of degradation and size distribution of long chain polymers. *Journal of Applied Physics* **12**:569–578.

Solomon, B. D., Barnes, J. R. and Halvorsen, K. E. (2007). Grain and cellulosic ethanol: history, economics, and energy policy, *Biomass and Bioenergy* **31**:416–425.

Sun, Y. and Cheng, J. (2002). Hydrolysis of lignocellulosic materials for ethanol production: a review. *Bioresource Technology* **83**:1–11.

Tanaka, M., Ikesaka, M. and Matsuno, R. (1988). Effect of pore size in substrate and diffusion of enzyme on hydrolysis of cellulosic materials with cellulases. *Biotechnology and Bioengineering* **32**:698–706.

Tsuchida, J. E., Rezende, C. A., de Oliveira-Silva, R., Lima, M. A., d'Eurydice, M. N., Polikarpov, I. and Bonagamba, T. J. (2014). Nuclear magnetic resonance investigation of water accessibility in cellulose of pretreated sugarcane bagasse. *Biotechnology for Biofuels* **7**:127.

Um, B.-H. and Bae, S.-H. (2011). Statistical methodology for optimizing the dilute acid hydrolysis of sugarcane bagasse. *Korean Journal of Chemical Engineering* **28**:1172–1176.

Weissberg, H. L. (1963). Effective diffusion coefficient in porous media. *Journal of Applied Physics* **34**:2636–2639.

Winkelhausen, E. and Kuzmanova, S. (1998). Microbial conversion of D-xylose to xylitol. *Journal of Fermentation and Bioengineering* **86**:1–14.

Yang, B. and Wyman, C. E. (2008a). Characterization of the degree of polymerisation of xylooligomers produced by flowthrough hydrolysis of pure xylan and corn stover with water. *Bioresource Technology* **99**:5756–5762.

Yang, B. and Wyman, C. E. (2008b). Pretreatment: the key to unlocking low-cost celluosic ethanol. *Biofuels, Bioproducts and Biorefining* **2**:26–40.

Zhang, J., Ma, X., Yu, J., Zhang X. and Tan, T. (2011a). The effects of four different pretreatments on enzymatic hydrolysis of sweet sorghum bagasse. *Bioresource Technology* **102**:4585–4589.

Zhang, R., Lu, X., Sun, Y., Wang, X. and Zhang, S. (2011b). Modeling and optimization of dilute nitric acid hydrolysis on corn stover. *Journal of Chemical Technology and Biotechnology* **86**:306–314.

Zhang, T. C. and Bishop, P. L. (1994). Evaluation of tortuosity factors and effective diffusivities in biofilms. *Water Research* **28**:2279–2287.

Zhao, X., Zhou, Y. and Liu, D. (2012). Kinetic model for glycan hydrolysis and formation of monosaccharides during dilute acid hydrolysis of sugarcane bagasse. *Bioresource Technology* **105**:160–168.

Zhou, W., Schuttler, H.-B., Hao, Z. and Xu, Y. (2009a). Cellulose hydrolysis in evolving substrate morphologies I: a general modeling formalism. *Biotechnology and Bioengineering* **104**:261–274.

Zhou, W., Hao, Z., Xu, Y. and Schuttler, H.-B. (2009b). Cellulose hydrolysis in evolving substrate morphologies II: numerical results and analysis. *Biotechnology and Bioengineering* **104**:275–289.

Ziff, R. M. and McGrady, E. D. (1986). Kinetics of polymer degradation. *Macromolecules* **19**:2513–2519.

CHAPTER 7

Hydrothermal liquefaction of lignin

Kameron G. Dunn and Philip A. Hobson

Centre for Tropical Crops and Biocommodities, Queensland University of Technology (QUT), Brisbane, Australia

7.1 Introduction

Oil, coal, and natural gas have provided a cheap and plentiful source of energy underpinning economic development since the industrial revolution. Though long before the dependency on these nonrenewable resources, man obtained energy and materials from biomass, a resource that is currently undergoing resurgence as a sustainable alternative to fossil fuels.

Biomass is a globally available resource and when managed sustainably can offer the prospect of reduced volatility in carbon-based energy markets and mitigate against carbon dioxide emissions linked to global warming. To provide a viable alternative to fossil fuels (at least in the shorter term), biomass must be transformed into products that are compatible with existing processing infrastructure and end-use technologies, as well as being produced on an equivalently large scale.

Second-generation biofuels utilizing lignocellulosic biomass such as bagasse, as the renewable feedstock, have the potential to produce fuels on a large enough scale to meet a significant fraction of demand while avoiding the use of feedstocks that compete with food production. Such biorefineries are expected to generate significant amounts of lignin or black liquor. Lignin is an amorphous branched aromatic polymer representing up to 22 wt% of the dry fiber component of bagasse, a waste product produced from the sugar cane industry. This represents a significant renewable resource of an aromatic product.

Lignin when depolymerized by a hydrothermal liquefaction process produces a range of phenolics. Such phenolics can be extracted as a monomer/oligomer phenolic-rich fraction (purified if required), and used as precursors or substitutes for epoxy or phenol–formaldehyde resins. Thus, production of phenolics has the potential to add significantly greater value to the original bagasse feedstock than conventional options such as bagasse combustion for the generation of export power. Furthermore, it is proposed that hydrothermal liquefaction provides an energy efficient economically favorable pathway for processing lignin to renewable fuels and commodity chemicals.

Sugarcane-Based Biofuels and Bioproducts, First Edition. Edited by Ian M. O'Hara and Sagadevan G. Mundree.

Hydrothermal liquefaction (sometimes referred to as simply, liquefaction) is a thermochemical process used for the depolymerization of both lignin and lignocellulosic material via hydrolysis and thus usually accomplished in an aqueous environment. As such, hydrothermal liquefaction does not require the drying of the lignocellulosic material before thermochemical conversion as generally required for other thermochemical processes, that is, gasification and pyrolysis.

Hydrothermal liquefaction requires significantly higher pressures and longer reaction times compared with gasification and pyrolysis processes. Table 7.1 summarizes the operating conditions required for the varying thermochemical technologies detailed.

Similar to pyrolysis, hydrothermal liquefaction also produces a multitude of products. The products contained in the biocrude produced from hydrothermal liquefaction of lignocellulosic material are generally less oxygenated and less reactive when compared with pyrolysis products, but are still not a petroleum analog given the degree of oxygenation (Elliott *et al.* 2015).

The conversion of lignin into a liquid via hydrothermal liquefaction has been studied extensively by many researchers using both basic and acid media and under supercritical and subcritical solvent conditions. Fang *et al.* (2008) proposed that the decomposition of lignin in water at elevated temperatures and pressures should be significantly different from that of pyrolysis because of additional hydrolysis reactions catalyzed by H^+ and OH^- ions and the associated phase behavior. The hydrolysis of lignin in either an acid or basic media has received ongoing consideration given the fewer simple degradation products obtained (Thring 1994).

Similar to oils, not all lignin is the same. The chemical structure of lignin is complex with structural variation evident between the different biomass lignocellulosic material, the environmental factors experienced by the plant, the morphological origin of the isolated lignin, and more importantly, the fractionation process used to recover the lignin (i.e., the delignification process).

Lignin does not exist in plant tissue as an independent polymer but it is bonded with cellulose and hemicellulose (i.e., carbohydrates) (Buranov and

Table 7.1 Summary of operating conditions for thermochemical processes.

	Temperature, °C	Pressure, bar	Reaction time	Drying	Solvent	Product
Gasification	600–1200	1–40	Seconds	Usually necessary	No	Solids and gases (syngas + CH_4)
Pyrolysis	400–630	1–5	Seconds to Minutes	Yes	No	Liquid (bio-oil), solids, and gases
Liquefaction	250–350	100–300	Minutes to hours	No	Yes	Liquid (biocrude + chemicals)

Mazza 2008). Lignin and carbohydrates are attached to each other by benzyl ether, benzyl ester, glycosidic, and acetal type bonds (Buranov and Mazza 2008).

Lignin from nonwoody or herbaceous crops (e.g., sugarcane, corn) can consist of three main monolignol monomers, methoxylated to various degrees: *p*-coumaryl alcohol, coniferyl alcohol, and sinapyl alcohol. These monolignols are present in lignin in the form of the phenylpropanoids *p*-hydroxyphenyl (H), guaiacyl (G), and syringyl (S) (Grabber 2005, Buranov and Mazza 2008, Hatfield *et al.* 2009). Figure 7.1 illustrates these three main monolignol structures.

For softwood lignin, coniferyl alcohol is the major monolignol, accounting for more than 95% of the lignin structure (Gosselink 2011). For hardwood species, both coniferyl alcohol and sinapyl are abundant, while the monolignol *p*-coumaryl alcohol is found predominately in herbaceous crops such as cereal straws and grasses (i.e., sugarcane). Grass lignins consist of lignin cores and peripheral units. The lignin cores are essentially lignins of the guaiacyl-syringyl type, while the peripheral units consist of *p*-hydroxycinnamic acid and ferulic acid groups that are bonded by ester linkages mostly to hydroxyl groups at C-γ in the lignin cores (Gratzl Josef and Chen 1999). Figure 7.2 is a representation of a possible polymer network of lignin and lignin linkage structures.

The most abundant interunit linkage or substructure in all lignin is the β-aryl ether bond (or the β-O-4 bond), and it is estimated that for softwoods this type of linkage represents up to 48% and for hardwoods 60% of the linkages between the monolignols (Sun *et al.* 2003). Figure 7.3 shows the common primary interunit linkages in native lignin.

The β-O-4 bond is the most easily chemically cleaved linkage in lignin forming the basis for various analytical techniques and chemical pulping processes. Other linkages such as the carbon–carbon bonds (5-5′, β-β, β-1, and β-5) and the α-O-5 bond linkage are more resilient to chemical and thermochemical degradation.

The sole goal for the depolymerization of lignin is for the recovery of monomeric phenolic products, which have application in a variety of industrial processes. During the process of hydrothermal liquefaction, lignin is converted

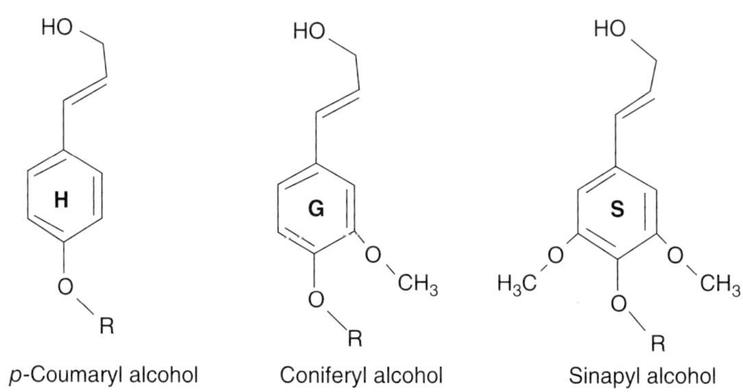

Figure 7.1 The monolignols.

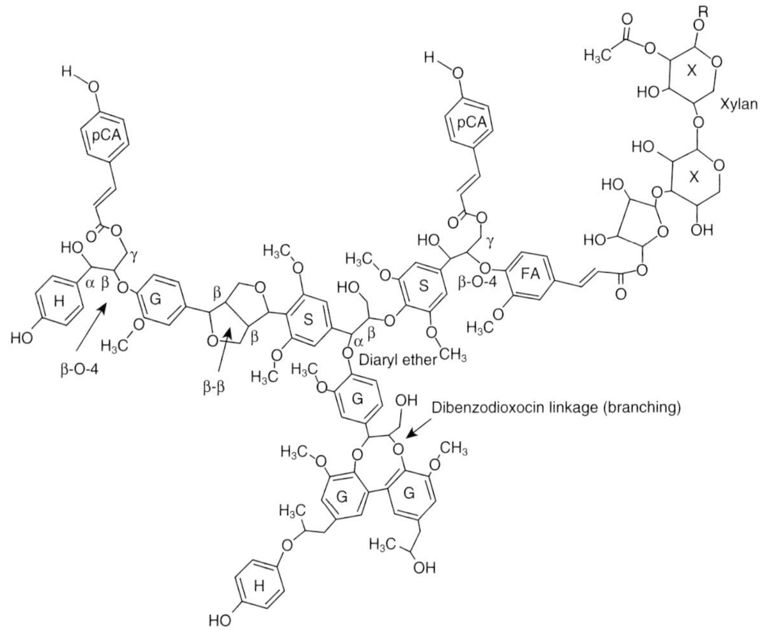

Figure 7.2 Schematic representation of lignin.

to monomeric phenolic products that can be extracted via conventional solvent extraction processes as a viscous liquid crude phenolic oil product. Gaseous products are also produced during hydrothermal liquefaction, as is char and a modified or residual solid lignin product. From a literature review undertaken by Dunn (2014) it was identified that the depolymerization via hydrothermal liquefaction of lignin to obtain such monomeric phenolic products suffers from

- the generation of a wide spectrum of chemical products, making product recovery and refining difficult;
- low yields of derived phenolic oils (<23.5% on dry lignin) or total monophenols (typically < 10% on dry lignin);
- the occurrence of competing repolymerization mechanisms, thus making the lignin structure less amenable to monomer production.

An evident and concerning feature of past research, that has somewhat inhibited improvements for the issues highlighted earlier, is the lack of an understanding of the initial structural attributes of the lignin, and how this structure is modified during liquefaction.

In recent years, extensive research has been undertaken (Sun *et al.* 2011, Vallejos *et al.* 2011, Wen *et al.* 2012, 2013a, b, del Río *et al.* 2015) on both the preparation techniques for lignin samples and in identifying structural attributes via analytical techniques such as two-dimensional nuclear magnetic resonance (NMR) (e.g., in particular heteronuclear single quantum correlation (HSQC)

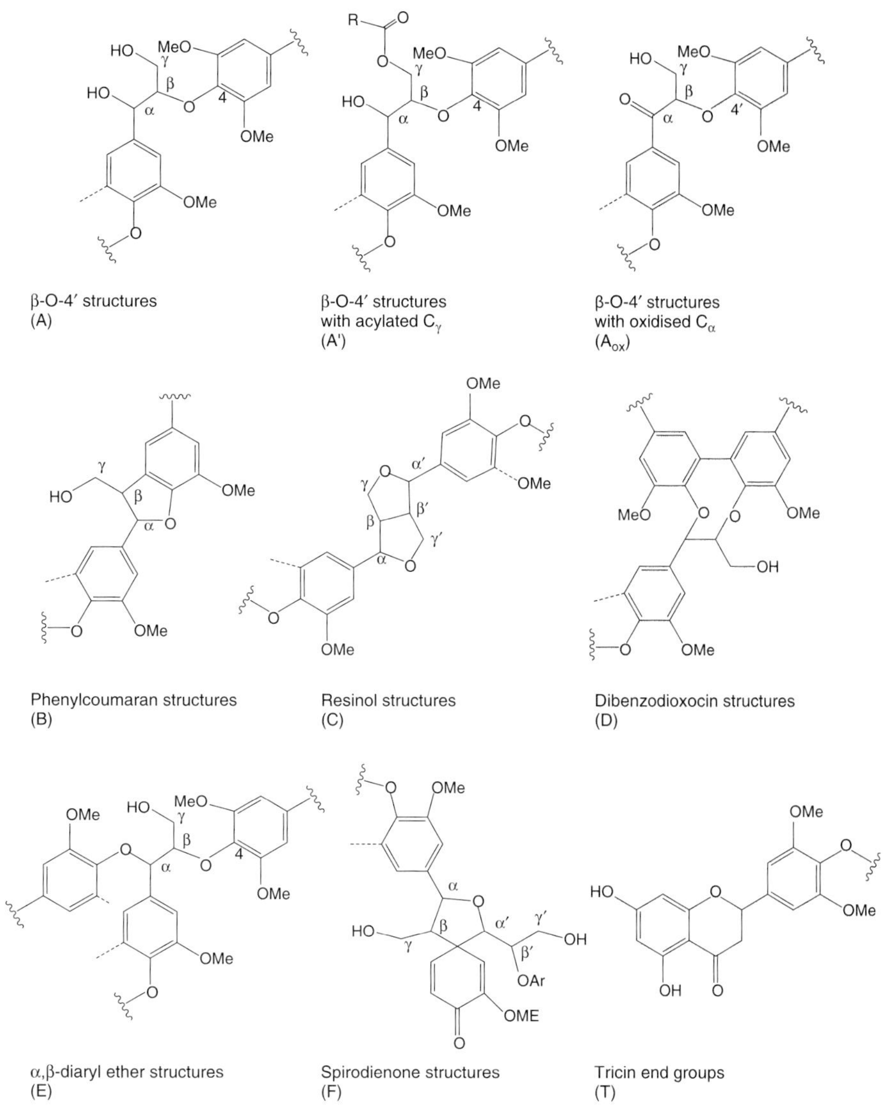

Figure 7.3 Primary interunit linkages.

experiments). These recent developments have enabled reliable predictions of the type of structures present within the native lignin samples and shown how these structures are modified during delignification or degradation processes, including, for example, the hydrothermal liquefaction of sugarcane soda lignin (Dunn 2014). It is these same analytical techniques that now provide a greater understanding of the chemical mechanisms at play during lignin hydrothermal liquefaction and hold the promise of obtaining greater yields of value-adding products recovered from lignin depolymerization.

7.2 A review of lignin alkaline hydrolysis research

Thring (1994) first studied the alkaline depolymerization of Alcell lignin (organosolv lignin) in a batch stirred reactor operating at target temperatures of 200–340 °C. Once the target temperature was achieved (30–45 min of heating), an extended reaction time of between 15 and 90 min was investigated. A 6 wt% lignin sample was reacted in an aqueous solution at varying concentrations of NaOH. Using a reaction ordinate (Ro) as presented in Equation 7.1, Thring (1994) assessed both the combined influence of reaction time and temperature into a single parameter, that is, a reaction severity parameter.

$$\mathrm{Ro} = \int_0^t \exp\left(\frac{T - 100}{14.75}\right)\mathrm{d}t \tag{7.1}$$

where t is reaction time in minutes and T is the reactor temperature in °C, at time t.

It was shown by Thring (1994) that at a high reaction ordinate (high reaction severity; 340 °C, 90 min), approximately 40 wt% of the lignin was recovered as an alkaline insoluble fraction (solid residue char), while at a low reaction ordinate (low reaction severity; 200 °C, 15 min) solid residue char was typically 8 wt%. The term char used by Thring (1994) referred to material that was filtered from the alkaline mixture before reducing the pH to precipitate and recover unreacted lignin. Reported oil yields by Thring (1994) using methylene chloride extraction peaked at 10 wt% of the initial lignin charge for a 0.5 wt% NaOH solution, although yields were more typically 7 wt%. A 5 wt% oil yield was reported when alkali was not added.

Monomer proportions reached a maximum of 4.4 wt% of the original lignin charge at a solution concentration of 0.5 wt% NaOH. Phenolic monomer products ranged from guaiacol and syringol at low severity reaction conditions to catechol and its derivatives under higher severity conditions. Although phenol yields reported by Thring (1994) were low, they were maximized at higher severity conditions reaching a value of 0.2 wt% lignin. Process conditions that maximized phenol yields gave corresponding catechol and guaiacol yields of 0.91 and 0.09 wt% of lignin, respectively. Syringol was present (1.05 wt% on lignin) only for lower-severity reactions. Char collected after filtering and without acidification of the reaction products revealed that char production reduced with increasing alkali concentrations. Char reduction was documented by Thring (1994) to be associated with a reduction in reploymerization reactions of the lignin fragments. Increasing concentrations of NaOH from 0 to 4 wt% increased liquid yields from 7 to 30 wt%, respectively.

Johnson *et al.* (2002) describes a patented process (Shabtai and Zmierczak 2001, Zmierczak and Miller 2006, 2008) for the conversion of lignin into a hydrocarbon for blending with gasoline. The initial stage of this process was a base-catalyzed depolymerization (BCD) process using a 1–5 wt% concentration of NaOH as a catalyst. The lower concentration of NaOH was preferred so as to reduce the associated cost of recovering the catalyst. Variables such as reaction time, pressure, type of alkali, alkaline concentration, and solvent (methanol) used were assessed.

It was claimed that high yields of depolymerized lignin of between 60 and 80 wt% (Zmierczak and Miller 2008), without methanol and using 5 wt% NaOH, were obtained at a temperature of 290 °C, residence times of 30 min, and with a reactor pressures of between 110 and 172 bar. The yield of depolymerized lignin was defined by Equation 7.2.

$$\text{Yield} = \frac{M_{\text{dl}}}{M_{\text{l}}} \times 100 \tag{7.2}$$

where M_{dl} is mass collected following acidification and solvent extraction of both the aqueous phase and the solid residue and M_{l} is the original mass of the lignin reactant

Johnson *et al.* (2002) noted that the aromatic methoxyl substituents of lignin are degraded during the BCD process via the loss of the methyl group, thus leading to the formation of hydroxyl groups attached to the aromatic structure of the monolignols. The largest identifiable component of the BCD process was found to be catechol. GC–MS spectra indicated that the quantities of phenol relative to catechol were small. Johnson *et al.* (2002) stated that elemental analysis indicated that all oxygen-containing functional groups in the lignin side chain were lost apart from a few that were converted to carbonyls.

Miller *et al.* (2002) reported on a collaborative study with the National Renewable Energy Laboratory (NREL) and the University of Utah to develop a scaled-up process for the conversion of lignin (Alcell mixed hardwoods) into valuable high-octane fuel additives based on the work of Shabtai and Zmierczak (2001). Miller *et al.* (2002) extensively assessed the impact of aqueous NaOH (290–330 °C) on the BCD process and showed that the primary factor governing the extent of lignin conversion is the Na to lignin ratio and that the NaOH concentration is a secondary factor. Maximum lignin conversions (i.e., minimum yields of insoluble lignin material) were achieved at NaOH-to-lignin molar ratios of 1.5–2. Such proportions represent quantities more appropriate to stoichiometric reactions rather than catalytic reaction processes. The effect of the water-to-lignin ratio for optimum Na-to-lignin ratios was shown to be minimal for concentrations of NaOH in the range of 1–20%.

Miller *et al.* (2002) found that an increase in reaction temperature resulted in a decrease in the yield of insolubles[1] (15% insolubles at 290 °C, 10% at 310 °C, and 8% at 330 °C). Increasing the NaOH concentration was found to

[1]Miller *et al.* defines the insoluble portion as the solid material remaining after acid precipitation, subsequent filtration, tetrahydrofuran (THF) extraction, and drying. The THF soluble component is considered an oligomer-rich fraction.

have an effect similar to that of temperature increase (34% insolubles at 1% NaOH (2 h), 23% at 5% NaOH (1 h), and 15% at 10% NaOH (0.4 h)). This NaOH-concentration-dependent trend was found to be modified by the reaction times; with extended reaction times, the reduction in the recovered insolubles trended to a constant value.

For a set of optimized experiments using NaOH, the maximum reported oil yield by Miller *et al.* (2002) of a phenolic monomer-rich oil fraction, extracted using diethyl ether was 23%, the yield of insolubles was 31%, and the oligomer-rich fraction recovered from the extraction of the precipitated solids was 29%.

The addition of CaO with the NaOH catalyst was found by Miller *et al.* (2002) to improve the reaction kinetics. The CaO was considered to assist the reaction by neutralizing the weak acids generated (e.g., the phenols and CO_2) and maintaining a high pH without inhibiting the degree of lignin conversion. The authors comment that the combination of CaO (i.e., $Ca(OH)_2$ when dissolved in water) and Na_2CO_3 (soda ash) has the same catalytic effect but offers cost savings compared with NaOH on a per moles of sodium basis.

Miller *et al.* (2002) also detailed how oxides of magnesium (i.e., MgO, Magox®) influence the conversion of lignin and insoluble proportions. It was shown that a combination of Magox and NaOH achieved a low insoluble lignin residue of 24%, outperforming all other combinations of catalysts at similar reaction conditions.

A process of recycling unconverted lignin was studied (Miller *et al.* 2002) and indicated that conversion yields would improve, but overall, less lignin would be converted than if the NaOH-to-lignin ratio was carefully controlled during the first pass. Extending the reaction time from 1 to 2 h did not improve the overall conversion yield.

Vigneault *et al.* (2006) assessed the BCD reaction of lignin derived from steam fractionated wood (*Populus tremuloides*) in a continuous flow reactor. Detail is presented (Vigneault *et al.* 2006) on the material balance of five extracts separated from the reaction mixture. These extracts included an aqueous extract containing lower-molecular-weight organics (methanol, acetic acid, and formic acid), a monomer-rich extract fraction (BCD oil), an oligomer-rich extract, an ether insoluble product (solid waste fraction), and a gaseous fraction. The authors (Vigneault *et al.* 2006) present typical percentage yields by mass (of the original lignin) in the BCD component products. These are reproduced in Table 7.2 for a BCD reaction with a lignin content of 10 wt%, reaction residence time of 5 min, NaOH-to-lignin mole ratio of 1.5 (lignin MW is 184.1 g/mol), and reaction temperatures of 290 and 315 °C. It is also noted from Table 7.2 that the higher temperature of 315 °C generated the least amount of solid residue.

Analysis by Vigneault *et al.* (2006) via ion-exclusion chromatography of the aqueous extract recovered from a BCD reaction operating at 315 °C revealed that methanol (59.5%), acetic acid (17.5%), and formic acid (18.5%) were the main

Table 7.2 Mass percentage of extracts from BCD (Vigneault *et al.* 2006).

Extract	Reaction temperature	
	290 °C	**315 °C**
Small organics (aqueous phase), %	7.0	11.4
Acid gas (CO_2), %	4.0	6.2
Monomer rich fraction, %	10.0	15.3
Oligomers rich fraction, %	38.3	51.6
Solid waste fraction, %	42.8	15.1
Total balance, %	102.0	99.7

organic components in the aqueous phase (6.8, 2, and 2.1 wt%, respectively, on lignin). The authors (Vigneault *et al.* 2006) state that methanol was most likely derived from the methoxyl substituent on the monolignol aromatic structure, while formic acid and acetic acid originated from the rupture of the aliphatic side chain of the lignin structure. The authors (Vigneault *et al.* 2006) also state that there is no evidence that BCD reactions cleave aromatic rings.

Both the monomer and oligomer-rich fractions were analyzed with reverse-phase high-performance liquid chromatography (HPLC) fitted with a UV detector (Vigneault *et al.* 2006). From the analysis, it was identified that approximately 37% (at 315 °C) to 45% (at 290 °C) of the oligomer-rich extract were identified as monomers with the remainder assumed to be oligomers (dimers and trimers). Similarly, the oligomer-rich extract was also shown to contain some monomers (1.4 wt% yield on lignin). Optimum monomer yields were reported at 9.8 wt% of the initial lignin, for a lignin feed concentration of 5 wt%.

The chemical constituents of the monomer-rich fraction extract were reported for varying reaction conditions for experiments designed to optimize monomer yield. The reaction variables assessed were reaction temperature (310–330 °C), reaction time (5–15 min), lignin content (5–15 wt%), molar ratio of NaOH to lignin (1–3), and pressure (122–150 atm). For low-severity reaction conditions (i.e., 310 °C and 5 min), vanillin (0.6 wt% yield on lignin) and syringol (2.1 wt% yield on lignin) were the most prevalent monomers. For moderate-severity reaction conditions, catechol (2.4 wt%), guaiacol (1.0 wt%), 4-methyl catechol (1.2 wt%), and 3-methoxy catechol (0.8 wt%) were the most prevalent monomers. Phenol at 1 wt% was the only monomer present in more than trace quantities at the high-severity (330 °C and 15 min) reaction conditions.

Vigneault *et al.* (2006) attribute the variation in chemical composition to the understanding that at low severity, the BCD reaction first breaks ether linkages and reduces the propyl side chains to one of two carbons, which is also accompanied by the formation of aldehydes and ketone structures. These structures are then hydrogenated to methyl and ethyl groups. For moderate-severity conditions, subsequent demethoxylation of the lower-severity products results in the formation of catechol and its derivatives. At higher-severity conditions, hydroxylation occurs to produce phenol.

Vigneault *et al.* (2006) deduced that the mass fraction of lignin in the feed had the greatest influence on the monomer yield, with higher lignin concentrations reducing the monomer yield. Furthermore, it was suggested that this behavior was a result of the competition between depolymerization and condensation reactions. To maximize monomer yield while also minimizing operating costs, the authors (Vigneault *et al.* 2006) suggested 10 wt% lignin as the most favorable of the concentrations tested (5, 10, and 15 wt% lignin).

From process simulation (using Aspen Plus) using a BCD reaction temperature of 315 °C, Vigneault *et al.* (2006) demonstrated that approximately 90% of the energy required for the BCD process was attributed to heating of the reaction products. This energy could be recovered to provide heat for coprocessing in an integrated biorefinery scenario. The remainder of the energy requirement was attributed primarily to maintaining the endothermic (950 kJ/kg of lignin) BCD reaction.

A CHO analysis was undertaken (Vigneault *et al.* 2006) and the heat of combustion calculated for the monomer-rich, oligomer-rich, and solid waste (insoluble) extracts. The monomer-rich extract had an estimated oxygen content of 32% (25.6 MJ/kg), the oligomer-rich fraction 22% (30.9 MJ/kg), and the solid waste 24% (27.8 MJ/kg).

Roberts (2008) published a thesis describing homogeneous and heterogeneous catalyzed hydrolysis of lignin during BCD. The lignin was obtained from the organosolv pulping of beech wood (a hardwood species). Roberts (2008) determined that the base-catalyzed liquid-phase hydrolysis of lignin depended primarily on the process conditions of temperature, pressure, catalyst concentration, and residence time. It was identified that depolymerization of lignin was limited due to consecutive addition and condensation reactions of the phenolic monomers to oligomers. Monomer yields were found to increase linearly with the base catalyst (NaOH) concentration; the formation of oil[2] increased strongly up to a base catalyst concentration of 2.5 wt% beyond which the rate of increase of oil yield diminished with higher catalyst concentrations.

For the BCD of organosolv lignin in a continuous reactor, oil yields peaked at 300 °C for the temperature range tested, that is, 240–340 °C (Roberts 2008). A residence time of 4.2 min showed the maximum oil yield, where residence

[2] "Oil" in the context of hydrothermal liquefaction refers to the product recovered from the aqueous reaction mixture following acidification and filtration to remove unconverted lignin from the reaction mixture. The remaining aqueous mixture is then extracted with an organic solvent to extract an oily product, which is then subsequently heated to remove the organic solvent and recover the oil extract.

times of up to 15 min were investigated. A pressure of 250 bar produced the greatest oil yield when pressures between 250 and 315 bar were tested. The highest yield determined across all conditions (pressure, temperature, and residence time) was 23.3 wt% of the original lignin. This was obtained from a 10 wt% lignin and 4 wt% NaOH aqueous solution. At this maximum oil yield, only 28% of the oil (i.e., 6.5 wt% lignin) was found to be monomer units. Increased residence times reduced the proportion of monomers and correspondingly increased the amount of oligomers in the recovered oil component. Increased concentrations of NaOH resulted in a corresponding linear increase in the yield of monomers. Decreased lignin concentration in solution also improved the monomer yield. At 2.5 wt% lignin in solution, the monomer yield was 9 wt% of the initial lignin, representing 43% of the oil extract designated as monomers.

At a reaction time of 40 min in a batch reactor operating with conditions similar to that of the continuous reactor (i.e., 300 °C, 10 wt% lignin, 4 wt% NaOH), Roberts (2008) found that the monomer fraction of the oil was mainly composed of catechol and its derivatives representing 65% of the known monomeric constituent fraction. The monomer yield was approximately 1 wt% of the initial lignin and the total oil yield about 23.5 wt% (i.e., 3.9% identified monomers in oil). Roberts (2008) suggests that without other means to reduce repolymerization this yield represents a maximum using the BCD process. The pressure associated with these optimum conditions was not detailed, though it would be expected to correspond to the saturation vapor pressure of water at 300 °C, that is, 85 bar. The batch autoclave reactors were set into a preheated aluminum block to ensure rapid heating rates.

In order to improve monomer yields, Roberts (2008) investigated the addition of boric acid as a blocking or capping agent to suppress addition and condensation (i.e., repolymerization) reactions subsequent to the formation of the primary products (i.e., the monophenols). Batch autoclaves were also placed into a preheated aluminum block for these trials. For a reaction temperature of 300 °C and reaction time of 40 min, the oil yield was 36 wt% compared with 23.5 wt% for the BCD process and 6 wt% for the noncatalyzed hydrothermal reaction. Higher ratios of boric acid to lignin increased the oil yield and reduced the number of products contained in the oil. However, these products tended to have higher molecular weights compared with those derived from the BCD process, with monomer components estimated to be only 2.8 wt% of the derived oil or 1 wt% of the initial lignin weight.

Roberts (2008) demonstrated that adjustment of the pH by the addition of NaOH (with the boric acid) increased oil yields to 52 wt% (pH = 13.0) and reduced the range of compounds contained in the product oil. Even when the concentration of NaOH was equal to that used in the BCD process with alkali alone (i.e., 4 wt%), oil yields were still slightly higher at 29 wt% with the addition of boric acid. Hence, it was shown that the capping agent increased lignin depolymerization and oil yield. ^{11}B NMR data presented by Roberts *et al.*

(2011) indicated the presence of coordinated boric esters in the oil extracts thus demonstrating that the addition of boric acid blocks the formation of higher-molecular-weight products by capping the phenolic hydroxyl groups and thereby preventing repolymerization. Furthermore, it is stated (Roberts *et al.* 2011) that phenolic hydroxyl groups are the primary active groups involved in the addition and condensation reactions, which lead to polymerization of the primary monomers. Based on LDI-TOF-MS results, the addition of boric acid reduces the complexity of the product oil produced.

Roberts *et al.* (2011) documented yield data on the products that remained in the aqueous phase, which were not recovered via solvent extraction. For one documented BCD reaction (10 wt% lignin, 4 wt% NaOH at 300 °C) the yield of solid residue following acidification of the BCD products was 80.2 wt% and the oil content extracted from the aqueous phase 11.5 wt%. For this particular reaction, the remaining aqueous soluble products not extracted equated to a total organic yield of 7.7 wt%. The authors (Roberts *et al.* 2011) reported that the remaining aqueous phase contained methanol, ethanol, and *o*-hydroxy phenolic compounds (i.e., most likely catechol), and that based on the close mass balance, aqueous phase reforming did not play a role under the conditions investigated. It is also evident from Roberts *et al.* (2011) that THF was used to dissolve the remaining solid residue in order to determine the oligomer proportion in the solid residue. The insoluble material following THF extraction is composed of ash and coke.

Roberts *et al.* (2011) postulate a mechanism for lignin depolymerization via the β-O-4 bond and the subsequent monomer product oligomerization. For the degradation of the β-O-4, it was postulated that the sodium cation and the hydroxide ion participate, with the sodium cation catalyzing the reaction by forming cation adducts with lignin that subsequently polarize the ether bond and reduce the energy required for heterolytic bond cleavage. Monomer polymerization is postulated to occur through phenolic and catecholic monomers that are negatively charged and in equilibrium with their corresponding phenolate and carbanion species.

Lavoie *et al.* (2011) describe the BCD of a lignin solution produced from base catalyzed steam treatment of agricultural hemp and softwood material. The lignin solution was adjusted to have a lignin content of 10 wt% and a NaOH content of 5 wt%. The experiments were carried out for 45 min at temperatures of 300, 310, 320, and 330 °C. The batch reactor was initially pressurized to 35 bar with nitrogen. Diethyl ether was used to extract the oil components from the aqueous BCD lignin product after the solution was first acidified to a pH of 1.7 with HCl. As with the approach adopted by Roberts *et al.* (2011), the solid residue was extracted with THF to recover oligomers (dimers and trimers).

The results (Lavoie *et al.* 2011) revealed that the yield of oil (a monomer-rich extract) increased with temperature (300–330 °C) from 8.4 to 10.9 wt% for hemp lignin and from 5.5 to 10.0 wt% for softwood lignin. The percentage of the oligomers reduced with increasing temperature (300–330 °C) from 56 to 47 wt% for hemp and from 70.3 to 60 wt% for softwood lignin.

Lavoie *et al.* (2011) also described the GC–MS results for the oil fraction and indicated that for increasing temperature phenol, catechol, 3-methyl catechol, cresols, and vanillin compounds increase, while most other compounds including guaiacol, syringol, and acetosyringone decrease in concentration. Interestingly, there was a large difference between the chemical compositions of the hemp BCD lignin oil compared with that of the softwood BCD lignin oil.

For the low-temperature BCD trials of Lavoie *et al.* (2011), GC–MS analysis of the oil extract revealed that guaiacol was the largest volatile component in the oil at 28% and 52% for a reaction temperature of 300 °C, for hemp and softwood, respectively. Phenol concentrations were also low at 2.6% and 1%, for hemp and softwood, respectively. It was noted that the proportion of phenol doubled when the reaction temperature was increased from 300 to 330 °C. Catechol was 7.5% and 1.8%, for hemp and softwood lignin, respectively, at the lower reaction temperature of 300 °C. These percentages increased dramatically to 30% and 32%, for hemp and softwood lignin, respectively, for the higher temperature BCD trials (330 °C). Syringol was not detected in any of the GC–MS results for the softwood lignin indicating that initial lignin composition plays an important role in the BCD products generated.

Olarte (2011) published a thesis, which assessed both the depolymerization of lignin in the presence of high temperature and base catalysts (NaOH, KOH, and NH_4OH) and investigated hydrodeoxygenation as a means of decreasing the oxygen-to-carbon ratio in lignin-derived model compounds. BCD of organosolv lignin (a hardwood mix) was undertaken in a 500 ml batch Parr reactor at temperatures ranging from 165 to 350 °C. Reactor heating times took between 45 and 60 min. Once heated to the desired temperature, reaction times were then continued for an additional 15–60 min. It was reported that no significant differences were observed in the ^{13}C NMR spectra of products from the 15 and 60 min reaction times, and thus the 15 min reaction time was adopted as the base case in the investigation for the Olarte (2011) study. The insensitivity to reaction time between the 15 and 60 min results is most likely attributed to the extended heat up times (45–60 min) required for the reactor vessel, as Dunn (2014) had later shown significant lignin structural modification during similar alkaline hydrothermal liquefaction reaction time variations in soda lignin from sugarcane bagasse.

^{13}C NMR results presented by Olarte (2011) of the aqueous mixture recovered from the reactor following alkaline hydrolysis with NaOH at 290 °C for both 15 and 60 min, indicated a notable increase in signal peaks downfield of 168 ppm (relative to that associated with the lignin feedstock material) denoting the presence of more oxidized functional groups (hence the oxidation of some bonds). A reduction in the aromatic carbons associated with methoxyl groups, a large reduction in peaks associated with syringyl carbons, and a significant decrease in actual methoxyl groups were also reported (Olarte 2011). The existence of a peak at 50 ppm was attributed to methanol in the crude aqueous solution. It was also suggested that syringaldehyde groups were more susceptible to the BCD reaction than guaiacyl groups.

Analysis as indicated by ^{13}C NMR spectra for BCD products from reaction temperatures of 165–350 °C with a 10% NaOH solution and a 15 min reaction time, revealed that increasing temperature

- increased the formation of carbonyl groups;
- reduced the prevalence of aromatic carbons associated with syringyls;
- produced large reductions in syringol aldehyde and compounds with methoxyl groups;
- promoted the formation of methanol and compounds with aliphatic groups.

^{13}C NMR spectra also showed that methanol proportions were always maximized at 290 °C. The greatest difference between the compositions of the product oil and the original lignin occurred at temperatures equal to or above 290 °C. At 350 °C, a 6.3 wt% solid residue was recovered from the crude alkaline mixture from the reactor when filtered.

As reported by Olarte (2011), analysis of the BCD products using KOH instead of NaOH as catalyst showed similar trends in terms of the distribution and signal intensities in ^{13}C NMR spectra. Inspection of the spectra presented suggests that KOH generated a higher proportion of alkyl or saturated structures for the conditions tested.

The alkali NH_4OH was also tested and compared with that of NaOH; the justification for using this alkali being the conversion during the BCD process of NH_4OH into NH_3 gas, which has the potential to assist with downstream separation of products. For BCD temperatures of 290 and 350 °C and using NH_4OH as a catalyst, the ^{13}C NMR spectra indicated different features in the spectra when compared with NaOH results. The difference in spectra was particularly associated with the carbonyl region and the aliphatic region of the spectra. Solid yields recovered from the crude alkali mix for the NH_4OH experiments were high at 37.2 and 51 wt% of the original lignin feed, for the 290 and 365 °C experiments, respectively. It was also reported by Olarte (2011) that nitrogen groups were present in the products. Observations of the relative proportions of the various structures as indicated by the ^{13}C NMR data presented, indicated that the aromatic proportions in the crude mix produced from the use of NH_4OH, appeared to have not been reduced to the same extent as that presented for the NaOH experiments for increasing reaction temperature. The high yields of solid residue removed before the ^{13}C NMR analysis for NH_4OH experiments may have influenced this result.

Gel permeation chromatography (GPC) analysis using THF as the solvent was undertaken to determine molecular weight. Acetylation was also done to ensure complete solubility in the solvent and to cap functional groups (carbonyl, phenolic, and hydroxyl) that can cause secondary interactions such as ion exclusion,

ion inclusions, and adsorption of elutes. Complete solubilization in THF of the acetylated solid material produced from NaOH and KOH alkaline experiments (recovered after evaporation and subsequent acetylation) was not possible, with Olarte (2011) estimating that 10% of the acetylated solid residue remained insoluble in THF.

GPC results revealed a reduction in the total molecular weight of the BCD products when compared with the lignin feedstock material. A greater proportion of molecular weight products in the range of 180–550 g/mol was observed for the lower temperature of 290 °C as compared with the higher temperature of 350 °C for all alkali. NaOH experiments showed the lowest proportion of products in the 2500–15,000 g/mol molecular weight range (estimated at <10%). For KOH, this figure was approximately 15% for the equivalent molecular weight range and for NH_4OH up to 30% estimated for the higher reaction temperature of 350 °C.

Olarte (2011) also presented the results of the identification and quantification of the monomeric products recovered with dichloromethane following acidification (pH of 2) of the crude solution recovered from the reactor. Products identified and quantified are syringaldehyde, 4-ethyl catechol, 2,6-dimethoxyphenol, 4-methyl catechol, 4-ethyl guaiacol, catechol, 4-methyl guaiacol, guaiacol, and phenol. Both NaOH and KOH catalyzed reactions were reported to show similar trends in monomer product formation and total detected monomer yields with changes in reaction temperature. Catechol was only detected at temperatures greater than 165 °C. A reduction in catechol yields at 350 °C was attributed to increased oxidation reactions. Syringaldehyde and syringol were also identified, and similar to catechol comprised a large proportion of the recovered monomeric fractions. Phenol proportions increased with reaction temperature. Total identified monomer yields were maximized at 290 °C for both NaOH and KOH catalyzed reactions. NH_4OH did not show this same trend as yield continued to increase for the reaction temperatures investigated.

Assessment of the identified compounds in the oil extracts recovered from the 290 °C experiments using the different catalysts revealed that the relative proportions of the identified compounds differ for the different catalysts. The following are the relative prevalence of BCD product compounds in order of decreasing yield as estimated from the graphs presented by Olarte (2011).

NaOH:

- Catechol ≫ guaiacol > 4-methyl catechol > phenol > syringaldehyde ≫ 4-ethyl catechol > 4-methyl guaiacol > 4-ethyl guaiacol > syringol

KOH:

- Catechol ≫ 4-methyl catechol ≫ 4-ethyl catechol > phenol > guaiacol > syringaldehyde ≫ 4-ethyl guaiacol > 4-methyl guaiacol > syringol

NH_4OH:

- Catechol > syringol > guaiacol > 4-methyl catechol > syringaldehyde > phenol > 4-methyl guaiacol

Zabaleta (2012) published a thesis on the hydrothermal liquefaction of lignin recovered from olive tree prunings. Lignin used in the liquefaction experiments was produced from an organosolv process following assessment of three different pulping or delignification processes, namely alkali treatment, autohydrolysis, and organosolv. The alkali pulping treatment was undertaken with a high molarity NaOH (7.5 wt%) solution and a low temperature of 90 °C for 90 min. The autohydrolysis delignification was undertaken at 180 °C for 30 min, and similarly the organosolv delignification using ethanol as the solvent was also undertaken at 180 °C for 30 min. The molecular weights of the lignin samples recovered and solubilized were 10,875, 3662, and 12,919 g/mol for the alkali, autohydrolysis, and organosolv lignin samples, respectively. Organosolv lignin extraction was then optimized through experimental design procedures to minimize lignin remaining in the pulp, while maximizing the dissolved lignin concentration. To maximize dissolved lignin, higher temperatures and low ethanol concentrations were recommended. The organosolv conditions (200 °C, 70 wt% ethanol, and 90 min) used to produce the lignin for subsequent hydrothermal liquefaction trials were a compromise between ensuring an adequate lignin quality and a reasonable yield of dissolved lignin in the black liquor product.

Lignin produced from the optimized organosolv process was used to test BCD conditions with a 4% alkali catalyst concentration, at a temperature of 300 °C and reaction time of 40 min. The reaction was carried out in a batch Parr reactor (series 5500). Oil was recovered by acidification with concentrated HCl (pH 1) with the precipitated solid fraction filtered from the oil and the aqueous fraction extracted with ethyl acetate to recover the phenolic compounds (oil). The filtered solid (residual lignin and char) was solubilized with THF and again gravity filtered to separate remaining solids (char) and the THF soluble component (termed lignin residue). Alkali catalysts tested were NaOH, KOH, LiOH, $Ca(OH)_2$, and K_2CO_3; a control experiment without catalyst was also undertaken. Reported oil yields for the experiments with the different catalysts were 19.7, 16.8, 10.2, and 6.9 wt%, for the NaOH, K_2CO_3, $Ca(OH)_2$, and the control experiment, respectively. KOH and LiOH produced oil yields of between 12 and 15 wt% based on the figures provided. Supported by the analysis of the product oils in GC–MS, the control experiment produced the highest proportions of guaiacol and syringol, while the hydroxides (NaOH, KOH, and LiOH) and the carbonate K_2CO_3 promoted the production of catechol (<2.5 wt% on lignin) and 4-methyl catechol. Similarly, the hydroxides and the carbonate showed small proportions of cresols not detected in the control and the $Ca(OH)_2$ experiments. Phenol yield in the oil was less than 1% (<0.2 wt% on lignin), with the hydroxide catalyst results showing the highest proportion of phenol.

The char yield was greatest for the control at 35 wt% on initial lignin, while KOH and NaOH showed the lowest yields of char at around 5 wt%. Residual lignin was highest for the hydroxides at approximately 45 wt%, while

Table 7.3 Results from Zabaleta (2012) of the BCD product trends and lignin feedstock molecular weight.

	Lignin ultrafiltration fraction					
	1	2	3	4	5	6
Molecular weight of feed lignin, g/mol	12,798	9302	7882	6555	5184	4527
Phenol wt% oil	2.71	3.03	6.09	3.22	4.18	7.35
Guaiacol wt% oil	0.42	0.44	0.52	0.43	0.53	0.56
Catechol wt% oil	8.08	8.36	11.44	13.25	19.11	22.82
4-Methyl catechol wt% oil	3.73	3.97	4.45	4.26	6.56	8.2
Total monomers wt% oil	16.15	17.42	25.31	22.53	31.97	41.72

the control was the lowest at 15 wt%. Molecular weight measurement using high-performance size-exclusion chromatography (HPSEC) of the residual lignin suggested that residual lignin had an increased molecular weight relative to the initial lignin feed. This result was more pronounced for the hydroxides and carbonate alkaline catalyst experiments. The lignin residue recovered from the NaOH catalyst experiment showed a threefold increase in the molecular weight relative to that of the initial lignin feed. HPSEC results from the study also suggest that a potassium-based alkali catalyst increased repolymerization.

Zabaleta (2012) also assessed the impact on lignin depolymerization based on a reaction ordinate (Ro) as presented previously in Equation 7.1 and used by Thring (1994). For a 4 wt% NaOH catalyst concentration, and a reaction temperature of 310 °C, and a reaction time of 30 min (i.e., representing a high reaction ordinate), an oil yield of 22.5 wt% was reported. At a lower reaction ordinate (250 °C and 90 min), the oil yield was lower at approximately 17.5 wt%.

GC–MS analysis of the oils produced indicated an increase in catechol(s), cresols, and phenol for more severe reactions, while syringol and guaiacol decreased as the reaction severity increased. Catechol and 4-methyl catechol as percentages of all the identified compounds were 42.6 and 35.3 wt%, respectively, for the severest reaction trialed. It was also reported that char yield was reduced with increasing severity (down to 1 wt% at a 310 °C and 30 min reaction), while residual lignin increased up to 60 wt% for the reaction conditions examined.

One of the most significant features of Zabaleta's (2012) work was the fact that BCD reactions were conducted with lignin that had been obtained from ultrafiltration using a membrane to produce six fractions with varying molecular weight ranges (Table 7.3). Oil yields for process conditions of 310 °C, 30 min, and 4 wt% NaOH (assumed) varied between 15.7 and 19.5 wt%. Total detected monomers increased from 16.15 to 41.72 wt% of oil, for decreasing molecular weight (12,800 to 4500 g/mol) of the lignin feedstock (Table 7.3). As observed from Table 7.3, phenol, catechol, and methyl catechol proportions in oil increased significantly with decreasing molecular weight of the lignin feedstock, while guaiacol remained relatively constant.

The results of the increased phenol content for lower-molecular-weight lignin oil extracts, though not discussed in the published thesis by Zabaleta (2012) is consistent with the findings of lignin structural analysis; that is, the H monolignol structures (considered to degrade to phenol in BCD reactions) are generally associated with terminal groups. Lower-molecular-weight lignin would have a higher proportion of these H monomer structures and consequently produce higher proportions of phenol in the derived oil.

Char formation for the ultrafiltrated lignin samples following BCD ranged between 6 and 8 wt%. The lignin residue yield ranged between 35 and 47 wt% with no clear trend evident relative to the ultrafiltrated fraction, that is, the initial lignin molecular weight.[3]

Similarly, there was no trend evident in the molecular weight of the solid residual lignin relative to the initial lignin molecular weight, other than an apparent increase in the molecular weights of the solid residue lignin. It was suggested by Zabaleta (2012) that unstable lignin fragments participate in reactions that do not follow any trend; however, it was also suggested that polymer depolymerization did occur as residual lignin polydispersivity was generally lower for the residual lignin fractions relative to the feed lignin material.

In order to improve monomer production from the hydrothermal depolymerization of lignin, Zabaleta (2012) and Toledano *et al.* (2014) investigated further the use of boric acid (as previously reported by Roberts (2008)) and phenol as capping agents. Mass ratios of lignin to boric acid investigated were 1:0.75, 1:1, and 1:2. These same values were used for the lignin-to-phenol ratios trialed. NaOH concentrations of 4 wt%, with a reaction temperature of 300 °C and reaction time of 40 min were used for the investigation of capping agents. For the boric acid trials, oil yield only increased when the ratio of lignin to boric acid reached 1:2, while for the phenol trials, oil yields increased linearly with increasing phenol concentration. It is also evident from the presented GC–MS results of the oil is that with the addition of phenol acting as a capping agent, high proportions of cresols are identified. Interestingly, compounds such as 4-hydroxybenzoic acid, 4-hyroxybenzaldehyde, and ferulic acid remained in the oil when phenol was added as a capping agent.

Zabaleta (2012) and Toledano *et al.* (2014) identified the addition of phenol as beneficial, as char production was typically about 7 wt% when phenol

[3]For comparison, the crude lignin (without ultrafiltration) at the same reaction severity (310 °C, 30 min, and 4 wt% NaOH) conditions produced an oil yield of 22.5 wt%, a char yield of 0.74 wt%, and a lignin residue of 60 wt%.

was added and residual lignin reduced to about 32 wt% with higher proportions of phenol. The molecular weight of the lignin residue for phenol addition generally showed a reduction in the molecular weight in comparison to the control experiment. The use of boric acid produced an increase in char to 21.24 wt% (compared with the control char yield of 5 wt%) and a decrease in the solid residue to approximately 25 wt% with increasing boric acid concentration. The use of boric acid showed no well-defined trend with respect to molecular weight of the lignin residue. From these results, the author concluded that phenol rather than boric acid was the preferred capping agent and worthy of further study.

Other investigators (Dorrestijn *et al.* 2000, Saisu *et al.* 2003, Man *et al.* 2005), reporting on the use of phenol to assist in the depolymerization of lignin, identified that char or tar reformation is reduced when homogeneous systems, that is, phenol and water mixtures are used. Saisu *et al.* (2003) showed that during solvolysis with phenol solutions in supercritical water, phenol participated in the depolymerization of lignin by potentially capturing the reactive species such as the methyl and ethyl groups and formaldehyde, which can polymerize the monophenols. Yields of the recovered oil from lignin hydrothermal treatment in supercritical water was generally higher in the presence of phenol than in its absence.

Man *et al.* (2005) presented similar results for the depolymerization of organosolv lignin in supercritical phenol–water solutions and cited the ability of phenol to suppress repolymerization by trapping and reacting with depolymerized low-molecular-weight fragments and monomeric compounds that would otherwise have oligomerized. Repolymerization of the monophenols was considered to lead to the generation of heavier fragments, that is, oligomers.

Beauchet *et al.* (2012) undertook the BCD processing of commercial Kraft lignin, Indulin AT, and Hemp lignin using a continuous flow reactor. Temperatures investigated ranged from 270 to 315 °C and the pressure set at 130 bar. Liquid hourly space velocities of 1.4–4 h^{-1} were investigated. The continuous reactor was 30 m long with an internal wall diameter of 0.66 cm. Solutions of 5 wt% NaOH were used to depolymerize lignin in a 10 wt% solution. Following the reaction, the products were quickly cooled with no solid present initially after cooling, and the contents acidified to a pH of 2 using HCl acid. The solid fraction (oligomer-rich fraction) produced following acidification was filtered from the solution, removed, and dried. The aqueous fraction was extracted with a solvent to produce the monomer-rich fraction. The remaining aqueous fraction contained low-molecular-weight organic compounds that were not extracted with the solvent.

Oil yields of the monomer-rich fraction from the Indulin lignin ranged from 11 to 19 wt%, with the higher yields achieved for the higher temperatures investigated (315 °C). The same trend existed for the lower-molecular-weight organics remaining in the aqueous fraction though yields ranged from 2.1 to 6.6 wt% on lignin for the conditions tested. Shorter liquid space velocities (i.e., longer residence times) showed an increase in monomer-rich oil yields and in the small organic compounds detected in the remaining aqueous fraction. The oligomer yield decreased with increasing temperature from about 72 to 49 wt% for the 270 and 315 °C conditions, respectively. Shorter liquid space velocities (longer residence times) showed a decrease in the oligomer fraction. The main gaseous product identified was CO_2 and traces of CH_4 (<0.2 wt%) was also detected. The yields of gas were estimated at between 13.2% and 25.5% based on a calculated mass difference; these high yields prompted Beauchet *et al.* (2012) to suggest that high CO_2 concentrations may inhibit the BCD reaction by reacting with NaOH to give Na_2CO_3.

Of the monomer-rich fraction produced from the Indulin lignin, 40–50% of the monomers were identified as monomeric compounds. This is equivalent to 8.4 wt% of the original lignin. The remaining fraction of the monomer-rich oil was composed of a wide range of other mono-aromatic compounds present in very low concentrations. At 315 °C and a liquid space velocity of 1.4 h^{-1}, the yield of catechol was up to 5 wt% of the original lignin and composed 25.8% of the oil fraction. For Hemp at 315 °C and a liquid space velocity of 2.6 h^{-1}, the catechol yield was approximately 2.5 wt% of the original lignin.

^{31}P NMR analysis of the oligomer fraction showed that almost all aliphatic hydroxyl groups of the original lignin (Indulin lignin) were removed during the alkaline treatment. The authors comment that the detection of acetic acid after the alkaline treatment confirmed the occurrence of C_α–C_β cleavage in the propane chains of the Indulin AT lignin.

Schuchardt *et al.* (1993) studied the alkaline hydrolysis of hydrolytic eucalyptus lignin using sodium formate in a batch and continuous reactor. The reactor (batch) was loaded with lignin and sodium formate to produce a 25% lignin in water or a 2.5% sodium formate in water solution. Following the reaction, the final pH of the solution was typically 8.3 and independent of the initial pH of the feed material. Reaction temperatures were varied between 200 and 300 °C. Temperatures of 270–300 °C produced the highest oil yields of 42.6–47.1%, respectively. Yield is based on a carbon wt% mass balance. Given the extraction approach, this oil represents both the monomer-rich and oligomer-rich fractions, and possibly residue lignin, as the reaction mixture pH was reduced to only 5 and the total contents extracted with a chloroform–ethanol (3:1 v/v) solvent. Microdistillation of the oil product produced a 42% mass fraction of colorless oil (i.e., 18.9% monomer-rich oil). GC–MS identified that a large portion of this oil was syringol and guaiacol, and that the total yields were 4 and 2 wt% on lignin, respectively. These yields were also compared with those reported for other studies (Schuchardt and Goncalves 1990) using alkaline hydrolysis in dioxane–water mixtures where yields of 5 wt% syringol and 8 wt% guaiacol were reported. Work undertaken in a continuous reactor using lower proportions of lignin (5%) and higher proportion of sodium formate (5–15%, pH = 9) produced an oil yield of 61% at an operating pressure of 120 bar. The temperature was reported (Schuchardt and Goncalves 1990) to be restricted to less than or equal to 300 °C, given concerns that above this temperature sodium formate would decompose into carbon dioxide and hydrogen.

Erdocia *et al.* (2014) studied the hydrothermal depolymerization BCD of organosolv lignin produced from olive tree prunings. The BCD reaction was carried out in a 4 wt% NaOH solution at a reaction temperature of 300 °C. Reaction times investigated were 20, 40, 60, 80, 90, and 100 min in a batch Parr reactor (series 4848) with a volume of 100 ml. Lignin loading was approximately 5 wt% in solution. Following the reaction, the pH was reduced to 1 using concentrated HCl, and residual lignin and coke filtered from solution. The remaining liquid faction was then extracted with ethyl acetate to recover an aromatic monomer-rich oil product. Solid residue was also washed with THF to separate coke from unreacted lignin. It was observed from this study that increasing reaction time from 20 to 80 min gradually increased the recovered oil yield from approximately 20 to 22.3 wt%. After 80 min reaction time, the oil yield findings showed a subsequent reduction in oil yield to 16.57 wt% for a reaction time of 100 min. This reduction in oil yields was attributed to depolymerization (recondensation reactions of the lignin/monomer fragments). This finding was somewhat confirmed by a gradual increase in char yields after an 80 min reaction duration. Overall the residual lignin was shown to gradually decrease during the first 60 min and subsequently level out at a yield of approximately 30 wt% for the remaining extended reaction durations.

Reported HPSEC chromatograms of the original lignin and residual lignin samples by Erdocia *et al.* (2014) revealed that the residual lignin samples comprised two main peaks, which were positioned off-centre in the HPSEC chromatogram to that of the original lignin. These two peaks in the residual lignin samples evidently revealed that the molecular weight of the residual samples showed both bulk smaller and larger molecular weight structures than that of the original lignin, thus identifying that residual lignin was not unconverted lignin or the same as the original lignin. The peak in the lignin residual samples reflecting larger molecular weight structures was attributed to the process of repolymerization reactions between unstable lignin fragments and the original lignin producing a lignin residual with higher molecular weight. The remaining peak representing the smaller molecular weight structures was attributed to depolymerization reactions. Lignin residual samples from the 90 and 100 min reactions contained the greatest proportion of the highest molecular weight structures; these two reactions also had the greatest coke yields. The 80-min lignin residue contained the largest proportion of lower-molecular-weight structures, which was consistent with the higher oil yields reported for this reaction.

7.3 Hydrolysis in subcritical and supercritical water without an alkali base

In 2003, Saisu *et al.* described the conversion of organosolv lignin at 400 °C in supercritical water and phenol water mixtures without a base catalyst (Saisu *et al.* 2003). The study involved a range of reaction times (10–65 min) and supercritical water densities (0 (pyrolysis), 0.1, and 0.5 g/ml) in a 10 ml batch reactor using 2 wt% lignin solutions. It was shown from this study that with supercritical water only at densities equivalent to 0.5 g/ml, the tetrahydrofuran-insoluble (TIS) yield fraction increased as reaction time increased up to 20 min and leveled out to a 20 wt% yield. Decreasing water density with similar reaction times tended to increase the TIS yield. It was also reported (Saisu *et al.* 2003) that the molecular weight of the tetrahydrofuran-soluble (TS) fraction (representing the monomers and oligomers) decreased with an increase in the water density.

It appeared from tabulated data that yields of hydrolysis products (mainly monomeric) in the TS fraction were greatest at 7.5 wt% for the initial stages of the reaction (i.e., <10 min), and as the reaction progressed these yields declined to 1.4% for 65 min reaction. Products such as syringol, guaiacol, catechol, and methoxy catechol were detected. The proportions of the guaiacol and syringol compounds decreased with time, catechol content peaked at 30 min, and methoxy catechol increased with increasing reaction time. Interestingly, compounds such as cresols, 4-ethyl phenol, and 2,2-dihydroxymethane were only detected when phenol was used in the solvent.

Wahyudiono *et al.* (2008) reported on the depolymerization of alkaline lignin in subcritical and supercritical water without a base catalyst in a batch reactor at temperatures of 350 and 400 °C for pressures of 250 and 400 bar. Reaction times from 5 to 240 min were explored. Reactor heating rates were approximately 120 °C/min. Wahyudiono *et al.* (2008) assessed the impact of water density, temperature, and reaction time on the generation of the monophenols *m*-cresol, *p*-cresol, *o*-cresol, phenol, and catechol, which remained dissolved in an aqueous/methanol solution following washing of the crude hydrothermal liquefaction product with methanol.

It was shown that the percentage of catechol decreased with increasing water density, while the percentages of the cresols and phenol increased gradually with increasing water density. The maximum proportion of catechol peaked at a reaction time of about 30 min for both temperatures investigated. Catechol was highest for the lower temperature of 350 °C reaching a value of 30% yield (quantitatively determined by HPLC of the aqueous methanol product). Phenol yield was shown to increase with an increase in the reaction time and for operation at the higher temperature of 400 °C. After 240 min at 400 °C, the yield of phenol reported was 11.75%. For a reaction time of 40 min and a 350 °C

Table 7.4 Maximum proportions (presented as yields) of compounds in the aqueous phase for varying reaction time (Wahyudiono *et al.* 2008).

Compounds in aqueous phase	Reaction temperature	
	350 °C	400 °C
Catechol, %	30.55	26.18
Phenol, %	3.31	11.75
m,p-Cresol, %	8.76	6.98
o-Cresol, %	3.65	3.95

Note: The values presented here are the maximum achieved proportions of the compounds for a varying set of experimental conditions, that is, for an experiment achieving high yields of catechol 30%, the corresponding phenol content would be low at approximately 1%, while *m,p*-cresol totaling 4% and *o*-cresol 1% yields.

reaction temperature the phenol yield was only 2%, while at 400 °C the phenol proportion was 6%. Table 7.4 provides details of the maximum proportions of compounds in the aqueous phase for varying reaction times.

The methanol insoluble fraction was considered to contain mainly char. Before the hydrothermal liquefaction treatment, the lignin was completely soluble in methanol. The char component increased with increased reaction times and was greatest (estimated at 25%) for the higher temperature of 400 °C and reaction time of 240 min.

Compound yields presented by Wahyudiono *et al.* (2008) are significantly higher than those presented to date for the BCD of similar types of lignin at similar operating temperatures. It is estimated from the data that the total identified monomer (catechol, cresols, and phenol) content contained in the aqueous product comprise at best approximately 35 wt% on lignin, with the lower temperature of 350 °C producing overall a greater yield from lignin of the identified monomers. The calculation of yield by Wahyudiono *et al.* (2008) assumes no evolution of gas.

Pińkowska *et al.* (2012) also report on subcritical and supercritical depolymerization of alkaline lignin and, similar to Wahyudiono *et al.* (2008), produced yield data that appears higher in comparison to that more generally reported for the BCD of lignin. Temperatures of between 280 and 390 °C were investigated with maximum monomeric yields (wt% lignin) reaching reported values of 11.23% (280 °C, 0 min), 11.11% (380 °C, 0 min), 4.21% (390 °C, 240 min), and 7.00% (390 °C, 240 min) for guaiacol, catechol, phenol, and cresol isomers, respectively. Insoluble residue yield behaved similar to that presented by Wahyudiono *et al.* (2008), which increased with increasing temperature for supercritical water.

Long *et al.* (2014) studied hydrothermal liquefaction of organosolv lignin produced from switchgrass at reaction temperatures ranging between 160 and

250 °C (subcritical) and for reaction times of between 2 and 8 h. It was reported that at the lower temperatures investigated (160 and 180 °C) and a reaction time of 2 h, no char was produced following THF extraction. Yield of oils (obtained from CH_2Cl_2 extraction) as expected were low at approximately 2–4 wt%. For a higher reaction temperature of 250 °C and reaction time of 2 h, oil yield increased to a 9.3 wt%, but residue char was also shown to drastically increase to 42.0 wt% for this same reaction.

An interesting observation reported by Long *et al.* (2014) was the variation in the proton NMR signals of the different THF soluble fractions collected and the initial lignin feedstock. It was reported from proton signal reductions in the proton spectra that signals associate with the H monomer were not present at the minimum temperature of 160 °C trialed. As the reaction temperature was raised, there was a reported increase in proton signals (1.76 and 2.43 ppm) typically associated with alkyl groups attached to aromatic rings. This prompted Long *et al.* (2014) to suggest breakage of the lignin structure. Based on the variations in the proton NMR signals relative to reaction temperature, Long *et al.* (2014) concluded that the H lignin structure monomer is more readily decomposed than that of the G and S lignin monomer structures, while the S lignin monomer is the most recalcitrant lignin monomer in an hydrothermal environment. Long *et al.* (2014) also presented the results of GC–MS of the oil extracts and FT-IR assessment of the char product, which were stated to confirm the findings of the proton NMR studies regarding the behavior of the lignin monomers during hydrothermal conditions.

7.4 Solvolysis with hydrogen donor solvent formic acid

A subset process of hydrothermal liquefaction termed solvolysis involves the use of solvents other than water. The solvent is generally selected for its ability to either donate hydrogen ions and/or suppress repolymerization of lignin.

Kleinert *et al.* (2009, Kleinert and Barth 2008a, 2008b, 2009) performed solvolysis on lignin types produced by strong acid, weak acid, enzymatic (Kleinert and Barth 2008a), and steam exploded delignification processes. Biomass feedstocks included birch, scots pine, Norway spruce, and aspen (Kleinert and Barth 2008b). Solvents investigated included ethanol, 2-propanol, or mixtures thereof, together with formic acid acting as the main molecular hydrogen donor. Solvent loadings varied between 6% and 20% (Kleinert and Barth 2008b). The reaction time varied between 4 and 17 h and reaction temperatures between 350 and 400 °C. The temperature of 350 °C represented a minimum limit to ensure satisfactory yields and low char. Optimum conditions were reported for reaction temperatures of 380 °C with reaction times of greater than 8 h. It was found that the addition of formic acid resulted in the production of less char and greater quantities of organic liquids, with a well-separated two-phase organic and aqueous system formed (unless methanol was used) during the reaction. Oil yields relative to lignin of between 60 and 99 wt%, and up to 130 wt% for isopropanol were reported as a result of the incorporation of alkyl groups. The aromatic products identified by GC–MS showed extensive alkylation (C_1–C_4 alkyl groups) and little to no methoxyl groups. Catechol and catechol derivates were not detected in the product oil. Noncyclic ketones and alkanes were also identified in the GC–MS spectra while polyaromatic hydrocarbon structures (PAHs) were either not present or only present in very small amounts. Char yields are reported as ranging between 2 and 15 wt%.

Studies of the behavior of guaiacol in a solvent system indicated that guaiacol was reduced to phenol and alkyl phenols with the assumption (given the fact that an aqueous layer is produced) that the oxygen was eliminated as water. Kleinert *et al.* (2009, Kleinert and Barth 2008a,b, 2009) attribute depolymerization and hydrodeoxygenation to the presence of formic acid or formate without which char yields and oxygen content are higher in the oil product. Formic acid was not detected in the product oil and was considered to be entirely consumed during the reaction.

Of interest is the degradation pathways of formic acid that are dependent upon the conditions employed. In a water-catalyzed high-temperature environment, two reaction pathways were proposed (Kleinert *et al.* 2009). The first was a dehydration pathway leading to CO_2, H_2, and H_2O and the second a decarboxylation pathway leading to the release of CO and H_2O. Without water (i.e., hydrogen bonding), formic acid decomposed into CO and H_2O or alternatively into CO_2 and H_2 via decarboxylation at both high temperature and high pressure; the later obviously being the preferred option.

Liguori and Barth (2011) also studied the depolymerization of a variety of different lignins (8 wt%) suspended in water and assisted by formic acid and solid acid catalyst, Nafion SAC-13 and a Pd-catalyst. In this study, temperature was maintained at 300 °C for 2 h and the pressure at 95 atm (calculated from the vapor pressure of water and the theoretical pressure from hydrogen obtained by the degradation of formic acid). Liguori and Barth (2011) stated that derived oil samples were stable at room temperature for at least 6 months and that separation of guaiacol from catechols could easily be achieved by the use of dichloromethane, which is selective for guaiacol. Though the use of solid catalysts has not been reported in this chapter, this work was included given the types of monomeric products reported (Liguori and Barth 2011). The main monomers identified were guaiacol, catechol, and resorcinol with their combined proportions equating up to 17.6 wt% of the recovered phenolic mixture produced from an acid hydrolysis lignin material containing initially 30% cellulose. The use of other solvents (ethanol, propenol) were also investigated (Liguori and Barth 2011) though after some preliminary tests were not considered further as they interfered with the reaction by generating highly alkylated monomeric products.

It was reported that the percentage yields of guaiacol and guaiacol derivatives were superior to other hydrodeoxygenation processes, however, inferior to yields reported using alkaline conditions in lignin depolymerization.

Gosselink (2011) assessed the ability of a supercritical CO_2/acetone/water system to depolymerize organosolv lignin produced from a mixed hardwood lignin sample and a wheat straw lignin sample. Formic acid was added to the system at specific intervals for the generation of *in situ* hydrogen by the decarboxylation of formic acid. Reaction temperatures of the experiments were 300 and 370 °C. Lignin was first dissolved in an 8:2 (v/v) acetone:water mixture at a concentration of 0.35 g/ml. For hardwood lignin depolymerization at 300 °C, a monomeric phenolic yield of 10 wt% on dry lignin was obtained when using formic acid. Without the addition of formic acid, the monomeric phenolic yield was 7 wt%. Individual monomeric yields relative to lignin were dependent upon the lignin material used; wheat straw lignin with the higher *p*-hydroxyphenyl (H) monolignol units gave rise to a substantially higher amount of H-derived phenolics, that is, phenol. Syringol showed the highest monomer yield derived from the depolymerized hardwood lignin feedstock with a yield of 3.6 wt% on dry lignin. For wheat straw lignin, the highest monomer yield was for syringic acid (2.0 wt% on lignin). Char yield was reasonably high with values ranging between 45 and 51.5 wt%. Gas yield was 6 wt% of the initial dried hardwood lignin and comprised portions of propane, ethane, methane, propylene, ethylene, and carbon monoxide. Gosselink (2011) makes reference to the benefits of removing monomeric products from the reaction system so as to avoid recondensation products such as oligomers and char. The evidence cited to support this is based on a comparison in monomeric yields of 1.7 wt% achieved without depolymerized product removal to that of 10 wt% achieved with depolymerized product removal.

Onwudili (2015) also assessed the use of formic acid to depolymerize alkaline lignin via a two-stage reaction process designed to validate if sufficient reaction time was first necessary, at a lower reaction temperature, in order to improve the overall oil yields produced from the depolymerization of alkali lignin. For one experiment (condition 1), the reaction was first held at 265 °C for 1 h and then heated to 350 °C and held for an additional 5 h. For the comparison experiment (condition 2), the reactor contents were first held at 265 °C for 5 h and then heated to 350 °C and held for an additional 1 h. Based on a reaction ordinate (log Ro), these two reaction conditions were identical with a log Ro of 6.97 determined. The results of the comparison between the two different experiments highlighted that a small benefit in oil yield (recovered by extraction with diethyl ether) was observed when allowing for an extended reaction time at the lower temperature when using formic acid or simply deionized water to depolymerize alkali lignin. For condition 1 with formic acid addition, the oil yields (ether extract mass relative to formic acid and lignin mass added) were 18.0 wt% and for condition 2, 20.6 wt%. In comparison, with deionized water alone, condition 1 produced a 5.96 wt% oil yield and condition 2 produced a 7.0 wt% oil yield. Water soluble product yields with formic acid addition were 18.4 and 18.0 wt% for conditions 1 and 2, respectively. This was considerably lower than the scenario without formic acid addition (31.5 and 35.8 wt% for conditions 1 and 2, respectively). What was evidently clear from the trials was that the addition of formic acid significantly reduced the amount of solid residual (char) material collected from the reaction in comparison to the deionized water alone experiments. An interesting observation from Onwudili (2015) is the drastically increased proportions of phenol for the condition 2 experiments above that of condition 1 experiments, a result that is independent of whether formic acid is added or not.

Rahimi *et al.* (2014) investigated the formic acid- and sodium formate-induced depolymerization of oxidized aspen lignin at a reaction temperature of 110 °C and a reaction duration of 24 h. Following the reaction and evaporation of formic acid, the liquid product remaining was extracted with ethyl acetate and a yield (relative to lignin content only) of 61.2 wt% reported for low-molecular-weight aromatic products. The insoluble residual lignin component remaining had a yield of 29.7 wt%. Studies from lignin model compounds (LMCs) also identified that the C_α ketone structure (oxidized lignin model compounds) is crucial to the success of the depolymerization reaction, with no cleavage products observed when the nonoxidized LMCs were subjected to formic acid/formate reaction conditions. This was confirmed when a nonoxidized aspen lignin was used in an identical reaction and a yield of only 7.2 wt% of the low-molecular-weight aromatic products was achieved.

Based on kinetic studies Rahimi *et al.* (2014) concluded that the reaction adopted for depolymerization of oxidized LMC is a redox neutral process that results in no net consumption of formic acid, and that the elimination of formic acid from the LMC structure is a rate-limiting step. The recovery of formic acid was cited as distinguishing the process from other lignin conversion methods as formic acid was not decomposed to produce a source of H_2. Based on the proposed reaction mechanism pathway, it was stated that the beneficial use of both formic acid combined with sodium formate resulted in the elimination of the rate-limiting step as a base (formate) was able to remove a proton from the C_β position, while the acid (formic acid) subsequently assisted in the loss of the formate as a leaving group.

7.5 Reported depolymerization pathways of lignin and lignin model compounds

It is known that thermochemical, photochemical, and enzymatic degradation of LMCs and technical lignin via reaction processes such as oxidation, hydrolysis, reduction, fragmentation, repolymerization, and rearrangement produces a wide spectrum of chemical products (Zakzeski *et al.* 2010, Pandey and Kim 2011). The composition and yield of such products are determined primarily by the nature of the monolignol (i.e., *p*-coumaryl alcohol, coniferyl alcohol, and sinapyl alcohol), the structural linkage bonds in lignin (i.e., β-O-4, β-5, α-O-4, 4-O-5, β-β), the solvent, the reaction time, the temperature, the pressure, and the catalyst used.

Zakzeski *et al.* (2010) reviewed the depolymerization of lignin model compounds by different degradation techniques and observed that the monolignol sinapyl alcohol, in comparison to other monolignols, was less likely to participate in repolymerization. This was attributed to the presence of the 3,5 methoxyl groups that occupy positions on the aromatic ring where 5-5′ bonds can form, thereby preventing repolymerization.

Although for the alkaline hydrolysis of syringol, Miller *et al.* (2002) showed that at 290 °C and a 2% concentrated solution of syringol, a material balance of the aromatics using GC–MS revealed only 75% of the aromatic content remained after a reaction time of 90 min. The aromatic loss was attributed to the reaction of syringol or syringol hydrolyzed products with each other to form higher-molecular-weight products (i.e., polymerization) that were not eluted from the GC column.

Miller *et al.* (2002) indicated that during the alkaline hydrolysis of syringol, two types of reactions occurred that involved the methoxyl substituents on the aromatic ring. These two reactions were identified as being the migration of the methyl group between phenolic groups and/or the loss of the methyl/methoxy groups from the aromatic ring. Migration reactions were cited as being significantly faster than demethylation reactions, but ultimately demethylation processes predominated. A proposed chemical pathway for syringol hydrolysis was detailed and is shown in Figure 7.4. Methanol and catechol were the main products from the hydrolysis of syringol as detected by GC–MS.

The reaction characteristics of a mixture of syringol and phenol present in aqueous NaOH as a function of time at 290 °C was also investigated by Miller *et al.* (2002). As for syringol alone, the main compounds in the product mix were methanol and catechol. High proportions of phenol were still present in the product mix. Syringol was observed to disappear much more rapidly than phenol and it was shown by a mass balance that 75% of the syringol and syringol hydrolysis products could not be accounted for by GC–MS analysis. Similarly, the reaction of anisole for the same conditions showed conversion to phenol and methanol. This indicated that phenol and anisole do not participate to the same extent as syringol in polymerization with monomeric compounds. It is expected that guaiacol given its methoxyl content would also participate in polymerization.

Roberts (2008) described the behavior of lignin model compounds in subcritical and supercritical water. A total of six model compounds were tested, which represented the types of common bond linkages seen in the native lignin structure. It was shown that product distribution from the BCD process (using

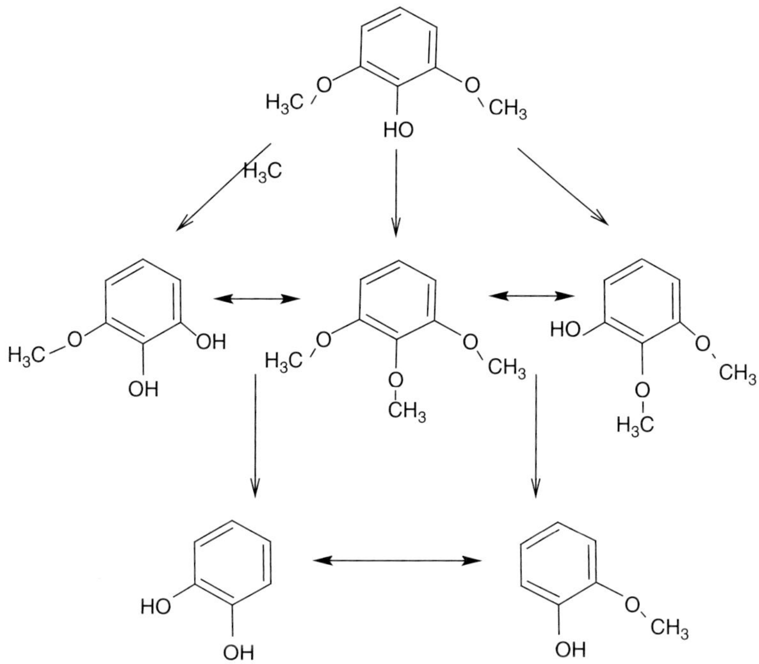

Figure 7.4 Syringol chemical degradation pathway proposed by Miller et al. (Adapted from Miller *et al.* 2002.)

different alkali salts) could be influenced by adjusting the density of the solvent, that is, water. An increase in solvent density shifted the reaction to an ionic reaction pathway and the formation of hydrolysis products (i.e., the monophenols) as opposed to a pyrolysis pathway where cracking and condensation reactions occur. It was also shown by using the model compounds that the use of alcohols and alcohol water mixtures as liquefaction mediums were undesirable, as hydrolysis reactions were inhibited and the product spectrum broadened due to reactions with the solvent. It was identified that based on the degradation results of lignin model compounds representing the aryl-alkyl ether bond of lignin, reaction conditions should not exceed 320 °C. Beyond this temperature, the hydrolysis products (monophenols) are degraded and form higher molecular pyrolysis type products induced by a radical coupling mechanism. The maximum yield of hydrolysis products from the model compound representing the aryl-alkyl ether bond was achieved for temperatures between 300 and 320 °C.

Roberts (2008) also showed that carbon–carbon bonds (aryl-aryl bonds) were not cleaved up to temperatures of 400 °C. The aryl-aryl ether bond could only be cleaved below a temperature of 400 °C when applying an appropriate base catalyst. The aryl-alkyl ether bond (i.e., the β-O-4 ether bond) was, however, readily cleaved above 270 °C.

7.6 The solid residue product

Residue produced from the alkaline hydrolysis of glycol lignin has been studied by Thring *et al.* (1990). Residue yields were shown to reduce with increasing severity (defined by Equation 7.1) for temperatures of between 180 and 340 °C. The methoxyl content, elemental oxygen, and hydrogen decreased in the solid residue with increasing reaction severity. Measurement of molecular weight with organic SEC showed that peak molecular weights (two peaks) reduced with increased reaction severity. A peak representing larger molecular weight products was shown to reduce more rapidly than that of a peak representing lower-molecular-weight products.

Thring *et al.* (1990) showed by FT-IR analysis that with increasing severity there was an increase in the carboxylic acid groups (1710 cm^{-1}) and a corresponding decrease in peak frequencies associate with syringyl and guaiacyl units. The syringyl and guaiacyl signals almost completely disappeared when higher severity conditions were employed. FT-IR analysis showed the generation of new bonds in the 3050–2800 cm^{-1} (representing aliphatics) and 900–800 cm^{-1} region (representing aromatic substitution).

^{13}C NMR investigation of the solid residue by Thring *et al.* (1990) revealed an increase in aliphatic carboxyl groups in the solid residue relative to the initial glycol lignin. Signals associated with primary and secondary alcoholic acetate structures decreased – a phenomenon attributed to extensive cleavage of the alcohol containing side chains of glycol lignin during alkaline hydrolysis. ^{13}C NMR also confirmed a reduction in aromatic signals of syringyl and guaiacyl, and a reduction in the methoxyl content relative to the initial glycol lignin. Carbohydrate signals detected in the glycol lignin and considered to be a lignin–carbohydrate complex (LCC) were absent in the solid residue samples. An LCC was assumed, as acid hydrolysis of the lignin showed sugars at < 1 wt%. New signals in the 10–34 ppm region and an increase in signals at 28.8 and 22.5 ppm at higher reaction severities prompted Thring *et al.* (1990) to suggest the formation of new bonds due to condensation reactions.

Beauchet *et al* (2012) studied the solid residue (oligomers and modified lignin) following alkaline hydrolysis using elemental analysis and ^{31}P NMR. Elemental analysis identified a significant reduction in oxygen content, which prompted Beauchet *et al* (2012) to suggest that the residue would facilitate subsequent hydrodeoxygenation reactions (reduced hydrogen requirement) for the purpose of upgrading to fuels. ^{31}P NMR showed that almost all aliphatic hydrogen groups of the original lignin material were removed during alkaline hydrolysis, and that catecholic structures increased with increasing residence time. This increase in catecholic groups was accompanied by a decrease in guaiacyl groups and an increase in *p*-hydroxyphenyl groups.

Vigneault *et al.* (2006) also investigated the solid residue following alkaline hydrolysis (BCD) using both ^{13}C NMR in the liquid and solid states. ^{13}C NMR spectra (Vigneault *et al.* 2006) showed the presence of carbonyl, aromatic, methoxyl, and aliphatic structures in the oligomer fraction. Vigneault *et al.* (2006) identified that the oligomer fraction contained large ^{13}C NMR signal intensities in the aliphatic region, which was attributed to nonaromatics, that is, potentially fatty acid structures, that may have been introduced from the original lignin material. CHO analysis confirmed that oligomer and solid waste fraction (char) had a higher weight percent carbon than that of the monomer-rich oil fraction recovered following BCD.

Fatty acid structures were also identified by Gellerstedt *et al.* (2004) as being chemically linked to Kraft lignin. Gellerstedt *et al.* (2004) proposed that these fatty acids were the origin of lignin condensation as indicated in the ^{13}C NMR spectra for structures associated with residual lignin following Kraft pulping.

Barbier *et al.* (2012) reported on the findings of the thermal depolymerization of lignin using supercritical water without alkali at 370 and 390 °C for reaction times of 5, 10, 20, and 40 min. Elemental analysis and ^{13}C NMR investigations were conducted on the water insoluble component (lignin residue). Elemental analysis identified an increase in the carbon content and a corresponding decrease for hydrogen and oxygen content with increasing reaction time. ^{13}C NMR investigations revealed a reduction in methoxyl carbons ($\approx$54 ppm) and an increase in aliphatic carbons ($\approx$33 ppm). A decrease in aromatic carbon attached to oxygen ($\approx$147 ppm) and an increase in aromatic carbon attached to carbon or vinylic carbons ($\approx$136 ppm) were also observed for increasing reaction time. The findings prompted Barbier *et al.* (2012) to suggest that ether linkages of lignin are broken under the hydrothermal conditions investigated.

7.7 Summary – strategies to increase yields of monophenols

It is evident from the findings in the literature (as shown in Tables 7.5, 7.6, and 7.7) that for alkaline hydrolysis of lignin alone, that is, without the addition of capping or hydrogen donating agents, the yield of identified monophenols is generally less than 10 wt% and the oil yield of the monomeric-rich fractions less than 23.5 wt%. Therefore, strategies are required to improve yields by enhancing depolymerization, reducing repolymerization (capping agents, hydrogen addition, and monomer extraction), and minimizing product distribution (catalytic upgrading). One depolymerization technique that shows some of the highest yields of monomer recovery from lignin is actually a structural characterization technique for lignin and addresses some of the criteria mentioned above. The Derivative Followed by Reductive Cleavage (DFRC) method for lignin analysis is an approach that incorporates the selective bond breakage of β-O-4 bonds between lignin aromatics. This technique has been shown to produce high yields of depolymerization monomers in the order of 95% from LMCs (Lu and

Table 7.5 Summary of the findings in the literature for the hydrolysis of lignin.

Reference	Lignin	Lignin %	Alkali concentration	Temperature, °C	Monomer rich oil yield%	Oligomer rich oil yield%	Solid residue	Maximum percentage of initial lignin feed				
								Phenol	Catechol	Guaiacol	Syringol	Identified monomers
Thring (1994)	Alcell (poplar, maple, and birch)	5	0.5 – 4	200 – 340	10 – 7			0.03–0.2	0.2–0.91	0.6–0.09	2.4	4.4–2.5
Miller *et al.* (2002)	Alcell mixed hardwoods	10	10	290–330			15–8					
Miller *et al.* (2002)	Alcell mixed hardwoods	10	1–10	290	23[a]	29	34[b]–15					
Vigneault *et al.* (2006)	Steam-exploded aspen	5–15	10	290	10.0	38.3	42.8	0.2	0.2	0.6	1.5	4.5
Vigneault *et al.* (2006)	Steam-exploded aspen	5–15	10	315	15.3	51.6	15.1	0.3	1.2	0.8	0.9	5.7
Vigneault *et al.* (2006)	Steam-exploded aspen	5	6.7	310–330				1.0	2.4	1.0	2.1	9.8
Roberts (2008)	Organosolv beech	10	4	300	23[c]							6.44
Roberts (2008)	Organosolv beech	2.5	1	320	21							9
Roberts (2008)	Organosolv beech	10	4	240–340						1.0	2.8	
Lavoie *et al.* (2011)	Kraft hemp	10	5	300–330	8.4–10.9	56–47						
Lavoie *et al.* (2011)	Kraft softwood	10	5	300–330	5.5–10	70.3–60						
Olarte (2011)	Organosolv hardwood mix	10	10	165–360								<5
Zabaleta (2012)	Organosolv olive	4.8	4	300	19.7	45	5	0.12	2.0	0.24	0	3.6

Zabaleta (2012)	Organosolv olive	4.8	4	250–310	17–22.5	35–60	7–1	0.03–0.75	0–9.58	0.9–0.27	0.65–0	1.65–19.27[d]
Zabaleta (2012)	Organosolv olive ultrafiltrated	4.8	4	310	19.5–15.7	37–47	6–8	0.46–1.25	1.37–3.87	0.07–0.1		2.75–7.1
Beauchet *et al.* (2012)	Kraft Indulin AT	10	5	270[e]	7.7–11	67.8–72.1		<0.1	1.3–0.3	2–1.5		4.5–3.0
Beauchet *et al.* (2012)	Kraft Indulin AT	10	5	290[f]	12.1–15	63.3–67.5		<0.1	3–1.3	1.5–2		6.1–4.9
Beauchet *et al.* (2012)	Kraft Indulin AT	10	5	315[f]	14.1–19.1	49.3–54.7		<0.1	5–3	0.3–1.3		8.4–6.3
Beauchet *et al.* (2012)	Hemp steam explosion	10	5	315					1–2.4	1–0.4		
Schuchardt *et al.* (1993)	Hydrolytic eucalyptus	5–25	2.5–15	240–300	42.6–61.0[f]					2	4	
Erdocia *et al.* (2014)	Organosolv olive	5	4	300	16.6–22.3	30	6–12.	0.1	1.5	0.02		4.3
Zmierczak and Miller (2008)	Organosolv	10	5	295	58		32					

[a]330 °C for 1 h, 2:1 lignin ratio to Na, but 3:1 ratio of OH^- to lignin, as CaO and NaOH where both used in the example.

[b]Time dependent.

[c]Maximised at 4.2 min.

[d]This figure is much higher than that typically reported and it is assumed that potentially GC-MS analysis undertaken assumed all oil produced was volatile and detectable.

[e]Undertaken for space velocities cf 1.4, 2.6, 4 h^{-1}.

[f]Yield based on a wt% carbon mass balance.

Table 7.6 Summary of the findings in the literature for the hydrolysis of lignin in subcritical and supercritical water without an alkali base.

Reference	Lignin	Lignin %	Alkali concen-tration	Tempera ture, °C	Monomer rich oil yield%	Oligomer rich oil yield%	Solid residue	Maximum percentage of initial lignin feed				
								Phenol	Catechol	Guaiacol	Syringol	Identified monomers
Saisu *et al.* (2003)	Organosolv	2	0	400			30		0.56	0.77	1.96	7.48
Wahyudiono *et al.* (2008)	Alkaline		0	350				7.17	5.5			
Wahyudiono *et al.* (2008)	Alkaline		0	400				7.96	2.1			
Pińkowska *et al.* (2012)	Alkaline	5.3	0	280			49–33[a]	1	3.91	11.23		11.5
Pińkowska *et al.* (2012)	Alkaline	5.3	0	390			25–13[b]	4.21	3			13.75
Long *et al.* (2014)	Organosolv switchgrass	2.5	0	160–250	2.4–9.3	62.5–19.8	0–42	<0.3		<0.3	<0.2	<6.7

[a]Time dependent.
[b]Addition of formic acid.

Table 7.7 Summary of the findings in the literature for the solvolysis of lignin with formic acid.

Reference	Lignin	Lignin %	Alkali concentration	Temperature, °C	Monomer rich oil yield%	Oligomer rich oil yield%	Solid residue	Maximum percentage of initial lignin feed				
								Phenol	Catechol	Guaiacol	Syringol	Identified monomers
Kleinert and Barth (2008a, 2008b, 2009), Kleinert *et al.* (2009)	A variety of types		0[a]	350–400	60–130[b]		2–15					
Gosselink (2011)	Organosolv hardwood		0	300	36.2		51.5					1.7[c]
Gosselink (2011)	Organosolv hardwood		0	300			40					7[d]
Gosselink (2011)	Organosolv hardwood		0[e]	300			45			1.6	3.6	10[e]
Gosselink (2011)	Organosolv straw		0[e]	300	45		47			1.6	1.8	12[e]
Onwudili (2015)	Alkaline		0	265–350	18–20.6		18					
Rahimi *et al.* (2014)	Oxidized aspen		0	110	61.2							

[a]Addition of formic acid.
[b]Yields are high as solvent assists in the alkylation of the aromatics.
[c]Without monomer removal.
[d]With monomer removal.
[e]Addition of formic acid (14% on lignin) at the start of reaction.

Ralph 1997) with minimal product compound variation. For real lignin samples, monomeric yields of between 17% and 36% are reported, with the yield dependent upon the type of lignin used and the lignin linkages present. Monomeric structural product variation is, however, fairly extensive (Lu and Ralph 1998) and structures with alkyl side chains still present.

To undertake DFRC of lignin, acetyl bromide is first reacted with the lignin at 50 °C for 3 h, a process where only α-ethers are cleaved in this reaction and all free OH (phenolic and aliphatic) are acetylated while benzylic positions are brominated. In the following reaction step, zinc powder is used as a reducing source and added to the system to ensure the rapid cleavage of the β-bromo ethers, a step that is also high yielding with both *cis* and *trans* (dominant at > 90%) lignin monomer isomers produced. Though the technique is not ideal for the generation of monophenols (catechol, phenol, etc.) or likely to be a commercially viable means of monophenol production, it does provide a demonstration of a reaction mechanism, which is able to selectively break bonds within lignin and is an important requirement for optimizing monophenol yield during the depolymerization of lignin.

Returning now to the practical alkaline depolymerization of lignin, there are several parameters that can be controlled and are known to influence monomer yield as reported in the literature. These parameters are reaction temperature, reaction time, reaction pressure, solvent loading, alkali concentration, lignin feed structure, and monomer separation.

7.7.1 Reaction temperature

It has been shown (Roberts 2008) that temperature is one approach to controlling selective bond breakage of LMCs (and potentially lignin). A temperature of 270 °C with extended reaction times (240 min) was shown to be suitable for cleavage of aryl-alkyl ether bonds (representing the β-O-4 bond), which are a common bond in sugarcane lignin (Sun *et al.* 2003) and considered the weakest bond between aromatic units in lignin. Temperatures in excess of 400 °C are required to cleave carbon–carbon bonds (i.e., representing 5-5′, β-1 and β-5 bonds), which are also present in small percentages in sugarcane lignin. The aryl-aryl ether bond could only be cleaved below a temperature of 400 °C by applying an appropriate base catalyst.

In addition, Roberts (2008) showed that temperatures greater than 320 °C were not suitable for lignin depolymerization as mentioned above this temperature hydrolysis products are degraded and reform pyrolysis products with higher molecular weights. Monomer-rich oil yields were also maximized at around 290 °C (Roberts 2008, Olarte 2011).

Temperatures of between 280 and 320 °C have consistently shown the greatest yield of monomeric-rich oil from the alkaline hydrolysis of lignin and that the higher the temperature, the greater the proportion of phenol.

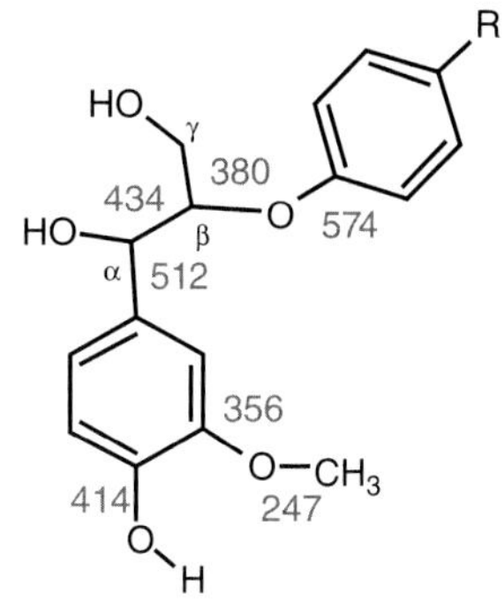

Figure 7.5 Approximation of the bond energies comprising the typical β-O-4 lignin linkage structure (highlighted bond energies are in units of kJ/mol).

Quantum calculations of the β-O-4 bond connectivity between LMC showed that the C_β-O bond (bond energy of 380 kJ/mol) was easier to crack than the C_α–C_β (434 kJ/mol) bond. The O-4 bond (574 kJ/mol) was the most stable bond and the 1-α bond the next (512 kJ/mol) (Liu *et al.* 2011). Catalytic hydrodeoxygenation studies of guaiacol revealed that the bond energy of the ArO-C bond was 247 kJ/mol, the C-OMe bond 356 kJ/mol, and the C-OH bond (hydroxyl) 414 kJ/mol (Bui *et al.* 2011). As an approximation to bond strength variations, Figure 7.5 shows the bond energies of the β-O-4 lignin linkage structure as detailed in the presented literature (Bui *et al.* 2011, Liu *et al.* 2011).

7.7.2 Reaction pressure

Pressure and temperature both influence the density of the solution and subsequently the yields and the reaction kinetics in the depolymerization of lignin. Approaching supercritical conditions, the solubility of alkali salts in solution reduces and the dielectric constant (extent of hydrogen bonding) also reduces allowing increased solubility of organics. In supercritical conditions, higher pressures imply higher densities. Higher densities were shown to improve alkali solubility, promote higher proportions of phenol, and overall yields from diphenyl ether and benzyl phenyl ether structures. For lignin, pressures above the water vapour saturation pressure corresponding to the saturation temperature of 320 °C showed a reduction in overall oil yield, as a result of a large reduction in the oligomer proportions of the oil (Roberts 2008).

7.7.3 Reaction time

It is evident from the literature that the shorter the reaction time, the greater the portion of monomeric products such as syringol and guaiacol formed. Longer reaction times favor the more saturated phenolic structures including phenol. Allowing the reaction to run for an extended periods of time should establish thermodynamic equilibrium products.

7.7.4 Lignin loading

It was identified (Roberts 2008) that the lower the concentration of lignin in solution the greater the portion of identified monophenols and the lower the oligomer fraction. However, in a commercial plant, there is an economic imperative to maximize solvent loading so as to reduce capital investment. A 10% lignin solution is typically used in the assessment so as to ensure complete lignin solubility. Vigneault *et al.* (2006) had also shown that recovered monomer content reduced significantly for steam exploded aspen lignin when the initial percent lignin in solution was greater than 10%.

7.7.5 Alkali molarity

Higher alkali molarity, assuming solubility limitations are not reached, favor higher yields of oil reaching a plateau at about 2.5 wt% of alkali (Roberts 2008) as well as reduced char or insoluble formation (Miller *et al.* 2002). High alkali concentrations are not extensively reported most likely because of the concerns associated with the requirement to recover the alkali for reuse.

7.7.6 Monomer separation

Gosselink (2011) identified the removal of monomers to be beneficial in increasing overall monomer yields by limiting oligomerization. In the context of alkaline hydrolysis, this could be achieved by a reactive distillation column (steam distillation via venting) or reaction in an organic counter current extraction system to remove monophenols as they are produced.

7.7.7 Lignin structure

It has been shown that lower-molecular-weight lignin has the potential to produce higher total monomer yields, including phenol and catechol yields (Zabaleta 2012), compared with higher-molecular-weight lignin depolymerized at the same conditions. Furthermore various papers have detailed the use of free-radical scavengers such as hydrogen, 2-naphthol, or phenol compounds to limit the extent of repolymerization and condensed products such as tars and char from being produced (Dorrestijn *et al.* 2000, Kleinert and Barth 2008a, Zakzeski *et al.* 2010). In addition, compounds that degrade to hydrogen (formate, formic acid, small organic acids) have also shown beneficial results in terms of increasing monomer yields. Hence both the initial lignin feedstock structure and the type of initial degradation products produced will influence the extent of overall lignin depolymerisation.

Finally another structural-based strategy to reduce product complexity and repolymerization is to genetically or chemically modify the monolignols in the source biomass to the same monolignol unit type before lignin depolymerization.

It is reasonable to expect that if there was only one monolignol present in the lignin structure then the depolymerization products would be less variable.

References

Barbier, J., Charon, N., Dupassieux, N., Loppinet-Serani, A., Mahé, L., Ponthus, J., Courtiade, M., Ducrozet, A., Quoineaud, A.-A. and Cansell, F. (2012). Hydrothermal conversion of lignin compounds. A detailed study of fragmentation and condensation reaction pathways. *Biomass and Bioenergy* **46**:479–491.

Beauchet, R., Monteil-Rivera, F. and Lavoie, J. M. (2012). Conversion of lignin to aromatic-based chemicals (L-chems) and biofuels (L-fuels). *Bioresource Technology* **121**:328–334.

Bui, V. N., Laurenti, D., Afanasiev, P. and Geantet, C. (2011). Hydrodeoxygenation of guaiacol with CoMo catalysts. Part I: promoting effect of cobalt on HDO selectivity and activity. *Applied Catalysis B: Environmental* **101**(3–4):239–245.

Buranov, A. U. and Mazza, G. (2008). Lignin in straw of herbaceous crops. *Industrial Crops and Products* **28**(3):237–259.

del Río, J. C., Lino, A. G., Colodette, J. L., Lima, C. F., Gutiérrez, A., Martínez, Á. T., Lu, F., Ralph, J. and Rencoret, J. (2015). Differences in the chemical structure of the lignins from sugarcane bagasse and straw. *Biomass and Bioenergy* **81**:322–338.

Dorrestijn, E., Laarhoven, L. J. J., Arends, I. W. C. E. and Mulder, P. (2000). The occurrence and reactivity of phenoxyl linkages in lignin and low rank coal. *Journal of Analytical and Applied Pyrolysis* **54**(1–2):153–192.

Dunn, K. G. (2014). Conversion of sugar cane lignin into aromatic products and fractionation of products for industrial use. Doctor of Philosophy, Queensland University of Technology.

Elliott, D. C., Biller, P., Ross, A. B., Schmidt, A. J. and Jones, S. B. (2015). Hydrothermal liquefaction of biomass: Developments from batch to continuous process. *Bioresource Technology* **178**:147–156.

Erdocia, X., Prado, R., Corcuera, M. Á. and Labidi, J. (2014). Base catalyzed depolymerization of lignin: influence of organosolv lignin nature. *Biomass and Bioenergy* **66**:379–386.

Fang, Z., Sato, T., Smith Jr., R. L., Inomata, H., Arai, K. and Kozinski, J. A. (2008). Reaction chemistry and phase behavior of lignin in high-temperature and supercritical water. *Bioresource Technology* **99**(9):3424–3430.

Gellerstedt, G., Majtnerova, A. and Zhang, L. (2004). Towards a new concept of lignin condensation in kraft pulping. Initial results. *Comptes Rendus Biologies* **327**(9–10):817–826.

Gosselink, R. J. A. (2011). Lignin as a renewable aromatic resource for the chemical industry. PhD Thesis, Wageningen University, Wageningen, Netherlands, 7 Dec 2011.

Grabber, J. H. (2005). How do lignin composition, structure, and cross-linking affect degradability? A review of cell wall model studies. *Crop Science* **45**(3):820.

Gratzl, J. S. and Chen, C.-L. (1999). Chemistry of pulping: lignin reactions. *Lignin: Historical, Biological, and Materials Perspectives*, American Chemical Society **742**:392–421.

Hatfield, R., Marita, J., Frost, K., Grabber, J., Ralph, J., Lu, F. and Kim, H. (2009). Grass lignin acylation: p-coumaroyl transferase activity and cell wall characteristics of C3 and C4 grasses. *Planta* **229**(6):1253.

Johnson, D. K., Chornet, E., Zmierczak, W. W. and Shabtai, J. S. (2002). Conversion of lignin into a hydrocarbon product for blending with gasoline. *Fuel Chemistry Division Reprints* **47**(1):380.

Kleinert, M. and Barth, T. (2008a). Phenols from lignin. *Chemical Engineering & Technology* **31**(5):736–745.

Kleinert, M. and Barth, T. (2008b). Towards a lignincellulosic biorefinery: Direct one-step conversion of lignin to hydrogen-enriched biofuel. *Energy & Fuels* **22**(2):1371–1379.

Kleinert, M. and Barth, T. (2009). One-step conversion of solid lignin to liquid products. WO patent. 2009/021733.

Kleinert, M., Gasson, J. R. and Barth, T. (2009). Optimizing solvolysis conditions for integrated depolymerisation and hydrodeoxygenation of lignin to produce liquid biofuel. *Journal of Analytical and Applied Pyrolysis* **85**(1–2):108–117.

Lavoie, J.-M., Baré, W. and Bilodeau, M. (2011). Depolymerization of steam-treated lignin for the production of green chemicals. *Bioresource Technology* **102**(7):4917–4920.

Liguori, L. and Barth, T. (2011). Palladium-Nafion SAC-13 catalysed depolymerisation of lignin to phenols in formic acid and water. *Journal of Analytical and Applied Pyrolysis* **92**(2):477–484.

Liu, J.-Y., Wu, S.-B. and Lou, R. (2011). Chemical structure and pyrolysis response of beta-O-4 lignin model polymer. *Bioresources* **6**(2):681.

Long, J., Xu, Y., Wang, T., Shu, R., Zhang, Q., Zhang, X., Fu, J. and Ma, L. (2014). Hydrothermal depolymerization of lignin: understanding the structural evolution. *BioResources* **9**(4):7162–7175.

Lu, F. and Ralph, J. (1997). Derivatization followed by reductive cleavage (DFRC Method), a new method for lignin analysis: protocol for analysis of DFRC monomers. *Journal of Agricultural and Food Chemistry* **45**(7):2590–2592.

Lu, F. and Ralph, J. (1998). The DFRC method for lignin analysis. 2. Monomers from isolated lignins. *Journal of Agricultural and Food Chemistry* **46**(2):547–552.

Man, X., Okuda, K., Ohara, S., Umetsu, M., Takami, S. a. and Adschiri, T. (2005). Disassembly of organosolv lignin in supercritical fluid - phenol as a suppressor for repolymerization. *Journal of the Japan Institute of Energy* **84**:486–490.

Miller, J. E., Evans, L., Mudd, J. E. and Brown, K. A. (2002). Batch microreactor studies of lignin depolymerization by bases. 2. Aqueous solvents. Sandia National Laboratories - SAND Report SAND2002-1318.

Olarte, M. V. (2011). Base-catalyzed depolymerization of lignin and hydrodeoxygenation of lignin model compounds for alternative fuel production. PhD dissertation, Georgia Institute of Technology.

Onwudili, J. A. (2015). Influence of reaction conditions on the composition of liquid products from two-stage catalytic hydrothermal processing of lignin. *Bioresource Technology* **187**:60–69.

Pandey, M. P. and Kim, C. S. (2011). Lignin depolymerization and conversion: a review of thermochemical methods. *Chemical Engineering & Technology* **34**(1):29–41.

Pińkowska, H., Wolak, P. and Złocińska, A. (2012). Hydrothermal decomposition of alkali lignin in sub- and supercritical water. *Chemical Engineering Journal* **187**(0):410–414.

Rahimi, A., Ulbrich, A., Coon, J. J. and Stahl, S. S. (2014). Formic-acid-induced depolymerization of oxidized lignin to aromatics. *Nature* **515**(7526):249–252.

Roberts, V. M. (2008). Homogeneous and heterogeneous catalyzed hydrolysis of lignin. PhD, Technische Univeritat Munchen.

Roberts, V. M., Stein, V., Reiner, T., Lemonidou, A., Li, X. and Lercher, J. A. (2011). Towards quantitative catalytic lignin depolymerization. *Chemistry – A European Journal* **17**(21):5939–5948.

Saisu, M., Sato, T., Watanabe, M., Adschiri, T. and Arai, K. (2003). Conversion of lignin with supercritical water in phenol mixtures. *Energy & Fuels* **17**(4):922–928.

Schuchardt, U. and Goncalves, R. (1990). Production of chemical and pharmaceutical feedstocks from hydrolytic eucalyptus lignin. In *Biomass for Energy and Industry*, vol. **2**, pp. 897–902. G. Grassi, G. Gosse, and G. dos Santos eds. Elsevier Applied Science: London, New York.

Schuchardt, U., Rodrigues, J. A. R., Cotrim, A. and Costa, J. L. M. (1993). Liquefaction of hydrolytic eucalyptus lignin with formate in water, using batch and continuous-flow reactors. *Bioresource Technology* **44**(2):123–129.

Shabtai, J. S. and Zmierczak, W. W. (2001). Process for the conversion of lignin to reformulated partially oxygenated gasoline. United states patent. 6,172,272.

Sun, J.-X., Sun, X.-F., Sun, R.-C., Fowler, P. and Baird, M. S. (2003). Inhomogeneities in the chemical structure of sugarcane bagasse lignin. *Journal of Agricultural and Food Chemistry* **51**(23):6719–6725.

Sun, J. X., Wang, H., Zhang, G., Fowler, P. and Rajaratnam, M. (2011). *Extraction and characterization of lignins from maize stem and sugarcane bagasse.* Journal of Applied Polymer Science **120**(6):3587–3595.

Thring, R. W. (1994). Alkaline degradation of ALCELL® lignin. *Biomass and Bioenergy* **7**(1–6):125–130.

Thring, R. W., Chornet, E., Bouchard, J. and Vindal, P. (1990). Characterization of lignin residues derived from the alkaline hydrolysis of glycol lignin. *Canadian Journal of Chemistry* **68**(1):82.

Toledano, A., Serrano, L. and Labidi, J. (2014). Improving base catalyzed lignin depolymerization by avoiding lignin repolymerization. *Fuel* **116**:617–624.

Vallejos, M. E., Felissia, F. E., Cruvelo, A. A. S., Zambon, M. D., Ramos, L. and Area, M. C. (2011). Chemical and physico-chemical characterisation of lignins obtained from ethanol-water fractionation of bagasse. *BioResources* **6**(2):1158–1171.

Vigneault, A., Johnson, D. K. and Chornet, E. (2006). Advance in the thermal depolymerisation of lignin via base-catalysis. *Science in Thermal and Chemical Biomass Conversion* **2**:1401–1419.

Wahyudiono, Sasaki, M. and Goto, M. (2008). Recovery of phenolic compounds through the decomposition of lignin in near and supercritical water. *Chemical Engineering and Processing: Process Intensification* **47**(9–10):1609–1619.

Wen, J., Xue, B. and Xu, F. (2012). Unveiling the structural heterogeneity of bamboo lignin. *Bioenergy Research* **5**:886–903.

Wen, J. L., Sun, S. L., Xue, B. L. and Sun, R. C. (2013a). Quantitative structural characterization of the lignins from the stem and pith of bamboo (Phyllostachys pubescens). *Holzforschung* **67**(6):613–627. ISSN (Online) 1437-434X, ISSN (Print) 0018-3830, DOI: 10.1515/hf-2012-0162.

Wen, J. L., Sun, S. L., Xue, B. L. and Sun, R. C. (2013b). Recent advances in characterization of lignin polymer by solution-state nuclear magnetic resonance (NMR) methodology. *Materials* **6**(1):359–391.

Zabaleta, A. T. (2012). *Lignin extraction, purification and depolymerisation study.* Thesis, University of the Basque Country.

Zakzeski, J., Bruijnincx, P. C. A., Jongerius, A. L. and Weckhuysen, B. M. (2010). The catalytic valorization of lignin for the production of renewable chemicals. *Chemical Reviews* **110**(6):3552–3599.

Zmierczak, W. W. and Miller, J. D. (2006). Process for the catalytic conversion of lignin to liquid bio-fuels. World patent. WO/2006/119357.

Zmierczak, W. W. and Miller, J. D. (2008). Process for the catalytic conversion of lignin to liquid bio-fuels and novel bio-fuels. United states patent. 11/932,408.

CHAPTER 8

Conversion of sugarcane carbohydrates into platform chemicals

Darryn W. Rackemann, Zhanying Zhang and William O.S. Doherty
Centre for Tropical Crops and Biocommodities, Queensland University of Technology (QUT), Brisbane, Australia

8.1 Introduction

Due to historically abundant supply and relatively low cost, the bulk of energy needs and vast majority of synthetic products and chemicals have been sourced from fossil-based resources. Increasing the supply of biobased products has gained importance in recent years as a step toward addressing the environmental issues associated with the use of fossil-based products. Biomass is the only renewable resource of fixed carbon, and can be used for the production of fuels and chemicals (Corma *et al.* 2007). Only 3–4% of the 150 billion tons of biomass produced each year by photosynthesis is used by humans for food and nonfood purposes (Wondu 2006). Low-value agricultural plants such as trees and grasses, energy crops, and food crop by-products are a preferred source of biomass for conversion into chemical products from both a techno-economic and socioeconomic point of view, as the biomass feedstock does not compete with food availability (Clark *et al.* 2006).

Worldwide, more than 1800 million tons of sugarcane is processed to produce over 250 million tons of raw sugar, which is mostly processed into refined sugar. The sugar manufacturing process also generates more than 40 million tons of molasses (>40% sugar content), which finds low value use in animal feed and in ethanol production. Bagasse, which amounts to more than 600 million tons (i.e., 25–30% of the sugarcane plant), is mainly used as fuel for combustion in boilers to generate process steam and electricity for the factory (Rein 2007). Surplus bagasse not used for the energy requirement of the factory finds use in cogeneration, and options for its potential use for the production of biofuels and commodity chemicals are currently being investigated around the world. The sugar industry has a major advantage over other agricultural-based industries because it has infrastructure in place for collecting and processing biomass.

Sugarcane-Based Biofuels and Bioproducts, First Edition. Edited by Ian M. O'Hara and Sagadevan G. Mundree.

8.1.1 Bagasse

The functionality of bagasse as a resource for conversion into fuels and chemicals depends upon the chemical constituents and physical properties. Bagasse contains varying amounts of key organic polymers (cellulose, hemicellulose, and lignin) and minor amounts of organic extractives (proteins, lipids) and inorganic materials (ash). The actual composition of the material primarily depends on sugarcane variety, age, and growth conditions of the plant.

In the production of sugar from sugarcane, the leaves and tops of sugarcane plant are separated from the stalk either by burning or during the mechanical harvesting process where the stalk is cut into smaller billets for transport to the factory. The sugarcane stalk consists of two main components, parenchyma cells and fibrovascular bundles. The parenchyma cells are thin-walled cells that store the sucrose that is synthesized in the leaves. The fibrovascular bundles transport nutrients, water, and sugars within the plant and provide structural strength.

In the raw sugar production process, the structural order of the fibers in the sugarcane plant is broken down by the sugarcane preparation and milling equipment in order to rupture the parenchyma cells to expose sucrose for extraction. At the end of the milling process, the resultant bagasse is a mixture of fiber components of varying length and composition (Pena *et al.* 2000). Bagasse from the sugarcane milling process generally contains 44–53% moisture, 1–2% soluble solids, 1–5% insoluble solids (dirt and ash), with the remainder being lignocellulosic fiber (Ebert 2008). Fiber composition of bagasse varies, but is typically 36–54% cellulose, 24–32% hemicellulose, and 16–28% lignin on a dry basis. Cellulose and hemicellulose can be hydrolyzed to monomer sugars mainly consisting of glucose, xylose, and arabinose.

8.1.2 Biorefining

Most organic chemicals produced from fossil-based resources can also be produced from biomass. Several studies have assessed the range of potential chemical products from biomass and, in fact, more than 300 potential chemical products have been identified (Werpy and Petersen 2004). These have the potential to further produce thousands of products similar to the petrochemical industry.

A sugar factory can be considered a biorefinery as the factory not only produces sugar and molasses, but other by-products of the sugar production process. For instance, bagasse and clarifier mud can be further converted into saleable products (Rein 2007). Typical products that are able to be produced in a sugarcane biorefinery are shown in Figure 8.1.

Molasses is commonly used as a feedstock for colocated alcohol distilleries but can also be used for animal feed and other food products. The main use of bagasse is as a fuel for the factory boilers, while clarifier mud can be converted into organic fertilizer or soil conditioner products. Bagasse that is surplus to the energy needs of the factory can be used to produce electricity (via cogeneration)

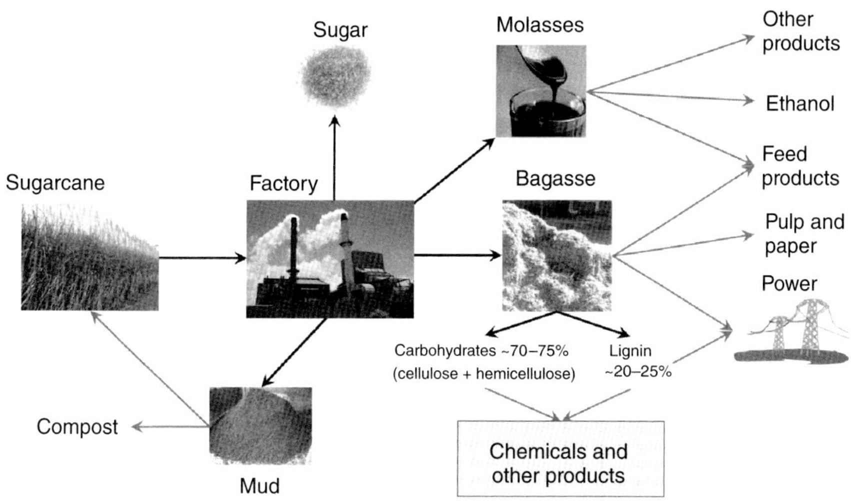

Figure 8.1 Current and potential products from sugarcane.

or as a feedstock for pulping and fiber products. However, product diversification in traditional sugar factories is limited with few high-value products being produced.

While bagasse is generally considered to be a relatively inexpensive feedstock, molasses and raw sugar can also be considered low cost if they can be transformed into higher-value products. The benefits of using molasses and raw sugar as feedstocks include (i) minimum pretreatment requirements as the sugars are more readily available, (ii) higher product yields, (iii) simpler recovery process of products, (iv) minimum product purification steps, and (v) less complicated processing technologies.

A substantial amount of research is currently being carried out worldwide to identify attractive chemical transformations of biomass into bulk organic chemicals or higher value products. A biorefinery approach with diversification into multiple products allows greater utilization of biomass resources. Biorefineries have a similar objective to petroleum refineries in that they (i) fractionate the feed source; (ii) convert intermediate fractions to products or chemical intermediates; and (iii) further process the chemical intermediates to higher-value products. In the case of biorefineries, various constituents of biomass can be made into useful platform chemicals through biochemical and thermochemical conversion steps. Platform chemicals are building blocks that have multitudes of uses in synthesis of nylons, synthetic rubbers, binders, resins, solvents, preservatives, nutraceuticals, herbicides, plasticizers, fuel additives, and pharmaceuticals (Kamm and Kamm 2004). The residues of biorefinery processing steps can be utilized for production of power and heat.

There are several routes for converting biomass into liquid fuels and chemicals. These are based on thermochemical or biochemical pathways. Over the past decade, a significant amount of research has focused on techniques that convert biomass to sugars (using acids or enzymes), which are subsequently fermented to ethanol in the presence of yeasts. One of the challenges in commercializing this technology is the high costs associated with lignocellulose pretreatment and saccharification.

Thermochemical processes (excluding combustion) can be broadly divided into three different conversion methods namely gasification, pyrolysis, and direct liquefaction. These processes are potentially highly efficient as most of the organic components in biomass are amenable to conversion at the relatively high temperatures involved. A significant component (20–30%) of lignocellulosic biomass is lignin, which is not readily amenable to microbial fermentation processes. Thermochemical processes generally require shorter reaction times (from seconds to minutes) compared with biochemical conversion processes (from days to weeks) and thus have several advantages over the latter.

Liquefaction refers to reactions that are typically carried out at temperatures between 250 and 450 °C and high pressures (50–200 bar). Liquefaction has energetic advantages as a thermochemical process since the operating pressure avoids phase changes and their associated enthalpic energy demands (Peterson *et al.* 2008). Liquefaction processes are typically suited to wet feedstocks such as grasses, energy crops, and food crop by-products. Lower temperature acid hydrolysis or solvolysis processes form a subset of liquefaction processes. One of the advantages of the incorporation of this technology into sugar factories is that much of the existing infrastructure and utilities currently available in a sugar factory (e.g., steam) can be used.

Many biorefinery concepts propose the use of lignin as a source of energy for combustion to generate process steam for heating duties (e.g., distillation) or power generation. Approaches currently being investigated to improve the economics of biorefineries include value adding to the lignin component not used in the primary process. Chemical hydrolysis processes are the focus of this chapter with regard to conversion of the carbohydrate content of lignocellulosic biomass such as bagasse to chemicals. While not directly amenable to conversion of lignin, this technology can be coupled with pretreatment/fractionation technologies to separate the lignin for value adding. Pretreatment technologies not only are able to remove lignin but can assist with making the carbohydrates more amenable to subsequent conversion. More details on derivation of valuable products from lignin can be found elsewhere (Bozell *et al.* 2007, Zakzeski *et al.* 2010, Doherty *et al.* 2011).

8.2 Platform chemicals

The National Renewable Energy Laboratory (Denver, USA) identified a number of key sugar-derived platform chemicals that can be produced from

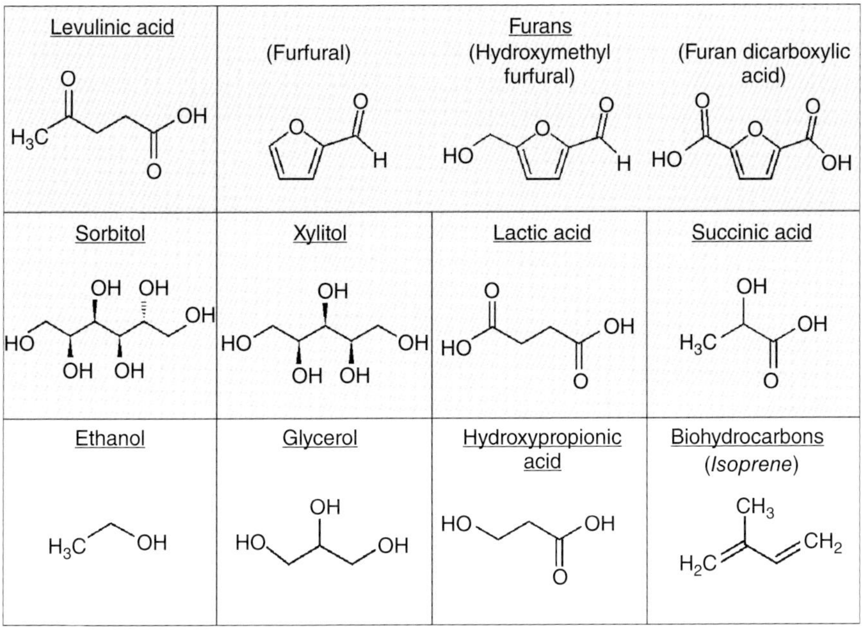

Figure 8.2 Top 10 chemicals produced from sugars.

carbohydrates (Werpy and Petersen 2004, Bozell and Petersen 2010). The product identification was proposed as a guide for research and reflects economics, industrial viability, size of markets, and the ability of a compound to serve as a platform for the production of derivatives. The list of target platform chemicals was refined in 2010 based on advances in technology developments (Bozell and Petersen 2010) and is shown in Figure 8.2. Many of the compounds (Figure 8.2) are produced via fermentation pathways, whereas those compounds that are shaded can be produced by thermochemical pathways. Furans and organic acids such as levulinic acid and formic acid can be produced through acid-catalyzed dehydration and hydrolysis of hexose and pentose sugars, and these platform chemical hydrolysis products are the focus of this chapter.

Polyols (sorbitol and xylitol) are mostly produced through catalytic hydrogenation of sugars but can also be produced via biochemical pathways. Lactic acid is a by-product from the thermochemical conversion of biomass but is also commercially produced by microbial fermentation. Similarly, succinic acid is industrially produced by the hydrogenation of maleic acid but can also be produced via fermentation of glucose.

The most prominent biofuel currently derived from carbohydrates is ethanol. As a fuel, ethanol is limited in utility due to its low energy content, potential for water contamination, and high production costs when produced from cellulosic sources. A further inherent disadvantage of cellulosic ethanol is that one-third of the available carbon is expelled as CO_2 during the production process. Ethanol may offer more promise as a solvent or platform chemical, particularly targeting higher value products or in the mitigation of CO_2 emissions (Rass-Hansen *et al.* 2007).

Although glycerol is a readily available by-product of biomass transformation to biodiesel, it can also be produced by microbial synthesis from sugars and has strong prospects as a primary building block in biorefineries. Isoprene is a high-value biohydrocarbon produced via biosynthetic routes and is used in rubber products. Hydroxypropionic acid is a carboxylic acid derived from microbial sources and used in the industrial production of acrylates, polymers, and plastics.

8.2.1 Furans

Furans are versatile chemicals and can be readily transformed into solvents, acrylate and polyester monomers, pharmaceuticals, and agrichemicals. Furans are heterocyclic organic compounds that consist of a five-membered aromatic ring with four carbon atoms and one oxygen atom. Furans are considerably more reactive than benzene in electrophilic substitution reactions, due to the electron-donating effects of the oxygen heteroatom.

In typical acid hydrolysis processes, cellulose is depolymerized and hydrolyzed to glucose while hemicellulose is hydrolyzed to hexose (e.g., glucose, mannose) and pentose sugars (e.g., xylose, arabinose). The hexose and pentose sugars are further converted to 5-hydroxymethylfurfural (HMF) and furfural as shown in Figure 8.3. The dehydration reactions of hexose and pentose sugars to HMF and furfural follow similar reaction pathways with protonation of three carbon atoms on the sugar ring to eliminate three water molecules (Agirrezabal-Telleria *et al.* 2014). Proposed reaction mechanisms include isomerization reactions between glucose and fructose, and between xylose and xylulose followed by either cyclic or acyclic dehydration.

Lignin components can also affect carbohydrate decomposition reactions with acid-soluble lignin-derived components polymerizing with sugars or other compounds present in the reaction mixture to form undesired insoluble polymeric materials (humins). Formation of humins also occurs from furfural and HMF degradation through aldol addition and condensation reactions (Patil and Lund 2011).

8.2.2 Furfural

Furfural is a versatile industrial solvent produced in large quantities worldwide from the acid-catalyzed dehydration of pentose sugars. Most furfural produced is converted to furfuryl alcohol and, while also a solvent, it is primarily used in the manufacture of foundry resins, adhesives, and wetting agents (Hoydonckx *et al.* 2000).

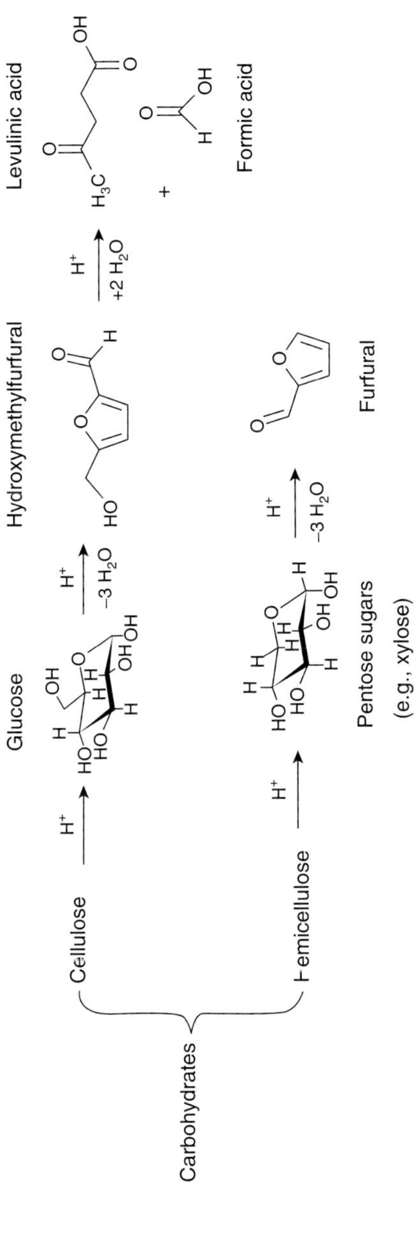

Figure 8.3 Acid-catalyzed decomposition of carbohydrates.

Other furfural derivatives include agricultural chemicals (herbicides, insecticides, and preservatives), perfumes and flavoring agents (furan, furanol, nitrofuran, furamone), plastics resins, and synthetic fiber (nylon), dyes, rubbers, and paints (Dias *et al.* 2010). Furfural can be converted to kerosene and diesel range biofuels through condensation reactions (Huber and Dumesic 2006) or hydrogenated to a range of solvents and fuel additives. These and other derivatives are shown in Figure 8.4.

8.2.3 HMF

In aqueous media, the major intermediate dehydration product of hexose sugars is HMF or its ester, ethoxymethylfurfural (EMF), when converted in alcoholic mediums. HMF is unstable under acidic aqueous conditions and readily hydrolyzes to levulinic acid and formic acid (Figure 8.3). With heterogeneous catalysts, HMF can be catalytically converted to energy-dense biofuels such as methylfuran (MF) and dimethylfuran (DMF) (Roman-Leshkov *et al.* 2007) or oxidized to highly functional monomers such as furan dicarbaldehyde (FDC) and furan dicarboxylic acid (FDCA) (Lewkowski 2001). FDCA is a monomer for the production of many polymers and has been touted as a green replacement for terephthalic acid and polyethylene terephthalate, principally used as a precursor to polyester for clothing and plastic bottles (Boisen *et al.* 2009).

Recent research has uncovered acid-catalyzed reactions for the selective production of halogenated furanics such as chloromethylfurfural (CMF) (Mascal and Nikitin 2010) and bromomethylfurfural (BMF) (Kumari *et al.* 2011). These furanics are hydrophobic molecules allowing biphasic systems to be used to extract the furanic product in high yields. Furthermore, the molecules are very reactive and the conversions of halogenated furanics into various valuable furan moieties and levulinates have been demonstrated (see Figure 8.5).

8.3 Organic acids

8.3.1 Levulinic acid

Levulinic acid is an ideal platform chemical that can be utilized to produce a number of biochemicals including succinic acid, resins, polymers, herbicides, pharmaceuticals and flavoring agents, solvents, plasticizers, surfactants, antifreeze agents, and biofuels/oxygenated fuel additives as shown in Figure 8.6 (Bozell *et al.* 2000).

Alternatively, production of higher-value levulinic acid derivatives has been studied broadly. Reactive extraction methods can be used to produce levulinate ester products that are more easily separated from the reaction mixture and can be used as food, flavoring, and fragrance agents (Ayoub 2008).

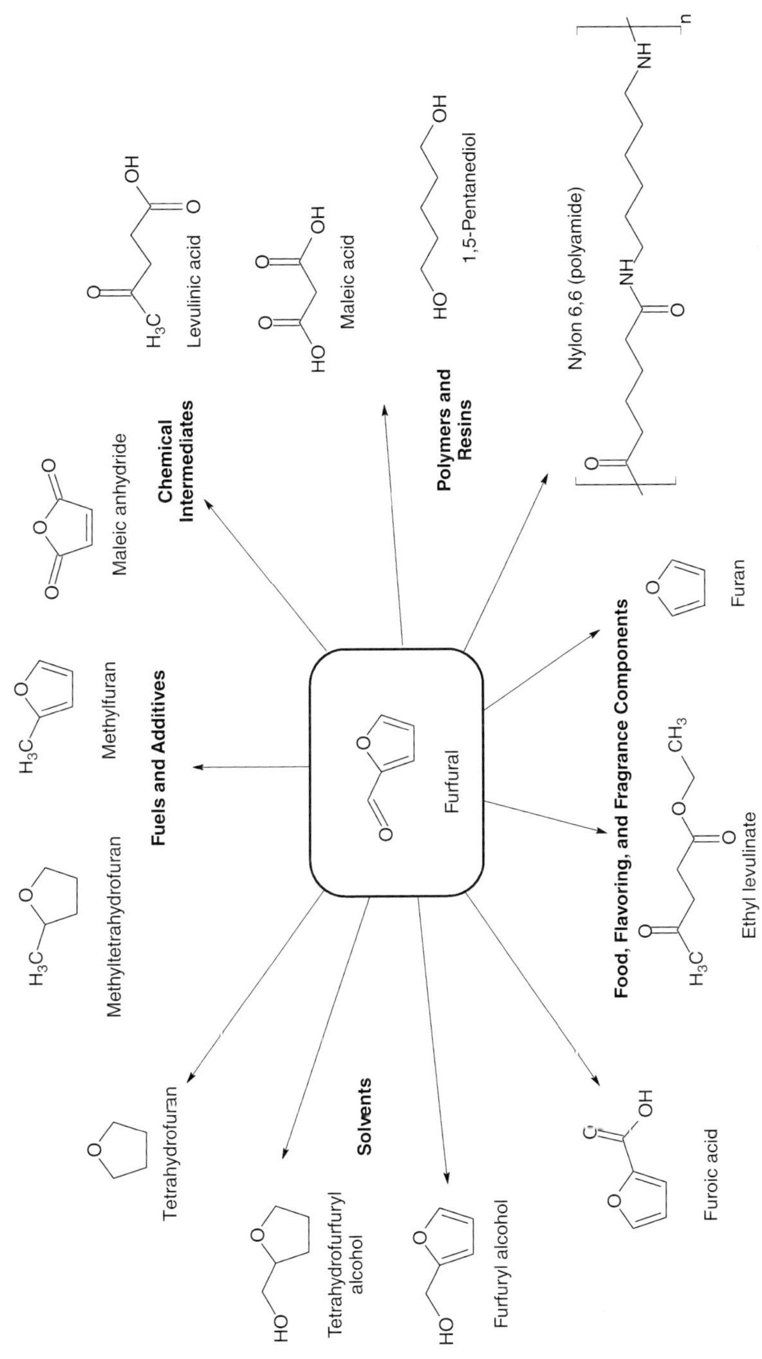

Figure 8.4 Furfural as a platform chemical.

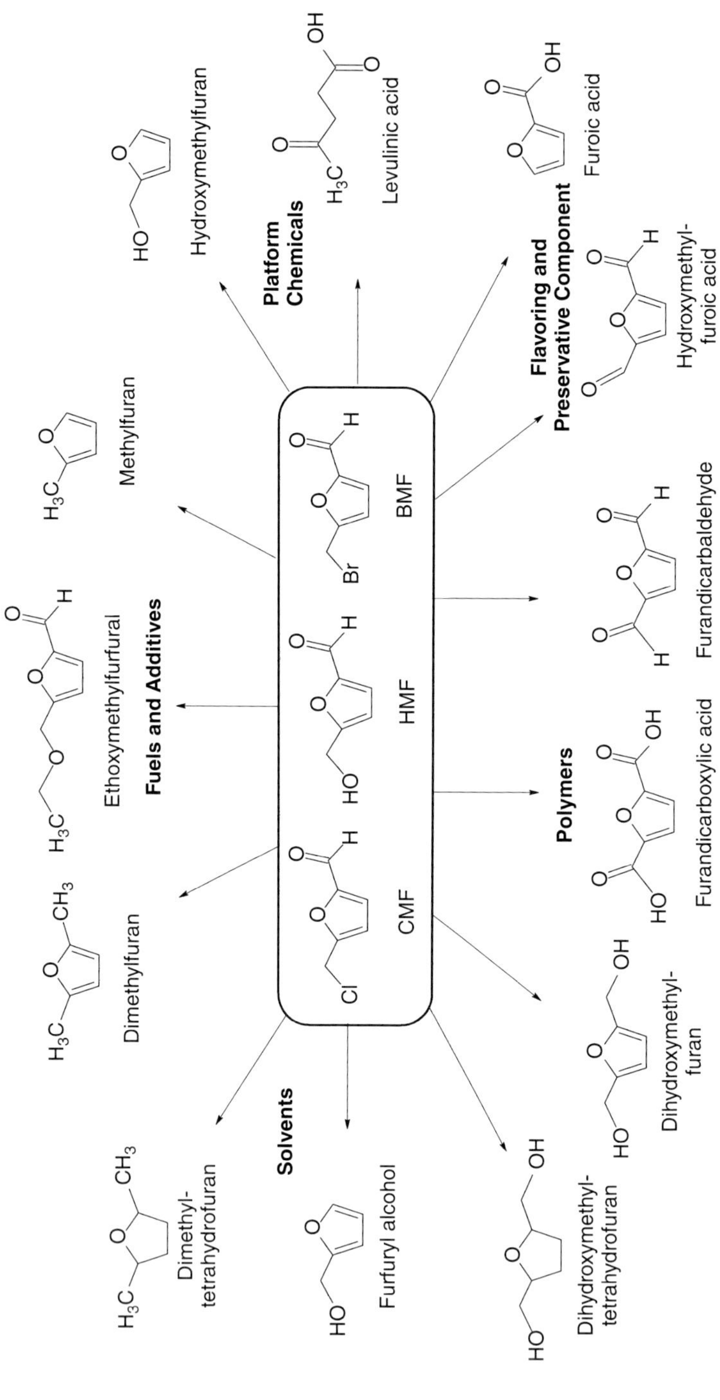

Figure 8.5 HMF/CMF/BMF as a platform chemical.

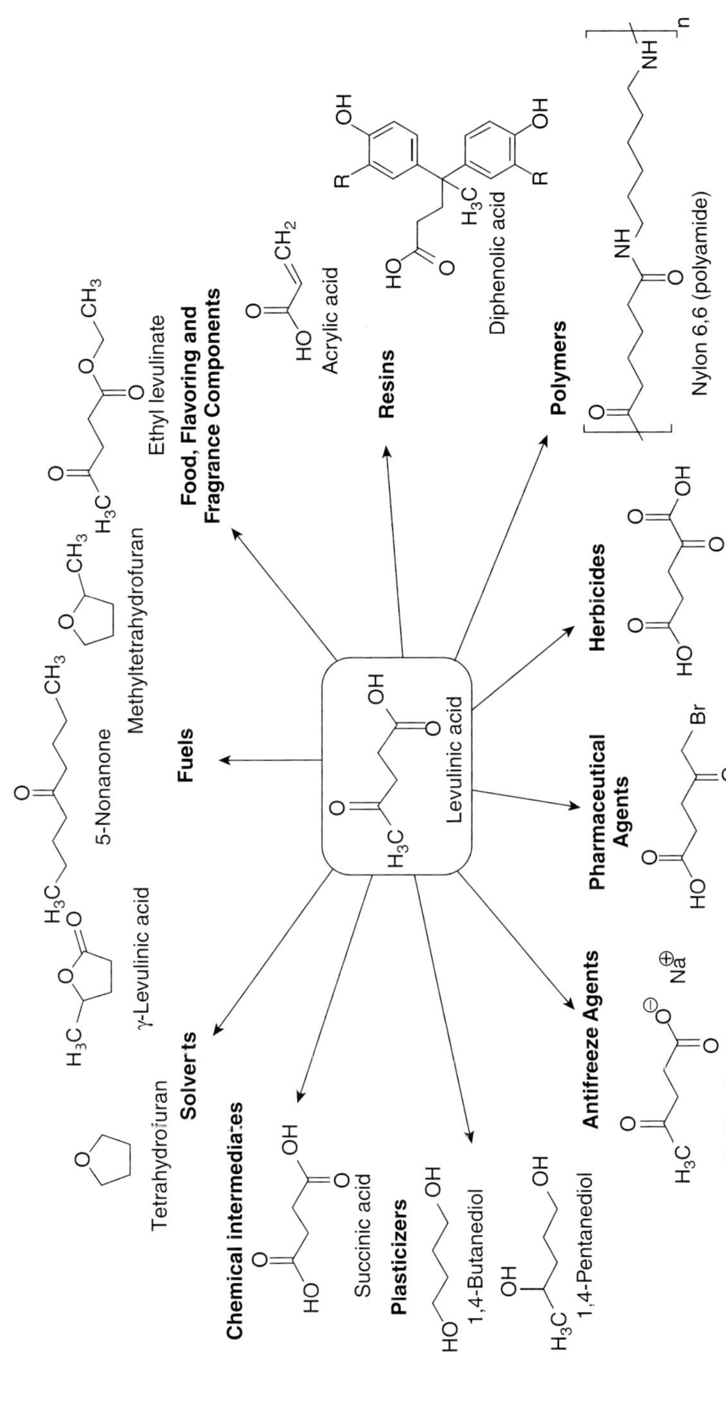

Figure 8.6 Levulinic acid as a platform chemical.

Ethyl levulinate is used as an oxygenated additive (Fitzpatrick 2006). Levulinic acid can also be condensed with aromatic alcohols to produce diphenolic acid, which is used in production of polymers, lubricants, fire-retardant materials, paints, and as a replacement for bisphenol A. Levulinic acid can be readily halogenated to yield organic halides such as 5-bromolevulinic acid that can be further converted to δ-aminolevulinic acid, which are useful pharmaceutical agents and herbicides.

Levulinic acid can be hydrogenated to γ-valerolactone (GVL) and methyltetrahydrofuran (MTHF) in very high (>90%) yields, and these additives can be readily blended with petroleum products to create cleaner-burning fuels with the advantage that they do not suffer from phase separation to become contaminated with water (cf. ethanol). Alternatively, GVL can be converted to liquid hydrocarbon fuels by decarboxylation to butene and oligomerization or through esterification and hydrogenation to 5-nonanone and subsequent conversion to alkanes (Alonso *et al.* 2010).

8.3.2 Formic acid

In acid-catalyzed conversion of hexose sugars to levulinic acid, formic acid is coproduced (Figure 8.3). Formic acid is a low-value commodity chemical used in the production of formaldehyde, rubber, plasticizers, pharmaceuticals, and textiles but in future may find increased utilization in fuel cell applications (Yu and Pickup 2008), as a feed source for hydrogen storage and production (Loges *et al.* 2008), or used in the transfer hydrogenation of levulinic acid to higher-value products.

8.4 Value of potential hydrolysis products

The widespread application of biobased products such as levulinic acid, furanics and their derivatives requires the production process to be economically viable, existence of sizable markets, and consumer acceptance of biobased products. To compete in chemical markets, biorefineries will require the manufacture of biobased products to directly or indirectly substitute or replace petroleum-based products or produce an intermediate that can be directly utilized in chemical production facilities (Fitzpatrick *et al.* 2010). Direct replacement implies that the same chemical is produced but from a different feedstock or using a new process to enable a competitive edge to be gained. Catalytic fast pyrolysis methods have been applied to biomass to produce benzene/toluene/xylene mixtures (BTX) and olefins (http://anellotech.com/tech.html), which are the building block chemicals of the petrochemical industry (Carlson *et al.* 2011).

On the other hand, indirect substitution requires the bioproduct to be functionally similar (i.e., biofuels, solvents, and biodegradable monomers)

Table 8.1 Value-added products derived from carbohydrates (O'Hara *et al.* 2013).

	Compound	Value
Feed	Bagasse	\$100/*t*[*]
	Cellulose pulp	~\$500/*t*
	Glucose	\$600–700/*t*
Platform Chemical Products	Ethanol	\$600–800/*t*
	Formic acid/acetic acid	\$900/*t*
	Furfural	\$1500/*t*
	Succinic acid	\$1800/*t*
	Polyols (sorbitol/xylitol)	\$2000–3000/*t*
	Levulinic acid/esters	\$2000–3000/*t*
	HMF/HMF esters	\$2000–4000/*t*
Upgraded Products	Fuel additives	\$900/*t*
	Terephthalic acid/FDCA	\$900/*t*
	Furfuryl alcohol	\$2000/*t*
	Bisphenol – A/diphenolic acid	\$2300/*t*
	Polymers and resins	~\$2500/*t*
	Solvents	\$2000–5000/*t*
	Herbicides	>\$2500/*t*
	Flavoring agents	>\$5000/*t*

[*]Based on dry weight.

to petroleum-based products placing importance on quality control. The biodegradability and biocompatibility properties of biobased products compared with hydrocarbon-derived products should improve their marketability (Yan *et al.* 2009). Furthermore, metrics for biomass conversion processes (i.e., life cycle assessment) should also be considered for comparison to conventional processes to meaningfully define whether biobased products are environmentally better (Sheldon 2011).

Table 8.1 shows the approximate values of various feedstocks and typical products from carbohydrates, as well as upgraded derivative products from these platform chemicals.

The production of biochemicals such as polymers and resins is attractive due to their higher value when compared with feedstock cost and fuels. However, the production of biofuels is still promising due to the large demand and immense market size. The production of transportation fuels requires biofuels to have compatible (noncorrosive) and suitable combustion and chemical properties. One issue surrounding application of biofuels is the lack of up-to-date data on toxicological properties, which is needed from a regulatory perspective and details on exhaust emissions and environmental impacts on the atmosphere and hydrosphere (Simmie and Würmel 2013).

8.5 Current technology for manufacture of furans and levulinic acid

The majority of research into the production of platform chemicals such as furfural and levulinic acid from biomass has been conducted with mineral acid catalysts. The first commercial production of levulinic acid from starch began in 1940 by A.E. Staley Company (Moyer 1942). Levulinic acid production was typically produced from starch and simple sugars and required greater reaction temperatures compared with the production of furfural. High feedstock costs led to the investigation of new feedstocks but low yields were achieved as slow heating rates and nonhexose sugar components of the feedstocks promoted side reactions that reduced levulinic acid yields (Ghorpade and Hanna 1997).

Furfural has been industrially produced from agricultural residues and was developed by Quaker Oats company in the 1920s (Zeitsch 2000). In the production of furfural, pressurized steam is not only used to heat but also to strip the product from the reaction mixture. The steam and furfural vapor extracted are condensed and then separated using decanting and distillation. Typical commercial furfural yields are low, <50% due to furfural's instability as auto-oxidation and decomposition from heat and acidic conditions contribute to loss reactions (Dunlop 1948). The high consumption of steam and the release of a large amount of wastewater are also major drawbacks of the Quaker-Oats process.

Although furfural has been industrially produced from agricultural residues since the 1920s, in recent years the majority of production occurs in low-cost countries such as China. The two largest furfural production plants in the world are constructed alongside sugarcane factories as large amounts of sugarcane bagasse are a by-product of sugar production from sugarcane (Watson and Connors 2008).

Technology advancements such as better reactor designs and continuous operation allowed for yields of levulinic acid or furfural to improve to more than 70% (Rackemann and Doherty 2011). However, a limitation of these processes was that they targeted conversion of only the hexose or pentose sugars with the remaining material treated as a waste and burned to produce energy. One reason for this is the yield dependence on processing conditions (Rackemann *et al.* 2014). High yields of levulinic acid can be achieved across a range of variables as long as two out of the three conditions are met: high acid catalyst concentration, long reaction time, or high temperature. This is because levulinic acid is relatively stable once formed. On the other hand, high furfural yields are achieved under milder conditions and using dilute feed concentrations as furfural degradation is enhanced under higher reactant loading and harsh operating conditions.

For this reason, multistage processes were developed, which offer the potential for the use of different operating conditions to enable optimized and selective production of a number of products. In the late 1960s, a batch process for the

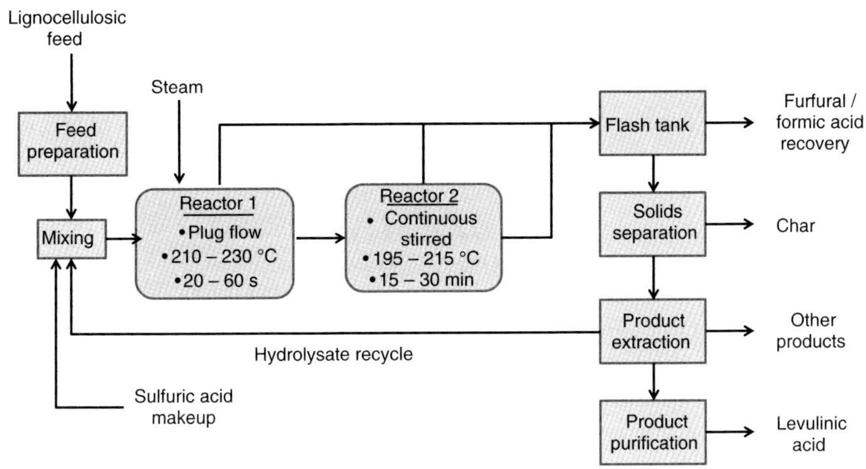

Figure 8.7 Schematic of Biofine process (Hayes *et al.* 2006).

concomitant production of levulinic acid and furfural from sugarcane residues was proposed using a two-step reaction process where furfural and other gases (acetic acid) were vented at lower temperatures before increasing the temperature to optimize levulinic acid production (Ramos-Rodriguez *et al.* 1968).

A recent and promising semi-commercial process utilizes the Biofine™ technology and involves a two-stage acid-catalyzed reaction process (Figure 8.7) to optimize process conditions (Hayes *et al.* 2006). In the first plug-flow reactor, the feed is dehydrated to HMF at between 210 and 230 °C (>3 MPa) for less than 30 s. Levulinic acid is produced in the second reactor at 195–215 °C (~1.5 MPa) for 15–30 min. The reaction conditions in the second reactor are such that formic acid and furfural are kept in the vapor phase to allow simple separation of these products. The variation in reactor volumes (residence times) reflects the fast cellulose conversion into sugars and subsequent slower sugar conversion into levulinic acid and the reactor system design minimizes conditions amenable to side-product formation. Fast heating rates assist the production of furfural, and the pressure reduction between reaction stages separates the furfural product into the vapor phase to reduce degradation reactions. In this way, 70–80% yields of both levulinic acid and furfural are achieved.

The small reactor volumes in the Biofine process minimize expensive, corrosion-resistant materials reducing the high equipment costs associated with mineral acid processes, but some waste disposal is still necessary and the process uses significant amounts of steam. Furthermore, processing difficulties have been experienced when working with some biomass feedstocks due to the presence of intractable materials (lignin) blocking the tubular reactors and the process also requires biomass of small particle size. These processing difficulties have, to date, restricted the commercial production of levulinic acid from biomass.

8.6 Technology improvements

There are a number of technological hurdles that historically have limited the economical manufacture of platform chemicals from carbohydrates including the following:

- *Poor yield from low cost feedstocks.* Lignocellulosics represent a low-cost source of pentose and hexose sugars but difficulties are encountered when processing lignocellulosics, which are related to the presence of lignin and formation of intractable materials. The nonselective nature of typical mineral catalysts used can also result in undesired side reactions (Bozell and Petersen 2010).
- *Inefficient product separation and recovery.* Distillation systems are typically used for separation and recovery. These systems are not economical due to the low product concentrations and the wide range of the boiling points of the products. Solvent extraction methods are complicated as by-products have similar solvation properties (polarity) and the method is limited in that small amounts of product left in the aqueous phase (low partitioning) require the use of large amounts of solvent to achieve high recovery.
- *Waste disposal and corrosion.* The predominant use of mineral acid catalysts leads to waste disposal issues and corrosion problems requiring expensive construction materials for both reactor and acid recovery plant.

To some extent, waste disposal can be mitigated through recycling the acid/solvent mixture at the expense of additional operating costs. Recycling will play a key role in the viability of the process and ensures zero effluent. Corrosion issues could be overcome by developing environmentally friendly processes utilizing less corrosive biodegradable acids, heterogeneous catalysts, or green solvent systems. Heterogeneous catalysts offer the greatest potential due to their simple and energy-efficient separation, reusable nature, limited corrosion, and waste disposal issues. However, a key technical barrier still remains the selectivity of new catalysts.

This section highlights emerging catalyst and solvent applications that are being used to produce platform chemicals and overcome the technical barriers discussed earlier. These technological advances can improve the method of product recovery by providing a processing environment that allows the use of biphasic systems and/or continuous extraction of products in high yield. Other methods that are applied to aid recovery include making use of the functional reactivity of the platform chemical. This may involve reactive extraction methods or converting the platform chemical to a nonaqueous product in cascade reactions to facilitate simpler product removal. One such example includes reacting aqueous hydrolysates of levulinic and formic acid with olefins to form hydrophobic fuel-like compounds that are easily separated (Fagan and Manzer 2006).

Other techniques involve the use of dual catalysts and solvents to enable "one-pot" reactions to produce water-immiscible products. For example, tetrahydrofuran derivatives (MTHF from pentose sugars and dimethyltetrahydrofuran from hexose sugars) can be produced directly from carbohydrates by a combined acid hydrolysis and hydrogenation reaction (Yang and Sen 2010).

New catalytic conversion technologies are commonly developed on simple sugar substrates with pentose sugars (xylose) more easily transformed than hexose sugars. While fructose is typically used as a model hexose sugar due to its facile conversion, it is not a common sugar derived from lignocellulosic biomass. Lower yields typically result from glucose and its polymers due to the comparatively stable (less reactive) nature of glucose resulting in less selective dehydration routes (van Dam *et al.* 1986) and polyglucans (starch and cellulose) require both a hydrolysis and dehydration step for conversion.

8.7 Catalysts

8.7.1 Homogeneous catalysts

The majority of research into the conversion of carbohydrates has been conducted with mineral acid (in particular, Brønsted acid) catalysts. Formic acid and acetic acid are also known catalysts that are able to hydrolyze hemicellulose and catalyze the dehydration of xylose to furfural in industrial processes\break (De Jong and Marcotullio 2010). However, these organic acid catalysts are less effective in the conversion of cellulose. One drawback of homogeneous acid-catalyzed processes is corrosive reaction conditions requiring special materials for reactor construction and chemical recovery systems increasing capital and operating costs. Also significant amounts of waste can be produced causing environmental pollution. This may be overcome by conducting the reaction in the vapor phase, which allows simpler catalyst recovery and product separation (Zeitsch 2001). Furfural yields of more than 95% were achieved from pentose sugars using gaseous HCl compared with typical yields of 50% obtained in a liquid acid-catalyzed process. While Zeitch only examined yields of furfural from monomer sugars, application of gaseous acid (HCl) stripping of bagasse was only able to achieve 60% furfural yields by other researchers (van Buijtenen *et al.* 2013). Gaseous acid catalysis using HCl has also been used to produce levulinic acid esters from biomass in alcohol media (Garves 1988a,b).

Steam stripping has also been utilized to isolate the furfural product from the aqueous acidic phase to stop homopolymerization and condensation reactions. This has been patented by the Suprayield process (Karinen *et al.* 2011) and Vedernikov process (De Jong and Marcotullio 2010), in which furfural yields of >80% are reported.

Sulfonic acids are strong, nonoxidizing, biodegradable acid catalysts that are more environmentally friendly, less corrosive, and minimize charring and other undesirable side reactions that occur in mineral acids such as sulfuric acid (Gernon *et al.* 1999). Common sulfonic acids include methanesulfonic acid (MSA) and *p*-toluenesulfonic acid (TSA). Both MSA and TSA have been used in the solvolysis liquefaction of cellulose and biomass, and were found to play an important role in reducing condensation reactions in comparison to mineral acids (Mun *et al.* 2006). Recently, MSA was shown to be an effective catalyst for the conversion of glucose and xylose mixtures to both levulinic acid and furfural (Rackemann *et al.* 2014).

Yields of levulinic acid and furfural of 60–80% from biomass are readily able to be achieved with mineral acid catalysts in aqueous systems under the right processing conditions.

8.7.2 Heterogeneous catalysts

Heterogeneous catalysts represent a viable alternative to homogeneous catalysts and offer an environmental advantage due to their selective and easy-to-handle nature, reduced equipment corrosion issues and relatively low cost if the catalyst can be easily separated and recycled (Lin and Huber 2009). One common issue with using heterogeneous catalysts in biomass conversion is the absorption of reactants, product and coke deposition onto the catalyst surface-limiting catalyst life (Gallo *et al.* 2013).

The function of Lewis and Brønsted acid sites in the production of furans from sugars was examined through their influence on catalyzing the dehydration reaction or the subsequent degradation of furan product. Product selectivity was found to increase with an increase in the Brønsted to Lewis acid site ratio of the catalyst (Weingarten *et al.* 2011). Indeed, heteropolyacids and solid super-acids are receiving much interest in biomass conversion not only because they possess high acidity but many are also bifunctional with both acid–base or Lewis–Brønsted acid sites (Chen *et al.* 2011). This second class of bifunctionality is interesting as typically Brønsted acids are more effective for the depolymerization of biomass, whereas Lewis acids are more effective for the conversion of sugars.

8.7.3 Levulinic acid

Until recently, low yields and reaction rates were obtained from lignocellulosic biomass using solid acid catalysts, and prolonged reaction times were required to achieve reasonable yields of levulinic acid (Rackemann and Doherty 2011). Water-soluble sulfonated hyperbranched polymers (WS-SPoly) were found to demonstrate superior catalytic performance in terms of activity and selectivity compared with conventional solid catalysts (Van de Vyver *et al.* 2011). These catalysts were able to produce yields of 70% and 76% levulinic acid from fructose and HMF feeds and were able to be recovered by ultrafiltration. However, the sulfonated polymer catalyst was not able to produce high yields of levulinic acid from more complex feeds and required extensive size reduction.

Chen and coworkers (2011) found a solid super-acid ($S_2O_4^{2-}/ZrO_2$–SiO_2–Sm_2O_3) could be utilized under optimal conditions (200 °C and 10 min) to obtain 70% yield of levulinic acid from steam-exploded rice straw. However, to ensure high catalytic activity, the biomass had to be ground to a very fine particle size, which represents an energy-intensive pretreatment step for an industrial process. Magnetic solid acids based on mesoporous silica and magnetic nanoparticles were able to produce levulinic acid in reasonable yields (54%) from 10% loading of cellulose (Lai *et al.* 2011). The magnetic property would aid recovery of the catalyst following reaction as other insoluble products (unreacted feed, humins, etc.) can complicate catalyst recovery.

The activity of heterogeneous catalysts can be restricted by mass transfer limitations when converting solid feedstocks, so a two-stage process has been proposed to first convert cellulose to soluble compounds and then to react the soluble compounds with Amberlyst-70 (Weingarten *et al.* 2012). While this approach allowed high cellulose loadings to be processed (~30%), yields of only 28% of levulinic acid were achieved.

Furans

Heterogeneous catalysts have had limited success in producing levulinic acid from lignocellulosic biomass, but more reports are available on their use to produce HMF and furfural. However, as furans are typically unstable in acidic aqueous solutions and ionic liquids, reactions are undertaken in nonaqueous media to suppress further conversion of the furan product (Agirrezabal-Telleria *et al.* 2014). While attractive, the use of ionic liquids in the production of biochemicals is likely to be a long-term prospect, as several key limitations need to be overcome including reducing their cost and developing efficient processes for separation and recovery of solvents and cocatalysts (Binder and Raines 2009). Therefore, the use of ionic liquids as reaction solvents is not discussed in this chapter but the reader is directed to recent review publications for more detail (Mukherjee *et al.* 2015).

Heterogeneous catalysts have had some success in producing HMF. Ion-exchange resins in the presence of dimethylsulfoxide (DMSO) and methyl isobutyl ketone (MIBK) were able to obtain 97% yield of HMF from fructose (Kuster 1990). Similarly, zeolites were able to obtain 98% yield of HMF from fructose in DMSO (Shimizu *et al.* 2009). HMF yields of ~60% from highly concentrated fructose feeds (30–40 wt%) at low temperatures (80 °C) were achieved using iron-based vanadyl phosphate catalyst (Carlini *et al.* 2004). Fixed-bed porous metal oxide–based (TiO_2 and ZrO_2) catalytic continuous processes have also been investigated for the production of HMF from carbohydrates (McNeff *et al.* 2010). With short residence times (2–3 min), yields of ~45% were achieved with concentrated sugars (30–50% fructose, glucose, and sucrose) at 180–200 °C and cellulose feeds at 260–280 °C. While relatively low, these yields are reasonable based on the high concentration of feed used.

Sulfonated carbon-based solid catalysts were able to convert fructose into HMF in >90% yields in DMSO at 130 °C after only 1.5 h (Wang *et al.* 2011). Other solid acid catalysts synthesized by carbonization and sulfonation of raw biomass materials (glucose, fructose, cellulose, lignin, and bamboo) were investigated (Guo *et al.* 2012), and the acid prepared from lignin was able to convert fructose into HMF with a yield of 84% in DMSO at 110 °C after only 10 min.

Metal and supported metal catalysts can also produce significantly improved reactivity of simple sugars but these catalysts have suffered from a lack of resistance to the harsh conditions and resultant products (acidic and chelating compounds) found in biomass processing (Rinaldi and Schuth 2009).

8.8 Solvolysis

Solvolysis involves a substitution or elimination reaction using a solvent that is typically an organic reagent. If water is used as the solvent, it is termed hydrolysis. Solvolysis has an advantage over hydrolysis in that solvents can be used to alter the properties of the reaction mixture to promote more selective reactions and due to their higher dielectric constants, allow reactions to occur at lower temperatures. Solvents can also be used in biphasic systems, which allow products to be selectively isolated from the reaction mixture preventing repolymerization reactions and improving recovery efficiency. Under laboratory conditions, such biphasic systems have been used to convert sugars to furanics in high yields with homogeneous catalysts and heterogeneous catalysts where the solvent is utilized to suppress subsequent conversion and degradation that can be significant under acidic conditions.

HMF was produced in 75% yield from fructose in DMSO with HCl catalyst and an organic phase of MIBK and 2-butanol used to enhance partitioning of the reactive species (Roman-Leshkov *et al.* 2006). The process was improved slightly (78% HMF yield) using tetrahydrofuran (THF) as the solvent with HCl catalyst and addition of an inorganic salt such as NaCl to improve partitioning into the extracting phase (Roman-Leshkov and Dumesic 2009). An 87% yield of HMF was produced from fructose in DMSO without HCl catalyst and using dichloromethane (DCM) as organic phase (Chheda *et al.* 2007). Other feed materials (glucose, sucrose, starch, and cellulose) were also tested but HMF yields were significantly lower.

DMSO is a dipolar aprotic solvent with electrophilic and nucleophilic ends that can react with hydroxyl groups and hydrogen bonding in sugar molecules to simultaneously promote the enolization step. This allows selective dehydration as a result of simple proton transfer and oxygen exchange with DMSO (James *et al.* 2010). The concerted catalysis of both electrophilic and nucleophilic ends reduces the chances of condensation or side reactions. Metal cations (i.e., salts) also aid the process by substituting for the electrophile, and chloride ions enhance hydrogen bonding strength of the solvent nucleophiles. Quantitative yields of HMF from fructose in DMSO using boron trifluoride acetate were produced at 100 °C in less than 1 h (Szmant and Chundury 1981). However, this method required very high catalyst loading (5:1 to feed) and did not transfer well to other feedstocks (yields of 42% from glucose).

Other solvents that have been tested include alcohols (methanol, propanol, pentanol, hexanol), ketones (acetone, butanone, pentanone, hexanone), and cyclic ethers (tetrahydropyran, MTHF). The use of alcohols was found to reduce polymerization reactions by protecting the carbonyl group through producing ethers and acetals of intermediate products (especially sugars and furanics), which diminished their predominance to oligomerization and polymerization reactions (Hu and Li 2011).

Furfural yields can also be improved through the use of organic solvents. THF in a single-phase cosolvent "one-pot" system able to produce 86% furfural, 21% HMF, and 40% levulinic acid from lignocellulosic biomass when catalyzed by sulfuric acid (Cai *et al.* 2013). However, combination of $FeCl_3$ with THF was able to enhance the catalytic performance and selectivity due to the Lewis acidity of the catalyst, which led to simultaneous yields of furfural (95%) and HMF (51%) directly from lignocellulosic biomass (Cai *et al.* 2014). Biphasic systems with MIBK have been utilized to produce furfural at yields of 85% from xylose (Weingarten *et al.* 2010).

The hydrogenated derivative of levulinic acid, GVL, is especially useful as a solvent due to its thermal stability (*c.f.* furfural) and solvation properties (Horvath *et al.* 2008). It is also water miscible and readily dissolves lignin and humins (acid insoluble products), which would allow reactor processing issues (fouling and blocking) to be overcome and its facile use in processing wet feedstocks (Alonso *et al.* 2013a,b).

The use of GVL and water (80:20) was able to produce levulinic acid from the cellulose fraction of corn stover in yields of 66% using sulfuric acid at 170 °C (Alonso *et al.* 2013a,b). Under other conditions, GVL with water as a cosolvent was able to produce furfural from the hemicellulose fraction of corn stover in yields of 96% (Alonso *et al.* 2013a,b). Heterogeneous acids (Amberlyst-70) were also able to convert cellulose to levulinic acid in yields of 69% in GVL–water systems (Alonso *et al.* 2013a,b). A small amount of water was found to be beneficial to the process but to achieve these high yields, reaction times of more than 4 h were employed. The catalyst lost activity with reuse unless a simple regeneration step (washing with dilute hydrogen peroxide) was employed. In addition, GVL swells the Amberlyst-70, thereby improving the diffusivity through the pores and increasing the catalytic activity of Amberlyst-70.

8.9 Other product chemicals

As an alternative to direct production of levulinic acid or furanic products, solvolysis reactions can produce other chemical products in either "one-pot" or sequential reactions. A number of potential products are discussed in the following sections.

8.9.1 Esters

Small chain-aliphatic alcohols (methanol, ethanol, and butanol) have been examined for the production of levulinic acid esters from biomass. Varying the process conditions (amount of water, type of solvent, and catalyst) can provide monophasic or biphasic conditions, which can allow the process to target specific products. The use of biphasic systems can improve carbon yield in the ester products by up to 90% (Tarabanko *et al.* 2008). The enhanced yields achieved are partially due to the suppression of humin formation caused by the solvent-modifying reaction pathways (Hu and Li 2011). The operating conditions also determine the amount of levulinic acid formed as the esterification is an equilibrium-limited reaction (Garves 1988a,b). Inorganic salts (Liu *et al.* 2012), heterogeneous acid catalysts (Peng *et al.* 2011 Saravanamurugan and Riisager 2012) as well as mixed acid catalysts (Tominaga *et al.* 2011) have also been used to produce levulinic acid esters from carbohydrates, albeit in lower yields compared with homogeneous catalysts. Amberlyst-70 as catalyst was also able to achieve high yields of methyl levulinate (>90%) from glucose but extension to cellulose and lignocellulosic biomass produced very low yields (<10%), which was attributed to the steric hindrance caused by feed structure affecting access to the catalyst (Hu *et al.* 2013).

Recently, the versatile process intensification technique of reactive distillation has been used for the continuous recovery and separation of organic acid esters. The technique was also shown to recover high purity levulinic acid as ethyl levulinate in almost quantitative yields (Kolah *et al.* 2012). Reactive distillation can improve selectivity and conversion, effectively utilize reaction heat, and avoid azeotropic issues. However, the *in situ* separation process in the reaction zone leads to complex interactions between vapor–liquid equilibrium, mass transfer, diffusion, and chemical kinetics.

8.9.2 Ketals

The semi-commercial production of levulinic ester ketals has recently been achieved by Segetis (http://www.segetis.com/technology.html). Levulinic ketals are another platform product that is considered more useful than levulinic acid due to broad solvency, compatibility, functionality, and exceptional hydrolytic and thermal stability, which are desirable traits for polymer-based products (Leibig *et al.* 2011). Levulinic ketals are readily made by a three-step process:

acid hydrolysis of biomass to levulinic acid; esterification of the carboxylic acid functional group; ketalization of the ketone functional group with glycerol with all steps catalyzed by similar acids. Ketalization of the levulinic ester is a very selective step (>95% yield).

8.9.3 Chloromethylfurfural

Halogenated furanics such as CMF can be produced in biphasic systems in high yield when a halogenated organic solvent and/or inorganic halide salt is employed. As CMF is hydrophobic, it readily sequesters into the organic solvent but effective mixing is required. Mascal and Nikitin (2010) were able to convert 10 wt% substrate loading of sugars and cellulose to CMF in yields of >90% using concentrated HCl and the low boiling organic solvent, dichloroethane (DCE) over a multiple-extraction 3 h cycle at 100 °C. Applying the process to corn stover was able to achieve an 88% total yield of CMF and levulinic acid (Mascal and Nikitin 2009a,b). The process also converted the xylan content of biomass to furfural in 40% yield, which was a significant reduction in the 68% yield from xylose) (Mascal and Nikitin 2009a,b).

A continuous biphasic flow process was also shown to produce 60–80% yields of CMF from fructose, glucose, and sucrose using DCM and DCE at temperatures of 100–120 °C and reaction times of ~1–5 min (Brasholz *et al.* 2011). Lower yields may have resulted from the lack of mixing in the continuous process reducing biphasic extraction efficiency as unreacted sugars were also observed in the aqueous fraction after separation. As DCE is a solvent with health and safety concerns, a range of other solvents (chloroform, toluene, cyclohexane, DCM, and DMF) were also tested with cyclohexane producing the next highest yield of CMF at 75% (Breeden *et al.* 2013).

CMF is very reactive and can be converted into HMF (86% + 10% levulinic acid), levulinic acid (91%), and ethyl levulinate (85%) (Mascal and Nikitin 2010). CMF can also be reduced to MF in yields up to 98% (Hamada and Fujisawa 1982).

8.9.4 GVL

Recently, a lot of attention has been given to GVL, as it has been identified as a useful solvent and platform molecule with many applications as a precursor for chemicals and fuels. GVL is produced in near quantitative yields by hydrogenation of levulinic acid over heterogeneous noble metal catalysts in vapor phase, liquid phase, or using supercritical fluids (Wright and Palkovits 2012, Zhang *et al.* 2012). Ruthenium-based catalysts provide the highest yields of GVL in liquid phase processing, which is relevant in cascade reaction systems following production of levulinic acid from biomass by acid hydrolysis. Dumesic and coworkers developed Ru/Rh and Ru–Sn catalysts supported on carbon where the bimetallic catalyst was active for the reduction of levulinic acid and simultaneous decomposition of formic acid to hydrogen with good stability in the presence of sulfuric acid (Wettstein *et al.* 2012). The use of mineral acids in the production of levulinic acid can poison the heterogeneous catalysts used in GVL upgrading process so a separation step is required. Other than an external gas, formic acid has also been successfully used as the hydrogen source for levulinic acid hydrogenation (Mehdi *et al.* 2008).

Dumesic's group has recently concentrated on lignocellulosic biomass conversion processes to levulinic acid using GVL as the solvent and subsequent conversion of levulinic acid to GVL. This approach for biomass conversion eliminates the need to separate the final product from the solvent (provided levulinic acid is not the target product), because the GVL product is the solvent.

8.10 Concluding remarks

Replacing hydrocarbon-based products with bioderived platform chemicals requires a high yielding and high efficiency manufacturing process with low operating and capital costs and simple product and catalyst recovery. To maximize value from biomass will require targeting a number of key products that are formed from the different biomass polymer components under different processing conditions. Multistage processes offer the potential for using different catalysts/solvents and/or operating conditions to enable optimized and selective production of a number of chemicals (i.e., furfural and levulinic acid) from biomass in high yields. Although multistage processes may be more complex, continuous processes and optimized reactor designs can be very effective in minimizing labor, maintenance, and energy costs associated with comparable batch operations. Pinch analysis technology combined with the process stream recycling methodologies adopted by the Biofine process also aids the minimization of energy consumption.

New technologies and processing environments that allow the use of biphasic systems and/or continuous extraction of products would increase reaction rates, yields, and product quality to provide the necessary step change for optimization of key reactions. The choice of solvent in an industrial process is influenced by polarity, boiling point, and partitioning ability as these provide a key measure on the recovery of product, catalyst, and solvent. In addition, development of more selective catalysts remains a key technical barrier to improve solvent-based systems. Heterogeneous catalysts offer the greatest potential due to their simple and energy-efficient separation, reusable nature, limited corrosion, and waste disposal issues.

The application of GVL as solvent (and target product) provides a significant advance to the development of effective biomass conversion processes particularly when coupled with heterogeneous catalysts. GVL can be used as a reaction solvent in a monophasic system or as an extraction solvent for levulinic acid and furfural in biphasic systems. Not requiring an external solvent (and external source of hydrogen due to in situ conversion of formic acid) allows purification and separation issues associated with industrial processes to be overcome. Although not optimized, this new technology shows potential in that it can produce similar overall product carbon yields to best practice cellulosic ethanol technology (86% conversion of cellulose and 64% conversion of hemicellulose to ethanol) (Dutta *et al.* 2010) and the products are of higher value than ethanol.

References

Agirrezabal-Telleria, I. et al. (2014). Heterogeneous acid-catalysts for the production of furan-derived compounds (furfural and hydroxymethylfurfural) from renewable carbohydrates: a review. *Catalysis Today* **234**:42–58.

Alonso, D. M. et al. (2010). Catalytic conversion of biomass to biofuels. *Green Chemistry* **12**:1493–1513.

Alonso, D. M. et al. (2013a). Direct conversion of cellulose to levulinic acid and gamma-valerolactone using solid acid catalysts. *Catalysis Science and Technology* **3**:927–931. DOI: 10.1039/c1032cy20689g.

Alonso, D. M. et al. (2013b). Integrated conversion of hemicellulose and cellulose from lignocellulosic biomass. *Energy and Environmental Science* **6**(1):76–80.

Ayoub, P. M. (2008). Process for the reactive extractive extraction of levulinic acid. US patent.

Binder, J. B. and Raines, R. T. (2009). Simple chemical transformation of lignocellulosic biomass into furans for fuels and chemicals. *Journal of American Chemistry Society* **171**:1979–1985.

Boisen, A. et al. (2009). Process integration for the conversion of glucose to 2,5-furandicarboxylic acid. *Chemical Engineering Research and Design* **87**(9A):1318–1327.

Bozell, J. J. et al. (2007). *Top Value-Added Chemicals from Biomass Volume II – Results of Screening for Potential Candidates from Biorefinery Lignin*, U.S. Department of Energy.

Bozell, J. J. et al. (2000). Production of levulinic acid and use as a platform chemical for derived products. *Resources Conservation and Recycling* **28**(3–4):227–239.

Bozell, J. J. and Petersen, G. R. (2010). Technology development for the production of biobased products from biorefinery carbohydrates; the US Department of Energy's "Top 10 revisited". *Green Chemistry* **12**(4):539–554.

Brasholz, M. et al. (2011). Highly efficient dehydration of carbohydrates to 5-(chloromethyl)furfural (CMF), 5-(hydroxymethyl)furfural (HMF) and levulinic acid by biphasic continuous flow processing. *Green Chemistry* **13**(5):1114–1117.

Breeden, S. W et al. (2013). Microwave heating for rapid conversion of sugars and polysaccharides to 5-chloromethyl furfural. *Green Chemistry* **15**(1):72–75.

Cai, C. M. et al. (2014). Coupling metal halides with a co-solvent to produce furfural and 5-HMF at high yields directly from lignocellulosic biomass as an integrated biofuels strategy. *Green Chemistry* **16**(8):3819–3829.

Cai, C. M. et al. (2013). THF co-solvent enhances hydrocarbon fuel precursor yields from lignocellulosic biomass. *Green Chemistry* **15**(11):3140–3145.

Carlini, C. et al. (2004). Heterogeneous catalysts based on vanadyl phosphate for fructose dehydration to 5-hydroxymethyl-2-furaldehyde. *Applied Catalysis A: General* **275**(1–2):111–118.

Carlson, T. R. et al. (2011). Production of green aromatics and olefins by catalytic fast pyrolysis of wood sawdust. *Energy and Environmental Science* **4**(1):145–161.

Chen, H. et al. (2011). Production of levulinic acid from steam-exploded rice straw via solid superacid. *Bioresource Technology* **102**(3):3568–3570.

Chheda, J. N. et al. (2007). Production of 5-hydroxymethylfurfural and furfural by dehydration of biomass-derived mono- and poly-saccharides. *Green Chemistry* **9**:342–350.

Clark, J. H. et al. (2006). Green chemistry and the biorefinery: a partnership for a sustainable future. *Green Chemistry* **8**:853–860.

Corma, A. et al. (2007). Chemical routes for the transformation of biomass into chemicals. *Chemical Reviews* **107**(6):2411–2502.

De Jong, W. and Marcotullio, G. (2010). Overview of biorefineries based on co-production of furfural, Existing concepts and novel developments. *International Journal of Chemical Reactor Engineering* **8**(1): (DOI:10.2202/1542-6580.2174).

Dias, A. S. et al. (2010). Furfural and furfural-based industrial chemicals. In *Ideas in Chemistry and Molecular Sciences*, pp. 165–186. B. Pignataro, ed. Wiley-VCH Verlag GmbH & Co. KGaA.

Doherty, W. O. S. et al. (2011). Value-adding to cellulosic ethanol: lignin polymers. *Industrial Crops and Products* **33**(2):259–276.

Dunlop, A. P. (1948). Furfural formation and behavior. *Industrial and Engineering Chemistry* **40**(2):204–209.

Dutta, A. et al. (2010). An economic comparison of different fermentation configurations to convert corn stover to ethanol using *Z. mobilis* and *Saccharomyces*. *Biotechnology Progress* **26**(1):64–72.

Ebert, J. (2008). *Furfural: Future Feedstock for Fuels and Chemicals. Biomass Magazine*. BBI International Media.

Fagan, P. J. and Manzer, L. E. (2006). Preparation of levulinic acid esters and formic acid esters from biomass and olefins. US patent.

Fitzpatrick, M. et al. (2010). A biorefinery processing perspective: treatment of lignocellulosic materials for the production of value-added products. *Bioresource Technology* **101**(23):8915–8922.

Fitzpatrick, S. W. (2006). The Biofine technology: a "bio-refinery" concept based on thermochemical conversion of cellulosic biomass. In *Feedstocks for the Future – Renewables for the Production of Chemicals and Materials*, pp. 271–287. J. J. Bozell and M. K. Patel, eds. American Chemical Society: Washington, US.

Gallo, J. M. et al. (2013). Production of furfural from lignocellulosic biomass using Beta zeolite and biomass-derived solvent. *Topics in Catalysis* **56**(18–20):1775–1781.

Garves, K. (1988a). Acid catalyzed cellulose degradation in alcohols. *Journal of Wood Chemistry and Technology* **8**(1):121–134.

Garves, K. (1988b). Synthesis of alkoxymethylfurfurals and alkyl levulinates from cellulose or lignocelluloses or starch and alcohols. German Patent (DE3621517).

Gernon, M. D. et al. (1999). Environmental benefits of methanesulfonic acid: Comparative properties and advantages. *Green Chemistry* **1**(3):127–140.

Ghorpade, V. M. and Hanna, M. A. (1997). Industrial applications for levulinic acid. In *Cereals: Novel Uses and Processes*, pp. 49–57. G. M. Campbell, C. Webb and S. L. McKee, eds. Plenum: New York, US.

Guo, F. et al. (2012). Conversion of fructose and glucose into 5-hydroxymethylfurfural with lignin-derived carbonaceous catalyst under microwave irradiation in dimethyl sulfoxide–ionic liquid mixtures. *Bioresource Technology* **112**:313–318.

Hamada, K. and Fujisawa, K. (1982). *Process for Producing 5-Methylfurfural*. Sumitomo Chemical Company Limited.

Hayes, D. J. et al. (2006). The Biofine process – production of levulinic acid, furfural, and formic acid from lignocellulosic feedstocks. In *Biorefineries – Industrial Processes and Products: Status Quo and Future Directions. Vol 1*, pp. 139–163. B. Kamm, P. R. Gruber and M. Kamm. John Wiley & Sons:Weinheim, Germany.

Horvath, I. T. et al. (2008). γ-Valerolactone – a sustainable liquid for energy and carbon-based chemicals. *Green Chemistry* **10**:238–242.

Hoydonckx, H. E. et al. (2000). Furfural and derivatives. *Ullmann's Encyclopedia of Industrial Chemistry*. G. Bellussi and M. Bohnet, eds. Wiley-VCH Verlag GmbH & Co KGaA.

Hu, X. and Li, C.-Z. (2011). Levulinic esters from the acid-catalysed reactions of sugars and alcohols as part of a bio-refinery. *Green Chemistry* **13**(7):1676–1679.

Hu, X. et al. (2013). Acid-catalyzed conversion of mono- and poly-sugars into platform chemicals: effects of molecular structure of sugar substrate. *Bioresource Technology* **113**:469–474.

Huber, G. and Dumesic, J. A. (2006). An overview of aqueous-phase catalytic processes for production of hydrogen and alkanes in a biorefinery. *Catalysis Today* **111**:119–132.

James, O. O. et al. (2010). Towards the conversion of carbohydrate biomass feedstocks to biofuels via hydroxylmethylfurfural. *Energy and Environmental Science* **3**:1833–1850.

Kamm, B. and Kamm, M. (2004). Principles of biorefineries. *Applied Microbiology and Biotechnology* **64**:137–145.

Karinen, R. et al. (2011). Biorefining: heterogeneously catalyzed reactions of carbohydrates for the production of furfural and hydroxymethylfurfural. *ChemSusChem* **4**(8):1002–1016.

Kolah, A. K. et al. (2012). Reactive distillation for synthesis of ethyl levulinate from biobased resources. 2012 AIChE Annual Meeting – Process Development Division – Process Intensification by Process Integration Pittsburgh, 28 Oct–2 Nov, 2012.

Kumari, N. et al. (2011). Synthesis of 5-bromomethylfurfural from cellulose as a potential intermediate for biofuel. *European Journal of Organic Chemistry* **2011**(7):1266–1270.

Kuster, B. F. M. (1990). 5-Hydroxymethylfurfural (HMF). A review focusing on its manufacture. *Starch* **42**(8):314–321.

Lai, D.-m. et al. (2011). Hydrolysis of biomass by magnetic solid acid. *Energy and Environmental Science* **4**(9):3552–3557.

Leibig, C. et al. (2011). Cellulosic-derived levulinic ketal esters: a new building block. In *Renewable and Sustainable Polymers*, pp. 111–116. American Chemical Society. **1063**.

Lewkowski, J. (2001). Synthesis, chemistry and applications of 5-hydroxymethylfurfural and its derivatives. *Arkivoc* **1**:17–54.

Lin, Y. C. and Huber, G. W. (2009). The critical role of heterogeneous catalysis in lignocellulosic biomass conversion. *Energy and Environmental Science* **2**:68–80.

Liu, J. et al. (2012). Conversion of fructose into 5-hydroxymethylfurfural (HMF) and its derivatives promoted by inorganic salt in alcohol. *Carbohydrate Research* **350**:20–24.

Loges, B. et al. (2008). Controlled generation of hydrogen from formic acid amine adducts at room temperature and application in H_2/O_2 fuel cells. *Angewandte Chemie International Edition* **47**(21):3962–3965.

Mascal, M. and Nikitin, E. B. (2009a). Dramatic advancements in the saccharide to 5-(chloromethyl)furfural conversion reaction. *ChemSusChem* **2**(9):859–861.

Mascal, M. and Nikitin, E. B. (2009b). Towards the efficient, total glycan utilization of biomass. *ChemSusChem* **2**(5):423–426.

Mascal, M. and Nikitin, E. B. (2010). High-yield conversion of plant biomass into the key value-added feedstocks 5-(hydroxymethyl)furfural, levulinic acid, and levulinic esters via 5-(chloromethyl)furfural. *Green Chemistry* **12**:370–373.

McNeff, C. V. et al. (2010). Continuous production of 5-hydroxymethylfurfural from simple and complex carbohydrates. *Applied Catalysis A: General* **384**:65–69.

Mehdi, H. et al. (2008). Integration of homogeneous and heterogeneous catalytic processes for a multi-step conversion of biomass: from sucrose to levulinic acid, γ-valerolactone, 1,4-pentanediol, 2-methyl-tetrahydrofuran, and alkanes. *Topics in Catalysis* **48**:49–54.

Moyer, W. W. (1942). Preparation of levulinic acid. US patent.

Mukherjee, A. et al. (2015). Review: sustainable production of hydroxymethylfurfural and levulinic acid: challenges and opportunities. *Biomass and Bioenergy* **72**:143–183.

Mun, S. P. et al. (2006). Effect of organic sulfonic acids as catalysts during phenol liquefaction of *Pinus radiata* bark. *Journal of Industrial and Engineering Chemistry* **12**(5):720–726.

O'Hara, I. M. et al. (2013). Prospects for the development of sugarcane biorefineries. International Society of Sugar Cane Technologists. M. Hogarth. Durban, Brazil. 28:(cd rom).

Patil, S. K. R. and Lund, C. R. F. (2011). Formation and growth of humins via aldol addition and condensation during acid-catalyzed conversion of 5-hydroxymethylfurfural. *Energy and Fuels* **25**(10):4745–4755.

Pena, C. G. et al. (2000). Bagasse. In *Handbook of Sugar Cane Derivatives*, 3rd Edition. S. R. Bugallo, ed. ICIDCA: Havana, Cuba.

Peng, L. et al. (2011). Conversion of carbohydrates biomass into levulinate esters using heterogeneous catalysts. *Applied Energy* **88**(12):4590–4596.

Peterson, A. A. et al. (2008). Thermochemical biofuel production in hydrothermal media: a review of sub- and supercritical water technologies. *Energy and Environmental Science* **1**:32–65.

Rackemann, D. W. et al. (2014). Methanesulfonic acid-catalyzed conversion of glucose and xylose mixtures to levulinic acid and furfural. *Industrial Crops and Products* **52**:46–57.

Rackemann, D. W. and Doherty, W. O. S. (2011). The conversion of lignocellulosics to levulinic acid. *Biofuels, Bioproducts and Biorefining* **5**(2):198–214.

Ramos-Rodriguez, E. et al. (1968). The simultaneous production of furfural and levulinic acid from bagasse. International Society of Sugar Cane Technologists. K.-C. Liu. Taipei, Taiwan. 13:1900–1911.

Rass-Hansen, J. et al. (2007). Bioethanol: fuel or feedstock? *Journal of Chemical Technology and Biotechnology* **82**(4):329–333.

Rein, P. (2007). Prospects for the conversion of a sugar mill into a biorefinery. International Society of Sugar Cane Technologists. M. Hogarth. Durban, South Africa. 26:44–59.

Rinaldi, R. and Schuth, F. (2009). Design of solid catalysts for the conversion of biomass. *Energy and Environmental Science* **2**(6):610–626.

Roman-Leshkov, Y. et al. (2007). Production of dimethylfuran for liquid fuels from biomass-derived carbohydrates. *Nature* **447**(7147):982–986.

Roman-Leshkov, Y. et al. (2006). Phase modifiers promote efficient production of hydroxymethylfurfural from fructose. *Science* **312**:297–303.

Roman-Leshkov, Y. and Dumesic, J. A. (2009). Solvent effects on fructose dehydration to 5-hydroxymethylfurfural in biphasic systems saturated with inorganic salts. *Topics in Catalysis* **52**:297–303.

Saravanamurugan, S. and Riisager, A. (2012). Solid acid catalysed formation of ethyl levulinate and ethyl glucopyranoside from mono- and disaccharides. *Catalysis Communications* **17**:71–75.

Sheldon, R. A. (2011). Utilisation of biomass for sustainable fuels and chemicals: molecules, methods and metrics. *Catalysis Today* **167**(1):3–13.

Shimizu, K.-I. et al. (2009). Enhanced production of hydroxymethylfurfural from fructose with solid acid catalysts by simple water removal methods. *Catalysis Communications* **10**(14):1849–1853.

Simmie, J. M. and Würmel, J. (2013). Harmonising production, properties and environmental consequences of liquid transport fuels from biomass – 2,5-dimethylfuran as a case study. *ChemSusChem* **6**(1):36–41.

Szmant, H. H. and Chundury, D. D. (1981). The preparation of 5-hydroxymethylfurfuraldehyde from high fructose corn syrup and other carbohydrates. *Journal of Chemical Technology and Biotechnology* **31**(1):135–145.

Tarabanko, V. E. et al. (2008). Sodium hydrosulfate as the catalyst for carbohydrate conversion into the levulinic acid and 5-hydroxymethylfurfural derivatives. *Journal of Siberian Federal University, Chemistry* **1**:35–49.

Tominaga, K.-i. et al. (2011). Mixed-acid systems for the catalytic synthesis of methyl levulinate from cellulose. *Green Chemistry* **13**(4):810–812.

van Dam, H. E. et al. (1986). The conversion of fructose and glucose in acidic media: formation of hydroxymethylfurfural. *Starch* **38**(3):95–101.

Van de Vyver, S. et al. (2011). Catalytic production of levulinic acid from cellulose and other biomass-derived carbohydrates with sulfonated hyperbranched poly(arylene oxindole)s. *Energy and Environmental Science* **4**(9):3601–3610.

van Buijtenen, J. et al. (2013). Furfural production by 'acidic steam stripping' of lignocellulose. *ChemSusChem* **6**:2132–2136.

Wang, J. et al. (2011). Efficient catalytic conversion of fructose into hydroxymethylfurfural by a novel carbon-based solid acid. *Green Chemistry* **13**(10):2678–2681.

Watson, L. J. and Connors, C. G. (2008). Furfural – a value adding opportunity for the Australian sugar industry. Australian Society of Sugar Cane Technologists. Townsville, Australia. 30:429–436.

Weingarten, R. et al. (2010). Kinetics of furfural production by dehydration of xylose in a biphasic reactor with microwave heating. *Green Chemistry* **12**:1423–1429.

Weingarten, R. et al. (2012). Production of levulinic acid from cellulose by hydrothermal decomposition combined with aqueous phase dehydration with a solid acid catalyst. *Energy and Environmental Science* **5**(6):7559–7574.

Weingarten, R. et al. (2011). Design of solid acid catalysts for aqueous-phase dehydration of carbohydrates: the role of Lewis and Brønsted acid sites. *Journal of Catalysis* **279**(1):174–182.

Werpy, T. and Petersen, G. (2004). *Top Value Added Chemicals from Biomass Volume I – Results of Screening for Potential Candidates from Sugars and Synthesis Gas*, U.S. Department of Energy.

Wettstein, S. G. et al. (2012). Production of levulinic acid and gamma-valerolactone (GVL) from cellulose using GVL as a solvent in biphasic systems. *Energy and Environmental Science* **5**(8):8199–8203.

Wondu. (2006). *Furfural chemicals and biofuels from agriculture*. Barton, Rural Industries Research and Development Corporation.

Wright, W. R. H. and Palkovits, R. (2012). Development of heterogeneous catalysts for the conversion of levulinic acid to γ-valerolactone. *ChemSusChem* **5**(9):1657–1667.

Yan, Z.-P. et al. (2009). Synthesis of γ-valerolactone by hydrogenation of biomass-derived levulinic acid over Ru/C catalyst. *Energy and Fuels* **23**(8):3853–3858.

Yang, W. and Sen, A. (2010). One-step catalytic transformation of carbohydrates and cellulosic biomass to 2,5-dimethyltetrahydrofuran for liquid fuels. *ChemSusChem* **3**(5):597–603.

Yu, X. and Pickup, P. G. (2008). Recent advances in direct formic acid fuel cells (DFAFC). *Journal of Power Sources* **182**(1):124–132.

Zakzeski, J. et al. (2010). The catalytic valorization of lignin for the production of renewable chemicals. *Chemical Reviews* **110**(6):3552–3599.

Zeitsch, K. J. (2000). *The Chemistry and Technology of Furfural and Its Many By-Products*. London, UK, Elsevier.

Zeitsch, K. J. (2001). Gaseous acid catalysis: an intriguing new process. *Chemical Innovation* **31**(1):40–44.

Zhang, J. et al. (2012). Advances in the catalytic production of valuable levulinic acid derivatives. *ChemCatChem* **4**(9):1230–1237.

CHAPTER 9

Cogeneration of sugarcane bagasse for renewable energy production

Anthony P. Mann

Centre for Tropical Crops and Biocommodities, Queensland University of Technology (QUT), Brisbane, Australia

9.1 Introduction

Increasing demands for energy, reducing stocks of fossil fuels, and concerns about global warming have made biomass an increasingly important source of energy for heating and electricity generation (IPCC 2012). Bagasse, a residue from sugarcane processing, is a significant biomass resource. Furthermore, sugar milling operations already have the infrastructure for harvesting and processing sugarcane and burning the bagasse produced. The heat produced from bagasse combustion is used to produce steam for factory operations and generate electricity; in most cases, the quantity of electricity produced is surplus to factory requirements. This simultaneous use of the heat produced by burning bagasse for factory operations and electricity generation is known as cogeneration.

In the past, however, there was little financial incentive for sugarcane factories to export electricity to the grid and for that reason, having surplus bagasse was seen as problem that had to be managed or avoided. The majority of sugarcane factories around the world were set up to avoid producing large quantities of surplus bagasse. Factory operations required large quantities of steam and sugar factory boilers were inefficient, which gave rise to the paradigm that sugar factories "waste energy very efficiently." This situation is changing with financial incentives being provided in many countries for producing electricity from renewable sources.

Bagasse is a renewable fuel and is essentially greenhouse gas neutral because the carbon dioxide produced during bagasse combustion is offset by the carbon dioxide absorbed by the photosynthesis of sugarcane in the field. With renewable electricity incentives in place, most new sugar factories are designed to be energy efficient with significant electricity export.

Sugarcane-Based Biofuels and Bioproducts, First Edition. Edited by Ian M. O'Hara and Sagadevan G. Mundree.

9.2 Background

The sugarcane crop can be considered to consist of sugarcane stalks, roots, tops, and leaves. In countries where sugarcane is mechanically harvested, the stalks are typically chopped into billets in the harvester. These billets, and some of the extraneous matter, usually between 5% and 10% of the mass of billets (Kent *et al.* 2003), form the sugarcane supply sent to a factory during normal crushing operations. This sugarcane supply consists of water, the dry insoluble component referred to as fiber, and the soluble solids component referred to as brix, which consists of sugar and soluble impurities. Most of the sugarcane stalk is water with the remainder of the stalk consisting of roughly equal proportions of insoluble and soluble solids (Figure 9.1).

The sugarcane juice extraction process extracts most of the water, sugar, and soluble impurities from the sugarcane supply. The remaining material, bagasse, primarily consists of fiber, water, and small amounts of brix (sugar and soluble impurities). For simplicity, it can be assumed (Kent 2010) that the soluble impurities in bagasse include equal proportions of reducing sugars, soluble ash, and other organic matter. Typical proportions of the bagasse constituents and their normal ranges are shown in Figure 9.2.

Most of the combustible material in bagasse is from the fiber (insoluble solids) in the sugarcane supply. The major components of this dry, ash- and brix-free combustible material are cellulose (45–55%), hemicellulose (20–30%), and lignin (15–26%) (Dixon 1988, Lamb 1979). Most of the soluble combustible components in bagasse (sugar, reducing sugars, and organic matter) have a lower heating value than the combustible fiber in bagasse. This can be taken into account using the relationship for bagasse heating values (h_b):

$$h_b = h_{\text{fibre, dabf}}(1 - 0.01(M + A_w)) - 0.01B_b(h_{\text{fibre, dabf}} - h_{\text{soluble solids}}) \tag{9.1}$$

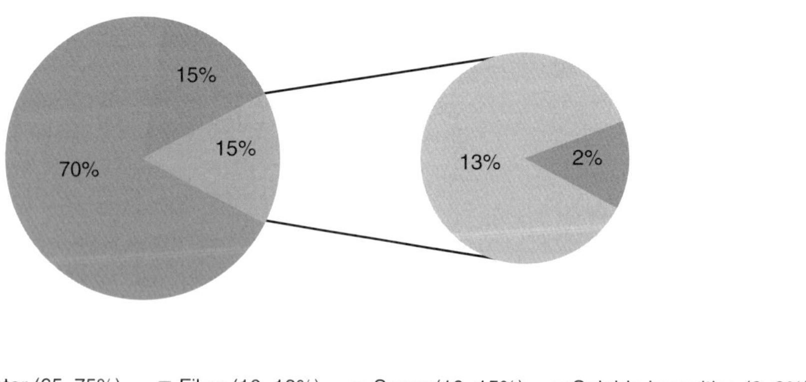

Figure 9.1 Typical sugarcane composition.

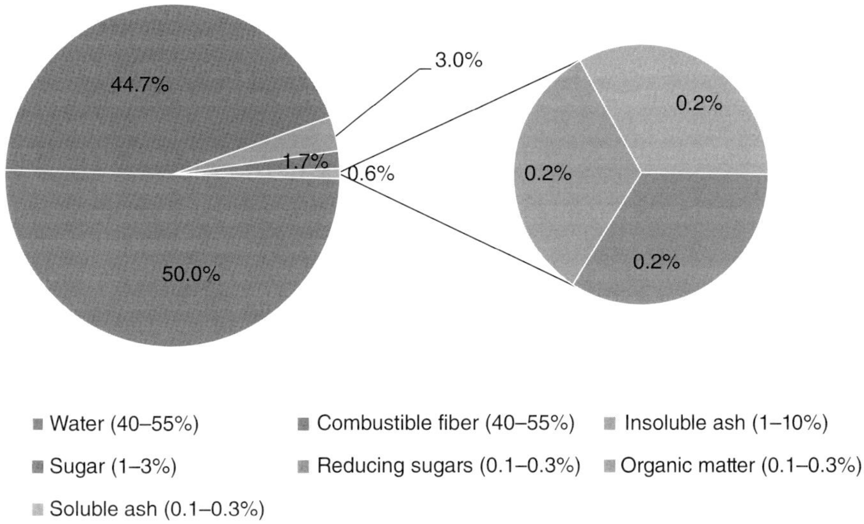

Figure 9.2 Typical bagasse composition.

where $h_{\text{fiber,dabf}}$ is the heating value of the dry, ash- and brix-free component of bagasse fiber, $h_{\text{soluble solids}}$ is the heating value of the soluble solids (mostly sucrose) in bagasse, M is the bagasse moisture content, A_w is the bagasse ash content (% wet basis), and B_b is the bagasse brix.

The heating values calculated in Equation 9.1 can be gross calorific values (GCVs) (which takes into account the latent heat of water vapor in combustion gases) or net calorific values (NCVs) (which ignores the latent heat of water vapor in combustion gases). Analyses reported by a number of researchers have found the heating value of dry, ash- and brix-free bagasse to be very consistent (Behne 1934, Don *et al.* 1977, Frew and James 1977). The GCV of dry, ash- and brix-free bagasse from a large number of bagasse samples was found to range between 19,400 and 19,800 kJ/kg (Don *et al.* 1977, Frew and James 1977). The expression for bagasse GCV proposed by Don *et al.* (1977) is equivalent to Equation 9.1 with $h_{\text{fiber,dabf}} - h_{\text{soluble solids}}$ equal to 3114 kJ/kg. This makes the GCV of the soluble solids in bagasse essentially the same as the sucrose GCV of 16,514 kJ/kg used by Frew and James (1977).

The NCV of bagasse ($h_{b,\text{net}}$) can be calculated from the bagasse GCV ($h_{b,\text{gross}}$) using

$$h_{b,\text{net}} = h_{b,\text{gross}} - 218.26H - 24.42M \tag{9.2}$$

where H is the hydrogen content of the bagasse (mass % as received). For typical bagasse with $H = 3\%$ and $M = 50\%$, the gross and NCVs differ by approximately 1876 kJ/kg.

The flow rate of bagasse available for use by a sugar factory boiler station ($\dot{m}_{b,\text{ available}}$) can be estimated using the following relationship:

$$\dot{m}_{b,\text{ available}} = \frac{\dot{m}_c(F - b_c)}{100 - M - B_b} \tag{9.3}$$

where $\dot{m}_c$ is the crushing rate, F is the fiber content of sugarcane, and b_c is the bagacillo content of sugarcane. Bagacillo is the term used to describe the small bagasse particles intentionally removed from the bagasse stream for use in the mud filters and the small bagasse particles remaining in the juice extracted during the crushing process. Normally, b_c is less than 0.5% but a study by Wright (2003) suggests that b_c could be much higher when there are high dirt levels in the sugarcane supply. The denominator of the right-hand side of Equation 9.3 is the fiber content of the bagasse. For typical factory conditions, the rate of bagasse production as a proportion of the crushing rate is approximately twice the fiber content of the sugarcane. Note that if the milling train performs poorly and the final bagasse has high moisture and brix contents, the quantity of bagasse that can be sent to the boiler station will increase according to Equation 9.3 but the heating value of this bagasse will reduce as reflected in Equation 9.1.

The quantity of bagasse required by the boiler station to generate steam for factory operations depends on the efficiency of the sugar factory boilers. Boiler efficiency is the thermal output expressed as a percentage of the thermal input (the energy associated with combustion of the fuel). If blowdown and changes in boiler heat storage due to changes in boiler drum level are ignored, then the efficiency of a bagasse fired boiler (η_b) can be defined by

$$\eta_b = \frac{100\dot{m}_s(h_s - h_f)}{\dot{m}_b h_b} \tag{9.4}$$

where $\dot{m}_s$ is the boiler steam flow, h_s is the enthalpy of the steam produced by the boiler, h_f is the enthalpy of boiler feedwater, and $\dot{m}_b$ is the boiler bagasse flow. On the right-hand side of Equation 9.4, the numerator is the boiler heat output and the denominator is the boiler heat input. If h_b is the bagasse GCV, then the corresponding boiler efficiency, $\eta_{b,i}$, will be the GCV basis boiler efficiency. Similarly, if h_b is the bagasse NCV, then the corresponding boiler efficiency, $\eta_{b,i}$, will be the NCV basis boiler efficiency.

The bagasse consumption of a sugar factory boiler station ($\dot{m}_{b,\text{ bs}}$) is the sum of the bagasse flows to each boilers ($\dot{m}_{b,\, i}$) plus an allowance for factory stoppages. The total bagasse loss associated with stoppages (and the associated start-ups) during a season can be divided by the crushing season length to give the corresponding rate of bagasse loss ($\dot{m}_{b,\text{ bs, stoppages}}$). Rearranging Equation 9.4 and including the bagasse loss due to stoppages gives the following equation for boiler station bagasse consumption:

$$\dot{m}_{b,\text{ bs}} = \sum_i \frac{100\,\dot{m}_{s,\,i}(h_{s,\,i} - h_{f,\,i})}{\eta_{b,\,i} h_b} + \dot{m}_{b,\text{ bs, stoppages}} \tag{9.5}$$

where the subscript i denotes the value for boiler i.

Once the boiler station bagasse consumption ($\dot{m}_{b,\,\mathrm{bs}}$) is known, the boiler station efficiency, η_{bs}, can be calculated from

$$\eta_{\mathrm{bs}} = \frac{\sum_i 100\,\dot{m}_{s,\,i}\left(h_{s,\,i} - h_{f,\,i}\right)}{\dot{m}_{b,\,\mathrm{bs}}\,h_b}. \tag{9.6}$$

The direct or input–output method of boiler efficiency determination (Equation 9.4) requires a measurement of the bagasse flow to the boiler, which is often not measured in practice. For this reason, the indirect or heat loss method of boiler efficiency determination is normally used where the losses per unit fuel flow are calculated from measured quantities. This method can be expressed as

$$\eta_b = 100\left(1 - \sum \frac{\mathrm{Losses}}{h_b}\right). \tag{9.7}$$

The losses in Equation 9.7 include those due to the sensible heat of the dry flue gas, fuel moisture, combustion moisture, air moisture, carbon monoxide (unburnt fuel volatile matter), unburnt solid fuel, and radiation. The dry gas loss is the sensible heat loss associated with the flue gas components other than water vapor. The fuel moisture loss is the sensible and latent heat of the water vapor in the flue gas that comes from evaporating the water in the fuel and the combustion moisture loss is the sensible and latent heat associated with the water produced by the reaction of hydrogen in the bagasse with oxygen. Combustion air contains water vapor; the moisture in this air is heated in the boiler and leaves the boiler at a higher temperature than ambient. The extra heat in this water vapor when it exits the final heat transfer component of the boiler is the air moisture loss. With complete combustion, all the carbon in the fuel should end up as carbon dioxide in the flue gas. Incomplete gas-phase combustion leaves carbon monoxide (CO) in the flue gas and the associated reduction in heat release is the carbon monoxide loss. Incomplete solid-phase combustion results in carbon leaving the boiler with the ash and the associated reduction in heat release is solid fuel loss. There is also a loss associated with the radiation from the boiler casing (radiation loss). To allow for the uncertainties associated with boiler efficiency testing, an unaccounted loss is normally included in boiler efficiency summaries.

Typical boiler efficiency loss components and the corresponding boiler effi ciencies based on the bagasse GCV and NCV are shown in Table 9.1. Note that the NCV basis fuel moisture, combustion moisture and air moisture losses in Table 9.1 do not include the latent heat in the water vapor. These calculations assume a final flue gas temperature of 280 °C, 50% excess air, 50% bagasse moisture content, and 1.7% bagasse ash content (wet basis). The losses in Table 9.1 are the energy losses per unit mass of bagasse divided by the appropriate calorific value and expressed as a percentage. For a bagasse-fired boiler, the major efficiency losses are those due to the sensible and latent heat of water vapor and the sensible heat of other gases exiting the final heat transfer surface of the boiler. Note that the moisture (fuel and combustion) losses are much smaller on

Table 9.1 Typical efficiency loss components and calculated efficiency for a sugar factory boiler with bagasse fuel.

Efficiency loss components (%)	GCV basis	NCV basis
Dry gas	12.77	15.91
Fuel moisture	15.32	3.12
Combustion moisture	8.29	1.68
Air moisture	0.3	0.38
Carbon monoxide	0.09	0.11
Unburnt solid fuel	0.67	0.84
Radiation	0.34	0.42
Unaccounted	1.00	1.00
Total losses	38.78	23.46
Boiler efficiency	61.22	76.54

an NCV basis, which shows that most of the moisture losses on a GCV basis are due to the latent heat of moisture in the flue gas. Carbon monoxide and unburnt solid fuel losses only become significant if combustion performance is poor.

Whether or not a factory produces surplus bagasse or requires auxiliary fuel depends on how the bagasse flow made available to the boiler station compares with the bagasse flow required by the boiler station. Equating the bagasse flow available to the boiler station (Equation 9.3) with the boiler station bagasse consumption (Equation 9.5) gives the following expression for the boiler station efficiency required for a bagasse balance:

$$\eta_{\mathrm{bs}} = \frac{\mathrm{SOC}(h_s - h_f)(100 - M - B_b)}{(F - b_c)h_b} \tag{9.8}$$

where the steam on cane (SOC) is the ratio of the factory steam requirements (equal to the quantity of steam produced by the boiler station) to the crushing rate, normally expressed as a percentage

$$\mathrm{SOC} = \frac{100\dot{m}_s}{\dot{m}_c}. \tag{9.9}$$

The SOC of most sugar factories that have not been adapted for high levels of electricity export ranges between 46% and 60% (Wright 2000). If the boiler station efficiency is greater than that calculated by Equation 9.8, the factory will produce a bagasse surplus; if the boiler station efficiency is less than that calculated by Equation 9.8, auxiliary fuel will be required. The boiler station efficiency required to balance the consumption and production of bagasse has been calculated for different sugarcane fiber levels and values of SOC as shown in Figure 9.3. The calculations assume a bagasse GCV of 9500 kJ/kg and bagasse moisture and brix contents of 50% and 2.3%, respectively. The assumed feedwater temperature is 95 °C and the assumed steam conditions are 280 °C and 1800 kPag.

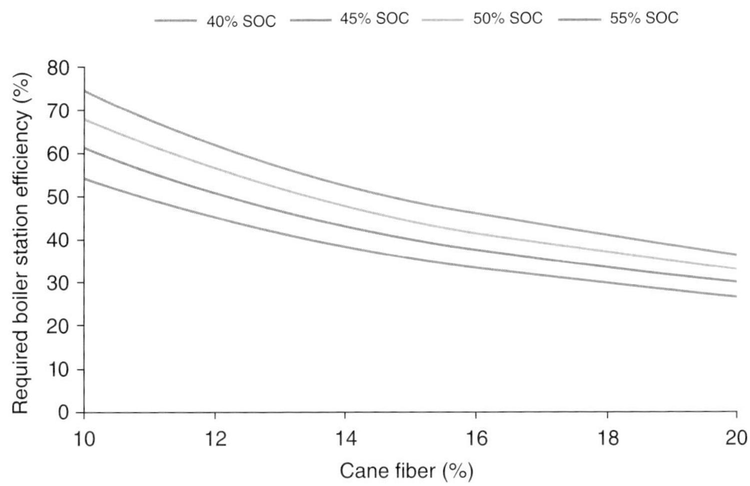

Figure 9.3 Calculated boiler station efficiency (% GCV basis) required to balance the production and consumption of bagasse in a sugar factory for different sugarcane fiber levels.

For a typical sugarcane fiber level of 14% and a typical factory SOC of 50%, a bagasse surplus would be expected if the boiler station efficiency exceeds 48%. If the sugarcane fiber levels are very low (<11%), the boiler station efficiency of a 50% SOC factory would need to be greater than 61% to avoid the use of auxiliary fuel. Normally, at the start of the crushing season, sugarcane fire levels are lower than average and auxiliary fuel is occasionally required; larger bagasse surpluses typically occur at the end of the season.

The curves in Figure 9.3 assume constant bagasse composition and heating value. In practice, variations in sugarcane fiber content are often the result of changes in the amount of dirt in the sugarcane supplied to the factory. High fiber levels do not always lead to large bagasse surpluses if a significant proportion of the measured fiber is ash. Figure 9.4 shows the calculated boiler station efficiencies required for a bagasse balance with varying bagasse ash contents. In these calculations, a constant sugarcane fiber level of 14%, constant bagasse moisture content of 50%, and constant bagasse brix of 2.3% were assumed. The bagasse ash content has a particularly large effect on the boiler station efficiency required for a bagasse balance in factories with high SOC.

9.3 Sugar factory processes without large-scale cogeneration

In a sugar factory that has not been modified to maximize electricity export via large-scale cogeneration, high-pressure (HP) steam produced by the boiler is used by back-pressure turbines to provide mechanical power to drive the shredder,

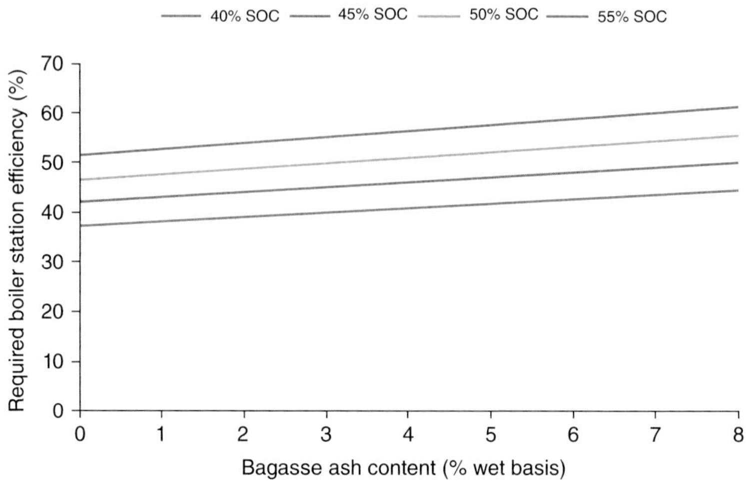

Figure 9.4 Calculated boiler station efficiency (% GCV basis) required to balance the production and consumption of bagasse in a sugar factory for varying bagasse ash contents.

milling train, boiler auxiliaries (forced draft fans, induced draft fans and feed-water pumps), and alternators to meet the electricity generation requirements of the factory. Lower-pressure (LP) steam is sourced primarily from the exhaust ports of these turbines and is used for process heating. The main users of LP steam are the factory juice heaters, juice evaporators, and crystallization pans. The process is a standard Rankine cycle with the evaporators, pans, and juice heaters performing the condensing role for the process. The return condensate from the evaporators, pans, and juice heaters is normally flashed to remove dissolved gasses before being sent to the boiler station (Wright 2001). A typical arrangement of a sugar factory steam cycle is shown in Figure 9.5.

Factories use multiple-effect evaporation to remove water from, and therefore increase the brix of, the sugarcane juice. In this process, typically between four and six evaporator vessels are arranged in series with the juice flowing sequentially through the vessels and becoming more concentrated (increasing the brix of the juice) from the first to the final vessel. LP steam is used to evaporate water from the juice in the first vessel, the vapor produced, known as vapor 1, is then used to evaporate water from the juice in the second vessel. The vapor produced from the second vessel (vapor 2) is then used for evaporating the juice in the third vessel. This process continues until the final vessel. The vapor produced by this final vessel then passes to a condenser. The temperatures and pressures of the vapors from each effect become progressively lower from the first to the final vessel. The difference in temperature between the LP steam and the final vapor in the condenser provides the driving force for multiple-effect evaporation. According to one of the principles established by Norbert Rillieux, the quantity of steam required to evaporate a unit mass of

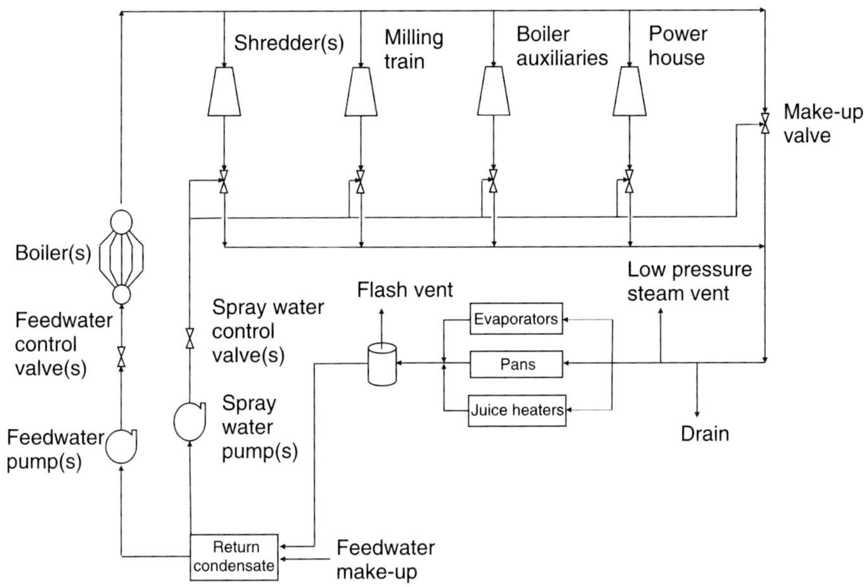

Figure 9.5 Typical arrangement of a sugar factor steam cycle without large-scale cogeneration.

water with multiple-effect evaporation is approximately equal to $\frac{1}{n}$ units where n is the number of vessels (Wright 1975, 2000). This means that a multiple-effect evaporator set with five vessels will use approximately 80% less LP steam to carry out the same evaporation duty as a single evaporator vessel supplied by the same LP steam. In a typical sugar factory without large-scale cogeneration, the evaporators consume roughly half the LP steam of the factory (Broadfoot 2001). Without multiple-effect evaporation, the SOC would exceed 100% and it would be impossible for sugar factories to operate without burning auxiliary fuel.

Typically, the quantity of LP steam required for process heating is more than the quantity of HP steam required to drive the shredder, milling train, boiler auxiliaries, and power house turbines. The total turbine exhaust is, therefore, insufficient to supply the factory's LP steam requirements. The balance of the LP steam requirements comes from HP steam that passes through a make-up valve. The exhaust steam from most factory turbines, and the steam exiting the make-up valve, is typically superheated. Desuperheating (spray) water is usually added to these streams so that the steam used by the evaporators, pans, and juice heaters is close to saturation.

The LP steam requirements of a sugar factory vary due to pan scheduling requirements, factory throughput variations, and operational changes. At times, the LP steam demand of the process may be less that the total exhaust from the factory turbines and LP steam may have to be vented. This wastes both energy and boiler feedwater.

The power supplied by the HP steam to a turbine ($\dot{W}_{\text{turb}}$) can be calculated using

$$\dot{W}_{\text{turb}} = \dot{m}_s(h_{\text{in}} - h_{\text{out}}) \tag{9.10}$$

where $\dot{m}_s$ is the steam flow, h_{in} is the enthalpy of the steam entering the turbine, and h_{out} is the enthalpy of the steam exiting the turbine. The power transferred to feedwater by a boiler feedwater pump ($\dot{W}_{\text{pump}}$) can be calculated using a similar expression

$$\dot{W}_{\text{pump}} = \dot{m}_f(h_{\text{out}} - h_{\text{in}}) \tag{9.11}$$

where $\dot{m}_f$ is the feedwater flow, h_{in} is the enthalpy of the feedwater entering the pump, and h_{out} is the enthalpy of the feedwater exiting the pump.

From Equation 9.10, the power supplied to a turbine and, therefore, turbine power output can be increased by reducing h_{out}. From Equation 9.11, the power transferred to the feedwater and, therefore, power required by a feedwater pump can be reduced by reducing h_{out}. The enthalpy of the outlet fluid depends on the isentropic efficiency (η_s), which for a turbine is defined as

$$\eta_{s,\text{ turb}} = \frac{100(h_{\text{in}} - h_{\text{out}})}{h_{\text{in}} - h_{\text{out},s}} \tag{9.12}$$

where $h_{\text{out},s}$ is what the enthalpy of the turbine outlet steam would be if the expansion in the turbine occurred isentropically. This is the lowest possible steam outlet enthalpy. For a pump, the isentropic efficiency is defined as

$$\eta_{s,\text{ pump}} = \frac{100(h_{\text{out}} - h_{\text{in}})}{h_{\text{out},s} - h_{\text{in}}} \tag{9.13}$$

where $h_{\text{out},s}$ is what the enthalpy of the pump outlet feedwater would be if the increase in feedwater pressure occurred isentropically.

The rate of heat transfer ($\dot{Q}$) from the LP steam to the evaporators, pans, and juice heaters can be calculated using

$$\dot{Q} = \dot{m}_s(h_{\text{in}} - h_{\text{out}}). \tag{9.14}$$

Table 9.2 summarizes the operating conditions for a typical factory with a sugarcane-crushing rate of 600 t/h and 50% SOC. For this factory, the LP steam requirements for process heating are slightly higher than the total exhaust flows from the factory turbines so that a small quantity of make-up steam is required. For these conditions, the HP and LP steam demands of the factory are well balanced so that neither a high quantity of steam flow through the make-up valve or continuous venting of LP steam occurs. The required turbine isentropic efficiencies to maintain such a balance in a factory with specified power and LP heating requirements depend on the boiler station steam conditions and the

Table 9.2 Crushing season conditions for a typical sugar factory.

Crushing rate (t/h)	600
Cane fiber (%)	14
LP steam demand (t/h)	300
Bagasse GCV (kJ/kg)	9,500
Boiler station steam pressure (kPag)	1,800
Boiler station steam temperature (°C)	280
Factory process heating steam pressure (kPag)	100
Factory process heating steam superheat (°C)	3
Boiler station feedwater temperature (°C)	100
Boiler station feedwater flow (t/h)	295
Boiler station blowdown (t/h)	0
Boiler station efficiency (% GCV basis)	55
Boiler station bagasse consumption (t/h)	144.5
Boiler station steam to bagasse ratio	2.04
Boiler station heat input (kW)	381,360
Boiler station heat output (kW)	209,748
Boiler station losses (kW)	171,612

Factory turbines (excluding power house)

	Shredder	Milling train	Boiler auxiliaries
Power input (kW)	4,300	3,100	1,300
Input steam flow (t/h)	58.8	65.5	24.3
Isentropic efficiency (%)	60.4	39.1	44.1
Spray water flow (t/h)	0.1	2.8	0.8
Power house turbine(s) power input (kW)			10,000
Power house turbine(s) input steam flow (t/h)			138.2
Power house turbine(s) isentropic efficiency (%)			59.7
Power house turbine(s) spray water flow (t/h)			0.4
Power house turbine(s) electrical conversion efficiency (%)			93
Power house turbine(s) electricity output (kW)			9,300
Factory electricity use (kW)			7,800
Electricity export (kW)			1,500
Make-up valve input steam flow (t/h)			8.1
Make-up valve spray water flow (t/h)			0.9
Total spray flow (t/h)			5.0

Process heating

	Evaporators	Pans	Juice heaters
Steam flow (t/h)	180	90	30
Heat transfer (kW)	110,373	55,186	18,395
Condensate flashing heat loss (kW)			7,094
Condensate flashing loss (t/h)			11.5
Crushing season length (h)			3,000
Crushing season bagasse usage (t)			433,547
Crushing electricity export (GWh)			4.5
Reduction in CO_2 emissions (t)			3,915

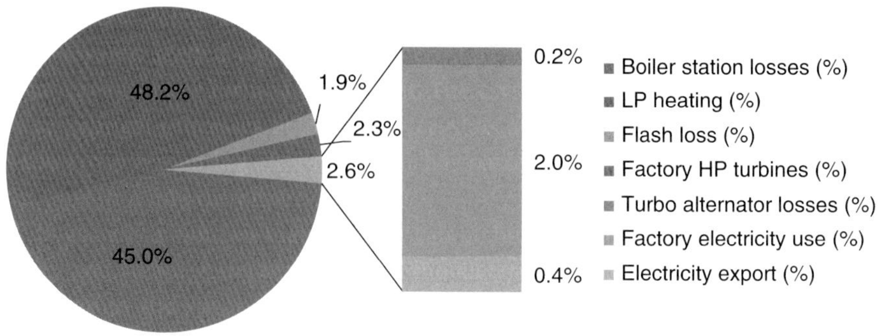

Figure 9.6 Crushing season bagasse energy use in a typical sugar factory.

LP steam mains pressure (Edwards 1982). The process heat transfer is almost 10 times the combined power supplied to all the turbines in the factory even though more than 95% of the steam produced by the boiler station is used by the turbines. The reason for this is that latent heat is extracted from the steam as it condenses when used for factory heating giving an enthalpy reduction of more than 2200 kJ/kg. No latent heat is extracted from the steam passing through the turbines; the enthalpy of steam passing through the turbines reduces by less than 300 kJ/kg.

Note that the rate of heat loss due to condensate flashing is more than the power supplied to most factory turbines. This heat loss can be minimized by installing a pressurized feedwater tank (Wright 2001) and by returning pressurized condensate directly to the feedwater pump (Mann and Dephoff 2006). The crushing season electricity export of 4.5 GWh is quite small and corresponds to an annual income of less than $0.5 million for an electricity price of $100/MWh.

The calculations summarized in Figure 9.6 show what happens to the energy produced when bagasse is burnt in a typical sugar factory with 55% boiler station efficiency and 50% steam on sugarcane. More than half the energy is used for process heating and most of the remaining energy is lost by the boiler station. The proportion of bagasse energy used in the factory turbines (shredder, milling train, and boiler auxiliaries) is very small by comparison (2.3%). Less than 3% of the energy from bagasse combustion is used to generate electricity. Most of this electricity is used by the factory leaving only 0.4% of the energy from bagasse combustion converted to export electricity. The calculations summarized in Figure 9.6 do not take into account energy losses due to venting of LP steam. It is likely that a typical factory without modifications for cogeneration and a good balance between HP and LP steam usage will have to vent LP steam due to pan stage steam requirements reducing more quickly than the boiler station can respond (Broadfoot and Mann 2014). If these vent losses are taken into account, the energy used for LP heating would reduce by the heat loss associated with this venting.

9.4 Sugar factory processes with large-scale cogeneration

It is clear from Figure 9.6 that to significantly increase the electricity export from sugar factories, minimizing the LP steam heating requirements and boiler losses should take priority. It is also important to increase the proportion of bagasse energy in the high-pressure steam that can be converted to electricity in the turbo alternator set and maximize the efficiency of this conversion.

9.4.1 Reducing LP steam heating requirements

Most factories that have not been modified for large-scale cogeneration use LP steam for all process heating except primary juice heating, which typically uses vapor 1 (Broadfoot 2001). Using vapor instead of LP steam for heating (vapor bleeding) reduces the LP steam requirements of a factory. The reduction in LP steam demand that can be achieved with vapor bleeding can be estimated using another of the principles established by Norbert Rillieux, which is that if a mass of vapor V is taken from the mth evaporator vessel to carry out a heating duty normally serviced by LP steam, the overall saving in process steam is approximately $\frac{m}{n}V$ units (Wright 2000). This means that using vapor from the later vessels in an evaporator set will give greater reductions in LP steam usage. However the vapor from the later evaporator vessels is lower temperature and may not be suitable for all factory heating applications. Furthermore, as the temperatures of vapors from evaporator vessels are lower than LP steam, heat transfer surface areas need to be increased to maintain the same heating duty. This additional surface area requirement increases capital costs. Options for reducing factory LP steam demand have been studied extensively (Broadfoot 2001, Lavarack *et al.* 2004, Rose *et al.* 2009, Wright 2000) and most of these studies propose extensive vapor bleeding.

Mann *et al.* (2015) carried out modeling of the different back-end heating arrangements summarized in Table 9.3. Arrangement VBA1 is typical of a factory that has not been modified for cogeneration, while VBA2 uses vapor 4 for primary juice heating. The remaining three vapor bleeding arrangements (VBA3, VBA4, and VBA5) have LP steam being progressively replaced by vapor 1 for secondary juice heating, pan heating, and FSJ heating duties. The final arrangement VBA5 has no process heating with LP steam.

The calculated SOC for the different vapor bleeding arrangements is shown in Figure 9.7. The SOC is 10% lower with vapor bleeding arrangement VBA5 than vapor bleeding arrangement VBA1. The largest reduction in SOC occurs when going from the VBA1 arrangement to the VBA2 arrangement, that is, replacing vapor 1 with vapor 4 for primary juice heating. Changes to the area allocation for the different evaporator stages or total area of the evaporator set are required as increased vapor bleed is incorporated because the heat transfer surface area

Table 9.3 Vapor bleeding arrangements and heating sources for primary, secondary, and ESJ heaters and pans (Mann *et al.* 2015).

Vapor bleeding arrangement	Heating source			
	Primary heaters	Secondary heaters	ESJ heaters	Pans
VBA1	Vapor 1	LP steam	LP steam	LP steam
VBA2	Vapor 4	LP steam	LP steam	LP steam
VBA3	Vapor 4	Vapor 1	LP steam	LP steam
VBA4	Vapor 4	Vapor 1	LP steam	Vapor 1
VBA5	Vapor 4	Vapor 1	Vapor 1	Vapor 1

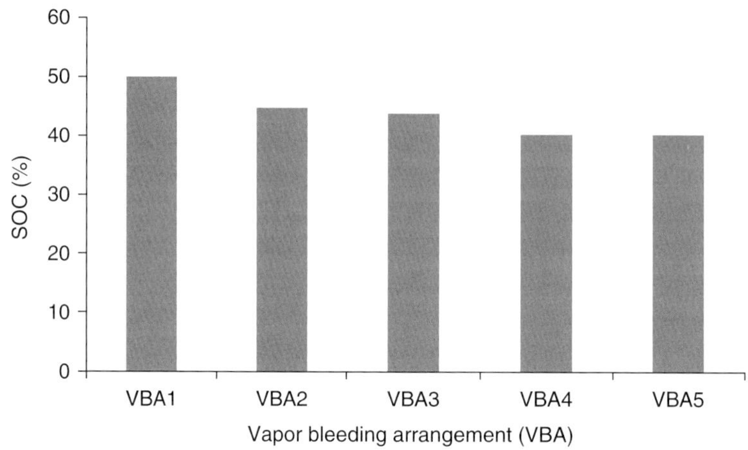

Figure 9.7 SOC for different vapor bleeding arrangements.

will have to be increased to maintain the same heat transfer if the vapor used for heating is at a lower temperature.

Other options for reducing factory LP steam demand include (Broadfoot 2001, Wright 2000) the following:

- Increasing the number of evaporator vessels. The steam consumption of an evaporator set is inversely proportional to the number of evaporator vessels. The incremental reduction in LP steam demand with each additional evaporator vessel decreases as the number of evaporator vessels increases. Due to the increased capital cost associated with additional vessels, few factories use more than six vessels in series.
- Increasing the temperature of the vapor from the final evaporator so that this vapor can be used for juice heating and by the pan stage. However, this option requires increased evaporator and pan heat transfer surface areas. The higher juice temperatures in the evaporators are also likely to cause increased sugar degradation and color formation.

- Using direct contact juice heating with vapor from the evaporator vessels (Tyson *et al.* 2010). This option requires increased evaporator surface area to handle the increased juice flow.
- Using vapor recompression around the pans or final evaporator vessel to increase vapor temperature and, therefore, reduce the quantity of vapor required for a give heating duty. However, it has been shown that vapor recompression requires more power than the extra power made available by the reduced LP steam demand (Wright 2000).
- Reducing the steam consumption of the evaporator set by using ESJ instead of process water in various parts of the factory (Attard 1989).

9.4.2 Reducing boiler station losses

Many boilers installed in sugar factories were designed to be inefficient to ensure complete disposal of the bagasse each season. Furthermore, as boilers age, their efficiency reduces as heat transfer surfaces foul and degrade and air leakage increases. Many older boilers may also be oversized so that they can be run in incineration mode to dispose of surplus bagasse. If a boiler is oversized, it will normally have to be run at higher excess air levels to maintain adequate suspension firing in the furnace and this further reduces boiler efficiency. Consequently, the efficiency of many boilers in the industry is comparable to or lower than that indicated in Table 9.1.

Many newer boilers installed in sugar factories for increased electricity export often have multiple air heater and/or economizer stages (Palmer *et al.* 2009) so that the final flue gas temperature is low (<130 °C). Furthermore, newer boilers can operate at much lower excess air levels due to improved boiler designs. Efficiency loss components and efficiencies (GCV basis and NCV basis) for a typical new cogeneration boiler are summarized in Table 9.4. The calculations summarized in Table 9.4 assume 50% bagasse moisture content and 1.7% bagasse ash content (wet basis).

The dry gas and moisture efficiency losses are much lower in a new high-efficiency cogeneration boiler due to the lower final flue gas temperature and the reduced operating excess air level (~25%). The largest remaining efficiency loss component is associated with fuel moisture. Reducing bagasse moisture will reduce the fuel moisture loss and, therefore, increase boiler efficiency.

One option for reducing bagasse moisture is to install a dewatering mill downstream of the milling train. However, previous studies (Downing *et al.* 2002) have found that even if a stand-alone dewatering mill could reduce the moisture content of the bagasse supplied to the boiler station by five percentage points it would not be economically viable for operation during the crushing season due to the high capital cost of a dewatering mill.

Table 9.4 Efficiency loss components and calculated efficiency for a new high-efficiency cogeneration boiler with bagasse fuel.

Efficiency loss components (%)	GCV basis	NCV basis
Dry gas	4.09	5.10
Fuel moisture	13.78	1.21
Combustion moisture	7.47	0.65
Air moisture	0.10	0.12
Carbon monoxide	0.04	0.05
Unburnt solid fuel	0.67	0.84
Radiation	0.34	0.42
Unaccounted	1.00	1.00
Total losses	27.49	9.39
Boiler efficiency	72.51	90.61

Bagasse drying using final flue gas can significantly increase boiler efficiency because the final flue gas temperature can be reduced to the dew point of the flue gas which for a boiler firing bagasse with no sulfur component is less than 80 °C (Edwards 1981). Unfortunately, bagasse dryers experience a range of operational problems such as fires, component wear, and increased bagasse dust emissions from the dryer bagasse. For these reasons, they have not been widely used in sugar factories.

A more attractive option with much lower capital cost than a dewatering mill or flue gas bagasse dryer is to exploit the phenomena of bagasse stockpile self-heating (Dixon 1988) during storage to reduce moisture content (Sexton *et al.* 2001). Past research indicates that most of the self-heating occurring in bagasse stockpiles is due to slow wet and dry oxidation reactions. If it is assumed that the stockpile is heated solely by the wet and dry oxidation reactions and that evaporation is the only way a stockpile loses heat, then the energy for evaporation of moisture from a stockpile must come from the oxidation of (and therefore loss of) combustible fiber in the stockpile.

The loss of combustible fiber can be calculated from the amount of moisture evaporated, the latent heat of evaporation of water at the stockpile temperature, and the heat liberated by the self-heating oxidation reactions (assumed here to be the GCV of combustible fiber in the bagasse). Calculations based on these assumptions were carried out by Mann (2010) and the results are summarized in Figure 9.8. The figure shows that the heat required to reduce the average moisture content in a bagasse stockpile from 50% to 40% corresponds to a combustible fiber loss of less than 5%. Further calculations in that work showed that the reduction in bagasse moisture content would increase boiler station efficiency (GCV basis) by more than five percentage points. The reduction in combustible fiber required by the boiler station was found to be approximately 1.7 times the loss of combustible fiber during storage. The net saving in combustible fiber

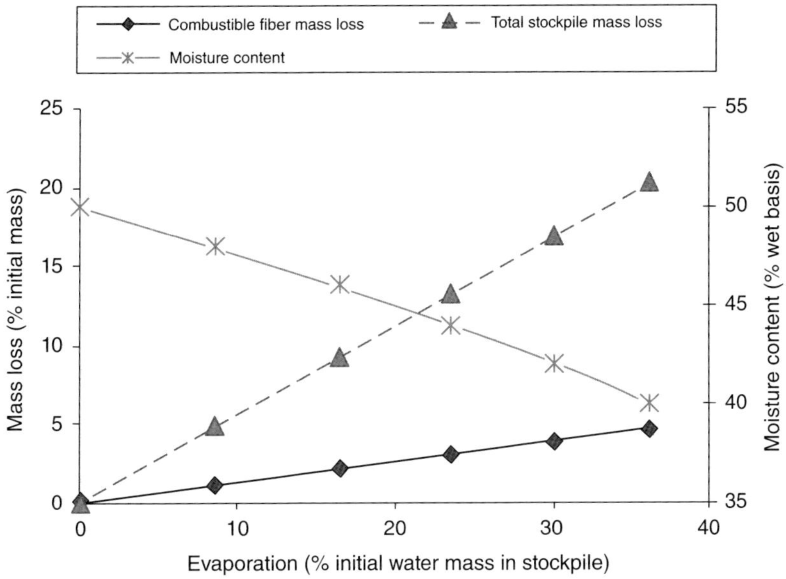

Figure 9.8 Calculated effect of evaporation on the loss of combustible fiber and loss of total stockpile mass. Calculations assume slow oxidation of combustible fiber supplies all the energy used for evaporation (Mann 2010).

occurs because in a stockpile the energy in the combustible fiber is used to evaporate moisture at the temperature of the stockpile while a boiler evaporates the moisture and heats the water vapor to the final flue gas temperature of the boiler.

9.4.3 Increasing power generation efficiency

From Equation 9.10, the power supplied to a turbine is proportional to the difference between the enthalpy of the inlet steam and the exhaust fluid(s). The enthalpy of the inlet steam can be increased by using high-pressure boilers that produce steam at higher temperature and, therefore, with higher enthalpy. A condensing turbine exhausts to a lower pressure (usually below atmospheric) and the exhaust fluid contains liquid; some of the latent heat of the inlet steam is recovered by the turbine and this reduces the enthalpy of the turbine exhaust. Therefore, for a given steam flow, significantly more power can extracted by a condensing turbine supplied with steam from a high-pressure boiler than from a back-pressure turbine supplied with steam from a low pressure boiler. The energy extracted per unit flow of steam can be increased further if the turbine has a high isentropic efficiency (Equation 9.12).

The efficiency of a Rankine cycle can be increased by increasing the average temperature at which heat is added to the working fluid. This can be done by feedwater heating. The feedwater heating can be accomplished in a shell and tube heat exchanger or in a deaerator where the heating steam and feedwater are mixed.

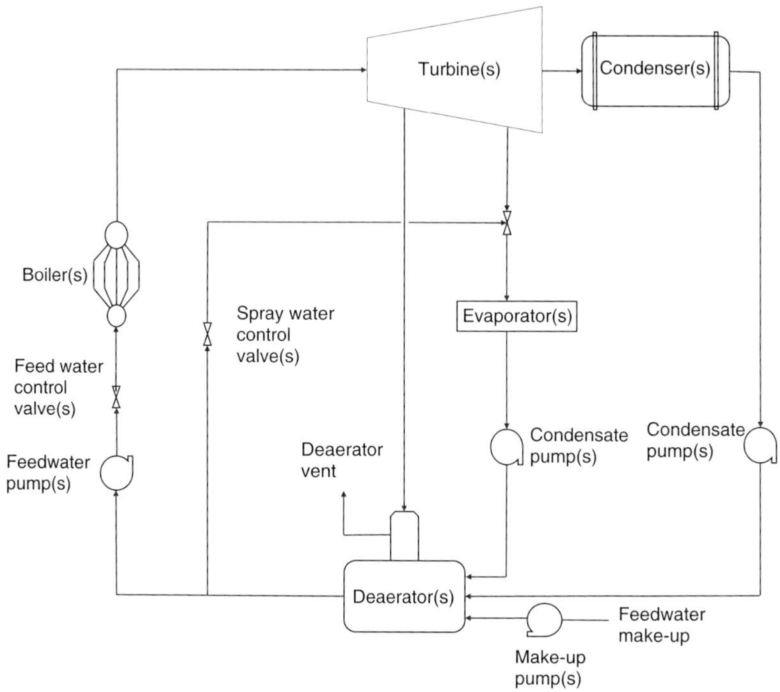

Figure 9.9 A possible arrangement of a steam cycle for a sugar factor steam setup for cogeneration.

9.4.4 A sugar factory cogeneration steam cycle

Figure 9.9 shows a possible arrangement of a sugar factory configured for increased electricity export. For this factory, all steam from the high-pressure boiler(s) is sent to high-pressure multistage turbine(s). These turbine(s) exhaust intermediate-pressure (IP) steam that is sent to a deaerator that removes dissolved gases from and heats the boiler feedwater, and low-pressure steam used by the factory evaporator set(s). All other heating in this factory (juice heating and pan heating) is carried out by vapors bled from the evaporator set(s). The exhaust from the turbine is sent to a condenser before it is returned via a deaerator to the boiler(s).

With this setup, all the factory prime movers are electrically driven so that no HP steam is used for factory operations. The LP steam extraction from the turbine(s) can adjust according to factory requirements.

Table 9.5 summarizes the operating conditions for a 600 t/h factory configured for cogeneration with the arrangement of Figure 9.9. The boiler station bagasse consumption for the conditions in Table 9.5 was chosen to match the boiler station bagasse consumption for the typical factory conditions of Table 9.2. This factory has high-pressure (7200 kPag) boilers with a boiler station efficiency of 65% and process side of the factory has been set up with extensive vapor bleeding so that the SOC is 40%. The condensing turbine exhausts to −89 kPag pressure.

Table 9.5 Crushing season conditions for a sugar factory setup for large-scale cogeneration.

Crushing rate (t/h)	600
Cane fiber (%)	14
LP steam demand (t/h)	240
Bagasse GCV (kJ/kg)	9,500
Boiler station steam pressure (kPag)	7,200
Boiler station steam temperature (°C)	520
Deaerator pass out pressure (kPag)	200
Deaerator vent rate (%)	1
Factory process heating steam pressure (kPag)	100
Factory process heating steam superheat (°C)	3
Turbine exhaust pressure (kPag)	−89
Boiler station feedwater temperature (°C)	133.7
Boiler station feedwater flow (t/h)	311.4
Boiler station blowdown (%)	1
Boiler station blowdown (t/h)	3.1
Boiler station steam output(t/h)	308.3
Boiler station efficiency (% GCV basis)	65
Boiler station bagasse consumption (t/h)	144.5
Boiler station steam to bagasse ratio	2.13
Boiler station heat input (kW)	381,360
Boiler station heat output (kW)	247,884
Boiler station losses excluding blowdown (kW)	133,476
Boiler station blowdown loss (kW)	626
Boiler station losses including blowdown (kW)	134,102
Turbine(s) high pressure stage isentropic efficiency (%)	85
Turbine(s) intermediate pressure stage isentropic efficiency (%)	80
Turbine(s) condensing stage isentropic efficiency (%)	75
Turbine(s) electrical conversion efficiency (%)	93
Deaerator steam extraction (t/h)	18.4
Deaerator vent flow (t/h)	3.0
Deaerator vent heat loss (kW)	2,249
LP steam extraction (t/h)	239.0
LP spray water flow (t/h)	1.0
Turbine exhaust flow (t/h)	51.0
Turbine(s) electricity output (kW)	62,374
Factory electricity use (kW)	14,100
Electricity export (kW)	48,274
Evaporator heat transfer (kW)	147,163
Condenser heat transfer (kW)	31,077
Crushing season length (h)	3,000
Crushing season bagasse usage (t)	433,547
Crushing electricity export (GWh)	144.8
Reduction in CO_2 emissions (t)	125,995

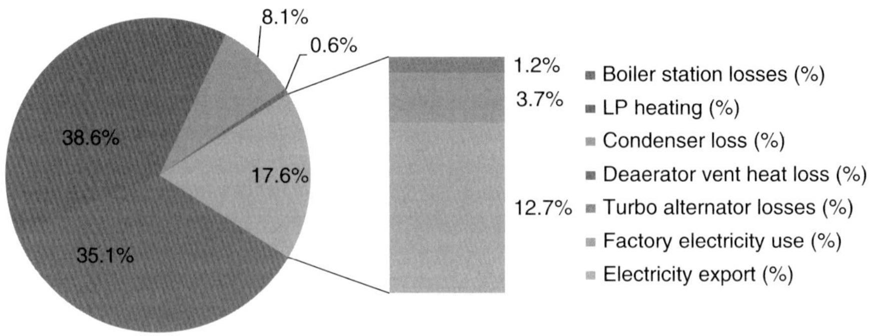

Figure 9.10 Crushing season bagasse energy use in a sugar factory setup for cogeneration.

The deaerator heats the feedwater to approximately 134 °C. Spray water is used to control the temperature of the steam sent to the evaporators. All stages of the turbine(s) have much higher isentropic efficiencies than the back pressure turbines used in a typical factory (Table 9.2). The higher boiler station efficiency for the conditions of Table 9.5 relative to the conditions of Table 9.2 more than compensates for the higher enthalpy of the steam produced by the boiler station so that the steam to bagasse ratio is higher for the cogeneration factory setup. The total electrical output of the factory turbo-alternator(s) is more than 62 MW; just over 14 MW of this used by the factory giving just over 48 MW electricity export. This gives just under 145 GWh electricity export over a 3,000 h crushing season. With an electricity price of $100 per MWh this gives annual revenues of $14.5 million – nearly 30 times the revenue due to electricity export of a typical sugar factory.

The calculations summarized in Figure 9.10 show the calculated bagasse energy usage distribution for a factory setup for cogeneration with the conditions of Table 9.5. Most of the bagasse energy is still used for LP heating or lost by the boiler station but electricity export (12.7%) accounts for a greater proportion of bagasse energy use than factory electricity use and all other losses combined.

9.5 Conclusions

Ever since the introduction of multiple-effect evaporation, the sugarcane industry has had the competitive advantage of producing a by-product, bagasse, which when burnt liberates more heat energy than what is required to carry out crushing operations. In the past, this surplus energy was wasted by designing and operating sugarcane factories to run inefficiently. In many parts of the factory, especially the boiler station, this mode of operation would increase maintenance costs due to excessive component wear.

Electricity exported by sugarcane factories will displace electricity generation by fossil fuels and will, therefore, mitigate global warming. With appropriate factory modifications, electricity export can also be a significant source of revenue for sugarcane factories. The promise of this revenue has helped justify the expansion and continued operation of existing factories and the construction of new factories around the world.

Cogeneration and the development of high-value by-products from bagasse are often thought of as competing options for utilizing bagasse. However, both cogeneration and bagasse value adding require factory improvements to reduce bagasse usage and infrastructure to manage bagasse storage and transport. The factory improvements and added infrastructure from cogeneration projects that increase bagasse surpluses provide more opportunities for bagasse value adding.

References

Attard, R. (1989). Substitution of evaporator supply juice (ESJ) for process water at North Eton sugar mill. *Proceedings of the Australian Society of Sugar Cane Technologists* **11**:199–206.

Behne, E. R. (1934). Thermal value of bagasse. *Proceedings of the Queensland Society of Sugar Cane Technologists,* **5**:175–177.

Broadfoot, R. (2001). Planning changes to the process sections of raw sugar factories for increased cogeneration. *Proceedings of the Australian Society of Sugar Cane Technologists,* **23**:395–402.

Broadfoot, R. and Mann, A. P. (2014). Management of a high grade pan schedule to reduce the steam consumption. *Proceedings of the Australian Society of Sugar Cane Technologists,* **36**:361–372.

Dixon, T. F. (1988). Spontaneous combustion in bagasse stockpiles. *Proceedings of the Australian Society of Sugar Cane Technologists,* **10**:53–62.

Don, C., Mellet, P., Ravno, B. and Bodger, R. (1977). Calorific values of South African bagasse. *Proceedings of the Annual Congress – South African Sugar Technologists Association,* **51**:169–173.

Downing, C. M., Hobson, P. A. and Kent, G. A. (2002). Is investment in a bagasse dewatering mill economically justifiable for cogeneration. *Proceedings of the Australian Society of Sugar Cane Technologists,* **24**:(CD-ROM).

Edwards, B. P. (1981). Bagasse drying. *Proceedings of the Australian Society of Sugar Cane Technologists,* **3**:203–206.

Edwards, B. P. (1982). The relationship connecting turbine efficiencies, electrical power exports, factory steam pressures, and the factory steam balance. *Proceedings of the Australian Society of Sugar Cane Technologists,* **4**:181–187.

Frew, R. and James, P. J. (1977). Combustion calculations for high ash content bagasse. *Proceedings of the Queensland Society of Sugar Cane Technologists,* **44**:327–334.

IPCC (2012) *Special Report on Renewable Energy Sources and Climate Change Mitigation,* Cambridge University Press, United Kingdom and New York, NY, USA.

Kent, G. A. (2010). Estimating bagasse consumption. *Proceedings of the Australian Society of Sugar Cane Technologists,* **32**:546–558.

Kent, G. A., Hoare, C. P., Miller, K. F. and Allen, W. J. (2003). Harvest and factory data to assist in evaluating the effect of extraneous matter on sugar production costs and revenues. *Proceedings of the Australian Society of Sugar Cane Technologists,* **25**: (CD-ROM).

Lamb, B. W. (1979). Combustion of bagasse, PhD, University of Sydney, Sydney.

Lavarack, B. P., Hodgson, J. J., Broadfoot, R., Vigh, S. and Venning, J. (2004). Improving the energy efficiency of sugar factories: Case study for Pioneer Mill. *Proceedings of the Australian Society of Sugar Cane Technologists,* **26**: (CD-ROM).

Mann, A. P. (2010). Predicting the effects of fuel properties on combustion performance. *Proceedings of the Australian Society of Sugar Cane Technologists*, **32**:629–637.

Mann, A. P. and Dephoff, R. M. (2006). Reducing bagasse consumption at Harwood Mill. *Proceedings of the Australian Society of Sugar Cane Technologists*, **28**:389–397.

Mann, A. P., Thaval, O. T., Broadfoot, R. and McFeaters, J. (2015). Improving factory performance using an integrated sugar factory model. *Proceedings of the Australian Society of Sugar Cane Technologists*, **37**:305–315.

Palmer, C., Farrell, R. M. and Hurt, J. (2009). The development of cogeneration projects at Condong and Broadwater mills. *Proceedings of the Australian Society of Sugar Cane Technologists*, **31**:401–409.

Rose, I., Scroope, P. and Broadfoot, R. (2009). Reducing factory steam consumption for cogeneration at Condong Mill. *Proceedings of the Australian Society of Sugar Cane Technologists*, **31**:420–429.

Sexton, M. J., Macaskill, C. and Gray, B. F. (2001). Self-heating and drying in two-dimensional bagasse piles. *Combustion Theory and Modelling*, **5**(4), 517–536.

Tyson, B., Drury, T. and Heck, D. (2010). Achieving low steam on cane with direct contact juice heaters. *Proceedings of the Australian Society of Sugar Cane Technologists*, **32**:517–521.

Wright, P. G. (1975). Simplified multi-effect evaporator calculations. *Proceedings of the Queensland Society of Sugar Cane Technologists*, **42**:185–193.

Wright, P. G. (2000). Sugar factory configurations for increased cogeneration. *Proceedings of the Australian Society of Sugar Cane Technologists*, **22**:347–354.

Wright, P. G. (2001). Increasing energy efficiency by conserving boiler feedwater temperature. *Proceedings of the Australian Society of Sugar Cane Technologists*, **23**:416–420.

Wright, P. G. (2003). The effect of dirt on bagasse quantity and heating value. *Proceedings of the Australian Society of Sugar Cane Technologists*, **25**, (CD-ROM).

CHAPTER 10

Pulp and paper production from sugarcane bagasse

Thomas J. Rainey[1] and Geoff Covey[2]

[1] *School of Chemistry, Physics and Mechanical Engineering, Science and Engineering Faculty, Queensland University of Technology (QUT), Brisbane, Australia*

[2] *Covey Consulting, Melbourne, Australia*

10.1 Background

The pulp and paper industry is very large and is now well in excess of $200 billion (FAO 2009). Estimates for the amount of bagasse used in the production of pulp and paper products vary but the general consensus is that it accounts for 2–5% of global production, making it one of the highest revenue earners for the global sugarcane industry. Currently, there are believed to be around 30 countries that use bagasse for paper manufacture. The paper industry remains a growth industry globally, with volumes forecast to increase 50% by 2035 (i.e., an additional 200 million tons) with most of the growth occurring in the packaging and tissue sectors. The growth will occur in emerging economies as the other economies will have net negative growth of 33 million tons by 2035 (Wilson 2013) amidst the rapid decline in newsprint consumption. As the growth is occurring in emerging economies, which constitutes the entirety of the bagasse-based paper sector, it is reasonable to presume that bagasse will grow in importance as a paper feedstock.

Bagasse is used for each of the four main paper categories, namely packaging and boxes, printing, writing and photocopier paper, tissues, and newsprint. Both chemical and mechanical pulping processes are used, and these processes are described in separate sections. This chapter focuses on the considerations for bagasse-based pulp and paper manufacture with only a broad overview of the general wood-based process. Numerous comprehensive texts have a more in-depth analysis of the science of pulp and paper manufacture (Smook 1992, Gullichsen and Fogelholm 1999, Paper Engineers Association 2007–2010).

Modern paper manufacturing is a highly capital-intensive industry. Wood -based chemical pulp mills typically cost in the order of billions of dollars, and cheaper mechanical pulp mills are in the order of $30–200 million. Although bagasse-based pulp mills are generally substantially smaller and cheaper than the large wood-based chemical pulp mills owing to the relatively mild pulping conditions, they are still a sizable investment.

Sugarcane-Based Biofuels and Bioproducts, First Edition. Edited by Ian M. O'Hara and Sagadevan G. Mundree.

10.2 History of bagasse in the pulp and paper industry

The first continuous paper machine was developed around the year 1800, allowing large quantities of paper to be produced for the first time. At this time, paper was produced from retired fabric, and its availability was becoming increasingly scarce as paper production grew. This necessitated a change in raw material and modern techniques for turning plant material (biomass) into pulp, the intermediate products for making paper, were not developed until the mid-19th century. Early investigations in Europe concluded that wood was the most suitable local raw material. Bagasse was brought back to Europe from the West Indies for investigation around this period.

Apart from the impracticality of using bagasse to meet European demand for paper, initial attempts to use bagasse for paper manufacture were unsuccessful primarily because the presence of pith is extremely detrimental to the papermaking process (Atchison 1962, 1971b, c, 1992c). In particular, bagasse pith severely reduces permeability and hence reduces paper productivity by around 30% when bagasse-based pulp is used instead of wood-based pulp (Rainey *et al.* 2008, 2009, 2010, 2011b, Rainey 2009). The short pith material blocks the holes in pores of the paper mat, preventing water from draining through it, reducing the production rate and various quality characteristics of the final paper product (Giertz and Varma 1979, Zanuttini 1997, Kasi Viswanathan and Gopalaratnam 1998). Initial problems with overcooking and supply-side issues were secondary factors, which also prevented the development of bagasse as a papermaking feedstock. For these reasons, interest in the suitability of bagasse for making paper stopped for more than 100 years.

Using wood for papermaking was appropriate for Europe and North America, which had abundant forestry resources. However, many emerging countries that did not have abundant forests such as India and China became dependent on imports of paper from developed countries. Many of these countries recognized the need for local paper production in the first half of the 20th century and the use of bagasse for paper production was revisited. Some of the early developmental work was funded by the United Nations programs in the 1950s and 1960s.

10.3 Depithing

10.3.1 The need for depithing

For bagasse papermaking, it is necessary to depith the bagasse. It has been unanimously agreed that removing 30% of the shortest bagasse fibers is essential for making pulp of acceptable quality (e.g., Atchison 1962, Giertz and Varma 1979, Paul and Kasi Viswanathan 1998, Covey *et al.* 2005, Rainey *et al.* 2010).

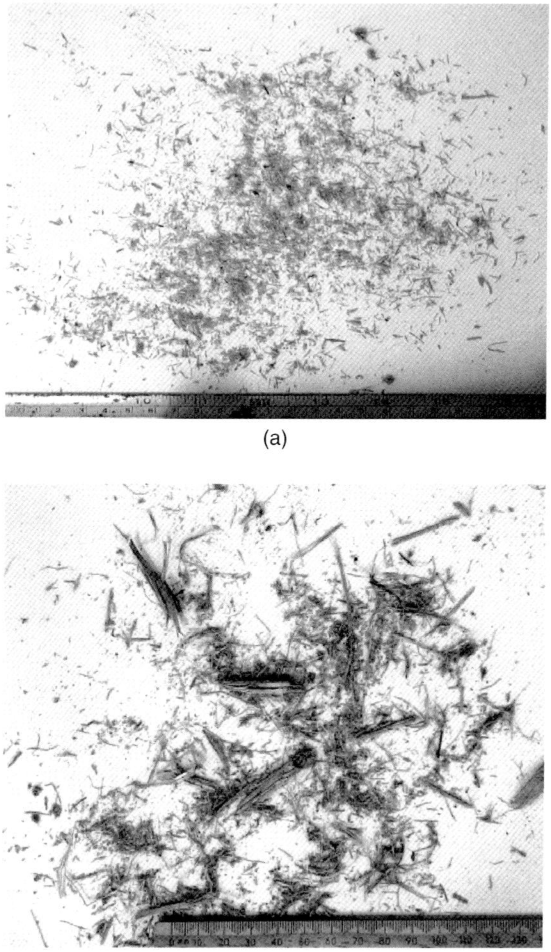

Figure 10.1 Photograph of (a) bagasse pith and (b) whole bagasse (Rainey 2009).

Most agricultural fibers require some kind of pretreatment before pulping. For bagasse, the very high amount of pith (Figure 10.1), which is mostly parenchyma material must be separated first. On a dry, water-insoluble basis, bagasse consists of "pith" from the center of the plant (30%), "fiber bundles" (65%; 50% from the rind and 15% from the internal material), and the epidermis (5%) (Atchison 1971b, c, Rao 1997).

The fiber bundles in bagasse are mostly sclerenchyma cells bound together. These cells are typically 1.0–1.2 mm in length and 20 μm in diameter (Paul and Kasi Viswanathan 1998, Rainey *et al.* 2010), with a wall thickness of 5 μm and a resulting lumen diameter of 10 μm (Rainey *et al.* 2010). The pith portion of bagasse is quite different anatomically from the other bagasse fibers. Pith is composed of thin-walled parenchyma cells that originate from the vascular bundles (Giertz and Varma 1979). The "pith" parenchyma material is much squarer (i.e., lower aspect ratio) than the sclerenchyma material with a maximum length in the order of 0.52 mm (Rainey *et al.* 2009) with a diameter of 140 μm (Paul and Kasi Viswanathan 1998).

Depithing is the process whereby the short pith material is separated from the rest of the bagasse. In practice, it is almost impossible to economically achieve extremely high efficiency of separation of pith at a sugar factory scale. The most efficient techniques involve cutting the cane lengthways and scooping out the central contents which has been previously achieved both for sugarcane (Moeltner 1980) and other pith bearing plants (Gominho *et al.* 2001, Reddy and Yang 2005). Instead, conventional, commercially available depithing equipment separates bagasse, which has been mechanically prepared, into pith and nonpith material based on size classification but there is still some cross-contamination of pith with nonpith material and vice versa. In one study, a commercial depithed bagasse was analyzed (Giertz and Varma 1979). Around 30% of the "pith" material was found to be non-parenchyma material.

Without depithing, it is virtually impossible to create a sheet of paper using normal industrial pulp washers and paper machines due to hydraulic holdup (i.e., extremely poor pulp freeness). Depithing is used to (Rao 1997, Lois-Correa 2012).

- improve pulp drainage in the pulp washers and at the paper machine;
- reduce cooking chemical usage;
- reduce foaming;
- reduce costs for handling and storage;
- reduce dirt count in the paper;
- improve tensile properties;
- improve black liquor quality.

Although depithing is mostly used for sugarcane bagasse, it has been investigated as a tool for improving the fiber quality of other plants such as cardunculus (Gominho *et al.* 2001) and sweet sorghum (Belayachi and Delmas 1995, Reddy and Yang 2005).

10.3.2 Depithing operation

Depithing of bagasse is usually undertaken in a two-stage process. The first stage is a "moist depithing" stage where bagasse (typically 48–50% moisture) is separated into two fractions using a hammermill and a screen (the pith material may be returned to the sugar factory boiler). The second stage is "wet depithing." Wet depithing most commonly occurs at the paper mill at a consistency of around 2.5% dry solids. The shortest 30% of bagasse fibers are removed.

Historically, there have been a number of designs for depithing equipment (Atchison 1962, Lois-Correa 2012), but modern depithers mostly have the

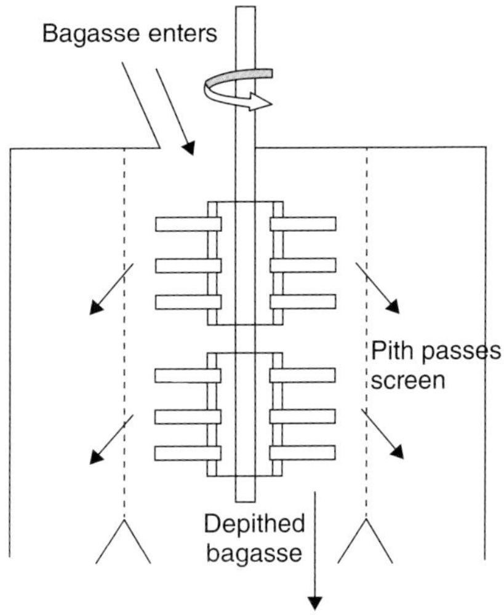

Figure 10.2 Sketch of a depither.

same operational principle–a hammer mill is mounted on a vertical shaft. The rotor can be hanging with only an upper support or with upper and lower supports (Caribe and Pallman models) and the rotor is sometimes split. The rotating assembly is mounted with straight hammers and operates at high speed, typically 1000–1800 rpm. The bagasse is fed into the top of the unit, and it is hit by the hammers, opening up clumps of bagasse. The pith is thrown through a screen while the longer bast material (i.e., the fiber bundles) falls down the center of the unit to a conveyor. The pith and bast fiber are conveyed away separately (Figure 10.2). Early model depithers worked on a similar principal but the rotor was mounted horizontally.

Due to the high wear on the hammers, depither hammers need to be periodically maintained/replaced. It was reported in one particularly problematic instance that the hammers had to be replaced every 3 days (Koen 1980). Some designs allow for isolating and replacing sets of hammers, rather than replacing the hammers individually.

Depithers are usually arranged in banks (conceptually shown in Figure 10.3). Bagasse enters from the sugar factory and feeds into the depithers through a series of chutes. The bagasse is commonly screw fed into the depithers (data not shown). The pith is separated from the bast fiber and conveyed away on communal conveyor belts.

Moist depithing occurs at either the sugar factory or the pulp and paper factory, and the pith is typically recovered for combustion. The depithed bagasse is normally conveyed to the pulp/paper factory for wet depithing and wet bulk storage.

Vibration can become an issue if depithing units are poorly designed or installed (Lois-Correa 1986, 2012; Diez and Lois 1989).

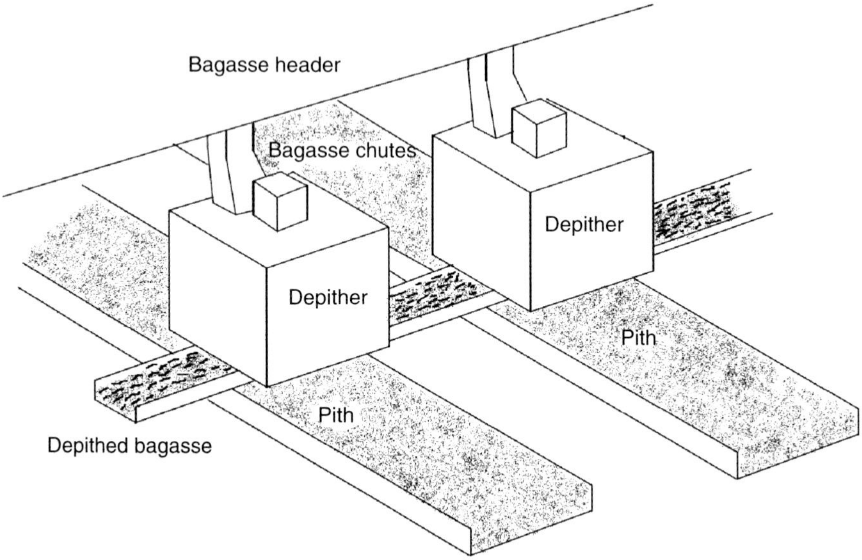

Figure 10.3 Typical arrangement of a bank of depithers.

10.3.3 Character of pith, depithed bagasse, and whole bagasse

The chemical character of raw bagasse, pith, and depithed bagasse has been reported by numerous authors using bagasse from around the world with fairly consistent findings (Lois and Suarez 1983, Lois-Correa 1986, Rao 1997, Rainey *et al.* 2009). The findings of these studies are summarized in Table 10.1, and average results are provided in Table 10.2. The consensus is that depithed bagasse has a higher α-cellulose content than whole bagasse (by 6%; refer to the data provided by Rao (1997) and Lois-Correa (1986) in Table 10.1), and that the ash content of the pith fraction is about three times higher, typically more than 5%. The ash content is higher because silica particles are relatively small and also pass through the depither screens with the pith.

Contrary to a commonly held belief, the majority of the literature surveyed reports that the lignin content of depithed bagasse is marginally lower than for the parent bagasse (Lois and Suarez 1983, Rao 1997, Rainey 2009). This suggests that pith has higher lignin content than rind material. Work by Lois and Suarez (1983) is in disagreement with this finding, for example.

10.3.4 Combustion of pith

For mechanically harvested cane, factories combusting pith have generally experienced 10–15% lower calorific value (4075 kcal/kg for 50% moisture) compared with whole bagasse (Atchison 1980, Rao 1997). During moist depithing operations, the ash is preferentially diverted to the pith fraction (Atchison 1980,

Table 10.1 Chemical composition of whole bagasse, depithed bagasse, and pith for Indian bagasse (Rao 1997), Australian bagasse (Rainey 2009), and Cuban bagasse (Lois and Suarez 1983, Lois-Correa 1986).

	Whole bagasse				Depithed bagasse				Pith	
	Rao (1997)	Rainey (2009)	Lois and Suarez (1983)	Lois-Correa (1986)	Rao (1997)	Rainey (2009)	Lois *et al.* (1983)	Lois-Correa (1986)	Rao (1997)	Lois-Correa (1986)
α-Cellulose	43.0	38.9	~39	43.0	45.3	40.5–45.6	~42	45.3	38.0	–
Pentosans	26.0	22.4	21	24.1	25.0	24.2–27.2	25	24.8	24.2	24.2
Total lignin	20.7	25.1	18	20.7	19.4	22.1–22.7	19	19.5	21.7	21.7
Ash	2.6	6.9		2.6	1.4	1.8–2.6		1.4	5.0	5.49
Extractives	2.8	0.48	2.5–4.5	2.8	2.4	0.34–0.49	3.6	2.4	2.9	2.95

Table 10.2 Average chemical composition of whole bagasse, depithed bagasse, and pith.

	Whole bagasse	Depithed bagasse	Pith
α-Cellulose	43.0	45.1	42.1
Pentosans	23.4	25.2	24.2
Total lignin	21.1	20.5	21.7
Ash	4.0	1.8	5.3
Extractives	3.3	2.8	2.9

Table 10.3 Fuel properties of pith and bagasse mixtures (Mann and O'Hara 2012).

	Moisture content (%)	Ash content (%)	Gross calorific value (kJ/kg)
Bagasse	50.00	4.60	8.929
Pith	53.00	8.46	7.580
Depithed bagasse	48.00	2.03	9.828
Bagasse/pith mixture	50.34	5.04	8.777

Zanuttini 1997, Paul and Kasi Viswanathan 1998, Rainey 2009). The ash content is higher, and the pith also has higher moisture content as shown in Table 10.3.

Fiber has a strong propensity to hydrogen bond with airborne water molecules. The higher moisture content of the pith is caused by its much higher surface area. Depithed bagasse has 43% lower water holding capacity than whole bagasse (Rainey *et al.* 2013). At the cellular level, pith *pulp* has a specific surface area of 14,100 cm^2/cm^3 (Rainey 2009). By comparison, depithed bagasse pulp has far lower specific surface area, ~2100 cm^2/cm^3. This very large difference in specific surface area makes it much easier for the pith to absorb ambient moisture than for rind material.

Converting a bagasse boiler to burn a higher proportion of pith reportedly requires "only minor changes" (Atchison 1980), namely increasing the amount of air and the position of entry into the boiler. Only "minor changes" are required provided cofiring of pith with whole bagasse is at a ratio of less than 1:3. However, 100% pith combustion occurs in some factories, as well as cofiring with fossil fuels. Cofiring of pith with fossil fuels is still commonly practiced because in most instances.

10.4 Storage of bagasse for papermaking

Rao (1997) notes that storage of bagasse for at least 2 months before papermaking is essential because (i) pith attached to bagasse fibers is loosened and the pith level is further reduced, improving processability at the paper machine; (ii) during storage the residual sugar in bagasse is destroyed and excessive foaming during the pulp cook is reduced; (iii) pulp quality is improved; and (iv) it is inevitable due to the seasonal nature of harvesting sugarcane and the year round operation of most pulp factories. If bagasse is not stored properly, it undergoes microbial attack and leads to increased color formation, loss of yield, and loss of pulp strength (Covey *et al.* 2006a,b).

In most instances, the depithed bagasse is stored for up to 12 months after *wet* depithing as this maximizes the quality of bagasse for paper manufacture. In these cases, the bagasse is stored in a pile with water recirculating through the pile. The recirculation liquor contains biologically active ingredients to assist with controlled degradation that improves pulp quality and pulping efficiency. For the best conservation of bagasse, it should be stored at its full water holding capacity to exclude air from the pile and to achieve maximum compaction (Rangan and Rangamannar 1998).

Figure 10.4 shows the impact of storage time on brightness for pulp made from bagasse taken from the surface of the pile. The decrease in brightness is mostly exhibited in pulp produced from bagasse taken from the surface.

For these reasons, moist depithed bagasse is rarely stored before papermaking. It is not known if there are currently any factories currently storing moist depithed bagasse; however, there are instances of this practice occurring in the past. In one article, it was reported that a factory produced moist depithed bagasse stacks 18 m high that had been compressed by bulldozers to a density of 150 kg/m^3 (bone dry basis) (Atchison 1971a).

Bagasse quality is also affected by its depth into the pile. Figure 10.5 compares pH, brightness, and yellowness of fresh bagasse with stored bagasse as a function of depth into the stockpile. Bagasse at the top of the pile showed the largest deterioration.

Bagasse is washed to reduce the level of silicates, which otherwise would dissolve in the pulping liquor and cause problems with process chemical recovery and evaporator scaling. The bagasse is dumped into a reclaim chest, made up to

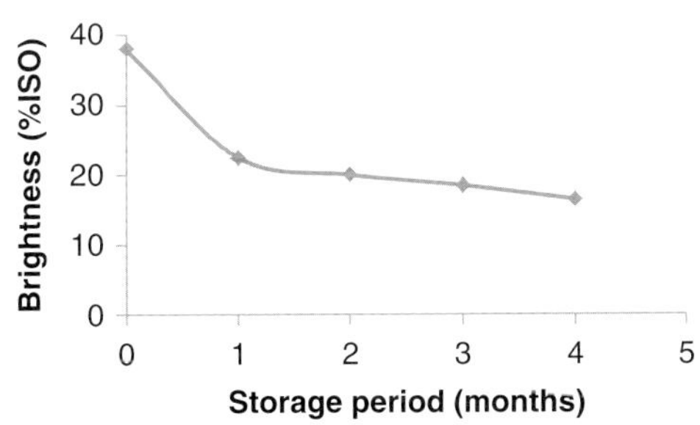

Figure 10.4 Effect of bagasse aging on brightness. (Adapted from Rangan and Rangamannar 1998.)

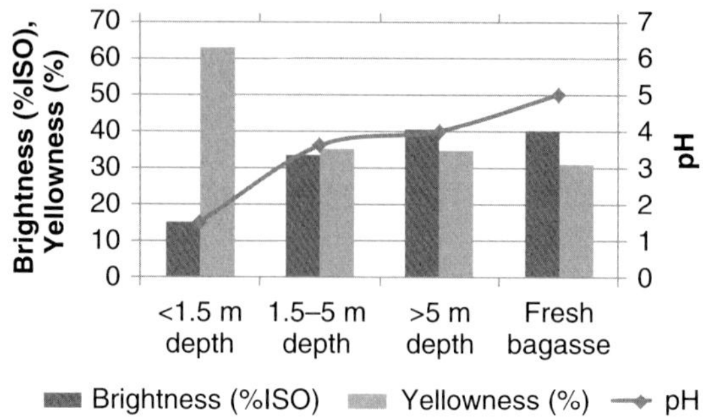

Figure 10.5 pH, brightness, and yellowness of stored bagasse through the stockpile. (Adapted from Rangan and Rangamannar 1998.)

~3% consistency and then may be pumped to a stone removal tank whereby the overflow is further processed to remove sand before going through an inclined screw to remove most of the water.

10.5 Chemical pulping and bleaching of bagasse

10.5.1 Digestion

In the digestion process, the fibrous material is broken down into individual pulp fibers which are the sclerenchyma cells in the case of bagasse and are typically 1.0–1.2 mm in length. For bagasse chemical pulp, the bagasse is normally loaded into a twin-screw horizontal digester and the fibers are exposed to chemicals and at elevated temperature and pressure. Soda, soda anthraquinone (AQ), and Kraft pulping are commonly used processes for producing bagasse chemical pulp.

Bagasse lignin is much more reactive than wood lignin, so the pulping conditions are very mild. For soda pulping the chemical charge is 12–16% of sodium hydroxide (as NaOH on dry fibers) at 170–175 °C and for Kraft pulping, it is 11–13% active alkali with 15–20% sulfidity. The pulping time is typically 10–12 min (Hurter 2007b) although residence times of up to 20 min are used in some factories (e.g., Rao 1997) (this is much milder than the 15–20% NaOH equivalent and total cooking time of 2.5–4 h for wood pulp). Pulp yield is typically in the range 48–55%, which varies with the extent of depithing, the general quality of the bagasse, and the variety of sugarcane (Gartside *et al.* 1983).

Bagasse digestion most commonly occurs in PanDdia style continuous digesters. Uniformity of cooking and ease of control results in better quality pulp, higher yield, lower chemical charge, and reduced energy consumption relative to wood pulping (Atchison 1992a, Darwesh *et al.* 1998). It allows a liquor-to-fiber ratio of 3.5:1 (Atchison 1992a), far lower than that required for batch digesters. PanDia digesters are a distinctive feature of bagasse pulping as this style is less common for wood digestion. A generic bagasse feeding

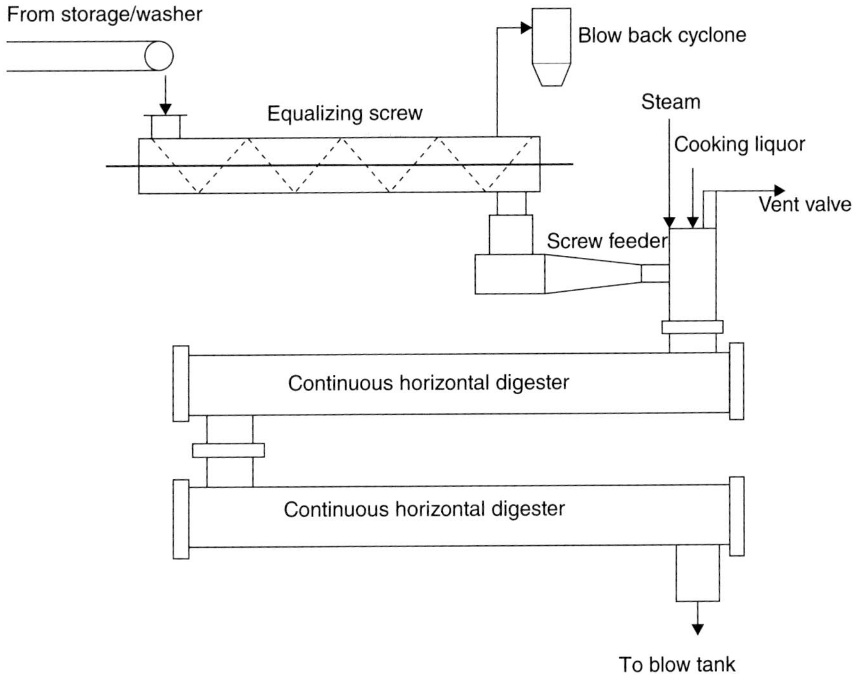

Figure 10.6 Bagasse digestion system for pulping agricultural fibers.

and continuous cooking system is shown in Figure 10.6. The wet bagasse is conveyed from the stockpile over a metering system and fed to the screw feeder. This removes excess water and creates a plug of bagasse, which allows feeding of the bagasse into the pressurized reactor. Steam and cooking liquor are subsequently injected. A blowback valve may be installed on the inlet chamber in the event that inadequate bagasse is metered to the digester (Darwesh *et al.* 1998, Nagarajan *et al.* 2003).

10.5.2 Black liquor

Similar to most agricultural residues, there is a high amount of silica present in bagasse compared with wood chips, and the amount varies widely (1–8%). The quantity depends on a range of agricultural factors such as manual or mechanical harvesting, soil and cane type, as well as climate (rainfall increases silica content). It is desirable to remove this silica otherwise there will be excessive equipment wear and it causes problems when concentrating the black liquor. The approaches to removing silica are generally to (i) remove silica from the raw material; (ii) remove silica from the black liquor or (iii) remove silica from the green liquor. Another approach is to install an extra black liquor evaporator so that one can always be offline for cleaning, mitigating the need for silica removal. The strategy employed reflects the extent of the problem.

Table 10.4 Bleaching sequences.

Chlorination	C	Reaction with chlorine in acidic medium
Alkaline extraction	E	Dissolution of reaction products with NaOH
Hypochlorite	H	Reaction with hypochlorite in alkaline solution
Chlorine Dioxide	D	Reaction with ClO_2 in acidic medium
Peroxide	P	Reaction with peroxides in alkali medium
Oxygen	O	Reaction with elemental oxygen at pressure in alkaline medium
Chelating agent	Q	A chelating agent is sometimes used for pulp derived from agricultural crops to remove transition metals which would otherwise consume peroxide
Ozone	Z	Reaction with ozone

The chemistry of removing silica is not particularly difficult but the physical separation of the silica without contamination can be problematic. One approach is to precipitate silica from black liquor by partial acidification with carbon dioxide using flue gas. Hindustan Newsprint Mills used a three-stage carbonation process across a range of nonwood materials. In this mill, 90% of the silica was removed by filtration but pH control and foaming issues were reported (Judt 1990, 1991). The two-stage causticizing process involves selectively precipitating silica in the first stage, and clean mud is generated in the second stage (Rao 1997). Sodium aluminate or bauxite has also been suggested to precipitate the silica (Bohmer 1984). Silica can be removed directly from bagasse by washing with a mild caustic solution (Gupta *et al.* 1997).

An entirely different approach has been to precipitate the silica onto the pulp (Tutus and Eroglu 2003), which reduces pulp strength and raises issues of market acceptance.

Although there are numerous approaches for removing silica, Gupta recently reported that the majority of Indian agricultural fiber pulp mills did not have a chemical recovery circuit (Gupta *et al.* 2001). It is understood that this practice appears to be declining and limited to India, and does not occur in larger factories with an export focus, reflecting its strengthening position in the world economy.

10.5.3 Bleaching

Pulp bleaching occurs by adding chemicals to the pulp, possibly in the presence of a gas or small amount of a secondary chemical, whereby the chemicals react with the pulp to increase its brightness. The pulp is then washed to recover the chemicals and pressed to reduce the moisture content. This sequence of chemical addition, reaction, washing, and pressing typically occurs 3–5 times. The order of chemicals used in each stage plays a role and so chemical treatments are traditionally given "short-hand" designations (Smook 1992) in order to describe bleaching sequences as shown in Table 10.4.

Stages that use more than one chemical are combined using subscripts (E_{OP}, e.g., refers to caustic extraction in the presence of oxygen and peroxide) or brackets.

Bagasse has traditionally been bleached using a CEH bleaching sequence (Hurter 2007a, i.e., chlorination followed by caustic extraction and hypochlorite) or CEHD, although a large range of sequences have been employed such as CEHH (Covey *et al.* 2006a,b). Similar to the broader paper industry, the bagasse paper industry is moving toward elemental-chlorine-free (ECF) bleaching and even total-chlorine-free (TCF) bleaching so as to minimize the environmental impact of using elemental chlorine. Some older bleach plants are being rebuilt to (ZD)(EO)D or (DQ)(PO)D (Ragnar 2009). A typical modern bleaching sequence is D–E–D with possible reinforcement of the *E* stage with oxygen or peroxide (E_O or E_{OP}) (Hurter 2007b).

Bagasse pulp bleaching requires relatively few chemicals compared with wood pulp bleaching reflecting the ease with which it is also digested. Metso advocated for OQ(PO) sequence (Lundahl 2002). Bagasse paper brightness is often a little lower than wood pulp brightness; 84% ISO brightness is fairly typical with a yield loss of 4–5% (cf. Figure 10.8 for comparison with literature values). Using multiple D bleaching stages, brightness levels over 90% ISO are achieved commercially. ECF and TCF sequences allow water recirculation, greatly reducing water demand compared with traditional bleaching sequences to $<10\,m^3/t$ pulp.

10.6 Mechanical and chemi-mechanical pulping

Although newsprint volumes are in decline in many countries, bagasse-based newsprint will continue to be produced for some time. Large-scale production of bagasse-based newsprint emerged during the 1980s as previous attempts had not used mechanical pulp (Atchison 1992b), and this was found to be essential to producing a good newsprint. Atchison (1992b) and Rangan and Rangamannar (1995) noted that newsprint requires: good optical properties, bulk, smoothness, and ink drying by absorption; bagasse mechanical pulp provides all of these.

Requirements for mechanical pulping of bagasse are very similar to those for wood, but one important point is to remove most of the sand/silica otherwise there will be excessive wear on refiner plates (e.g., Rangan and Rangamannar 1995). Zanuttini (1997) notes that thorough depithing will assist sand removal and that a chelant added during depithing can remove some metal ions, which otherwise hinder brightness development. It was also noted that excessively vigorous depithing will damage the longer fibers, so a compromise is required.

Rangan and Rangamannar (1995) noted that one Indian newsprint mill adds sodium hydroxide and sodium sulfite to the eye of the refiner to give pH 8–9 and this results in 15–20% less power (c.f. TMP or RMP), reduced stiff fibers, improved strength, better bleaching response, and less variable pulp quality.

Mechanical bagasse fiber has a number of useful properties but is rather weak; therefore, a portion of stronger "reinforcing" fibers must be used in the furnish. For example, Forest Research Institute (2001) noted that Tamil Nadu Newsprint and Paper Mills uses 40% mechanical bagasse, 30% chemical bagasse, 15% CTMP, and 15% broke for newsprint.

A number of mills also make semi-chemical pulp from bagasse. El-Saied *et al.* (2001) describes a process whereby bagasse is subjected to a mild chemical treatment (10% NaOH, 180 min at 90 °C followed by refining) to produce a pulp that is much stronger than mechanical pulp, but with inferior optical properties, which is useful for packaging grades.

As with wood pulping, experiments have been undertaken with white-rot fungus as a precursor for mechanical and chemical pulping (e.g. Ramos *et al.* 2001). However, as with wood pulping, the reduction in refining energy and improvement in pulp strength is offset by the long processing time and loss in brightness.

10.7 Papermaking

The formation of a fiber mat is a filtration process whereby a suspension of fibers are deposited one layer at a time and water drains through the formed mat (Parker 2001). The result is that the flexible fibers overlay one another forming a sheet of paper. For bagasse pulp, the high content of short pith fiber impedes drainage, which invariably reduces paper production rate when compared with wood pulp.

The two main types of paper formers processing pulp into paper products are the traditional Fourdrinier formers, which are in decline as a proportion of the industry generally, and twin-wire formers. Pulp fibers are suspended in water at 0.1–1% (depending on the grade) before being sprayed onto the moving fabric, which operates at speeds from 50 to 2000 m/min, depending on the product being made. As with most equipment used in the bagasse paper industry, these formers were originally designed for wood pulp rather than bagasse pulp.

Following paper formation, most of the water is drained from the fibers, and the moist wet sheet is mechanically pressed, dried, and reeled. Once a continuous dry paper sheet is produced, it is slit, cut, and packaged ready for distribution.

10.7.1 Fiber morphology

Bagasse pulp more closely resembles hardwood eucalypt pulp rather than softwood pulp as its fiber length is around 1.0–1.2 mm in diameter; refer Figure 10.7 (softwood is typically around 3 mm in length). Bagasse pulp is typically around

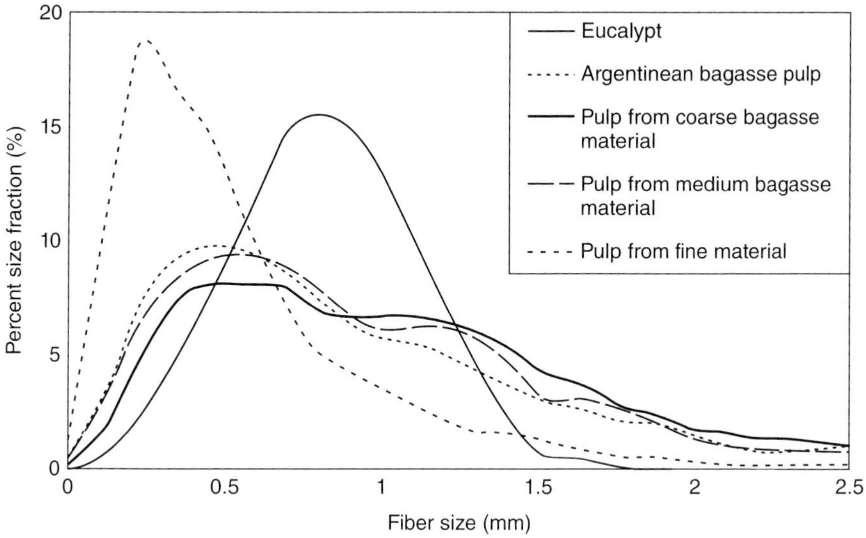

Figure 10.7 Bagasse pulp fiber length distribution (Rainey *et al.* 2009; with permission from the authors).

20 μm in thickness and has a thicker cell wall than eucalypt, making it stiffer. This attribute makes it ideal for applications such as corrugating medium and folded paper toweling. A comparison of its morphology is given in Table 10.5, where two types of depithed bagasse pulp were analyzed alongside a commercial depithed bagasse pulp (from Argentina), pith pulp, and eucalypt pulp. In the figure, the pulp from the coarse fraction is from the coarsest 25% of the raw bagasse (according to sieve size), and the medium fraction is the depithed fraction smaller than this.

10.7.2 Suitability of bagasse for various paper grades

As for any paper grade, bagasse pulp is blended to meet the required properties of the finished product. Chemical pulp is used for applications where expensive high-quality pulp is required and mechanical pulp is used to lower production costs while keeping within the limits of the required physical property parameters. Some examples of bagasse-based furnishes are provided in Table 10.6.

Both Hurter (2007c) and Rao (1997) provide overviews of the suitability of bagasse for various paper grades, namely tissue, toweling, printing, and writing papers. For these applications, bagasse pulp can supply 80–100% of the furnish in combination with the usual supplementary long fiber to improve strength. Bagasse makes acceptable quality newsprint although a bleached chemi-mechanical pulp is often required for part of the furnish. The use of bagasse pulp for linerboard and sack Kraft paper is also commonplace (Hurter 2007c).

Table 10.5 Bagasse pulp morphology (Rainey 2009, Rainey *et al.* 2010).

	Bagasse pulp derived from the coarse bagasse fraction	Bagasse pulp derived from the medium bagasse fraction	Eucalypt pulp
Fiber width (μm)	20.2	20.7	18.6
Fiber thickness (μm)	13.9	12.7	11.6
Wall thickness (μm)	5.13	4.72	4.19
Fiber area (μm^2)	214	200	159
Fiber perimeter (μm)	68.5	67.6	60.9
Wall area (μm^2)	186	169	132
Lumen area (μm^2)	31.0	27.9	26.3
Collapse ratio (–)	1.48	1.66	1.63
Minimum wall thickness (μm)	3.46	3.18	2.75
Maximum wall thickness (μm)	7.03	6.70	6.09

Table 10.6 Typical furnishes for newsprint and printing and writing paper.

	Newsprint		Printing and writing paper	
	Factory 1	Factory 2	Factory 1	Factory 2
Chemical bagasse pulp	~60%	35–40%	75%	70–80%
Chemi-mechanical		40–55%		
Mechanical bagasse pulp	~40		25	
Long fiber wood pulp		15–25%		10–30%
Filler		0–8%		10–20%

10.7.3 Physical properties

Figure 10.8 shows the properties of a Kraft bagasse pulp in comparison to wood-based pulp. Apart from the initial freeness measurement, the other properties were measured at 40° SR after beating. Bagasse pulp fibers are best suited for applications where hardwood pulp might be used in large proportions, such as photocopier paper. The main characteristics of bagasse pulp are

- poor initial freeness impeding drainage (caused by the residual pith content);
- low tear and tensile strength;
- lower opacity;
- high stiffness (caused by the relatively thick cell walls);
- high smoothness.

Apart from depithing, the general quality of the prepared fiber also significantly affects the physical strength properties. Increasing the milling extent causes the bagasse pulp to have lower strength properties (Gartside and Langfors

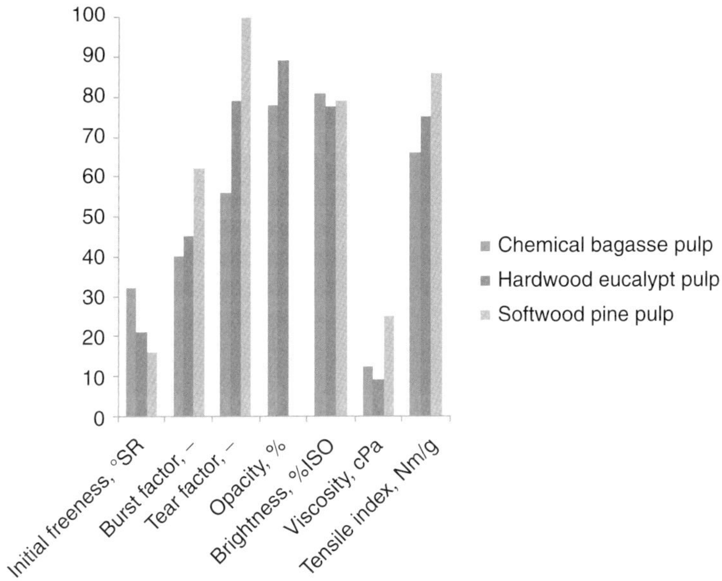

Figure 10.8 Pulp and handsheet properties of Kraft bagasse pulp compared with Kraft wood pulp. (Adapted from Kasi Viswanathan and Gopalaratnam 1998.)

1981, Gartside *et al.* 1982, 1983). This may potentially result in a trade-off needing to be struck between sugar extraction and acceptable pulp quality.

10.7.4 Effect of pith on paper production

If bagasse pulp directly replaces an equivalent wood pulp, paper machine speeds reduce by about 30%. Although the effect of pith on paper physical properties is well known (Paul and Kasi Viswanathan 1998), the effect on machine speed is less certain. A series of studies were undertaken to determine the effect of increasing the extent of depithing from 30% to 40% on the permeability and compressibility properties of bagasse pulp. Permeability and compressibility properties are key properties that affect machine speed. Although increasing the level of depithing reduces the amount of pulp that can be made from a given quantity of bagasse, there are significant productivity gains at the paper machine.

For the common situation of an integrated sugar, pulp, and paper factory, the economic considerations for an appropriate level of depithing include capacity limitations at the pulp mill or the paper mill. A higher depithing level does not necessarily increase the price of the bagasse as the extra pith is returned to the sugar mill boiler with the rest of the pith for steam and electricity production, although this may affect the maximum capacity of the pulp mill.

Permeability experiments elucidated that by pulping the coarsest 25% of bagasse, the permeability of bagasse pulp will increase by 100–120% for the conditions investigated by Rainey (2009). Flocculant systems, which are effective for bagasse pulp, did not affect the compressibility of pulp pads under steady-state conditions, but under dynamic conditions that are similar to those experienced on a paper machine, the dewatering effect of flocculants increased as the amount of pith increased (Rainey *et al.* 2010). Pulp washers also benefit from increased depithing.

Further experiments were conducted to simulate the highly turbulent conditions of a paper machine using a rapidly draining dynamic Britt Jar under vacuum. These showed that pith helped bagasse pulp to consolidate and dewater more quickly. Although formation using very highly depithed bagasse may conceivably be marginally slower *at the paper machine* with lower pith content, there are lower energy costs to dry the bagasse paper sheet and tear strength is significantly increased leading to reduced downtime particularly for tissue grades (Rainey 2012).

10.8 Alternate uses of bagasse pulp

Bagasse pulp is a commodity in its own right and it is exported around the world owing to its unique properties. Traditionally, it was sold at a discount to hardwood pulp, but this has not been the case for at least 5 years and occasionally bagasse pulp can command a small premium.

Dissolving pulps are made by chemical processes to produce high levels of α-cellulose. Dissolving pulps are converted into products such as rayon and cellulose acetate. The viscose rayon industry is the largest user of dissolving pulps. The cellulose is causticisized, aged, and dissolved in carbon disulfide to form xanthate. Pure cellulose is regenerated by extruding the xanthate into a sulfuric acid bath (Smook 1992). The main issue with using bagasse for dissolving pulp is the high hemicellulose content compared with wood and hence extra purification is required. Alternate uses of dissolving pulp are in pharmaceuticals and biotechnology applications (Villamil *et al.* 1995).

There has been some work on using bagasse in nonwoven fabrics, in particular at Louisiana State University and Agricultural and Mechanical College (Chiparus and Chen 2003). This study found that bagasse fibers made by atmospheric treatment of bagasse with 20–40 g/l NaOH were suitable for use in a wide range of nonwoven materials at bagasse loading of up to 70% (the limit was stiffness of bagasse fibers). The nonwoven materials were suitable for various agricultural uses such as biodegradable plant pots, poultry bedding, and aquaculture. It is not clear whether these products have been made commercially.

Bagasse has also attracted interest for use in composites with polymers. Although for many applications long reinforcing fibers are preferred, there are benefits in using short fibers in some applications particularly where a high modulus is required.

In an extensive review, Verma *et al.* (2012) noted work using bagasse fibers in composites with a wide range of polymers including phenolic resins, thermoplastics, and polyester. Work has also been performed on esterification of the bagasse to a thermoformable material.

In a number of cases, bagasse has been mixed with other fibers (natural and synthetic) to further extend the range of properties that can be developed. For example, (Saw *et al.* 2011, 2012) showed that using bagasse and jute (ratio 3:1) or coir (ratio 1:1) fibers together gave improved dimensional stability and mechanical properties and also improved thermal stability.

Work has also been undertaken on the use of bagasse fiber in cement composites (Karade 2010, Onesippea *et al.* 2010). The addition of mechanical and chemical bagasse pulp to concrete reduces its thermal conductivity, specific heat, and density and makes it a better insulator for use in energy-efficient buildings, but there can be issues in achieving surface compatibility between the components.

As far as the authors are aware, none of these composites is in commercial production. However, there is commercial production of particle board and fiber board in India (Karade 2010).

References

Atchison, J. E. (1962). Bagasse becoming a major raw material for manufacture of pulp and paper – background, present status, and future possibilities. Proceedings of the ISSCT Conference: 1185–1211.

Atchison, J. E. (1971a). Modern methods of purchasing and handling of bagasse – major advances in the sixties. Tappi Non-Wood Plant Fibers Committee: 5–29.

Atchison, J. E. (1971b). Review of bagasse depithing. Proceedings of the ISSCT Conference.

Atchison, J. E. (1971c). Review of progress with bagasse for use in industry. Proceedings of the ISSCT Conference.

Atchison, J. E. (1980). The fuel value of bagasse pith and developments in pith burning at leading bagasse pulp and paper mills and/or sugar mills. Proceedings of the ISSCT Conference.

Atchison, J. E. (1992a). The global choice for pulping bagasse – the rapid cooking horizontal tube continuous digester with screw feeder. Sugar y Azucar(February): 94–102.

Atchison, J. E. (1992b). Making the right choices for successful bagasse newsprint production: part 1. *Tappi Journal* **75**(12): 63–68.

Atchison, J. E. (1992c). Making the right choices for successful bagasse newsprint production: part 2. *Tappi Journal* **76**(1): 63–68.

Belayachi, L. and Delmas, M. (1995). Sweet sorghum: a quality raw material for the manufacturing of chemical paper pulp. *Biomass and Bioenergy* **8**(6): 411–417.

Bohmer, V. J. (1984). Silica removal from soda nonwood pulping. *Tappi Journal* **67**(11): 116.

Chiparus, O. I. and Chen, Y. (2003). An image method to evaluate bagasse fiber dimensions. *Bioresources Technology* **90**: 305–309.

Covey, G., Rainey, T. and Shore, D. (2006a). The potential for bagasse pulping in Australia. *Appita Journal* **59**(1): 17–22.

Covey, G., Rainey, T. J. and Shore, D. (2005). The potential for bagasse pulping in Australia. Appita Conference Proceedings, Auckland, New Zealand.

Covey, G., Rainey, T.J., and Shore, D. (2006b). A new opportunity to pulp bagasse in Australia. Proceedings of the Australia Society of Sugar Cane Technology, Mackay.

Darwesh, H. R., Mahendra, S. and Khan, P. A. (1998). An experience with Pandia continuous digester on agricultural residues. *IPPTA Journal* **10**(3): 67–73.

Diez, F. and Lois, J. A. (1989). The development of Cuban depithers and the results obtained. Proceedings of the ISSCT Conference.

El-Saied, H., El-Sakhawy, M., Ismai, I. and El-Shawadefy, M. (2001). Bagasse SEmichemical pulp by alkali treatment. *IPPTA Journal* **13**(4): 39.

FAO (2009), *State of the World's Forests, 2009*, Food and Agriculture Organization of the United Nations, Rome, Italy p. 70.

Forest Research Institute (2001). High yield pulping and bleaching of plantation grown wood species and bagasse using alkaline peroxide mechanical pulping/bleaching (APMP) process. Retrieved 30-11-2015, from http://www.dcpulppaper.org/gifs/report8.pdf.

Gartside, G. and Langfors, N. G. (1981). Developing the potential of Australian bagasse as a paper resource. Proceedings of the Australia Society Sugar Cane Technology: 233–238.

Gartside, G., Langfors, N. G. and Miller, K. (1982). An evaluation of the damage to sugarcane fibre during milling. *Holzforschung* **36**: 23–28.

Gartside, G., Langfors, N. G. and Miller, K. (1983). The effect of milling on the papermaking properties of several Australian sugarcane varieties. Proceedings of the Tappi Pulping Conference: 275–282.

Giertz, H. W. and Varma, R. S. (1979). *Studies on the pulping of bagasse and the influence of pith on paper properties: non-wood plant fiber pulping progress report*, pp. 53–69. Tappi Press: Atlanta.

Gominho, J., Fernandez, J. and Pereira, H. (2001). *Cynara cardunculus* L. – a new fibre crop for pulp and paper production. *Industrial Crops and Products* **13**: 1–10.

Gullichsen, J. and Fogelholm, C. (1999). *Chemical Pulping*. Helsinki: Finland, Fapet Oy.

Gupta, A., Mathur, R. and Kulkarni, A. (2001). Non-process elements in nonwood black liquors; their effect on the operation and energy efficiency of chemical recovery systems. *Appita Journal* **54**(6): 518.

Gupta, H. K., Shivare, P., Roy, T. K. and Rao, N. J. (1997). Alkaline extraction of bagasse to get rid of silica. *IPPTA* **9**(2): 29–35.

Hurter, R. (2007a). *Bagasse Fibre Processing Overview. Developments in Pulp and Paper Production from Sugarcane Bagasse*. Queensland University of Technology: Brisbane, Australia.

Hurter, R. (2007b). *Developments in Pulp and Paper Production from Sugarcane Bagasse*. Queensland University of Technology: Brisbane, Australia.

Hurter, R. (2007c). *Introduction to Bagasse Products. Developments in Pulp and Paper Production from Sugarcane Bagasse*. Queensland University of Technology: Brisbane, Australia.

Judt, M. (1990). Straw – opportunities and innovations: successful operation of industrial desilication plant in India. Peterborough (UK): Session 7, Paper 2.

Judt, M. (1991). Desilication problems can be overcome. PPI Feb: 65.

Karade, S. (2010). Cement-bonded composites from lignocellulosic wastes. *Construction and Building Materials* **24**(8): 1323–1330.

Kasi Viswanathan, K. S. and Gopalaratnam, N. (1998). Effect of bagasse pulp on quality – a study of various Indian paper grades. North America Nonwood Fiber Symposium Proceedings, Tappi Press.

Koen, A. (1980). Reducing maintenance costs of shredder hammers and cane knives. Proceedings of the South African Sugar Cane Technologists Association.

Lois-Correa, J. A. (1986). Depithing and storage of bagasse in Cuba: experiences and future development. *Proceedings of the ISSCT Conference*.

Lois-Correa, J. A. (2012). Depithers for efficient preparation of sugar cane bagasse fibers in pulp and paper industry. *Ingeniería Investigación y Tecnología* **13**(4): 417–424.

Lois, J. and Suarez, R. (1983). Experimental evaluation of bagasse storage in bale systems. Proceedings of the ISSCT Conference.

Lundahl, Y. (2002). *Nonwood Pulping the Metso Way*. Metso Technology Days.

Mann, A. P. and O'Hara, I. (2012). *Predicting the Effects of Bagasse Depithing Operations on Boiler Combustion Performance*. Australian Society of Sugar Cane Technologists: Palm Cove.

Moeltner, H. G. (1980). The sugar cane separation process and composition board manufacture from "Comrind". Proceedings of the ISSCT Conference.

Nagarajan, V., Balasubramaniam, R., Govind, S. and Kasi Viswanathan, K. S. (2003). Maintenance improvements in continuous digester for bagasse pulping. *IPPTA* **15**(3): 51–54.

Onesippea, C., Passe-Courtina, N., Toro, F., Delvasto, S., Bilba, K. and Arsene, M. A. (2010). Sugarcane bagasse fibres reinforced cement composites: thermal considerations. *Composites Part A* **41**: 549–556.

Paper Engineers Association (2007–2010). *Papermaking Science and Technology*. Helsinki: Paperii ja Puu Oy.

Parker, I. H. (2001). *Headbox and Forming Section. APPI Principles of Papermaking Short Course*. Australian Pulp and Paper Institute: Melbourne.

Paul, S. K. and Kasi Viswanathan, K. S. (1998). Influence of pith on bagasse pulp, paper and black liquor properties. *IPPTA Journal* **10**(3): 1–8.

Ragnar, M. (2009). *Environmentally Benign Bleaching Technologies: Current Technology Needs of Paper Industry*. Central Pulp and Paper Research Institute: Saharanpur.

Rainey, T. J. (2009). *A Study into the Permeability and Compressibility Properties of Australian Bagasse Pulp. Faculty of Built Environment and Engineering*. Queensland University of Technology: Brisbane.

Rainey, T. J. (2012). A comparison between highly depithed and conventionally depithed bagasse pulp. *Appita Journal* **65**(2): 178–183.

Rainey, T. J., Doherty, B., Martinez, D. M., Brown, R. and Kelson, N. A. (2010). The effect of flocculants on the filtration properties of bagasse pulp. *Tappi Journal* **9**(5): 7–14.

Rainey, T. J., Doherty, W. O. S., Brown, R. J., Martinez, D. M. and Kelson, N. A. (2008). Determination of the permeability parameters of bagasse pulp from two different sugar extraction methods. Tappi EPE Conference, Portland, OR, USA.

Rainey, T. J., Doherty, W. O. S., Brown, R. J., Martinez, D. M. and Kelson, N. A. (2009). An Experimental Study of Australian Sugarcane Bagasse Pulp Permeability. *Appita Journal* **62**(4): 296–302.

Rainey, T. J., Doherty, W. O. S., Martinez, D. M., Brown, R. J. and Kelson, N. A. (2011). Pressure filtration of bagasse pulp. *Transport in Porous Media* **86**(3): 737–751.

Rainey, T. J., O'Hara, I., Mann, A. P., Bakır, C. H. and Plaza, F (2013). Effect of depithing on the safety and environmental aspects of bagasse stockpiling. *Process Safety and Environmental Protection* **91**: 5, 378–385.

Ramos, J., Gonzalez, M., Ramirez, F., Young, R. and Zunga, V. (2001). Biomechanical and biochemical pulping of sugarcane bagasse with *Ceriporiopsis subvermispora* fungal and xylanase pretreatments. *Journal of Agricultural and Food Chemistry* **49**: 1180–1186.

Rangan, S. G., and Rangamannar, G. (1998). Development in bagasse fiber preparation for the manufacture of fine papers and newsprint – a review of facilities in two successful Indian mills. Proceedings of the Tappi Pulping Conference: 775–789.

Rangan, S. G. and Rangamannar, G. (1995). Development in mechanical pulping of bagasse for manufacture of newsprint. Proceedings of the Tappi Pulping Conference: 583–592.

Rao, M. (1997). *Industrial Utilization of Sugar Cane and Its Coproducts*. ISPCK Publishers and Distributors: New Delhi, India.

Reddy, N. and Yang, Y. (2005). Biofibers from agricultural byproducts for industrial applications. *Trends in Biotechnology* **23**(1): 22–27.

Saw, S., Sarkhel, G. and Choudhury, A. (2011). Dynamic mechanical analysis of randomly oriented short bagasse/coir hybrid fibre-reinforced epoxy novolac composites. *Fibers and Polymers* **12**(4): 506–513.

Saw, S. K., Sarkhel, G. and Choudhury, A. (2012). Effect of layering pattern on the physical, mechanical, and thermal properties of jute/bagasse hybrid fiber-reinforced epoxy novolac composites. *Polymer Composites* **33**(10): 1824–1831.

Smook, G. A. (1992). *Handbook for Pulp and Paper Technologists*. Tappi Press: Atlanta.

Tutus, A. and Eroglu, H. (2003). A practical solution to the silica problem in straw pulping. *Appita Journal* **56**(2): 111–115.

Verma, D., Gope, P. C., Maheshwari, M. K. and Sharma, R. K. (2012). Bagasse fiber composites – a review. *Journal of Material and Environmental Science* **3**(6): 1079–1092.

Villamil, G., Islen, P. and Perez, M. (1995). Added value of third generation bagasse by-products. Proceedings 22nd Conference of the International Society of Sugarcane Technologists: 403–406.

Wilson, R. (2013). Challenges pulp and paper producers face entering new markets. *Appita Journal* **66**(3): 170–171.

Zanuttini, M. (1997). Factors determining the quality of bagasse mechanical pulps. *Cellulose Chemistry Technology* **31**: 381–390.

CHAPTER 11
Sugarcane-derived animal feed

Mark D. Harrison
Centre for Tropical Crops and Biocommodities, Queensland University of Technology (QUT), Brisbane, Australia

11.1 Introduction

Sugarcane cultivation in the tropics and subtropics has been a matter of record for as long as written and pictorial records have survived the ravages of time, and it is undoubtedly true that the crop was cultivated by primitive humans long before. It is also undoubtedly true that cultivated sugarcane has been part of the diet of domesticated animals for a similar period of time and, to this day, crop residues from cultivated sugarcane are an important source of forage or supplementary fodder in the tropics and subtropics (Peacock 1996, Aregheore 2005).

In contrast, commercial sugarcane production now occurs in more than 100 countries in an area of more than 26 million hectares (FAOSTAT 2014), and its cultivation generates the highest annual tonnage of biomass of any crop, approximately 1.84 billion tons (FAOSTAT 2014). With an increasing global demand for meat, milk, dairy, and other products derived from domesticated animals, there is an increasing demand for animal feed and an increasing need to integrate existing crop residues and processing products into animal feed. Given the prominence of sugarcane in global plant biomass production, it is clear that sugarcane has a significant role to play in increasing global animal production. Furthermore, modeling of integrated sugarcane and animal production shows that the approach can have substantial economic advantages (Gradiz *et al.* 2007).

Sugarcane has significant advantages as a forage crop in the tropics and subtropics including (i) adaptation to tropical and subtropical conditions, (ii) reduced sensitivity to poor soil fertility, heat and humidity, and insect and disease pressures compared with other crops, (iii) existing technology for production and processing at industrial scale, (iv) high biomass yields can be achieved, and (v) the unusual ability to maintain consistent quality as a standing crop in the field (Pate *et al.* 2002). Indeed, there are examples of sugarcane cultivars that have been specifically bred to provide animal feed and thrive under region-specific environmental conditions (Suzuki *et al.* 2014).

This chapter provides an overview of how sugarcane crop residues, the products of sugarcane processing, and the products of raw sugar production are

---

Sugarcane-Based Biofuels and Bioproducts, First Edition. Edited by Ian M. O'Hara and Sagadevan G. Mundree.

being used to support animal production and describes technologies available to enhance them for this purpose.

11.1.1 The anatomy of the sugarcane plant

Sugarcane is a perennial grass (family Poaceae) grown commercially in the tropics and subtropics, and accounts for approximately two-thirds of the world's sugar production. As with other members of Poaceae family (such as maize, wheat, rice, and bamboo), the stalk of the adult sugarcane plant (known also by the botanically correct term, "culm") is composed of a series of repeating phytomeric units, each of which consists of a node, an attached leaf, and an internode. The leaves associated with each phytomeric unit undergo age-related senescence, and both green and senescent leaves remain attached to the stalk at maturity. The internodes in the mature stalk of adult sugarcane plants are typically between 150 and 200 mm in length but are significantly shorter at both the top of the stalk (where internodes are elongating) and the base of the stalk where new shoots (known as "tillers") are being formed.

The immature and expanding internodes at the uppermost region of the stalk are enclosed by overlapping leaf sheathes and are collectively known as "tops," and there is a natural breaking point at which the tops will separate from the remainder of the stalk. The leaves on the sugarcane plant not associated with the tops are collectively known as "trash."

Unlike other perennial grasses, sugarcane typically accumulates between 8 and 17 g of sucrose per 100 g of stalk (wet weight). The internodes contain the cells and subcellular structures responsible for the transport and storage of sucrose. Sucrose constitutes the majority (>90%) of the total soluble sugar in sugarcane stalk with the remainder consisting of a mixture of glucose and fructose (both of which are "reducing" sugars because they contain an aldehyde group or are capable of forming one via isomerization in solution). These reducing sugars represent approximately 10% of the total soluble sugars in tops but diminish down the stalk to only a few percent in mature internodes. The reader is directed to detailed reviews of sugarcane anatomy for more information (James 2004, Rae *et al.* 2013).

11.1.2 Sugarcane production, processing, and sugar refining

Sugarcane is typically harvested in the cooler, dryer months to ensure maximum sucrose content in the stalk and the length of the crushing season is dependent, principally, upon crop size, processing rate, and frequency and intensity of rain. As a result, sugarcane harvesting typically coincides with the season in which pasture forage for grazing animals (such as cattle and sheep) is at its most limited.

Traditionally, mature sugarcane is harvested after burning to remove leaves, weeds, and trash (Figure 11.1), and this is seen as an effective way to maintain the highest possible rates of harvest and to deliver "clean" sugarcane to

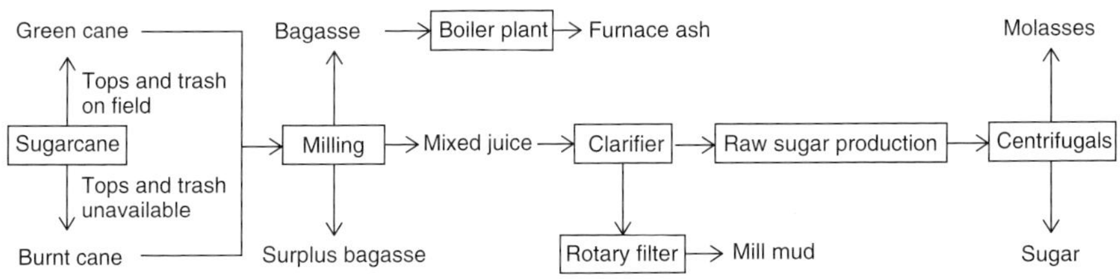

Figure 11.1 General schematic of sugarcane processing and sugar production.

sugar mills. Removal of trash also can prevent delays in sugarcane regrowth in cooler climates, which leads to a reduction in sugar yields (Viator *et al.* 2005). However, there are significant disadvantages associated with burning of sugarcane, including atmospheric pollution and soil water losses in field. Commercial harvesting of green sugarcane is becoming more widespread (Environment Directorate 2011), particularly in those growing regions where power generation from trash is economic and/or concern over the environmental impact of sugarcane burning has resulted in legislative change. Although the decision to burn sugarcane has impacts upon subsequent sugarcane processing and sugar production (Rein 2007), the use of burnt sugarcane obviously has a profound impact on the availability of trash and tops for animal feed applications.

Sugarcane is either harvested manually or mechanically (Figure 11.1) and, while there are industries such as those in Australia and the United States of America where the entire crop is mechanically harvested, greater than 50% of the world's sugarcane crop is harvested manually. It should be noted that preharvest burning of sugarcane is a common practice where manual harvesting predominates and these two agricultural processes are inextricably linked. In most cases, manual harvesting leads to lower losses of sucrose in the field, reduced soil compaction, reduced damage to the roots of the sugarcane plant (known as the "stool"), and delivery of higher quality sugarcane to the sugar factory. Mechanical harvesting can be undertaken so as to either retain the full stalk length or chop the stalk into 150–300 mm long "billets." Chopper harvesters, while being associated with sugar losses, are the only practical mechanical option for harvesting when sugarcane tonnage is high, sugarcane has lodged, or when sugarcane is being harvested green. Although it is true that the proportion of sugarcane being manually harvested is falling, it will remain important in commercial sugarcane production for the foreseeable future. Regardless of the harvesting method employed, however, it is critical that harvested sugarcane is crushed promptly to minimize sugar deterioration and microbial growth.

Sugarcane delivered to the factory is processed through a number of key steps to produce sugarcane juice from which raw sugar can be prepared (Figure 11.1). Sugarcane is typically delivered by road or rail and transferred to the sugarcane conveyor by either feeder or spiller tables. The sugarcane conveyor then transfers the sugarcane to a sugarcane cleaning system. Sugarcane cleaning has significant advantages for the factory (e.g., reduced wear, increased factory capacity, reduced energy consumption, and lower losses of sugar in mill mud, bagasse, and molasses) and is typically undertaken using dry cane. Although sugarcane can be cleaned wet, the observed sugar losses and downstream effects on processing efficiency limit its utility.

Sugarcane is then prepared for milling by either the use of rotating knives or a swing hammer shredder, and then typically enters a series of milling units (known as a "mill train") that separate the fiber (known as bagasse) and sugarcane juice (Figure 11.1). Hot water is supplied to the last milling unit in the mill train and is fed counter to the flow of sugarcane through the milling train to ensure efficient removal of sugar. The bagasse is then transported for stockpiling and then either stored as surplus or burned in the boiler plant to generate electricity or as a means of disposal (Figure 11.1).

The sugarcane juice from the milling trains is referred to as mixed juice, being a mixture of juice from the first and second milling units. Juice is screened to remove small pieces of bagasse, heated, and then clarified using lime or lime saccharate. The goal of this process is to lower the color and turbidity of the sugarcane juice before production of raw sugar. The byproduct from this process is mill mud, a mixture of fine bagasse particles, water, juice, and flocculent (Figure 11.1). The mill mud is filtered to recover residual sugar, and the dewatered solid exits the factory as a by-product. The clarified juice is evaporated to concentrate the juice to 65–68% dissolved solids. The syrup is then further concentrated by heating under vacuum and the resulting "mother liquor" is seeded with crystalline sugar to start the process of bulk sugar crystallization.

The mixture of mother liquor and sugar crystals is called massecuite and the sugar crystals are separated from the residual liquid (molasses) by centrifugation (Figure 11.1). The "first molasses" or "A molasses" is typically subjected to two more rounds of sugar crystallization to maximize overall sugar yield. Each of these subsequent crystallization steps produces sugar in decreasing yield, as well as "B molasses" and "C molasses" (also known as "blackstrap molasses"). The reader is directed to detailed review of sugarcane processing for more information (Rein 2007).

11.1.3 Scope of the chapter

The solid or liquid outputs from sugarcane production, processing, and sugar production that cannot be used directly for animal feed (such as mill mud and boiler ash) fall outside the scope of this chapter. As a result, fermentation products for animal feed derived from bagasse, molasses, or sucrose are not considered herein. Furthermore, while raw or white sugar has been used in the past in animal feed (so-called "fodder sugar") (Pate *et al.* 1971), the addition of sugar to animal feed has long been uneconomic and will not be considered herein. As a result, the chapter focuses only on whole sugarcane, tops and trash, bagasse, sugarcane juice, and molasses.

While the cost of animal feed (either purchased or produced) typically represents 50–80% of the total cost of animal production, it is important to note that when considering the potential "value" of biomass derived from sugarcane in animal feed (particularly tops, trash, and bagasse) numerous factors need to be taken into consideration; these factors include scale of animal production (i.e., subsistence or commercial), mode of animal production (i.e., forage-fed, feedlot, or a combination thereof), seasonal variation in forage availability and quality, the presence of chemical residues, animal age and breeding status, desired outcome (i.e., rapid weight gain, maintenance, or survival), availability of water, and access to technologies and chemicals to enhance nutritional value.

11.2 Crop residues and processing products

11.2.1 Whole sugarcane

Whole sugarcane (i.e., stalk, juice, tops, and trash) is used extensively in the tropics and subtropics as a roughage source for the production of cattle (Santos *et al.* 2011, Hulshof *et al.* 2012, Pedreira *et al.* 2012, Rotta *et al.* 2014), pigs (Xande *et al.* 2009, 2010), sheep (Magana *et al.* 2009, Cutrim *et al.* 2012), and goats (Cabral *et al.* 2009). Although not for human consumption, it is interesting to note that sugarcane (either processed using cryogenic grinding or pelleting) has been assessed as an additive to the diets of mice (Wang *et al.* 2007, 2012) and cats (Fischer *et al.* 2012) to improve their digestive and, hence, general health.

The dry matter (DM) content and chemical composition of whole sugarcane used for animal feed varies significantly (Table 11.1) depending upon cultivar, climate, season, and method of preparation (e.g., removal of brown leaves, drying, grinding). While whole sugarcane contains significant amounts of soluble sugars that can support rapid microbial growth in the rumen (Xande *et al.* 2010), the overall nutritional value of whole sugarcane is relatively low compared with other animal feed ingredients (such as corn meal, soybean meal, and rice bran) (Cutrim *et al.* 2012) because of relatively low levels of protein and minerals (Kutty and Prasad 1980), and relatively low digestibility of the polysaccharides in the fibers from the leaves and stalk (Patil *et al.* 1999, Xande *et al.* 2009). However, despite this, the value of whole sugarcane in global animal production cannot be understated; for example, Brazil has the largest commercial cattle herd in the world with more than 217 million cattle (FAOSTAT 2014) and, given its preeminence as a sugarcane producer, it is no surprise that fresh chopped sugarcane is used as the primary source of roughage in finishing diets in a significant proportion (~15%) of Brazilian feedlots (Oliveira and Millen 2014). Fresh chopped sugarcane was once the primary roughage source used by the majority of Brazilian feedlots (Millen *et al.* 2009) but increasing feedlot size has reduced the economic value of roughage in finishing diets and increased the use of grains and concentrates.

Table 11.1 Dry matter (g/kg) and chemical composition (g/kg DM) of whole sugarcane used in animal feed.

	Xande *et al.* (2010)	Santos *et al.* (2011)	Rotta *et al.* (2014)	Lascano *et al.* (2012)	Santiago *et al.* (2013)	Cutrim *et al.* (2012)	Average	SD[a]
Dry matter	258	292	282	276	283	216	268	28
Ash	15	nd[b]	nd	nd	40	19	25	13
Crude protein ($N \times 6.25$)	12	27	28	24	25.6	27	24	6
NDF[c]	403	443	459	576	422	547	475	70
ADF[d]	266	258	nd	217	258	366	273	55
GE[e] (MJ/kg DM)	17.9	nd	nd	nd	nd	nd	nd	nd

[a]Standard deviation.
[b]Not determined.
[c]Neutral-detergent fiber (i.e., cellulose, hemicellulose, and lignin).
[d]Acid-detergent fiber (i.e., cellulose and lignin).
[e]Gross energy (i.e., the total energy in feed).

Increasing the proportion of fresh chopped sugarcane in feedlot diets of growing-finishing steers results in a corresponding decrease in the rate of weight gain, feed intake, and carcass quality (Pate *et al.* 2002), and this arises primarily from the lower energy value of sugarcane compared with the other dietary components that it is replacing. While fresh chopped sugarcane is equivalent to corn silage as a roughage in a high-concentrate diet, it has only ~70% of the value of corn silage when used as the major component of the diet (Pate *et al.* 2002).

Compound animal diets to promote growth (as opposed to subsistence compound diets used during drought periods where weight gain is undesirable) that are based on whole sugarcane require supplementation with significant amounts of nitrogen (Suzuki *et al.* 2010, Hulshof *et al.* 2012, Lascano *et al.* 2012, Santiago *et al.* 2013). Protein is the preferred source of nitrogen in animal feeds but in regions where sugarcane forms the basis of the diet, it is likely that such protein is expensive. As a result, less expensive sources of nitrogen such as urea are more economic.

It has been demonstrated that, in feedlot cattle, urea can supply all of the nitrogen required in sugarcane basal diets without causing harmful effects but with slower and less efficient weight gains than those obtained with protein as the nitrogen source (Pate *et al.* 2002). Furthermore, despite the relatively limited number of studies on the feeding of whole sugarcane to animals, a mechanistic model of sugarcane digestion has been developed and used to identify supplements that improve milk production in dairy cattle (Kebreab *et al.* 2001).

11.2.2 Tops and trash

Sugarcane tops and trash often represent an untapped source of roughage in the diet of ruminants (e.g., cattle, water buffalo, sheep, and goats), and monogastric

Table 11.2 Dry matter (g/kg) and chemical composition (g/kg DM) of sugarcane tops.

	Galina *et al.* (2007)	Ortiz-Rubio *et al.* (2008)	Gendley *et al.* (2009)	Grieve (1976)	Naseeven (1986)	de Lima *et al.* (2014)	Butter-worth (1963)	Average	SD[a]
Dry matter	869	932	nd[b]	817	290	310	256	nd	nd
Ash	58	nd	78	65	85	51	62	67	13
Crude protein ($N \times 6.25$)	37	54	44	55	59	63	63	54	10
NDF[c]	781	722	696	nd	633	731	nd	713	54
ADF[d]	407	411	408	nd	431	458	nd	423	22
ME[e] (MJ/kg DM)	9.16	8.97	nd	nd	9.12	nd	9.12[f]	9.10	0.08

[a] Standard deviation.
[b] Not determined.
[c] Neutral-detergent fiber (i.e., cellulose, hemicellulose, and lignin).
[d] Acid-detergent fiber (i.e., cellulose and lignin).
[e] Metabolizable energy (i.e., the difference between the total energy in feed and the energy lost in waste) in cattle.
[f] Butterworth (1964).

herbivores (e.g., horses). Grazing of mature, standing sugarcane in commercial animal production systems is recommended only in an emergency (Pate *et al.* 2002). Despite these issues, sugarcane tops and trash can be an effective source of roughage if supplemented with nitrogen (either as protein or as urea) (Pate *et al.* 2002, Galina *et al.* 2004, 2007, Karbhari *et al.* 2007, Ortiz-Rubio *et al.* 2008) and other nutrients (Gendley *et al.* 2003, 2009, Sharma *et al.* 2007), and are routinely used in both subsistence and commercial agriculture in Asia (Reddy and Prasad 1982a,b, Chen *et al.* 1983, Baconawa 1986, Reddy *et al.* 1992, Yuangklang *et al.* 2005), Africa (Grieve 1976, Mahala *et al.* 2013), South America (Thiago *et al.* 1986, Bravo *et al.* 1996, de Lima *et al.* 2014, Oliveira and Millen 2014), and North America (Pate *et al.* 2002).

The dry matter content of tops is highly variable and depends on the season and whether the tops have been freshly harvested (Rein 2007) or, as is often the case, air-dried before use in animal feed trials (Table 11.2). Chemical composition of tops is also variable (Table 11.2) but average compositions can be determined.

It should be noted that where intensive sugarcane production is undertaken and chemicals are applied care should be taken to ensure that chemical residues on sugarcane tops and trash do not enter the diets of ruminant or nonruminant animals. The occurrence of such residues will depend on a number of factors including (i) the type of chemicals applied, (ii) whether application was undertaken in accordance with manufacturer's instructions, (iii) the interval between application and harvest, (iv) whether the tops and trash were subjected to ensilage and the application of inoculants, if any, to enhance ensilage, and (v) the proportion of sugarcane tops and trash in the animals diet. In Australia, two chemicals used in sugarcane production (picloram and terbutryn) have been deemed to require further investigation and the development of risk management strategies to minimize the likelihood of unacceptable residues in export meat and offal from cattle and sheep fed sugarcane tops and trash are recommended (MacLachlan 2010).

11.2.3 Bagasse

Bagasse consists primarily of the insoluble polysaccharides (cellulose and hemicellulose) and aromatic polymers (lignin) that make up the primary and secondary cell walls in the sugarcane stalk. Therefore, bagasse contributes little in the way of protein, lipids, and minerals to animal feed, and there are relatively few studies describing the incorporation of bagasse as roughage into animal feed without some form of processing to enhance digestibility (Marshall and Vanhorn 1975, Romanponce *et al.* 1975, Hegde *et al.* 1982, Briceno *et al.* 1987, Brondani *et al.* 1991, Singh *et al.* 2004). Despite this, while only a small proportion (<7.5%) of cattle in Brazil spend any significant period of time in feedlots, the size of the industry means that the sugarcane bagasse used as the primary source of roughage in finishing diets in a significant proportion (approximately 20%) of Brazilian feedlots represents a substantial amount (Oliveira and Millen 2014).

Bagasse also has value in the diet of sheep (Brondani *et al.* 1991, Ferreira *et al.* 2014) and nonruminants, such as chickens (Hegazy and el-Faramawy 2001). Bagasse has also been investigated as an additive to the diet of chickens at a low (8 wt%) inclusion rate (Hegazy and el-Faramawy 2001). Furthermore, although untreated sugarcane bagasse is considered to have limited utility in aquaculture feed (Ulloa *et al.* 2004), untreated sugarcane bagasse has been used as a substrate to promote the growth of microbial biofilms to serve as a source of food in aquaculture production of fringe-lipped carp (Mridula *et al.* 2003) and can assist in bioremediation of ammonia from shrimp farm waste water (Krishnani *et al.* 2006).

11.2.4 Molasses

Sugarcane molasses is a valuable component in animal feed and is often the cheapest source of energy for animals produced in proximity to a sugar mill (Tillman *et al.* 1951). Molasses typically contains 25–33% water (Caldwell 2001) and provides 8–9 MJ of metabolizable energy per kilogram to cattle (Wiedmeier *et al.* 1992), both of which will vary depending upon the level of refinement (see Section 1.2). While this is approximately two-thirds that of cereal grains (assuming molasses contains 25% water and grains contain 10% water) (Table 11.3), the carbohydrates and minerals within molasses are readily accessible to gut microbes. However, the low level of nitrogen in molasses (Table 11.3) limits the ability of rumen microbes to fully utilize the sugars in molasses and so the addition of nitrogen (e.g., protein, urea) is required to provide maximum nutritional benefit from molasses in the diets of ruminants, such as cattle.

Table 11.3 Chemical composition of sugarcane of molasses.

Nutrient	% DM range	Average %DM	Nutrient	Abundance range (mg/kg)	Average abundance (mg/kg)
Crude protein	1.2–10.6	3.5	Iron	145–640	297
Ash	5.9–13.4	8.5	Zinc	7.5–37.3	13.1
Total sugars	44.9–50.2	47.2	Copper	6.6–68.4	14.0
Calcium	0.02–0.84	0.57	Manganese	2.1–67.1	28.3
Chlorine	0.33–2.27	1.19	Cobalt	1.5–4.23	2.45
Magnesium	0.12–0.58	0.33			
Phosphorous	0.05–0.48	0.10			
Potassium	1.24–2.64	1.98			
Sodium	0.05–0.48	0.21			
Sulfur	0.34–1.31	0.86			

Source: Adapted from (Caldwell 2001).

Molasses, in addition to providing energy and minerals, also improves the palatability of animal feed, acts as an excellent carrier of microingredients, functions as a binder in total mixed rations (which reduces ration "sorting," particularly in cattle), aids in pellet production, and reduces physical degradation of animal feed pellets (i.e., formation of dust).

The proportion of molasses in an animal diet varies widely (from liquid feeds consisting almost exclusively of molasses to solid compound feeds containing less than 5% molasses and even solid feed blocks) and presents an equally wide range of both opportunities and challenges. Liquid feed consisting of molasses and a nitrogen source (such as urea) makes an excellent crisis feed during drought when little dry feed is available (Onwuka 1999). However, intake of molasses supplemented with nitrogen as a liquid feed under these conditions needs to be carefully monitored to prevent molasses toxicity (i.e., "molasses drunkeness," a condition that arises (ultimately) from vitamin B deficiency) and bloat.

It should be noted that phosphorous deficiency in animals is widespread in the tropics (McDowell *et al.* 1984) and, as a result of its relatively high phosphorous content (Table 11.3), molasses can be a particularly valuable source of phosphorous in supplemental feed for ruminants grazing on low-quality forage in these regions (Leng 1990). Wet-mix pig diets where the majority of the digestible energy is supplied by molasses have been studied since the late 1960s (Preston *et al.* 1967).

There can be substantial challenges in maintaining live weight gains of above 1.0 kg/day in cattle fed high-molasses diets (Sundstrom and Palmer 1977, Lindsay *et al.* 1998) but recent research suggests that, through careful dietary formulation and supplementation, diets containing >30% (dry matter) as molasses can achieve live weight gains of >1.5 kg/day over a 140-day period (Hunter 2012). Such studies underscore the opportunities for integration of animal and sugarcane production.

Molasses is typically included in grain-based feedlot cattle diets at rates ranging from 8% to 10% dry matter (Sparke 1997). Furthermore, molasses (in conjunction with a nitrogen source) has been long recognized as valuable additive to diets based on low-quality roughages (including sugarcane bagasse) for cattle production (Brown 1990, 1993, Stateler *et al.* 1995, Royes *et al.* 2001) but care needs to be taken to balance the addition of molasses with nitrogen so that fibers digestibility, and hence weight gain (Sparke 1997), is not compromised (reviewed in (Pate 1983)).

Molasses is relatively high in sulfur (Table 11.3) and so a sulfur source should not be used in molasses-based supplements. Indeed, the high level of sulfur in molasses-based diets can result in decreased copper absorption due to the formation of ruminal thiomolybdates (Mason 1990). Despite the value of molasses in grain-based feedlot cattle diets there are specific circumstances during animal production in which molasses supplementation can be detrimental. For example, a key issue in feedlot cattle production is their performance in the period immediately after they enter the feedlot; the 3 weeks immediately after arrival establishes their feeding pattern, immune competence, and (ultimately) their potential performance (Lofgreen and Kiesling 1985, Duff and Galyean 2007). Feed supplementation with urea-molasses during this period does not significantly improve health and live weight gain, and may make cattle without permanent incisors more susceptible to diseases, such as bovine respiratory disease (Croft *et al.* 2014).

11.2.5 Sugarcane juice

Sugarcane juice has long been recognized as an additive to animal feed in the tropics and subtropics (Preston *et al.* 1968) and is widely used to replace grain as the main energy source in pig feed, particularly in the Caribbean (Mena 1981, Diaz *et al.* 1992, Xande *et al.* 2009). The majority (i.e., 83%) of the organic matter in sugarcane juice consists of soluble sugars (i.e., sucrose, glucose, and fructose), while the remainder consists of chlorophyll, tannins, wax, and fibers (Rostagno 2005). Given its high-soluble sugar content, the energy in sugarcane juice is almost 100% accessible to pigs (Xande *et al.* 2010).

11.3 Processing sugarcane residues to enhance their value in animal feed

Sugarcane crop residues and processing by-products have substantial value as animal feedstocks, particularly in ruminants that have the capacity to utilize biomass polysaccharides because of the consortium of anaerobic bacteria, protozoa, and fungi present in their rumen. However, in those sugarcane crop residues containing intact or partially intact cell walls (i.e., tops, trash, and bagasse), the high lignin content, low level of soluble carbohydrates and relatively low levels of fermentable nitrogen and bypass protein limit the efficiency with which ruminants and nonruminants can release the metabolic energy from these materials. This limitation arises from the inherent complexity and resilience of the plant cell wall and the connections between its three principal components (i.e., cellulose, hemicellulose, and lignin). Physical, chemical, physicochemical, and biological treatment can substantially improve the nutritional value of sugarcane crop residues (Walker and Kohler 1981).

11.3.1 Ensilage/microbial conditioning

Ensilage is an ancient forage crop preservation method that is particularly valuable in the tropics for the storage of sugarcane without adversely affecting nutritional content (Montanez-Valdez *et al.* 2013). Fresh chopped sugarcane consists of ~75% water, while commercially produced sugarcane bagasse has a water content of ~50%. As a result, excessive seepage losses can occur during the ensilage of both fresh-chopped sugarcane and bagasse. Ensiling sugarcane is also a convenient storage solution when large quantities are harvested after events such as fire or frost (Pedroso *et al.* 2006) and plays an important role in both subsistence and commercial tropical agriculture. For example, approximately 22% of all dairy farms in Brazil use sugarcane silage as a source of energy and fibers for their cattle (Bernardes and do Rego 2014) either throughout the year or part of the year in production systems that include pasture (Costa *et al.* 2013). The composition of sugarcane silage (Table 11.4) varies significantly because of the same factors effecting the composition of whole sugarcane (Table 11.1).

Table 11.4 Dry matter (g/kg) and chemical composition (g/kg DM) of sugarcane silage

	Santana *et al.* (2012)	**Santos *et al.* (2011)**	**Roman *et al.* (2011)**	**Suzuki *et al.* (2014)**	**Reyes *et al.* (2014)**	**Average**	**SD[a]**
Dry matter	26	25	24	21	28	25	3
Ash	43	73	40	73	nd[b]	75	18
Crude protein ($N \times 6.25$)	85	35	41	36	87	57	28
NDF[d]	700	536	709	718	616	656	78
ADF[d]	445	329	473	445	473	433	60
ME[e] (MJ/kg DM)	11.17	nd	nd	nd	8.54	9.86	nd

[a]Standard deviation.
[b]Not determined
[c]Neutral-detergent fiber (i.e., cellulose, hemicellulose, and lignin).
[d]Acid-detergent fiber (i.e., cellulose and lignin).
[e]Metabolizable energy in cattle (i.e., the difference between the total energy in feed and the energy lost in waste).

Epiphytic microorganisms (such as lactobacteria) ferment the free sugars present in the biomass (McDonald *et al.* 1991) during optimal ensilage and rapidly generate an acidic environment that limits the growth of microorganisms that degrade plant cell wall polysaccharides (Thompson *et al.* 2005). Water content and the concentration of free sugars are critical to ensure that those microorganisms that reduce pH dominate the silage microbiome. In the absence of free sugars or if free sugar concentration in sugarcane silage is very high, pH increases and significant losses of cellulose and hemicellulose occur (Kung and Stanley 1982, Thompson *et al.* 2005), resulting in the loss of as much as 30% of the dry mass (Alli and Baker 1982, Kung and Stanley 1982). Under these conditions, the major end-product of microbial fermentation is ethanol, rather than lactic acid, primarily through the growth of yeast (Avila *et al.* 2010, Roth *et al.* 2010). The presence of ethanol in silage can reduce palatability (Buchanansmith 1990), increase acetate and caproate during ruminant digestion (Durix *et al.* 1991), and alter the flavor of the milk produced by ruminants (Randby *et al.* 1999). It should also be noted that ethanol is metabolized to acetate in the rumen with concomitant release of methane (Pedroso *et al.* 2008), an atmospheric pollutant.

The ensilage of sugarcane (whole sugarcane, tops, or bagasse) can be enhanced through the addition of microbial inoculants to improve the process and generate better quality silage (Roth *et al.* 2010, Schmidt *et al.* 2014). As a result, there are numerous commercial microbial inoculants for ensilage and lactic acid bacteria are the main microorganisms used for this purpose (Avila *et al.* 2014). However, the results of microbial inoculation on silage quality can vary widely and is influenced by how the bagasse is stored, how inoculant is applied (Apas *et al.* 2008), and whether or not the sugarcane has been burnt (Roth *et al.* 2010). Despite this, the capacity of the microorganisms to utilize the free sugars present in the silage and to convert them into organic acids remains the key factor in determining the success of inoculant application for silage production (Avila *et al.* 2009).

Inoculants containing lactic acid bacteria (which also produce acetic or propionic acids) not only reduce the growth of yeast, but also reduce the growth of fungi because of the fungicidal effect of these acids (Moon 1983) and secrete antimicrobial peptides called bacteriocins (O'Shea *et al.* 2013). However, not all lactic acid bacteria inoculants produce high-quality sugarcane silage and it has recently been observed that differing strains of the same species of lactic acid bacteria produce silages of variable quality (Avila *et al.* 2014), underscoring the need for careful selection of the microbe or microbes in commercial inoculation products. The addition of plant biomass from other species, such as *Atriplex nummularia* (also known as "old man saltbush"), has also been shown to improve the quality of the resulting sugarcane silage (da Silva *et al.* 2014).

Ensiled sugarcane is typically included in animal diets as an ingredient in compound feed (Briceno *et al.* 1987, Roman *et al.* 2011, Santana *et al.* 2012, Daniel *et al.* 2013, Reyes *et al.* 2014, Suzuki *et al.* 2014) but can be delivered as a supplemental feed (Gomez-Vazquez *et al.* 2011). Sugarcane silage is a viable alternative to whole-crop maize silage in compound feed for beef cattle produced in feedlots but whole-crop maize silage provides superior finishing and dressing percentage (i.e., the percentage of live weight that remains in the dressed carcass) over a 96-day period (Roman *et al.* 2011). While sugarcane silage has been shown to enhance total dry matter intake, total gain, and average daily gain in beef steers grazing on stargrass (Gomez-Vazquez *et al.* 2011), it was not clear if the positive benefits arose from the presence of sugarcane silage in the supplemental feed or from the presence of ground corn grain, urea, and mineral premix.

11.3.2 Chemical conditioning

Chemical treatment at ambient temperature (also known as chemical conditioning or chemical ensilage) can significantly improve the quality of ensiled sugarcane bagasse, trash, and tops, and the need to overcome the relatively low nutritive value of crop residues, such as cereal straws, using such treatment has long been recognized (Lehmann 1891). The majority of these well-studied processes utilize alkali, oxidants, acids, or combinations thereof.

Alkaline conditioning

Treatment of low quality of roughage with sodium hydroxide (NaOH) (known as the Bekmann method) was devised during World War I (Homb 1984) and was in widespread use during periods of acute fodder shortage in Europe until the late 1960s (Owen and Jayasuriya 1989). This process was superseded with "dry treatment" of roughage using sodium hydroxide (Lampila 1963, Wilson and Pigden 1964), which overcame the need for high water content and the resulting loss of nutrients via leaching from the Beckmann method.

Semi-dry roughage alkaline conditioning partially dissolves lignin and silica, hydrolyzes the chemical linkages between lignin and hemicellulose, and causes tightly packed cellulose microfibrils to swell, all of which enhance microbial enzyme access to cellulose and hemicellulose (Jackson 1977). The technique began to be applied commercially, both on farm and at industrial facilities, in Europe during the 1970s. The effect of NaOH on a wide range of roughages has been studied extensively (Capper *et al.* 1977, Jackson 1977, Nicholson 1981) and is still used today at a commercial-scale and for laboratory studies (Pires *et al.* 2006), primarily because of its effectiveness and the ready availability of NaOH.

The treatment of sugarcane tops with 10 wt% NaOH increases rumen dry matter digestibility to 86%, compared with only 56% in untreated sugarcane tops (Chandra and Jackson 1971). At this level of NaOH, the rumen dry matter digestibility of sugarcane tops is similar to that of maize cobs, wheat straw, paddy straw, or sorghum stover (Chandra and Jackson 1971). The treatment of sugarcane bagasse with 2% (w/w dry fiber) NaOH before inclusion in complete rations significantly enhances feed consumption, daily weight gain, and daily milk production by lactating dairy cows compared with untreated bagasse (Randel *et al.* 1972). The *in vitro* digestibility of the organic matter in raw fresh bagasse is ~30% but increases to 55% with the addition of 7.5% NaOH (w/w/ dry fiber) (Tudor *et al.* 1985), which upgrades bagasse to a metabolizable energy content equivalent to that of medium-quality hay (~7 MJ/kg dry matter).

Conditioning of sugarcane bagasse with 5% NaOH (w/w dry fiber) improves average live weight gain and efficiency of feed conversion in lambs at both 19 and 40 wt% in complete rations (Jacobs and Vanniekerk 1985). Side-by-side comparison of NaOH with a wide range of other conditioning chemicals revealed that chemical conditioning of bagasse with 9% NaOH results in the greatest increase in *in vitro* organic matter digestibility (Ibrahim and Pearce 1983).

These studies confirmed that NaOH-treated bagasse can be readily converted to a feed that can maintain weight of livestock during drought or, if supplemented with nitrogen, allow significant weight gain. During NaOH conditioning of low-quality roughage, excess alkalis are not typically neutralized, nor are sodium or other cations removed before feeding. The use of NaOH alone for the chemical conditioning of low-quality roughage, therefore, increases the amount of sodium in the feed and an upper limit of 6% NaOH (w/w dry fiber) is recommended to prevent impairment of kidney function in cattle (Jackson 1977). As a result, there was considerable interest in the mid-1970s in developing alkali mixtures to reduce the level of sodium in the final product (Tudor *et al.* 1985) and in the use of alternative alkaline treatment chemicals.

Ammoniation of crop residues using either anhydrous ammonia gas (Jose *et al.* 2004), aqueous ammonium hydroxide (NH_4OH) (Waiss *et al.* 1972), or urea (which is degraded by bacterial enzymes during ensilage into gaseous ammonia) (Kobeisy *et al.* 2005, Ahmed *et al.* 2013) was developed as an alternative to the Beckmann method and, as recently as 1995, the FAO recommended ammoniation to improve the value of fibrous agricultural residues for animal production (Preston 1995). Ammoniation using animal urine as the source of ammonia has been investigated (Sundstøl 1993) but this approach is not used commercially.

Ammoniation has been studied extensively (Nicholson 1981) and the process is used commercially to improve the nutritive quality of low-quality forage. Ammonia is a slow reacting chemical (e.g., Waiss *et al.* (1972) determined optimal conditions for rice straw were 5% added ammonia with 30% added water and 30 days treatment at ambient temperature) but the rate of chemical conditioning can be sped up at elevated temperature. However, ammoniation of plant biomass at temperatures >90 °C can produce toxic compounds that cause health problems in cattle and these compounds can affect both calves and humans drinking milk from intoxicated cattle (Perdok and Leng 1985). Rumen degradation of neutral detergent fiber (i.e., cellulose, hemicellulose, and lignin) increases from 46% in raw bagasse silage to 57% in silage supplemented with 10 wt% urea (Ahmed *et al.* 2013). Ammonia is particularly valuable for the preservation of stored material at moisture content above that which is typically considered to be safe (i.e., 16%).

NaOH and ammonia are the primary alkalis used for the chemical conditioning of sugarcane residues but a number of alternatives have been investigated, including lime ($Ca(OH)_2$) (Gandi *et al.* 1997), sodium sulfide (Na_2S) (Jose *et al.* 2004, Pires *et al.* 2004), and calcium oxide (CaO) (Carvalho *et al.* 2009, Murta *et al.* 2011, Santos *et al.* 2011). While the benefits of conditioning with hydrated lime ($Ca(OH)_2$) have been amply demonstrated, the benefits from the use of CaO are not as clear (Carvalho *et al.* 2009, Murta *et al.* 2011) and, in fact, daily weight gain in cattle feeding on bagasse ensiled with 2.5% Na_2S is reduced slightly compared with cattle feeding on untreated bagasse silage (Jose *et al.* 2004).

Oxidative conditioning

Wood pulping and delignification (i.e., pulp bleaching) using ozone has been well characterized (Osawa *et al.* 1963, Secrist and Singh 1971) and ozone treatment of wood and other lignocellulosic materials (including sugarcane bagasse) to improve digestion by ruminant microorganisms is well established (Neely 1984). Ozone is attractive as a treatment chemical for a number of reasons: (i) it does not leave acidic, basic, or toxic residues in pretreated biomass, (ii) it is inherently unstable and can, therefore, be easily decomposed back to oxygen using a simple catalyst bed for removal from residual gaseous effluent (Neely *et al.* 1975), (iii) ozone can be generated on-site, which can avoid issues of chemical supply in remote areas, and (iv) most ozonation reactions take place at near-ambient temperature and pressure, resulting in the use of simplified, lower cost equipment.

The rumen fluid digestibility of bagasse increases from 15% in untreated bagasse to 40% in bagasse after conditioning with 3% ozone at 30% water content (Neely 1984). The cost of ozone treatment limits its utility as an economic treatment for the production of animal feed if wood is the feedstock but, where biomass is available at a central location as a low-value by-product (e.g., sugarcane bagasse), there is the potential to generate cost-effective ruminant feed using ozone treatment.

Alkaline/oxidative conditioning

Prior success in the use of mixtures of alkaline and oxidizing agents to condition wheat and barley straw (Lewis *et al.* 1988, Bhargava *et al.* 1989) led Amjed *et al.* to determine the effect of conditioning sugarcane bagasse or sugarcane pith with alkaline hydrogen peroxide in a submerged system on cell wall composition and digestibility (Amjed *et al.* 1992). While the *in vitro* digestibility of both sugarcane bagasse and pith increased from ~19% to at least 49% after 24 h of conditioning, organic matter (>30%) and crude protein (>79%) losses seriously limit the utility of this method for the production of animal feed. That said, it is possible that an alternative method of alkaline hydrogen peroxide delivery (i.e., spraying of a concentrated solution onto bagasse followed by mixing) would improve digestibility without the loss of such important components of the starting material.

Acid conditioning

Acid conditioning of sugarcane bagasse has been assessed using both gaseous (Crosthwaite *et al.* 1984) and liquid (Crosthwaite *et al.* 1984) acids. The cellulose digestibility and *in vitro* dry matter digestibility of oven-dried and milled bagasse after soaking in sulfuric acid and conditioning for 9 months increased by 87% and 65%, respectively (Crosthwaite *et al.* 1984).

11.3.3 Physical processing (baling, pelletization, depithing)

Mechanical processing of forage using grinding or pelleting typically results in an increase in daily intake but reduced digestibility of the dry matter (Nicholson 1981). However, the products of digestion are utilized more efficiently so that there is an increase in the net energy value. Pelleted sugarcane bagasse can replace cotton seed meal as roughage at 25% in compound feed for dairy cattle without significantly impacting upon milk production, milk fat content, and live weight gain (Marshall and Vanhorn 1975).

In cattle, feeding diets based upon finely ground roughage can lead to digestive disorders resulting in cessation of rumination, bloat, parakeratosis, depressed butter fat content in milk, reduced appetite, and lower gastrointestinal fill (Nicholson 1981). These effects can be effectively countered by the addition of a small proportion of unmilled roughage in the diet. The effect of chemical and physical processing tend to be additive and can have indirect benefits. For example, baled NaOH-treated sugarcane bagasse can be stored for up to 6 months with no significant effect on quality or acceptability to cattle (Tudor and Inkerman 1989).

11.3.4 Pretreatment

Pretreatment, in the present context, is defined as any physicochemical process at elevated temperature (>100 °C) and/or pressure. The stability of, and interconnections between, the major components of plant cell walls makes pretreatment of plant biomass essential for efficient and cost-effective enzymatic saccharification of its carbohydrate component to fermentable sugars *in vitro*. While a range of both physical and chemical changes occur during pretreatment, effective pretreatment of plant biomass is (ultimately) dependent upon achieving significant lignin solubilization, lignin modification, hemicellulose solubilization, defibrillation, particle size reduction, or combinations thereof.

A notable physical change during plant biomass pretreatment at high pressure (with or without the addition of a chemical catalyst) occurs when the pressure in the reaction vessel is rapidly reduced back to atmospheric; the resulting "steam-explosion" both defibrillates and reduces particle size (Harrison *et al.* 2013). Currently, the fermentable sugars produced by enzymatic saccharification of plant biomass *in vitro* are primarily used to supply energy to yeast for the production of ethanol, a so-called second-generation biofuel (Sun and Cheng 2002).

The goal of all pretreatment technologies for biofuel production is to improve access of hydrolytic enzymes to cell wall polysaccharides, while minimizing degradation of sugars and formation of yeast fermentation inhibitors (Hendriks and Zeeman 2009). The reader is directed to some of the multiplicity of excellent papers in the area for further information (Hsu 1996, Wyman *et al.* 2005, Chandra *et al.* 2007, Yang and Wyman 2008, Chen *et al.* 2009, Kumar *et al.* 2009, Agbor *et al.* 2011, Galbe and Zacchi 2012).

The goal of sugarcane bagasse, trash, and tops pretreatment for the production of animal feed depends upon the digestive capacity of the target animal. In the case of ruminants, the goal is similar to that of pretreatment technologies to produce biofuels and enhancing hydrolytic enzyme access to cell wall polysaccharides allows ruminant microbes (which includes not only bacteria, but also protozoa and fungi) to more rapidly and completely utilize the constituent monosaccharides to fuel their growth (Neely 1984, Jung and Bernardo 2012). There is also a need to enhance rumen microbial adherence to the fibers (Akin 1979), minimize the formation of compounds that negatively impact upon microbial growth, and minimize the formation of compounds that have a negative impact upon either live weight gain or general animal health.

In the case of nonruminants (i.e., animals that cannot digest the cellulose, hemicellulose, and lignin in plant biomass), the goal of pretreatment is to saccharify cellulose and hemicellulose so that the resulting monosaccharides can be taken up directly by the animal to provide energy for growth. As with ruminants, there is also a need to minimize the formation of compounds that have a negative impact upon live weight gain or general animal health. As a result, there is unlikely to be a single pretreatment technology that significantly enhances the nutritional value of sugarcane bagasse, trash, and tops for both ruminants and nonruminants. It should be noted that the majority of studies described in the following sections focus on sugarcane biomass pretreatment to enhance ruminant production.

Steam pretreatment

Steam pretreatment of sugarcane crop residues is attractive because no additional chemical catalyst is required. Steam pretreatment of plant biomass undertaken at elevated temperature (170–220 °C) and pressure results in "autohydrolysis," a process that modifies fiber structure and composition. During autohydrolysis, acetic and uronic acids are released from hemicelluloses, which lowers the pH of the extract (Brasch and Free 1965). The hemicelluloses are then dissolved as oligomers after acid hydrolysis, and the hemicellulose oligomers in solution are partially hydrolyzed to monosaccharides and sugar degradation products (including hydroxymethylfurfural and furfural) (Chen *et al.* 2010).

Lignin in sugarcane bagasse has a glass transition temperature of ~130 °C (Mousavioun *et al.* 2010), which means that any pretreatment undertaken at or above that temperature will result in structural modification of the lignin therein.

Furthermore, steam pretreatment offers the opportunity to further disrupt the fibers using steam-explosion.

Steam pretreatment of sugarcane bagasse at 180–190 °C and 1667 kPa for 5 min reduces the hemicellulose and lignin content by ~75% and ~15%, respectively, which significantly increases *in vitro* dry and organic matter digestibility (Manzano *et al.* 2000). Likewise, steam pretreatment of sugarcane pith at 19 bar (~210 °C) for 3 min increases *in vitro* dry matter digestibility compared with untreated sugarcane pith (Chaji *et al.* 2010).

The *in vitro* dry and organic matter digestibility of sugarcane bagasse increases after pretreatment using steam at ~210 °C for 4 min (Pate 1979). It should be noted that, in this study, the steam pretreated bagasse was ensiled for 2 months before analysis and the pH of the ensiled bagasse was 3.5. Furthermore, this study also demonstrated that inclusion of steam pretreated bagasse at ≥32% (dry matter) in compound feed reduced palatability to cattle and potentially limits the use of steam pretreated bagasse to intermediate levels in cattle feed (Pate 1979). This was supported by the observation that steam-pretreated bagasse improved average live weight gain and efficiency of feed conversion in lambs at an inclusion rate of 19 wt% in complete rations but reduced these parameters at an inclusion rate of 40 wt% (Jacobs and Vanniekerk 1985).

A number of additional animal feeding studies using steam pretreated sugarcane have also been undertaken. Steam pretreatment of sugarcane bagasse increases total digestible nutrition by up to 60% and steam pretreated sugarcane pith can be successfully used as a substitute for sugar beet pulp in feeding of buffaloes (Eslami *et al.* 2007). Pelleted steam-pretreated sugarcane bagasse is equivalent to the nongrain portion of corn silage for maintaining milk yield in dairy cattle (Harris *et al.* 1983). There is a decrease in the cost of lamb production if steam-pretreated sugarcane pith is used a partial substitute for barley in compound feed (Hashemipour *et al.* 2009). Unfortunately, steam pretreatment conditions (e.g., time, temperature, pressure, rate of steam application, rate of pressure release) were not supplied by the authors, so it is difficult to correlate improvements in animal performance with the physical and chemical changes in the sugarcane biomass.

As described previously (Section 3.1), however, care must be taken to ensure that any positive effects on performance and carcass characteristics are attributed to the addition of steam-treated sugarcane and not additives included in the diet to counteract the inherently low nitrogen content of sugarcane bagasse (Hashemipour *et al.* 2009).

Alkaline pretreatment

Ammonia fiber expansion (AFEX) is a physicochemical pretreatment in which plant biomass is exposed to aqueous ammonia under pressure (200–400 psi) at moderate temperatures (80–150 °C) for 5–30 min, followed by explosive decompression. AFEX depolymerizes cellulose, partially solubilizes hemicellulose, and modifies lignin, all of which enhance subsequent enzymatic saccharification

(Teymouri *et al.* 2004). AFEX treatment using 1.5 g NH_3/g dry matter for 30 min at 150 °C improves *in vitro* digestibility of neutral detergent fiber in sugarcane bagasse by 68% and increases the nitrogen content by approximately fivefold (Bals *et al.* 2010). Of particular note is the observation that while AFEX treatment solubilizes some fiber, the fact that the only products from AFEX treatment are a solid residue and exhaust gas (primarily ammonia), which means that much of the solubilized fiber may be available as a source of energy for animals. Also, the substantial increase in nonprotein nitrogen content has the potential to significantly enhance the capacity of rumen microbes to more effectively utilize both the oligosaccharides and polysaccharides in AFEX-treated sugarcane bagasse.

Pretreatment of bagasse with mixtures of Na_2S and NaOH significantly increased both *in vitro* dry and organic matter digestibility (Manzano *et al.* 2000). Subsequent incorporation of bagasse pretreated with 4% Na_2S and 6% NaOH at 160 °C for 8 min into compound feed resulted in significantly increased dry matter, organic matter, neutral detergent fiber, and acid detergent fiber digestibility in sheep compared with compound feed incorporating steam-treated bagasse (Manzano *et al.* 2000). Unfortunately, no direct comparisons with either N_2S or NaOH alone were provided by Manzano *et al.* (2000) and so it is not possible to determine which catalyst had the greatest impact on digestibility or the degree of synergy (if any).

The combination of alkaline pretreatment and steam explosion has been demonstrated to be a particularly effective approach to increase the *in vitro* organic matter digestibility of sugarcane bagasse (Playne 1984). Of particular note in this study was the inclusion of a high-level technoeconomic assessment of alkaline steam explosion for sugarcane bagasse pretreatment based upon the cost of the alkaline pretreatment catalysts (i.e., NaOH, anhydrous NH_3, $Ca(OH)_2$, Na_2CO_3, and combinations thereof) and expressed as cost of catalyst per kilogram additional digestible organic matter. While lime was identified as the least expensive alkaline catalyst, it should be noted that alkali is also able to serves as a neutralizing agent during acidogenic fermentation and ammonia provides organic nitrogen for microbial growth and can be recycled. As a result, any technoeconomic assessment of any sugarcane pretreatment using a catalyst (regardless of the nature of the catalyst) needs to take into account the effect of the catalyst (either positive or negative) on palatability, microbial fermentation in the rumen, and digestion in monogastric animals.

Acid pretreatment

Mild acid pretreatment of sugarcane bagasse can efficiently solubilize hemicellulose and modify lignin, which enhances enzyme accessibility to cellulose (Harrison *et al.* 2013, Zhang *et al.* 2013). Mineral acids (e.g., hydrochloric acid, sulfuric acid) are most often used as catalysts in such processes, and the use of strong mineral acids can have drawbacks including increased sugar losses, production of inhibitors of microbial growth, and higher capital costs because of the need for corrosion resistant equipment. While the changes in fiber composition and structure from such process correlate with enhanced digestibility by rumen microbes (Jung and Bernardo 2012), there is a dearth of published reports describing the incorporation of sugarcane pretreated using mineral acid catalysis for enhanced nutrition in animal feed.

The use of weaker inorganic or organic acids as catalysts can alleviate some of the negative impacts associated with mineral acid catalysts during plant biomass pretreatment and where the acid has nutritional value, there may be opportunities to further enhance the overall value of the pretreated sugarcane biomass in animal feed. For example, *in situ* rumen degradation of sugarcane bagasse can be increased from ~35% in untreated bagasse to ~70% by pretreatment with 2.9% (w/w) phosphoric acid at 197 °C (Fontana *et al.* 1995). However, partial lignin and hemicellulose removal during phosphoric acid catalyzed pretreatment of sugarcane bagasse increases the lag phase during ruminant digestion unless the solubilized C5 and C6 sugars from hemicellulose are retained to fuel the initial burst of microbial growth (Deschamps *et al.* 1996). It should be noted that, in such a situation, there is a requirement that pretreatment conditions need to be optimized to balance fiber pretreatment/hemicellulose solubilization with the production of toxic by-products or cattle growth inhibitors.

11.4 Conclusions

The products of sugarcane farming and sugar processing already have a substantial role to play in supporting animal production for both subsistence and commercial farmers. The size and scope of the global sugarcane industry means that greater integration of these existing products into animal feeds will result in (ultimately) nutritional and economic benefits for subsistence farmers, the majority of whom live in the tropics and subtropics, as well as improve the profitability of commercial animal production in these countries. A variety of technologies to improve the nutritional value of whole sugarcane, tops, and bagasse have been developed and some are used commercially. The challenges are to further develop these technologies such that they can be made more readily available to farmers (regardless of scale), to apply emerging biomass-processing technologies to enhance the value of sugarcane for animal feed, and to develop new technologies to (ultimately) enhance the economics of animal weight gain using the products from the world's most abundant biomass crop.

References

Agbor, V. B., Cicek, N., Sparling, R., Berlin, A. and Levin, D. B. (2011). Biomass pretreatment: fundamentals toward application. *Biotechnology Advances* **29**(6): 675–685.

Ahmed, M. H., Babiker, S. A., Elseed, A. E. M. A. and Mohammed, A. M. (2013). Effect of urea-treatment on nutritive value of sugarcane bagasse. *ARPN Journal of Science and Technology* **3**(8): 834–838.

Akin, D. E. (1979). Microscopic evaluation of forage digestion by rumen microorganisms – a review. *Journal of Animal Science* **48**(3): 701–710.

Alli, I. and Baker, B. E. (1982). Studies on the fermentation of chopped sugarcane. *Animal Feed Science and Technology* **7**(4): 411–417.

Amjed, M., Jung, H. G. and Donker, J. D. (1992). Effect of alkaline hydrogen peroxide treatment on cell wall composition and digestion kinetics of sugarcane residues and wheat straw. *Journal of Animal Science* **70**(9): 2877–2884.

Apas, A. L., Arena, M. E., Draksler, D. and Gonzalez, S. N. (2008). Utilization of sugarcane industrial residues as animal food and probiotic medium. *Journal of Bioscience and Bioengineering* **106**(4): 363–367.

Aregheore, E. M. (2005). *Feeds and forages in Pacific Island farming systems.* FAO Grassland and Pasture Crops , South Pacific Special

Avila, C. L., Bravo Martins, C. E. and Schwan, R. F. (2010). Identification and characterization of yeasts in sugarcane silages. *Journal of Applied Microbiology* **109**(5): 1677–1686.

Avila, C. L., Carvalho, B. F., Pinto, J. C., Duarte, W. F. and Schwan, R. F. (2014). The use of *Lactobacillus* species as starter cultures for enhancing the quality of sugar cane silage. *Journal of Dairy Science* **97**(2): 940–951.

Avila, C. L. S., Pinto, J. C., Figueiredo, H. C. P. and Schwan, R. F. (2009). Effects of an indigenous and a commercial *Lactobacillus buchneri* strain on quality of sugar cane silage. *Grass and Forage Science* **64**(4): 384–394.

Baconawa, E. T. (1986). Case study – prospects for reconversion of sugarcane into animal feeds in the Philippines. Proceedings of an FAO Expert Consultation, Santo Domingo, Dominican Republic, FAO.

Bals, B., Rogers, C., Jin, M. J., Balan, V. and Dale, B. (2010). Evaluation of ammonia fibre expansion (AFEX) pretreatment for enzymatic hydrolysis of switchgrass harvested in different seasons and locations. *Biotechnology for Biofuels* **3**.

Bernardes, T. F. and do Rego, A. C. (2014). Study on the practices of silage production and utilization on Brazilian dairy farms. *Journal of Dairy Science* **97**(3): 1852–1861.

Bhargava, P. K., Orskov, E. R. and Walli, T. K. (1989). Effect of soaking, ensilage and hydrogen-peroxide treatment of barley straw on rumen degradability. *Animal Feed Science and Technology* **22**(4): 295–303.

Bracch, D. J. and Free, K. W. (1965). Prehydrolysis-kraft pulping of *Pinus radiata* grown in New Zealand. *Tappi* **48**: 245–248.

Bravo, M., Lasso, M., Esnaola, M. A. and Preston, T. R. (1996). Preliminary studies on the use of chopped sugar cane stalks as the basal diet for fattening pigs. *Livestock research for rural development* **8**.

Briceno, J. V., Van Horn, H. H., Harris, B., Jr., and Wilcox, C. J. (1987). Effects of neutral detergent fiber and roughage source on dry matter intake and milk yield and composition of dairy cows. *Journal of Dairy Science* **70**(2): 298–308.

Brondani, A., Towns, R., Chou, K., Cook, R. M. and Barradas, H. (1991). Effects of isoacids, urea, and sulfur on ruminal fermentation in sheep fed high-fiber diets. *Journal of Dairy Science* **74**(8): 2724–2727.

Brown, W. F. (1990). Ammoniation or cane molasses supplementation of tropical grass hay. *Journal of Production Agriculture* **3**(3): 377–381.

Brown, W. F. (1993). Cane molasses and cottonseed meal supplementation of ammoniated tropical grass hay for yearling cattle. *Journal of Animal Science* **71**(12): 3451–3457.

Buchanansmith, J. G. (1990). An investigation into palatability as a factor responsible for reduced intake of silage by sheep. *Animal Production* **50**: 253–260.

Butterworth, M. H. (1963). Digestibility trials on forages in Trinidad and their use in prediction of nutritive value. *Journal of Agricultural Science* **60**: 341–346.

Butterworth, M. H. (1964). The digestible energy content of some tropical forages. *Journal of Agricultural Science* **64**: 319–321.

Cabral, A. D., Batista, A. M., Mustafa, A., de Carvalho, F. F., Guim, A., Monteiro, P. S. and Lucena, R. B. (2009). Performance of dairy goats fed whole sugarcane. *Tropical Animal Health and Production* **41**(3): 279–283.

Caldwell, D. (2001) *Molasses in Feeds. Advances in Equine Nutrition II, 1998–2000.* Kentucky Equine Research.

Capper, B. S., Morgan, D. J. and Parr, W. H. (1977). Alkali-treated roughages for feeding ruminants – review. *Tropical Science* **19**(2): 73–88.

Carvalho, G. G. P., Cavali, J., Fernandes, F. E. P., Rosa, L. O., Olivindo, C. S., Porto, M. O., Pires, A. J. V. and Garcia, R. (2009). Chemical composition and dry matter digestibility of sugarcane bagasse treated with calcium oxide. *Arquivo Brasileiro De Medicina Veterinaria E Zootecnia* **61**(6): 1346–1352.

Chaji, M., Naserian, A. A., Valizadeh, R., Mohammadabadi, T. and Mirzadeh, K. (2010). Potential use of high-temperature and low-temperature steam treatment, sodium hydroxide and an enzyme mixture for improving the nutritional value of sugarcane pith. *South African Journal of Animal Science* **40**(1): 22–32.

Chandra, R. P., Bura, R., Mabee, W. E., Berlin, A., Pan, X. and Saddler, J. N. (2007). Substrate pretreatment: the key to effective enzymatic hydrolysis of lignocellulosics? *Biofuels* **108**: 67–93.

Chandra, S. and Jackson, M. G. (1971). Study of various chemical treatments to remove lignin from coarse roughages and increase their digestibility. *Journal of Agricultural Science* **77**: 11.

Chen, M., Zhao, J. and Xia, L. M. (2009). Comparison of four different chemical pretreatments of corn stover for enhancing enzymatic digestibility. *Biomass and Bioenergy* **33**(10): 1381–1385.

Chen, M. C., Chen, C. P., Lee, C. F. and Yi, J. Z. (1983). Comparisons of the feeding value of fresh, wafered and pelleted sugarcane tops for cattle. *Journal of the Agricultural Association of China* (121): 66–77.

Chen, X., Lawoko, M. and Heiningen, A. (2010). Kinetics and mechanism of autohydrolysis of hardwoods. *Bioresour Technol* **101**(20): 7812–7819.

Costa, J. H., Hotzel, M. J., Longo, C. and Balcao, L. F. (2013). A survey of management practices that influence production and welfare of dairy cattle on family farms in southern Brazil. *Journal of Dairy Science* **96**(1): 307–317.

Croft, I., Clayton, E. and Cusack, P. (2014). Health and production of feedlot cattle following supplementation with urea-molasses in starter pens. *Australian Veterinary Journal* **92**(5): 166–170.

Crosthwaite, C., Ishihara, M. and Richards, G. N. (1984). Acid-aging of lignocellulosics to improve ruminant digestibility – application to bagasse, wheat and rice straw and oat hulls. *Journal of the Science of Food and Agriculture* **35**(10): 1041–1050.

Cutrim, D. O., Alves, K. S., Oliveira, L. R., da Conceicao dos Santos, R., da Mata, V. J., do Carmo, D. M., Gomes, D. I., Mezzomo, R. and de Carvalho, F. F. (2012). Elephant grass, sugarcane, and rice bran in diets for confined sheep. *Tropical Animal Health and Production* **44**(8): 1855–1863.

da Silva, J. G., Queiroz, M. A. A., Araujo, G. G. L., da Silva, B. G., da Cunha, J. A. and Rodrigues, P. H. M. (2014). Fermentation characteristics of sugarcane silages with saltbush. *Ciencia Rural* **44**(3): 555–560.

Daniel, J. L., Amaral, R. C., Goulart, R. S., Zopollatto, M., Santos, V. P., Toledo Filho, S. G., Cabezas-Garcia, E. H., Lima, J. R., Santos, M. C. and Nussio, L. G. (2013). Short-term effects of silage volatile compounds on feed intake and digestion in beef cattle. *Journal of Animal Science* **91**(5): 2321–2331.

de Lima, J. A., Gavioli, I. L. D., Barbosa, C. M. P., Berndt, A., Gimenes, F. M. D., Paz, C. C. D. and da Cunha, E. A. (2014). Soybean silage and sugarcane tops silage on lamb performance (vol 43, pg 1478, 2013). *Ciencia Rural* **44**(4): 757–757.

Deschamps, F. C., Ramos, L. P. and Fontana, J. D. (1996). Pretreatment of sugar cane bagasse for enhanced ruminal digestion. *Applied Biochemistry and Biotechnology* **57–58**: 171–182.

Diaz, C. P., Elias, A., Savon, L., Rodriguez, Y. and Achang, J. (1992). Inclusion of protein sugar-cane juice in the diets for young-pigs. *Cuban Journal of Agricultural Science* **26**(1): 71–77.

Duff, G. C. and Galyean, M. L. (2007). Board-invited review: recent advances in management of highly stressed, newly received feedlot cattle. *Journal of Animal Science* **85**(3): 823–840.

Durix, A., Jeanblain, C., Sallmann, H. P. and Jouany, J. P. (1991). Use of a semicontinuous culture system (rusitec) to study the metabolism of ethanol in the rumen and its effects on ruminal digestion. *Canadian Journal of Animal Science* **71**(1): 115–123.

Environment Directorate, J. M. o. t. C. C. a. t. W. P. o. C., Pesticides, and Biotechnology (2011). Consensus document on compositional considerations for new varieties of sugarcane (*Saccharum* ssp. hybrids): key food and feed nutrients, anti-nutrients, and toxicants. Series on Safety of Novel Foods and Feeds, Organisation for Economic Co-operation and Development: 1–43.

Eslami, M., Mashyekhee, M. and Kasiri, H. M. (2007). Evaluation of feeding steam treated bagasse pith on milk production and blood parameters of dairy buffaloes. *Italian Journal of Animal Science* **6**: 469–471.

FAOSTAT (2014). FAOSTAT.

Ferreira, E. M., Pires, A. V., Susin, I., Gentil, R. S., Gilaverte, S., Parente, M. D. M., Biehl, M. V. and Ribeiro, C. V. D. (2014). Lamb performance, milk production and composition from ewes supplemented with soybean oil partially replaced by fish oil blend. *Livestock Science* **163**: 51–61.

Fischer, M. M., Kessler, A. M., de Sa, L. R., Vasconcellos, R. S., Filho, F. O., Nogueira, S. P., Oliveira, M. C. and Carciofi, A. C. (2012). Fiber fermentability effects on energy and macronutrient digestibility, fecal traits, postprandial metabolite responses, and colon histology of overweight cats. *Journal of Animal Science* **90**(7): 2233–2245.

Fontana, J. D., Ramos, L. P. and Deschamps, F. C. (1995). Pretreated sugar-cane bagasse as a model for cattle feeding. *Applied Biochemistry and Biotechnology* **51–52**: 105–116.

Galbe, M. and Zacchi, G. (2012). Pretreatment: the key to efficient utilization of lignocellulosic materials. *Biomass & Bioenergy* **46**: 70–78.

Galina, M. A., Guerrero, M. and Puga, C. D. (2007). Fattening Pelibuey lambs with sugar cane tops and corn complemented with or without slow intake urea supplement. *Small Ruminant Research* **70**(2–3): 101–109.

Galina, M. A., Hummel, J. D., Sanchez, M. and Haenlein, G. F. W. (2004). Fattening Rambouillet lambs with corn stubble or alfalfa, slow intake urea supplementation or balanced concentrate. *Small Ruminant Research* **53**(1–2): 89–98.

Gandi, J., Holtzapple, M. T., Ferrer, A., Byers, F. M., Turner, N. D., Nagwani, M. and Chang, S. S. (1997). Lime treatment of agricultural residues to improve rumen digestibility. *Animal Feed Science and Technology* **68**(3–4): 195–211.

Gendley, M. K., Singh, P. and Garg, A. K. (2003). Performance of crossbred cattle fed solely on chopped green sugarcane tops and supplemented with concentrate mixture or urea molasses liquid diet. *Indian Journal of Animal Sciences* **73**(9): 1061–1065.

Gendley, M. K., Singh, P., Garg, A. K., Tiwari, S. P., Kumari, K. and Dutta, G. K. (2009). The studies on nutrient balances in crossbred cattle bulls fed chopped green sugarcane tops supplemented with some agro industrial by-products. *Tropical Animal Health and Production* **41**(6): 943–949.

Gomez-Vazquez, A., Pinos-Rodriguez, J. M., Garcia-Lopez, J. C., de la Cruz-Lazaro, E., Luna-Palomera, C. and Sanchez-Hernandez, R. (2011). Nutritional value of sugarcane silage enriched with corn grain, urea, and minerals as feed supplement on growth performance of beef steers grazing stargrass. *Tropical Animal Health and Production* **43**(1): 215–220.

Gradiz, L., Sugimoto, A., Ujihara, K., Fukuhara, S., Kahi, A. K. and Hirooka, H. (2007). Beef cow-calf production system integrated with sugarcane production: simulation model development and application in Japan. *Agricultural Systems* **94**(3): 750–762.

Grieve, D. G. (1976). Nutritive value of rice straw, sugar-cane tops and sorghum tops fed to goats and sheep. 1. Dry matter intake and digestibility. *Ghana Journal of Agricultural Science* **9**: 103–109.

Harris, B., Vanhorn, H. H., Manookian, K. E., Marshall, S. P., Taylor, M. J. and Wilcox, C. J. (1983). Sugarcane silage, sodium hydroxide-treated and steam pressure-treated sugarcane bagasse, corn-silage, cottonseed hulls, sodium-bicarbonate, and aspergillis-oryzae product in complete rations for lactating cows. *Journal of Dairy Science* **66**(7): 1474–1485.

Harrison, M. D., Zhang, Z. Y., Shand, K., O'Hara, I. M., Doherty, W. O. S. and Dale, J. L. (2013). Effect of pretreatment on saccharification of sugar cane bagasse by complex and simple enzyme mixtures. *Bioresource Technology* **148**: 105–113.

Hashemipour, E., Dabiri, N. and Hashemipour, H. (2009). Effect of diets containing different levels of steam treated pith bagasse on performance and carcass characteristics of fattening Arabi lambs. *Journal of Animal and Veterinary Advances* **8**(11): 2139–2142.

Hegazy, R. A. and el-Faramawy, A. A. (2001). Substitution of sugar cane bagasse in the chicken diet and immune response. *Nahrung* **45**(5): 364–367.

Hegde, S. N., Rolls, B. A. and Coates, M. E. (1982). The effects of the gut microflora and dietary fibre on energy utilization by the chick. *British Journal of Nutrition* **48**(1): 73–80.

Hendriks, A. and Zeeman, G. (2009). Pretreatments to enhance the digestibility of lignocellulosic biomass. *Bioresource Technology* **100**(1): 10–18.

Homb, T. (1984). Wet treatment with sodium hydroxide. In *Straw and Other Fibrous By-Products as Feed*. F. Sundstol and E. Owen, eds. Elsevier: Amsterdam.

Hsu, T.-A. (1996). Pretreatment of biomass. In *Handbook on Bioethanol: Production and Utilization*, pp. 179–212. C. E. Wyman, ed. Taylor & Francis: Washington, DC.

Hulshof, R. B., Berndt, A., Gerrits, W. J., Dijkstra, J., van Zijderveld, S. M., Newbold, J. R. and Perdok, H. B. (2012). Dietary nitrate supplementation reduces methane emission in beef cattle fed sugarcane-based diets. *Journal of Animal Science* **90**(7): 2317–2323.

Hunter, R. A. (2012). High-molasses diets for intensive feeding of cattle. *Animal Production Science* **52**(9): 787–794.

Ibrahim, M. N. M. and Pearce, G. R. (1983). Effects of chemical pretreatments on the composition and invitro digestibility of crop by-products. *Agricultural Wastes* **5**(3): 135–156.

Jackson, M. G. (1977). Review article: the alkali treatment of straws. *Animal Feed Science and Technology* **2**: 105–130.

Jacobs, G. A. and Vanniekerk, B. D. H. (1985). Sodium and steam-treated sugarcane bagasse in low and high-fiber fattening diets for lambs. *South African Journal of Animal Science-Suid-Afrikaanse Tydskrif Vir Veekunde* **15**(4): 166–168.

James, G. L. (2004). An introduction to sugarcane. In *Sugarcane*, pp. 1–18. G. L. James. Blackwell Science Ltd.: Oxford, UK.

Jose, A., Pires, V., Garcia, R., Valadares, S. D., Pereira, O. G., Cecon, P. R., da Silva, F. F., Silva, P. A. and Veloso, C. M. (2004). Diets with of sugarcane bagasse treated with anhydrous ammonia and/or sodium sulfate fed heifers. *Revista Brasileira De Zootecnia-Brazilian Journal of Animal Science* **33**(4): 1078–1085.

Jung, H. J. G. and Bernardo, R. (2012). Comparison of cell wall polysaccharide hydrolysis by a dilute acid/enzymatic saccharification process and rumen microorganisms. *Bioenergy Research* **5**(2): 319–329.

Karbhari, P. S., Balakrishnan, V. and Murugan, M. (2007). Substitutional feeding value of ensiled sugarcane tops and its effects in crossbred heifer's/cow's reproductive performance. *Asian Journal of Animal and Veterinary Advances* **2**: 21–26.

Kebreab, E., Assis, A. G., Dijkstra, J. and France, J. (2001). Evaluating sugarcane diets for dairy cows using a digestion model. *Tropical Animal Health and Production* **33**(2): 127–139.

Kobeisy, M. A., Zenhom, M., Salem, A. and Hayder, M. (2005). Sugarcane bagasse silage treated with different levels of urea for improvement sheep production. II. Body weight changes and ewes' reproductive performance. Annual Conference of the European Federation of Animal Science, Uppsala.

Krishnani, K. K., Parimala, V., Gupta, B. P., Azad, I. S., Meng, X. and Abraham, M. (2006). Bagasse-assisted bioremediation of ammonia from shrimp farm wastewater. *Water Environment Research* **78**(9): 938–950.

Kumar, R., Mago, G., Balan, V. and Wyman, C. E. (2009). Physical and chemical characterizations of corn stover and poplar solids resulting from leading pretreatment technologies. *Bioresource Technology* **100**(17): 3948–3962.

Kung, L. and Stanley, R. W. (1982). Effect of stage of maturity on the nutritive-value of whole-plant sugarcane preserved as silage. *Journal of Animal Science* **54**(4): 689–696.

Kutty, K. P. A. and Prasad, D. A. (1980). Studies on improving nutritive-value of sugarcane tops with urea or dried poultry waste by ensiling techniques. *Indian Journal of Animal Sciences* **50**(2): 189–193.

Lampila, M. (1963). Experiments with alkali straw and urea. *Annales Agriculture Fenniae* **2**: 105–108.

Lascano, G. J., Velez, M., Tricarico, J. M. and Heinrichs, A. J. (2012). Short communication: Nutrient utilization of fresh sugarcane-based diets with slow-release nonprotein nitrogen addition for control-fed dairy heifers. *Journal of Dairy Science* **95**(1): 370–376.

Lehmann, F. (1891). Der Nahrwert der Cellulose. *Landwirtschaftliche Versuchsstationen* **38**: 337–338.

Leng, R. A. (1990). Factors affecting the utilization of 'poor-quality' forages by ruminants particularly under tropical conditions. *Nutrition Research Reviews* **3**(1): 277–303.

Lewis, S. M., Montgomery, L., Garleb, K. A., Berger, L. L. and Fahey, G. C., Jr., (1988). Effects of alkaline hydrogen peroxide treatment on in vitro degradation of cellulosic substrates by mixed ruminal microorganisms and *Bacteroides succinogenes* S85. *Applied and Environmental Microbiology* **54**(5): 1163–1169.

Lindsay, J. A., Cooper, N. J. and Batterham, I. (1998). A molasses based production feeding system for Brahman cattle. *Animal Production in Australia* **22**: 119–121.

Lofgreen, G. P. and Kiesling, H. E. (1985). Effects of receiving and growing diets on compensatory gains of stressed calves. *Journal of Animal Science* **61**(2): 321–328.

MacLachlan, D. (2010). Pesticide risk profile for the feeding of sugarcane tops and trash to animals. A. Q. a. I. Service. Canberra, Australian Government Department of Agriculture, Fisheries, and Forestry: 1–29.

Magana, R., Aguirre, J., Aguirre, A., Gomez, A., Martinez, S., Lemus, C., Ulloa, R. and Ly, J. (2009). Entire sugar cane or sugar cane residues for feeding sheep: performance of hairless sheep. *Livestock Research for Rural Development* **21**(2).

Mahala, A. G., Mokhtar, A. M. S., Amasiab, E. O. and AttaElmnan, B. A. (2013). Sugarcane tops as animal feed. *Global Journal of Agricultural Research and Reviews* **1**(1): 53–57.

Manzano, R. P., Fukushima, R. S., Gomes, J. D. F. and Garippo, G. (2000). Digestibility of sugar cane bagasse treated with chemical reagents and steam pressure. *Brazilian Journal of Animal Science* **29**(4): 1196–1204.

Marshall, S. P. and Vanhorn, H. H. (1975). Complete rations for dairy-cattle. 2. Sugarcane bagasse pellets as roughage in blended rations for lactating cows. *Journal of Dairy Science* **58**(6): 896–900.

Mason, J. (1990). The biochemical pathogenesis of molybdenum-induced copper deficiency syndromes in ruminants – towards the final chapter. *Irish Veterinary Journal* **43**(1): 18–21.

McDonald, P., Henderson, P. N. and Heron, S. (1991). *The Biochemistry of Silage*. Chalcombe Publications: Marlow, UK.

McDowell, L., Conrad, J. H. and Ellis, G. L. (1984). Mineral deficiencies and imbalances and their diagnosis. In *Herbivore Nutrition in the Subtropics and Tropics*. F. M. C. Gilchrist and R. I. Mackie. The Science Press: Craighall, South Africa.

Mena, A. (1981). Sugar cane juice as an energy source for fattening pigs. *Tropical Animal Production* **6**: 338–344.

Millen, D. D., Pacheco, R. D., Arrigoni, M. D., Galyean, M. L. and Vasconcelos, J. T. (2009). A snapshot of management practices and nutritional recommendations used by feedlot nutritionists in Brazil. *Journal of Animal Science* **87**(10): 3427–3439.

Montanez-Valdez, O. D., Reyes-Gutierrez, J. A., Guerra-Medina, C. E. and Salem, A.-F. Z. M. (2013). Rumen dry matter degradability of fresh and ensiled sugarcane. *African Journal of Biotechnology* **12**(19): 2743–2747.

Moon, N. J. (1983). Inhibition of the growth of acid tolerant yeasts by acetate, lactate and propionate and their synergistic mixtures. *Journal of Applied Bacteriology* **55**(3): 453–460.

Mousavioun, P., Doherty, W. O. S. and George, G. (2010). Thermal stability and miscibility of poly(hydroxybutyrate) and soda lignin blends. *Industrial Crops and Products* **32**(3): 656–661.

Mridula, R. M., Manissery, J. K., Keshavanath, P., Shankar, K. M., Nandeesha, M. C. and Rajesh, K. M. (2003). Water quality, biofilm production and growth of fringe-lipped carp (*Labeo fimbriatus*) in tanks provided with two solid substrates. *Bioresource Technology* **87**(3): 263–267.

Murta, R. M., Chaves, M. A., Pires, A. J. V., Veloso, C. M., da Silva, F. F., Neto, A. L. R., Eustaquio, A. and dos Santos, P. E. F. (2011). Performance and nutrients apparent digestibility in sheep fed diets containing sugar cane bagasse treated with calcium oxide. *Brazilian Journal of Animal Science* **40**(6): 1325–1332.

Naseeven, M. R. (1986). Sugarcane tops as animal feed. Proceedings of an FAO Expert Consultation – Sugarcane as Feed, Santo Domingo, Dominican Republic, FAO.

Neely, W. C. (1984). Factors affecting the pretreatment of biomass with gaseous ozone. *Biotechnology and Bioengineering* **26**(1): 59–65.

Neely, W. C., West, A. D. and Hall, T. D. (1975). Ozone removal from arc lamp cooling air. *Journal of Physics E-Scientific Instruments* **8**(7): 543–543.

Nicholson, J. W. G. (1981). Nutrition and feeding aspects of the utilization of processed lignocellulosic waste materials by animals. *Agriculture and Environment* **6**(2–3): 205–228.

O'Shea, E. F., Cotter, P. D., Ross, R. P. and Hill, C. (2013). Strategies to improve the bacteriocin protection provided by lactic acid bacteria. *Current Opinion in Biotechnology* **24**: 130–134.

Oliveira, C. A. and Millen, D. D. (2014). Survey of nutritional recommendations and management practices adopted by feedlot cattle nutritionists in Brazil. *Animal Feed Science and Technology* **197**: 64–75.

Onwuka, C. F. I. (1999). Molasses blocks as supplementary feed resource for ruminants. *Archivos de Zootecnia* **48**: 89–94.

Ortiz-Rubio, M. A., Orskov, E. R., Milne, J. and Galina, H. M. A. (2008). Effect of different sources of nitrogen on in situ degradability and feed intake of Zebu cattle fed sugarcane tops (*Saccharum officinarum*). *Animal Feed Science and Technology* **140**(1–2): 139–154.

Osawa, Z., Erby, W. A., Sarkanen, K. V., Carpenter, E. and Schverch, C. (1963). The action of gaseous reagents on cellulosic materials II. *Pulping of wood with ozone Tappi* **46**: 84–88.

Owen, E. and Jayasuriya, M. C. N. (1989). Use of Crop Residues as Animal Feeds in Developing-Countries. *Research and Development in Agriculture* **6**(3): 129–138.

Pate, F. M. (1979). *Improving the Nutritive Value of Sugarcane Bagasse with Steam-Pressure Treatment*. Agricultural Research and Education Centre, Department Animal Sciences, University of Florida: Belle Glade.

Pate, F. M. (1983). *Molasses in Beef Nutrition. Molasses in Animal Nutrition*. National Feed Ingredients Association: West Des Moine, Iowa, USA.

Pate, F. M., Alvarez, J. H., Phillips, J. D. and Eiland, B. R. (2002). Sugarcane as a cattle feed: production and utilization. In *Florida Sugarcane Handbook*, pp. 1–21. R. W. Rice, ed. Institute of Food and Agricultural Sciences.

Pate, F. M., Beardsle. D. W., Hayes, B. W., Palmer, A. Z. and Carpente, J. W. (1971). Value of raw sugar in steer-finishing rations. *Journal of Animal Science* **32**(2): 389.

Patil, N. V., Kharadi, V. B., Desia, P. M. and Deai, B. M. (1999). Comparative nutritional value of fresh and sundried sugarcane tips in buffalo calves. *Indian Journal of Animal Nutrition* **6**: 259–266.

Peacock, C. (1996). *Basic Nutrition. Improving Goat Production in the Tropics: A Manual for Development Workers*, pp. 65–91. OXFAM in association with FARM-AFRICA: Oxford, UK.

Pedreira, M. D., Berchelli, T. T., Primavesi, O., de Oliveira, S. G., Frighetto, R. and de Lima, M. A. (2012). Influence of different supplements and sugarcane (*Saccharum officinarum* L.) cultivars on intake, digestible variables and methane production of dairy heifers under tropical conditions. *Tropical Animal Health and Production* **44**(7): 1773–1778.

Pedroso, A. D., Nussio, L. G., Loures, D. R. S., Paziani, S. D., Ribeiro, J. L., Mari, L. J., Zopollatto, M., Schmidt, P., Mattos, W. R. S. and Horii, J. (2008). Fermentation, losses, and aerobic stability of sugarcane silages treated with chemical or bacterial additives. *Scientia Agricola* **65**(6): 589–594.

Pedroso, A. D. F., Nussio, L. G., Barioni, W., Rodrigues, A. D., Loures, D. R. S., de Campos, F., Ribeiro, J. L., Mari, L. J., Zopollatto, M., Junqueira, M., Schmidt, P., Paziani, S. D. and Horii, J. (2006). Performance of Holstein heifers fed sugarcane silages treated with urea, sodium benzoate or *Lactobacillus buchneri*. *Pesquisa Agropecuaria Brasileira* **41**(4): 649–654.

Perdok, H. B. and Leng, R. A. (1985). Hyperexcitability in cattle fed (thermo-) ammoniated rice straw or wheat crop. Proceedings of the 3rd Asian-Australasian Association of Animal Production Societies Symposium Feeding Systems of Animals in Temperate Areas.

Pires, A. J. V., Garcia, R., Valadares, S. D., Pereira, O. G., Cecon, P. R., da Silva, F. F., Silvan, P. A. and Itavo, L. C. V. (2004). Degradability of sugarcane bagasse treated with anhydrous ammonia and/or sodium sulfate. *Brazilian Journal of Animal Science* **33**(4): 1071–1077.

Pires, A. J. V., Reis, R. A., de Carvalho, G. G. P., Siqueira, G. R. and Bernardes, T. F. (2006). Sugarcane bagasse treated with sodium hydroxide. *Brazilian Journal of Animal Science* **35**(3): 953–957.

Playne, M. J. (1984). Increased digestibility of bagasse by pretreatment with alkalis and steam explosion. *Biotechnology and Bioengineering* **26**(5): 426–433.

Preston, T. R. (1995). Technologies for improving the use of renewable natural resources. In *Tropical Animal Feeding – A Manual for Research Workers*, pp. 1–14. A. Speedy and R. Sansoucy, eds. FAO: Rome.

Preston, T. R., Elias, A., Willis, M. B. and Sutherla, T. M. (1967). Intensive beef production from molasses and urea. *Nature* **216**(5116): 721.

Preston, T. R., Macleod, N. A., Lassota, L., Willis, M. B. and Velazque, M (1968). Sugar cane products as energy sources for pigs. *Nature* **219**(5155): 727.

Rae, A. L., Martinelli, A. P. and Dornelas, M. C. (2013). Anatomy and morphology. In *Sugarcane: Physiology, Biochemistry & Functional Biology*, pp. 19–34. P. H. Moore and F. C. Botha, eds. John Wiley & Sons: Somerset, NJ, USA.

Randby, A. T., Selmer-Olsen, I. and Baevre, L. (1999). Effect of ethanol in feed on milk flavor and chemical composition. *Journal of Dairy Science* **82**(2): 420–428.

Randel, P. F., Ramirez, A., Valencia, I. and Carrero, R. (1972). Alkali-treated and raw sugarcane bagasse as roughages in complete rations for lactating cows. *Journal of Dairy Science* **55**(10): 1492.

Reddy, M. S., Prasad, D. A. and Krishna, N. (1992). Effect of nitrogen fortification on protein degradability of silages based on sugarcane tops in fistulated steers. *Indian Journal of Animal Sciences* **62**(1): 81–82.

Reddy, R. R. and Prasad, D. A. (1982a). Evaluation of complete rations consisting of enriched sugarcane tops (30-percent-Dpw on Dmb) silage and an economic concentrate mixture in varying ratios. *Indian Journal of Animal Sciences* **52**(9): 754–757.

Reddy, R. R. and Prasad, D. A. (1982b). Studies on improving nutritive-value of sugarcane tops with urea (1.5-percent) or dried poultry waste (30-percent or 40-percent on Dmb) by ensiling techniques and development of complete rations for growing Nellore lambs. *Indian Journal of Animal Sciences* **52**(7): 524–529.

Rein, P. (2007). *Cane Sugar Engineering*. Verlag: Berlin, Germany.

Reyes, J., Montanez-Valdez, O., Guerra, C. and Palma, J. (2014). Effect of sugarcane silage on productive parameters of replacement Holstein-Friesian heifers. *Revista Mvz Cordoba* **19**(1): 3962–3969.

Roman, J., Jobim, C. C., de Resende, F. D., Siqueira, G. R., de Faria, M. H. and Neto, R. A. D. (2011). Performance of finishing beef cattle fed different diets containing whole-crop maize silage or sugarcane silage. *Brazilian Journal of Animal Science* **40**(3): 682–689.

Romanponce, H., Vanhorn, H. H., Marshall, S. P., Wilcox, C. J. and Randel, P. F. (1975). Complete rations for dairy-cattle. 5. Interaction of sugarcane bagasse quantity and form with soybean-meal, urea, and starea. *Journal of Dairy Science* **58**(9): 1320–1327.

Rostagno, H. (2005). *Brazilian tables for poultry and swine: composition of feedstuffs and nutritional requirements*. Universidade Federal de Vicosa: Vicosa, Minas Gerais, Brazil, **186**.

Roth, A. P. D. P., Reis, R. A., Siqueira, G. R., Roth, M. D. P., de Resende, F. D. and Monteiro, R. R. (2010). Sugarcane silage production treated with additives at different times post burning. *Brazilian Journal of Animal Science* **39**(1): 88–96.

Rotta, P. P., Filho, S. C., Engle, T. E., Costa, E. S. L. F., Sathler, D. F., Prado, I. N., Bonafe, E. G., Zawadzki, F. and Visentainer, J. V. (2014). The impact of dietary sugarcane addition to finishing diets on performance, apparent digestibility, and fatty acid composition of Holstein x Zebu bulls. *Journal of Animal Science* **92**(6): 2641–2653.

Royes, J. B., Brown, W. F., Martin, F. G. and Bates, D. B. (2001). Source and level of energy supplementation for yearling cattle fed ammoniated hay. *Journal of Animal Science* **79**(5): 1313–1321.

Santana, M. H. A., Rossi, P., Almeida, R. and Cucco, D. C. (2012). Feed efficiency and its correlations with carcass traits measured by ultrasound in Nellore bulls. *Livestock Science* **145**(1–3): 252–257.

Santiago, A. M. F., Campos, J. M. D., de Oliveira, A. S., Valadares, S. D. C., Santos, S. A., de Souza, S. M. and Santiago, I. F. (2013). Urea in sugarcane-based diets for dairy cows. *Brazilian Journal of Animal Science* **42**(6): 456–462.

Santos, S. A., Valadares, S. D., Detmann, E., Valadares, R. F. D., Ruas, J. R. D. and Amaral, P. D. (2011). Different forage sources for F1 Holstein x Gir dairy cows. *Livestock Science* **142**(1–3): 48–58.

Schmidt, P., Nussio, L. G., Queiroz, O. C. M., Santos, M. C., Zopollatto, M., de Toledo, S. G. and Daniel, J. L. P. (2014). Effects of *Lactobacillus buchneri* on the nutritive value of sugarcane silage for finishing beef bulls. *Brazilian Journal of Animal Science* **43**(1): 8–13.

Secrist, R. B. and Singh, R. P. (1971). Kraft pulp bleaching II. Studies on the ozonation of chemical pulps. *Tappi Journal* **54**: 583–584.

Sharma, J., Singh, M., Kumar, R., Kumar, A. and Singh, V. (2007). Utilization of sugarcane tops in growing crossbred heifers. *Indian Journal of Animal Nutrition* **24**(1): 65–66.

Singh, P., Garg, A. K., Tiwari, R. K. and Agrawal, D. K. (2004). Influence of feeding raw bagasse pith/urea treated bagasse pith on intake and utilization of nutrients in crossbred bulls. *Indian Journal of Animal Sciences* **74**(12): 1223–1226.

Sparke, E. J. (1997). Expanded use of sugarcane by-products for intensive beef cattle feeding. Final Report. Sydney, Australia.

Stateler, D. A., Kunkle, W. E. and Hammond, A. C. (1995). Effect of protein level and source in molasses slurries on the performance of growing cattle fed hay during winter. *Journal of Animal Science* **73**(10): 3078–3084.

Sun, Y. and Cheng, J. Y. (2002). Hydrolysis of lignocellulosic materials for ethanol production: a review. *Bioresource Technology* **83**(1): 1–11.

Sundstøl, F. (1993). Urinary urea and other urinary compounds as feed supplement. In *Urine – A wasted, Renewable Natural Resources*, pp. 36–41. F. Sundstøl and E. Owen, eds. NORAGRIC: Norway, **12**.

Sundstrom, B. and Palmer, W. A. (1977). Molasses based fattening rations for beef cattle. *Technical Bulletin* **17**.

Suzuki, T., Sakaigaichi, T., Kamiya, M., Kamiya, Y., Hattori, I., Sato, K., Terauchi, T. and Tanaka, M. (2014). Feeding of fodder-sugarcane silage to Holstein cows. *Jarq-Japan Agricultural Research Quarterly* **48**(2): 183–193.

Suzuki, T., Sakaigaichi, T., Terajima, Y., Matsuoka, M., Kamiya, Y., Hattori, I. and Tanaka, M. (2010). Chemical composition and in situ degradability of two varieties of sugarcane at different growth stages in subtropical Japan. *Grassland Science* **56**(3): 134–140.

Teymouri, F., Laureano-Perez, L., Alizadeh, H. and Dale, B. E. (2004). Ammonia fiber explosion treatment of corn stover. *Applied Biochemistry and Biotechnology* **113**: 951–963.

Thiago, L. R. L. D., Costa, F. P., Dasilva, J. M., Baroli, E. A., Figueiredo, G. R. and Curvo, J. B. E. (1986). The use of sugarcane tops for fattening steers in feedlots, supplemented with corn, sorghum, urea and pigeon-pea forage. *Pesquisa Agropecuaria Brasileira* **21**(3): 317–321.

Thompson, D. N., Barnes, J. M. and Houghton, T. P. (2005). Effect of additions on ensiling and microbial community of senesced wheat straw. *Applied Biochemistry and Biotechnology* **121–124**: 21–46.

Tillman, A. D., Singletary, C. B., Kidwell, J. F. and Bray, C. I. (1951). Methods of feeding cane molasses and urea to beef cattle. *Journal of Animal Science* **10**(4): 939–946.

Tudor, G. D. and Inkerman, P. A. (1989). Alkali treated bagasse – potential as feed for ruminants. Retrieved 01-10-2014, from http://livestocklibrary.com.au/handle/1234/19600.

Tudor, G. D., Martin, P. R., Inkerman, P. A., Bromage, E. K. and Foster, D. H. (1985). Alkali-treated bagasse as a medium quality roughage for ruminants. Proceedings of the Australian Society of Sugar Cane Technologies: 59–63.

Ulloa, J. B., van Weerd, J. H., Huisman, E. A. and Verreth, J. A. (2004). Tropical agricultural residues and their potential uses in fish feeds: the Costa Rican situation. *Waste Management* **24**(1): 87–97.

Viator, R. P., Johnson, R. and Richard, E. P. J. (2005). Challenges of post-harvest residue management in the Louisiana sugar-cane industry. *Proceedings – International Society of Sugar Cane Technologists* **25**(2): 228–234.

Waiss, A. C., Walker, H. G., Garrett, W. N., Guggolz, J. and Kohler, G. O. (1972). Improving digestibility of straws for ruminant feed by aqueous ammonia. *Journal of Animal Science* **35**(1): 109.

Walker, H. G. and Kohler, G. O. (1981). Treated and untreated cellulosic wastes as animal feeds – recent work in the United-States-of-America. *Agriculture and Environment* **6**(2–3): 229–243.

Wang, Z. Q., Yu, Y., Zhang, X. H., Floyd, Z. E., Boudreau, A., Lian, K. and Cefalu, W. T. (2012). Comparing the effects of nano-sized sugarcane fiber with cellulose and psyllium on hepatic cellular signaling in mice. *International Journal of Nanomedicine* **7**: 2999–3012.

Wang, Z. Q., Zuberi, A. R., Zhang, X. H., Macgowan, J., Qin, J., Ye, X., Son, L., Wu, Q., Lian, K. and Cefalu, W. T. (2007). Effects of dietary fibers on weight gain, carbohydrate metabolism, and gastric ghrelin gene expression in mice fed a high-fat diet. *Metabolism* **56**(12): 1635–1642.

Wiedmeier, R. D., Tanner, B. H., Bair, J. R., Shenton, H. T., Arambel, M. J. and Walters, J. L. (1992). Effects of a new molasses by-product, concentrated separator by-product, on nutrient digestibility and ruminal fermentation in cattle. *Journal of Animal Science* **70**(6): 1936–1940.

Wilson, R. K. and Pigden, W. J. (1964). Effect of sodium hydroxide treatment on utilization of wheat straw + poplar wood by rumen microorganisms. *Canadian Journal of Animal Science* **44**(1): 122.

Wyman, C. E., Dale, B. E., Elander, R. T., Holtzapple, M., Ladisch, M. R. and Lee, Y. Y. (2005). Coordinated development of leading biomass pretreatment technologies. *Bioresource Technology* **96**(18): 1959–1966.

Xande, X., Mourot, J., Archimede, H., Gourdine, J. L. and Renaudeau, D. (2009). Effect of sugarcane diets and a high fibre commercial diet on fresh meat and dry-cured ham quality in local Caribbean pigs. *Meat Science* **82**(1): 106–112.

Xande, X., Regnier, C., Archimede, H., Bocage, B., Noblet, J. and Renaudeau, D. (2010). Nutritional values of sugarcane products in local Caribbean growing pigs. *Animal* **4**(5): 745–754.

Yang, B. and Wyman, C. E. (2008). Pretreatment: the key to unlocking low-cost cellulosic ethanol. *Biofuels Bioproducts & Biorefining-Biofpr* **2**(1): 26–40.

Yuangklang, C., Wanapat, M. and Wachirapakorn, C. (2005). Effects of pelleted sugarcane tops on voluntary feed intake, digestibility and rumen fermentation in beef cattle. *Asian-Australasian Journal of Animal Sciences* **18**(1): 22–26.

Zhang, Z., Wong, H. H., Albertson, P. L., Doherty, W. O. and O'Hara, I. M. (2013). Laboratory and pilot scale pretreatment of sugarcane bagasse by acidified aqueous glycerol solutions. *Bioresource Technology* **138**: 14–21.

PART III
Systems and sustainability

CHAPTER 12

Integrated first- and second-generation processes for bioethanol production from sugarcane

Marina O. de Souza Dias[1,2], Otávio Cavalett[2], Rubens M. Filho[2,3] and Antonio Bonomi[2,3]

[1] *Instituto de Ciência e Tecnologia (ICT), Universidade Federal de São Paulo (UNIFESP), São Paulo, Brazil*

[2] *Laboratório Nacional de Ciência e Tecnologia do Bioetanol (CTBE), Centro Nacional de Pesquisa em Energia e Materiais (CNPEM), Campinas, Brazil*

[3] *Faculdade de Engenharia Química, Universidade Estadual de Campinas (FEQ/UNICAMP), Campinas, Brazil*

12.1 Introduction

Efficient conversion of biomass into energy resources remains one of the greatest challenges faced by humanity in the search for a sustainable energy future. By the end of this century, the global population is expected to reach 10 billion people, which along with improvements in standard of living requires an increase in food and energy supply (Azadi *et al.* 2013). This is even more critical when considering that the human development index (HDI) appears to be directly related with energy consumption (Dale and Ong 2012), and fossil sources of energy are becoming more expensive.

Bioethanol is one of the most promising biofuels (Cardona and Sánchez 2007; Dantas *et al.* 2013; Rabelo *et al.* 2011), currently produced from first-generation feedstocks such as sugarcane and corn, but may also be produced from lignocellulosic feedstocks such as sugarcane bagasse, straw, forest residues, and other agricultural residues that are abundant, cheap, and not used as a primary food source. Efficient technologies for the production of lignocellulosic (or second-generation) ethanol, however, are still under development (Barakat *et al.* 2014; Yu *et al.* 2013) and challenges concerning both its technical and economic feasibility remain yet to be solved. Even though the use of lignocellulosic feedstocks for ethanol production may present advantages in terms of feedstock availability, issues such as capital costs and enzyme price still hinder second-generation ethanol economic feasibility (Stephen *et al.* 2012).

Sugarcane-Based Biofuels and Bioproducts, First Edition. Edited by Ian M. O'Hara and Sagadevan G. Mundree.

In the search for an optimized process configuration, different authors have shown that the integration of first- and second-generation ethanol production processes can be more economical and efficient than stand-alone second-generation processes (Čuček *et al.* 2011, Dias *et al.* 2012a, Palacios-Bereche *et al.* 2013). As a result, integrated first- and second-generation ethanol production can improve the feasibility of lignocellulosic ethanol and its industrial implementation.

Sugarcane is currently the most efficient feedstock for ethanol production, with reported net energy ratios much higher than ethanol from corn (Srirangan *et al.* 2012), which is the other major feedstock for ethanol production globally. As a result of first-generation ethanol production from sugarcane juice, large amounts of sugarcane bagasse and straw (sugarcane tops and leaves) are produced as byproducts of the agricultural and industrial processes. These byproducts are lignocellulosic materials, which are currently discarded or used as fuel to supply thermal and electric energy for the process, but may be converted to ethanol and other biofuels through second-generation processes. In fact, process improvements and the use of more efficient high-pressure boilers may lead to simultaneous production of electricity and surplus bagasse for second-generation processes.

In many countries, sugarcane straw is usually burnt before harvest, but since sugarcane burning is being eliminated in countries such as Brazil (Seabra *et al.* 2010), the amount of sugarcane straw available for use as an energy source or as feedstock for second-generation processes will be potentially larger.

Production of second-generation ethanol integrated in a first-generation sugarcane biorefinery possesses several advantages over a stand-alone second-generation plant, such as the ready availability of lignocellulosic feedstock (sugarcane bagasse) at the factory site and the possibility of sharing infrastructure (Dias *et al.* 2009). However, sugarcane bagasse is currently used as fuel in cogeneration systems for production of steam and electricity, so second-generation ethanol production from sugarcane bagasse may compete with its current use as a solid fuel (Dias *et al.* 2011a), unless some changes in the process are carried out in order to increase the amount of surplus bagasse available. These changes can be advantageous for both first- and second-generation ethanol production from sugarcane (decreasing losses and energy consumption, for instance), even though they may require additional investment. Process improvements in first-generation ethanol production can lead to significant gains in overall integrated first- and second-generation ethanol economics, as shown by Dias *et al.* (2012b).

In addition to these process improvements, integration of first- and second-generation ethanol production in a sugarcane biorefinery may also contribute to greater diversification of biorefinery revenues. In integrated processes, sugarcane bagasse and straw may either be used as a solid fuel for cogeneration or as feedstock for ethanol production, if the plant has some flexibility of operation. Thus, if electricity prices are favorable, such as in the case of an electricity shortage, the plant may divert more lignocellulosic material to the boilers, while if ethanol prices are advantageous, all surplus lignocellulosic materials can be used as feedstocks for ethanol production (Dias *et al.* 2013a).

Different processes may be used to convert lignocellulosic feedstocks into second-generation biofuels. The main routes are based on biochemical (pretreatment and hydrolysis) and thermochemical processes (pyrolysis or gasification followed by synthesis). Each route has its advantages when integrated with first-generation processes. In the biochemical route, hydrolyzed liquor may be fermented in a mixture with sugarcane juice, thus decreasing the eventual inhibitory effects of compounds generated during pretreatment on the fermenting microorganism (Dias *et al.* 2012b). The thermochemical process, on the other hand, offers large potential for thermal integration and a consequent decrease in energy consumption, since operations are carried out under high temperatures (Čuček *et al.* 2011).

In this chapter, the integrated production of first- and second-generation ethanol from sugarcane, including its technical, economic, and environmental aspects are discussed. The biochemical route for second-generation ethanol production, comprising of steam explosion pretreatment and enzymatic hydrolysis, is considered as an example.

12.2 Process descriptions

In this section, a brief description of both first- and second-generation ethanol production processes from sugarcane is provided, highlighting the features that these processes should present in order for their integration to become feasible. Cogeneration is discussed in a separate section. A scheme of an integrated first- and second-generation ethanol production process is illustrated in Figure 12.1.

12.2.1 First-generation ethanol production

Ethanol production from sugarcane in a first-generation autonomous distillery, in which all sugarcane juice is used for ethanol production, is essentially comprised of the following main steps: sugarcane reception, cleaning, and preparation; juice extraction, treatment, concentration, and fermentation; distillation; dehydration; and cogeneration. During juice extraction, sugarcane bagasse is separated from the juice and sent to the cogeneration boiler, where it is burnt for the production of steam, supplying the thermal energy for the process and generating electricity in steam turbines. If sugarcane straw is collected from the fields, this is also used as fuel in the boilers.

In order to integrate first- and second-generation ethanol production, steam consumption must be reduced, so that more lignocellulosic material is available

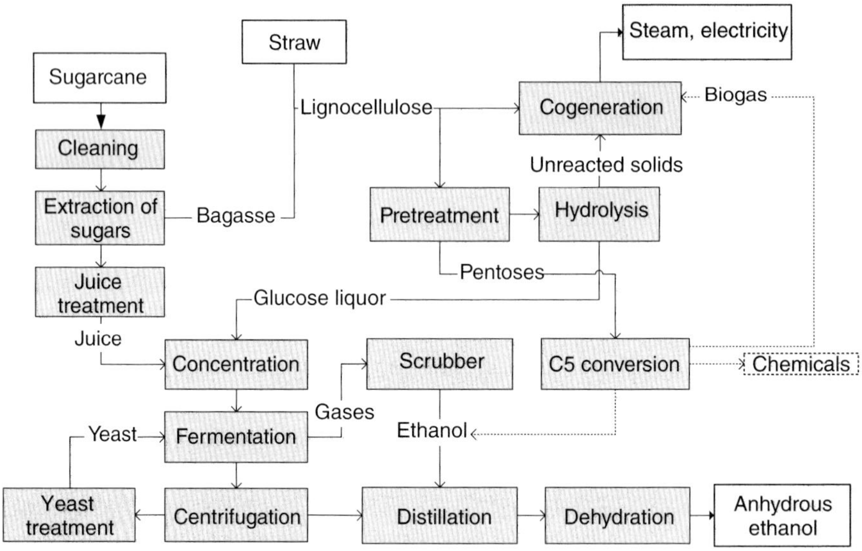

Figure 12.1 Simplified block flow diagram of the integrated first- and second-generation ethanol production process.

for use as an ethanol production feedstock. Several process improvements can be implemented in the first-generation ethanol production process in order to decrease its energy consumption and the most relevant are explained later.

During sugarcane preparation and juice extraction, large amounts of energy are required to drive shredders and mills. These items are the most common equipment used for juice extraction in Brazil, which is the largest sugarcane producer in the world. Both shredders and mills are usually driven by high-pressure steam but electric drives on these items offer greater overall steam efficiency and tend to be used in new or retrofitted mills in order to reduce process steam consumption (Ensinas *et al.* 2007). An alternative to the use of mills for sugarcane juice extraction, the use of diffusers, offers the possibility to decrease steam consumption, since the amount of mechanical energy required to operate diffusers is significantly smaller than the energy required for mills (10 kWh/t of sugarcane for diffusers compared with 15 kWh/t of sugarcane for mills, according to Pellegrini and Oliveira Júnior (2011)).

Juice treatment is comprised of the addition of chemicals (phosphoric acid, lime, and flocculant polymer), air removal, settlement, filtration of mud, and screening of clarified juice, to remove impurities. The clarified juice undergoes concentration before fermentation, in order to produce a juice with adequate sugar content. Most autonomous distilleries in Brazil use single-stage evaporation for juice concentration, but steam consumption may be significantly reduced if multiple effect evaporators, such as the ones used in sugar production, are employed (Dias *et al.* 2012b).

After concentration, juice is sent to fermentation, which is usually undertaken through fed-batch (Melle-Boinot) or continuous processes. These processes take place at temperatures around 30–34 °C and produce wine with relatively low ethanol content (up to 10 °GL) due to yeast inhibition related to substrate, product, and temperature. In order to produce fuel ethanol, the wine must be purified to at least 95 °GL; thus, the low ethanol content of the wine requires a significant input of energy in the subsequent purification step. Alternative fermentation processes such as low temperature or extractive fermentation allow production of wine with higher ethanol contents and consequently lead to lower energy consumption in distillation (Dias *et al.* 2012c). Product recovery through the use of membranes can also be used to produce high concentration wines, but technological issues, such as membrane fouling remain to be solved (Abels *et al.* 2013).

Because water and ethanol form an azeotrope with ethanol concentration of around 95.6 °GL under atmospheric pressures, wine is usually purified in a series of distillation and rectification columns followed by a dehydration process. In Brazil, the most common dehydration processes are azeotropic distillation with cyclohexane, extractive distillation with monoethyleneglycol, and adsorption using molecular sieves.

The conventional distillation system used in Brazil has relatively high-energy consumption; however, operation of the distillation columns under different pressure levels (double-effect distillation) allows thermal integration between columns reboilers and condensers and can provide a significant reduction in steam consumption of this system (Dias *et al.* 2011b).

Better technologies for dehydration systems can also promote important reduction in steam consumption if efficient processes are employed. For example, pervaporation systems have been developed and promote a considerable reduction in process steam consumption, but have not yet been widely implemented in the industry due to the high cost of membranes and the difficulties associated with fouling and long-term performance. Recent developments have suggested that pervaporation systems may constitute the entire purification process from fermented wine to anhydrous ethanol, eliminating energy-intensive distillation processes (Abels *et al.* 2013).

12.2.2 Second-generation ethanol production

As previously mentioned, the focus of this section is on the aspects of the second-generation ethanol production process required to provide a feasible integration to first-generation ethanol production from sugarcane. The biochemical process comprised of lignocellulosic pretreatment and hydrolysis is discussed, although there are opportunities for the thermochemical route to integrate with first-generation process also.

Surplus sugarcane bagasse and straw (the amount exceeding the quantity of bagasse required for cogeneration to supply the thermal and electrical energy demand of the process, both first and second generations) is used as the feedstock in the second-generation process. In the biochemical route, lignocellulosic feedstock is pretreated and hydrolyzed to produce fermentable sugars. Among several process options, pretreatment based on steam explosion has received a lot of attention. In this process, the lignocellulosic feedstock is subjected to steam treatment, in which the hemicellulose is converted to pentoses (C5 sugars). The process may or may not be catalyzed by the addition of acids, but in both cases steam is required, and cellulose losses may be significant.

A solid–liquid separation may be used to separate the solids (mainly cellulose and lignin) and produce a pentose-rich liquor stream, which can be used for a wide variety of applications such as fermentation to ethanol, biogas, or production of chemicals (Dias *et al.* 2013b, Mariano *et al.* 2013, Rabelo *et al.* 2011). Pentoses fermentation to ethanol cannot be achieved using the same microorganism (*Saccharomyces cerevisiae*) employed in conventional alcoholic fermentation of C6 sugars such as glucose. Engineered microorganisms are being developed to promote simultaneous fermentation of C5 and C6 sugars, but their future application in industrial processes in sugarcane biorefineries is currently uncertain as the need for a sterile environment and relatively long fermentation time, among other technical risks, may hinder their application. Assuming pentoses fermentation to ethanol is viable, the energy demand of the process with pentoses fermentation will be relatively higher than hexoses fermentation alone due to fermentation of liquor with lower sugar concentrations and greater energy requirements for the subsequent ethanol purification stage. In addition, the amount of lignocellulosic material available for use as feedstock is significantly smaller when pentoses fermentation to ethanol takes place compared with pentoses biodigestion (Dias *et al.* 2013b). Biogas production has the advantage of producing an energy source for combustion (methane) that can further reduce the bagasse required for process energy, resulting in a greater yield of ethanol from glucose, which is much more readily fermented to ethanol. The use of pentoses for chemicals production, on the other hand, may increase the revenues of the process through diversification of the production portfolio with higher value products. Production of butanol from pentoses in the integrated first- and second-generation ethanol production process was shown to be more profitable than production of biogas, using current technology (Mariano *et al.* 2013).

Other pretreatment processes such as liquid hot water, AFEX, and dilute-acid processes can also be considered for biochemical ethanol production. In the integrated process, energy consumption is one of the most important factors, since it impacts the amount of feedstock available for second-generation processes. Further details about the range of pretreatment processes are widely available in the literature (Alvira *et al.* 2010, Balat *et al.* 2008, Mora-Pale *et al.* 2011, Ruiz *et al.* 2013, Sathitsuksanoh *et al.* 2013, Tao *et al.* 2011, Zhao *et al.* 2012).

The solids obtained after steam explosion pretreatment contain unreacted hemicellulose, cellulose, and lignin. Even though the steam explosion pretreatment process may promote some delignification (Balat *et al.* 2008), an additional delignification step may be carried out, to further remove lignin and increase cellulose accessibility in the subsequent enzymatic hydrolysis step. This removal of lignin eliminates issues associated with the adsorption of cellulases onto lignin (Zhao *et al.* 2012).

Alkaline delignification and other delignification processes subject cellulignin to the action of solvents and heat (e.g., 10% NaOH (m/V), 100 °C, 1 h), promoting lignin solubilization and its consequent separation in a filter or another solid–liquid separation operation. While increasing cellulose accessibility to enzymes, this operation can promote a significant increase in process energy consumption, thus decreasing the amount of feedstock available for second-generation processes. Under specific process conditions, alkaline delignification after pretreatment did not increase overall ethanol production in the integrated first- and second-generation ethanol production process with pretreatment by steam explosion because of the increase in energy demand (Dias *et al.* 2013c).

The next step in second-generation processes is hydrolysis or saccharification, in which cellulose is converted to glucose under the action of enzymes or acids. Research has focused on the use of enzymes, since acid hydrolysis requires corrosion-resistant equipment and may lead to cellulose and hemicellulose degradation to hydroxymethylfurfural and furfural, respectively. These components are fermentation inhibitors and may also be produced during pretreatment, depending on process conditions (Andrade *et al.* 2013).

Enzymatic hydrolysis in the integrated first- and second-generation process should be balanced in terms of yield and reaction duration. Long reaction times require relatively large equipment, thus increasing capital costs. Since unreacted solids are used as fuels in the cogeneration system, it may not be economical to use extremely long reaction times to achieve minor increases in yields (Giese *et al.* 2013, Pereira *et al.* 2011, Wu *et al.* 2011). In addition, most hydrolysis studies are carried out using low solids loading (Giese *et al.* 2013, Silva *et al.* 2011), because hydrolysis yield and productivity are strongly affected by the solids content in hydrolysis reactors (Lau and Dale 2009). Despite leading to higher yields, the use of low solids loadings requires more water for dilution, generates more wastes, and requires more energy for concentration and purification, since it produces low ethanol titers. Low solids loading, therefore, increases production costs (Caspeta *et al.* 2014) and, most importantly, reduces the internal rate of return of the plant. In the integrated process, increased energy consumption leads to reduced amounts of lignocellulosic material available for product production. A comparison between hydrolysis at 5–20 wt% in the integrated first- and second-generation ethanol production process from sugarcane showed that the lowest overall ethanol production was obtained from hydrolysis with low solids loadings, even though this led to higher glucan yields (Dias *et al.* 2013c).

Higher enzyme loadings also promote an increase in hydrolysis yields but enzyme costs are one of the main constraints for second-generation ethanol feasibility. In the integrated process, hydrolysis residues are used as fuel, thus total carbohydrate conversion is not necessarily required and lower enzyme loadings may be used in the hydrolysis reactor.

After cellulose hydrolysis, glucose fermentation to ethanol takes place. In the integrated process, fermentation can be carried in a mixture with sugarcane juice, that is, in the same reactors used for first-generation ethanol production using *S. cerevisiae*. Andrade *et al.* (2013) showed that high yields (98–100% of the theoretical) could be obtained for the fermentation of sugarcane bagasse hydrolysate (pretreated with hydrogen peroxide) in a mixture with sugarcane molasses, even in the presence of inhibitors generated during pretreatment such as acetic acid. The authors showed that lower yields (approximately 75%) were obtained for the fermentation of pure hydrolysate due to the large amount of fermentation inhibitors, showing that integrating first- and second-generation allows a more efficient fermentation process.

After fermentation, ethanol produced in the integrated process is purified in the set of rectification and distillation columns and dehydrated using the same processes employed in first-generation ethanol production.

12.2.3 Cogeneration in integrated first- and second-generation ethanol production from sugarcane

The cogeneration in a sugarcane-processing facility is based on burning sugarcane bagasse and straw to produce steam and electricity to supply the energy demand of the process. In the integrated first- and second-generation production process, since the feedstock for combustion is also the feedstock for second-generation processes, the cogeneration system can play a significant role in the assessment of the sustainability indicators for the facility, since it is directly related to the production of second-generation ethanol and electricity.

Modern, efficient boilers must be used in these integrated processes, since they require less fuel (bagasse) to produce the same amount of steam, when compared with older boilers such as those used in most of the existing sugarcane facilities in Brazil. These facilities were mostly built when electricity could not be sold to the grid, so the cogeneration system was designed to ensure disposal of sugarcane bagasse guaranteeing only energetic self-sufficiency. Only since the early 2000s have greenfield mills been designed to produce large quantities of surplus electricity for sale to the grid (Leal *et al.* 2013a).

Low pressure (e.g., 22 bar) boilers cost less and require less bagasse to produce steam than high-pressure (e.g., 90 bar) boilers. Thus, if there is no market for electricity, the mill may continue to use these low-pressure boilers and hence use surplus bagasse to produce more second-generation ethanol, at the expense of producing small electricity surpluses. An analysis of the integrated first- and second-generation ethanol production process in Brazil showed that processes with 22 bar boilers produce 5% more ethanol than ones with 82 bar boilers; however, even though the investment required is smaller, the revenues obtained with electricity sale in the mill with high-pressure boilers outweigh the smaller gains in ethanol production, since the difference in electricity production is more than 300% higher (Dias *et al.* 2013b).

The cogeneration system in the integrated first- and second-generation ethanol production process is comprised of boiler and back pressure steam turbines. The minimum amount of sugarcane bagasse and straw that is required to be used as fuel in the cogeneration system is the one required to generate the thermal energy to supply the process. Condensing steam turbines allow combustion of all the lignocellulosic material available, producing more surplus electricity at the expense of burning more bagasse and straw. If the production process is to have some flexibility to divert lignocellulosic material either for electricity or second-generation ethanol production, the cogeneration system must have condensing steam turbines (Dias *et al.* 2013a). Results for an analysis of the Brazilian flexible plant with first- and second-generation ethanol production showed that the flexible plant has better economic indicators such as internal rate of return, but is more sensitive to changes in ethanol price than in electricity price.

12.2.4 Some aspects of the process integration

The amount of lignocellulosic material (sugarcane bagasse and straw) available for use as a feedstock in second-generation ethanol production depends on the energy consumption of the integrated process, both first and second generations. In addition, residues (unreacted cellulose and lignin) from the second-generation process may be used as solid fuels for combustion, thus displacing more lignocellulosic feedstock for second-generation ethanol production. Therefore, the exact amount of lignocellulosic feedstock available for use in second generation is a function of the energy consumption of both first- and second-generation processes, the configuration of the second-generation process and the extent of pretreatment and hydrolysis reactions, which have an impact on the amount of residues that may be used as fuels.

The fate of pentoses obtained from the hemicellulosic fraction of sugarcane bagasse is important as well. Different authors have considered pentoses biodigestion to produce biogas, which can be an additional fuel for the process (Dias *et al.* 2013a, Rabelo *et al.* 2011). The configuration of the integrated process highlighting the different interactions between process steps is illustrated in Figure 12.2.

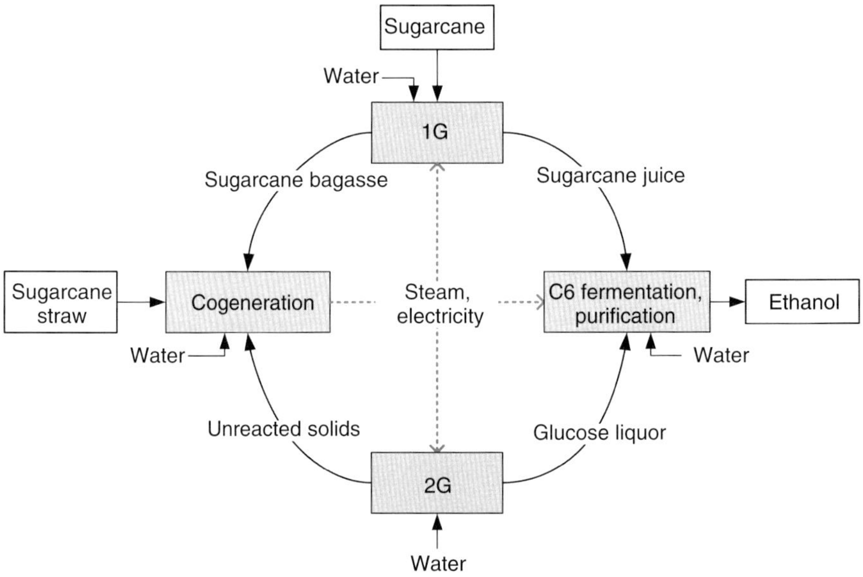

Figure 12.2 Scheme of the interaction among the main steps of the integrated first- and second-generation ethanol production process.

Water consumption is also a critical factor. As indicated in Figure 12.2, water is required in all process steps, especially in second-generation process, with hydrolysis and pretreatment requiring large amounts to produce streams with low solids loading. Since one of the main components of sugarcane is water, mass integration in integrated first- and second-generation ethanol production allows recovery and reuse of water from the sugarcane plant and this availability of process water may increase the feasibility of second-generation processes (Mosqueira-Salazar *et al.* 2013).

Another issue that requires consideration is the production of stillage (or vinasse, the residue obtained after ethanol purification) in the integrated process. Vinasse currently produced in first-generation ethanol production is used for fertigation of the sugarcane plantation (Cavalett *et al.* 2012), providing water and nutrients for the crop. Since the amount of vinasse produced is proportional (approximately 11 l vinasse/l anhydrous ethanol) to the amount of ethanol (Dias *et al.* 2012c), in the integrated process there will be more vinasse to be transported to the field, increasing transportation costs. In addition, vinasse produced in second-generation processes does not contain the necessary nutrients such as potassium (Rabelo *et al.* 2011), so besides being produced in larger quantities, the quality of the vinasse of the integrated process will be diminished.

The results of process simulation analyses of different process configurations for ethanol and electricity production in first- and second-generation processes are illustrated in Figure 12.3. 1G is the optimized first-generation

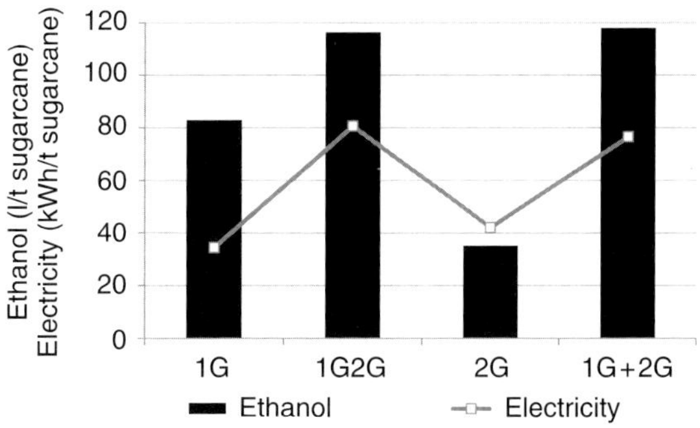

Figure 12.3 Results for ethanol and electricity production. 1G: optimized first-generation autonomous distillery producing surplus lignocellulosic material for the stand-alone second-generation plant; 1G2G: integrated first- and second-generation sugarcane biorefinery; 2G: stand-alone second-generation process; 1G + 2G: independent first- and second-generation ethanol from sugarcane.

autonomous distillery producing surplus lignocellulosic material for use as feedstock in a stand-alone second-generation plant; 1G2G is the integrated first- and second-generation sugarcane biorefinery, producing second-generation ethanol from surplus bagasse and trash through steam explosion and enzymatic hydrolysis followed by C5 and C6 fermentation; 2G represents the stand-alone second-generation process; 1G + 2G represents the results obtained for 1G and 2G, or the production of first- and second-generation ethanol from sugarcane in plants that are not integrated. Parameters adopted in the calculations are the same used by Dias *et al.* (2012a).

Results in Figure 12.3 show that the technical parameters (ethanol and electricity production) of the first- and second-generation ethanol production process from sugarcane are similar for an integrated (1G2G) and independent (1G + 2G) first- and second-generation processes, considering a first-generation (1G) plant supplying the feedstock for the stand-alone (2G) plant.

12.3 Economic aspects of first- and second-generation ethanol production

As previously mentioned, second-generation ethanol production processes are not currently considered to be economically feasible although early-stage commercialization of the technology is occurring. Some of the impediments to more widespread commercialization include capital expenditures and enzyme costs.

Capital costs may be high in equipment that must support high pressures or acid concentrations, such as in the pretreatment step, and in the large reactors required in hydrolysis due to low solids loading (usually up to 10%) and long reaction times (between 24 and 72 h). As mentioned previously, most of the hydrolysis reaction takes place within the first 24 h, and the incremental gains obtained from 24–48 h and 72 h hydrolysis times are much smaller (Banerjee *et al.* 2012, Giese *et al.* 2013). Since in integrated processes the unreacted material is ultimately burnt in the boilers, displacing additional lignocellulose for use as feedstock, there may be significant advantages to be gained from low hydrolysis reaction times for integrated processes.

Enzyme costs represent one of the most important components of the cost of second-generation ethanol production (Stephen *et al.* 2012). In spite of enzyme developments (enzyme activity and production cost), enzyme use in second-generation ethanol production can be optimized using low enzyme loadings, since it is not necessary to achieve a complete conversion of cellulose. This also avoids the requirement for much more effective and costly pretreatments, which can minimize issues associated with enzymes adsorption onto lignin. Nevertheless, all aspects of enzymatic hydrolysis must be considered in the complete first- and second-generation ethanol production in order to obtain the optimum process conditions.

On-site enzyme production has the advantage of allowing the production of enzyme mixtures suitable for the specific pretreatment and feedstock of the process (Delabona *et al.* 2012). In addition, on-site enzyme production has other advantages over external production, such as the possibility of using a fraction of the sugarcane juice or hydrolysate as sugar source, the production of a less purified enzyme mixture since there would be no need to transport it over long distances and potential to decrease costs, since purchasing commercial enzymes is a great economical constraint against second-generation ethanol feasibility (Sørensen *et al.* 2011).

While the second-generation feedstock in the sugarcane process is not expected to constitute a significant fraction of the costs, since it is already available at the sugarcane-processing site (in the case of sugarcane bagasse), transportation of sugarcane straw to the industrial plant using current technologies may hinder its use due to its low density. Different researchers have been investigating efficient means of straw collection and transportation from the field to the plant (Leal *et al.* 2013b, Macedo *et al.* 2001); however, it has been shown that the recovery of a fraction of straw is fundamental to the economics of integrated first- and second-generation ethanol production. Without the processing of straw, the amount of second-generation ethanol produced using only surplus bagasse as feedstock is greatly reduced (Dias *et al.* 2011a) and more focus needs to be given to enhancing the viability of straw recovery and transport.

In a similar manner to the technical results displayed in Figure 12.3, Figure 12.4 shows the results of an economic comparison of the integrated and independent first- and second-generation ethanol production from sugarcane, using data provided by Dias *et al.* (2012a) for a facility processing 2 million metric tons of sugarcane per year. Results show that the integrated process

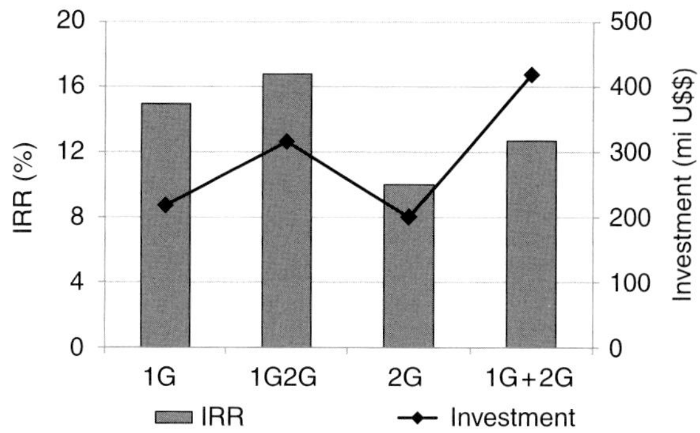

Figure 12.4 Internal rate of return (IRR) and investment required for first- and second-generation ethanol production from sugarcane. 1G: optimized first-generation autonomous distillery producing surplus lignocellulosic material for the stand-alone second-generation plant; 1G2G: integrated first- and second-generation sugarcane biorefinery; 2G: stand-alone second-generation process; 1G + 2G: independent first- and second-generation ethanol from sugarcane.

(1G2G) has a smaller investment cost and much higher internal rate of return than the independent (1G + 2G, comprised by 1G and stand-alone 2G) first- and second-generation ethanol production processes. Results were calculated for the Brazilian scenario, but are expected to apply more generally for cases where no subsides are included for either first- or second-generation ethanol production.

12.4 Environmental aspects of first- and second-generation ethanol production

In order to evaluate environmental aspects of sugarcane biorefineries, the life cycle assessment (LCA) methodology has been used as a key tool (Cavalett *et al.* 2012, Dias *et al.* 2012a, b, c, 2013a, b, Junqueira *et al.* 2012, Luo *et al.* 2009, 2010, Alvarenga *et al.* 2013, Ometto *et al.* 2009, Amores *et al.* 2013, González-García *et al.* 2010, 2012, Renouf *et al.* 2011, Melamu and von Blottnitz 2011). LCA is a recognized method for quantitative estimation of the environmental impacts of a product (good or service) during its entire life cycle, from extraction of raw materials through manufacturing, commercialization logistics, use, and final disposal or recycling. In LCA, substantially broader environmental aspects can be covered, ranging from greenhouse gases (GHG) emissions and fossil resource depletion to acidification, toxicity, water and land use aspects; hence, it is an appropriate tool for quantifying environmental impacts of a product system. The ISO 14040 series provides a technically rigorous framework for carrying out LCAs (ISO 2006a,b).

Regarding the possibilities of evaluating different environmental impact categories by applying different life cycle impacts assessment (LCIA) methods,

Cavalett *et al.* (2013) showed that the use of different LCIA methods can lead to different comparative environmental impacts of ethanol and gasoline, mainly when single-score indicators are applied. Results also showed that first-generation ethanol presents better environmental performance than gasoline in important categories such as global warming, fossil depletion, ecotoxicities, and ozone layer depletion but worse environmental performance than gasoline in the categories acidification, eutrophication, photochemical oxidation, and agricultural land use.

Similarly, González-García *et al.* (2010) showed that the production and use of ethanol derived from different lignocellulosic feedstock as a liquid fuel would reduce fossil fuels dependence and GHG emissions but would increase acidification, eutrophication, and photochemical smog, compared with production and use of gasoline as liquid fuel. Luo *et al.* (2009) also found similar results regarding the comparison of environmental impacts of ethanol and gasoline using the CML LCIA method.

Regarding the methods for evaluating global warming potential in the ethanol production in Brazil, Galdos *et al.* (2013) showed the importance of including black carbon emissions when calculating global warming and human health environmental indicators. This study quantitatively shows a technological trend, considering past, present, and future scenarios for ethanol production in Brazil, of decreasing environmental impacts. Preharvesting burning phase-out will significantly decrease emissions of black carbon and GHG in the sugarcane production phase. In addition, increases in yield of sugarcane per hectare, and of ethanol per ton of sugarcane, will decrease environmental impacts per unit of ethanol produced. Results showed that second-generation ethanol production plays a key role in increasing the amount of ethanol produced per unit of area. The study showed that the Brazilian sugarcane sector presents a trend of using resources more efficiently, as well as promoting good management practices that reduce its environmental impacts over time.

Impact allocation to multiple products produced from sugarcane is an important issue (Amores *et al.* 2013, Renouf *et al.* 2011). When coproducts are included (e.g., molasses, electricity, ethanol), economic allocation is less useful due to uncertainty in the mechanism for assigning impacts. To enable the consistent representation of impact across the full range of sugarcane products, mass allocation combined with energy allocation for the energy obtained from bagasse combustion and cogeneration is preferable (Amores *et al.* 2013). However, economic allocation is still the most used method for allocation of environmental impacts in the LCA of sugarcane biorefineries (Cavalett *et al.* 2012, Dias *et al.* 2012a, b, c, 2013a, b, Junqueira *et al.* 2012).

Several studies have evaluated environmental aspects of first-generation ethanol production using a life cycle approach. However, many of these studies use only certain aspects of the overall LCIA methodologies and focus only on specific impacts such as GHG emissions and energy balances (Macedo *et al.* 2008, Bordonal *et al.* 2012, Capaz *et al.* 2013).

The agricultural production is responsible for the majority of environmental impacts in the ethanol production system (Cavalett *et al.* 2012, Ometto *et al.* 2009, Amores *et al.* 2013). The influence of the sugarcane-processing system and sugarcane-growing variability suggests that representing sugarcane products on a generic or average basis should be avoided. Instead, results should be specific to the sugarcane-growing region and the processing methods used (Renouf *et al.* 2011, Amores *et al.* 2013).

Environmental impacts calculated using the LCA methodology indicate that optimization technologies such as high-pressure boilers, system integration for decreasing steam consumption, use of molecular sieves for ethanol dehydration, and use of straw for power generation, among others, have great potential to contribute to a significant decrease in the environmental impacts (around 25%) in comparison to present sugarcane biorefineries (Cavalett *et al.* 2012). Therefore, integrating second- and first-generation ethanol production process from sugarcane has the potential to reduce environmental impacts, since process improvements are required.

Environmental impacts calculated using the LCA methodology showed that current integrated first- and second-generation ethanol production scenario characterized by use of pentoses for biogas production and no alkaline delignification for lignin removal after steam explosion pretreatment in the second-generation hydrolysis step has the potential to decrease the environmental impacts in relation to the optimized first-generation ethanol production (Dias *et al.* 2012a).

Also, integrated first- and second-generation ethanol presented lower impacts than first generation for acidification and eutrophication. These impact categories are mainly related to fertilizer use in the agricultural stage. Since more ethanol is produced per unit of area on integrated scenarios, this process present better results for these categories (Dias *et al.* 2012a, b). Also, stand-alone second-generation ethanol production presents higher environmental impacts in comparison to the integrated first- and second-generation ethanol production process (per liter of ethanol produced) (Dias *et al.* 2012a) because lower yields (per ton of sugarcane) and more inputs are necessary for the production of second-generation ethanol in comparison to the first-generation ethanol. However, it is important to highlight that these results are strongly affected by the allocation criteria used for first-generation ethanol and bagasse (to be used in the stand-alone second-generation ethanol process). The results previously cited are obtained using economic allocation.

Melamu and von Blottnitz (2011) evaluated different second-generation ethanol production scenarios diverting bagasse to second-generation ethanol and replacing it with coal in sugarcane refineries. They concluded that diverting bagasse into second-generation ethanol in the absence of energy improvements on the sugarcane mill results in an increase in the GHG emissions. Similar increased impacts are observed for other impact categories such as nonrenewable energy use, aquatic eutrophication, and terrestrial acidification.

As previously mentioned, Dias *et al.* (2013a) evaluated a flexible biorefinery with the possibility of diverting a fraction of the lignocellulosic material (sugarcane bagasse and straw) either for electricity production or as a feedstock for second-generation ethanol production. Both fixed and flexible plants operating at maximum ethanol production lead to the highest avoided carbon dioxide emissions when usage phase and displacement of equivalent fossil products are considered (avoided emissions due to the products).

However, biorefineries producing more ethanol present higher environmental impacts per unit of ethanol produced than the configuration with maximum electricity production due the high impacts of inputs (sodium hydroxide, enzymes, and sulfuric acid, for instance) used in the second-generation process. Therefore, diverting lignocellulosic material to the cogeneration system presents lower environmental impacts than producing second-generation ethanol, even though emissions in the cogeneration system are higher. Luo *et al.* (2009) found similar results considering the global warming potential but the comparison did not include a flexible plant.

Another study from González-García *et al.* (2012) evaluating ethanol production and power generation systems involving *Salix* spp. biomass showed that ethanol production is the best choice from an energy perspective, whereas the electricity production seems to be a more suitable alternative when global warming is the decisive factor.

12.5 Final remarks

In this chapter, the integration of first- and second-generation ethanol production from sugarcane was discussed. The main aspects of the process were highlighted and some remarks on its economic and environmental indicators were provided. Integrated first- and second-generation processes present advantages over stand-alone second-generation ethanol production from sugarcane when considered from both a technical and economic perspective, while environmental performance depends on the methods and allocation criteria adopted in the analysis.

References

Abels, C., Carstensen, F., Wessling, M. (2013). Membrane processes in biorefinery applications. *Journal of Membrane Science* **444**:285–317.

Alvarenga, R.A.F., Dewulf, J., De Meester, S., Wathelet, A., Villers, J., Thommeret, R., Hruska, Z. (2013). Life cycle assessment of bioethanol-based PVC. Part 1: Attributional approach. *Biofuels, Bioproducts and Biorefining* **7**:386–395.

Alvira, P., Tomás-Pejó, E., Ballesteros, M., Negro, M. J. (2010). Pretreatment technologies for an efficient bioethanol production process based on enzymatic hydrolysis: A review. *Bioresource Technology* **101**:4851–4861.

Amores, M. J., Mele, F. D., Jiménez L., Castells, F. (2013). Life cycle assessment of fuel ethanol from sugarcane in Argentina. *The International Journal of Life Cycle Assessment* **18**:1344–1357.

Andrade, R. R., Rabelo, S. C., Maugeri Filho, F., Maciel Filho, R., Costa, A. C. (2013). Evaluation of the alcoholic fermentation kinetics of enzymatic hydrolysates from sugarcane bagasse (*Saccharum officinarum* L.). *Journal of Chemical Technology & Biotechnology* **88**:1049–1057.

Azadi, P., Inderwildi, O. R., Farnood, R., King, D. A. (2013). Liquid fuels, hydrogen and chemicals from lignin: A critical review. *Renewable and Sustainable Energy Reviews* **21**:506–523.

Balat, M., Balat, H., Öz, C. (2008). Progress in bioethanol processing. *Progress in Energy and Combustion Science* **34**:551–573.

Banerjee, G., Car, S., Liu, T., Williams, D. L., Meza, S. L., Walton, J. D., Hodge, D. B. (2012). Scale-up and integration of alkaline hydrogen peroxide pretreatment, enzymatic hydrolysis, and ethanolic fermentation. *Biotechnology and Bioengineering* **109**:922–931.

Barakat, A., Chuetor, S., Monlau, F., Solhy, A., Rouau, X. (2014). Eco-friendly dry chemo-mechanical pretreatments of lignocellulosic biomass: Impact on energy and yield of the enzymatic hydrolysis. *Applied Energy* **113**:97–105.

Bordonal, R. O., Figueiredo E. B., La Scala Jr, N. (2012). Greenhouse gas balance due to the conversion of sugarcane areas from burned to green harvest, considering other conservationist management practices. *GCB Bioenergy* **4**:846–858.

Capaz, R. S., Carvalho, V. S. B., Nogueira, L. A. H. (2013). Impact of mechanization and previous burning reduction on GHG emissions of sugarcane harvesting operations in Brazil. *Applied Energy* **102**:220–228.

Cardona, C. A., Sánchez, O. J. (2007). Fuel ethanol production: Process design trends and integration opportunities. *Bioresource Technology* **98**:2415–2457.

Caspeta, L., Caro-Bermúdez, M. A., Ponce-Noyola, T., Martinez, A. (2014). Enzymatic hydrolysis at high-solids loadings for the conversion of agave bagasse to fuel ethanol. *Applied Energy* **113**:277–286.

Cavalett, O., Junqueira, T. L., Dias, M. O. S., Jesus, C. D. F., Mantelatto, P. E., Cunha, M. P., Franco, H. C. J., Cardoso, T. F., Maciel Filho, R., Rossell, C. E. V., Bonomi, A. (2012). Environmental and economic assessment of sugarcane first generation biorefineries in Brazil. *Clean Technologies and Environmental Policy* **14**:399–410.

Cavalett, O., Chagas, M.F., Seabra, J.E.A., Bonomi, A., (2013). Comparative LCA of ethanol versus gasoline in Brazil using different LCIA methods. *The International Journal of Life Cycle Assessment* **18**:647–658.

Čuček, L., Martín, M., Grossmann, I.E., Kravanja, Z. (2011). Energy, water and process technologies integration for the simultaneous production of ethanol and food from the entire corn plant. *Computers and Chemical Engineering* **35**:1547–1557.

Dale, B. E., Ong, R. G. (2012). Energy, wealth, and human development: why and how biomass pretreatment research must improve. *Biotechnology Progress* **28**:893–898.

Dantas, G. A., Legey, L. F. L., Mazzone, A. (2013). Energy from sugarcane bagasse in Brazil: An assessment of the productivity and cost of different technological routes. *Renewable and Sustainable Energy Reviews* **21**:356–364.

Delabona, P. D. S., Farinas, C. S., da Silva, M. R., Azzoni, S. F., Pradella, J. G. C. (2012). Use of a new *Trichoderma harzianum* strain isolated from the Amazon rainforest with pretreated sugar cane bagasse for on-site cellulase production. *Bioresource Technology* **107**:517–21.

Dias, M. O. S., Ensinas, A. V., Nebra, S. A., Maciel Filho, R., Rossell, C. E. V., Wolf-Maciel, M. R. (2009). Production of bioethanol and other bio-based materials from sugarcane bagasse: Integration to conventional bioethanol production process. *Chemical Engineering Research and Design* **87**:1206–1216.

Dias, M. O. S., Cunha, M. P., Jesus, C. D. F., Rocha, G. J. M., Pradella, J. G. C., Rossell, C. E. V., Maciel Filho, R., Bonomi, A. (2011a). Second generation ethanol in Brazil: can it compete with electricity production? *Bioresource Technology* **102**:8964–8971.

Dias, M. O. S., Modesto, M., Ensinas, A. V., Nebra, S. A., Maciel Filho, R., Rossell, C. E. V. (2011b). Improving bioethanol production from sugarcane: evaluation of distillation, thermal integration and cogeneration systems. *Energy* **36**:3691–3703.

Dias, M. O. S., Junqueira, T. L., Cavalett, O., Cunha, M. P., Jesus, C. D. F., Rossell, C. E. V., Maciel Filho, R., Bonomi, A. (2012a). Integrated versus stand-alone second generation ethanol production from sugarcane bagasse and trash. *Bioresource Technology* **103**:152–161.

Dias, M. O. S., Junqueira, T. L., Jesus, C. D. F., Rossell, C. E. V., Maciel Filho, R., Bonomi, A. (2012b). Improving second generation ethanol production through optimization of first generation production process from sugarcane. *Energy* **43**:246–252.

Dias, M. O. S., Junqueira, T. L., Jesus, C. D. F., Rossell, C. E. V., Maciel Filho, R., Bonomi, A. (2012c). Improving bioethanol production – Comparison between extractive and low temperature fermentation. *Applied Energy* **98**:548–555.

Dias, M. O. S., Junqueira, T. L., Cavalett, O., Pavanello, L. G., Cunha, M. P., Jesus, C. D. F., Maciel Filho, R., Bonomi, A. (2013a). Biorefineries for the production of first and second generation ethanol and electricity from sugarcane. *Applied Energy* **109**:72–78.

Dias, M. O. S., Junqueira, T. L., Cavalett, O., Cunha, M. P., Jesus, C. D. F., Mantelatto, P. E., Rossell, C.E.V., Maciel Filho, R., Bonomi, A. (2013b). Cogeneration in integrated first and second generation ethanol from sugarcane. *Chemical Engineering Research and Design* **91**:1411–1417.

Dias, M. O. S., Junqueira, T. L., Rossell, C. E. V., Maciel Filho, R., Bonomi, A. (2013c). Evaluation of process configurations for second generation integrated with first generation bioethanol production from sugarcane. *Fuel Processing Technology* **109**:84–89.

Ensinas, A. V., Nebra, S. A., Lozano, M. A., Serra, L. (2007). Analysis of process steam demand reduction and electricity generation in sugar and ethanol production from sugarcane. *Energy Conversion and Management* **48**:2978–2987.

Galdos, M., Cavalett, O., Seabra, J. E. A., Nogueira, L. A. H., Bonomi, A. (2013). Trends in global warming and human health impacts related to Brazilian sugarcane ethanol production considering black carbon emissions. *Applied Energy* **104**:576–582.

Giese, E. C., Pierozzi, M., Dussán, K. J., Chandel, A. K., Silva, S. S. (2013). Enzymatic saccharification of acid-alkali pretreated sugarcane bagasse using commercial enzyme preparations. *Journal of Chemical Technology Biotechnology* **88**:1266–1272.

González-García, S. Iribarren, D., Susmozas, A., Dufour, J., Murphy R. J. (2012). Life cycle assessment of two alternative bioenergy systems involving Salix spp. biomass: Bioethanol production and power generation. *Applied Energy* **95**:111–122.

González-García, S., Moreira, M. T., Feijoo, G. (2010). Comparative environmental performance of lignocellulosic ethanol from different feedstocks. *Renewable and Sustainable Energy Reviews* **14**:2077–2085.

ISO. (2006a). *ISO International Standard 14040: Environmental management — Life cycle assessment — Principles and framework*, p. 28. International Organization for Standardization, Geneva, Switzerland.

ISO. (2006b). *ISO International Standard 14044: Environmental management — Life cycle assessment — Requirements and guidelines*, p. 54. International Organization for Standardization, Geneva, Switzerland.

Junqueira, T. L., Dias, M. O. S., Cavalett, O., Jesus, C.D.F., Cunha, M.P., Rossell, C. E. V., Maciel Filho, R., Bonomi, A. (2012). Economic and environmental assessment of integrated 1st and 2nd generation sugarcane bioethanol production evaluating different 2nd generation process alternatives. *Computer Aided Chemical Engineering* **30**:177–181.

Lau, M. W., Dale, B. E. (2009). Cellulosic ethanol production from AFEX-treated corn stover using *Saccharomyces cerevisiae* 424A(LNH-ST). *Proceedings of the National Academy of Sciences of the United States of America* **106**:1368–73.

Leal, M. R. L. V., Walter, A., Seabra, J. E. A. (2013a). Sugarcane as an energy source. *Biomass Conversion and Biorefinery* **3**:17–26.

Leal, M. R. L. V., Galdos, M. V., Scarpare, F. V., Seabra, J. E. A., Walter, A., Oliveira, C. O. F. (2013b). Sugarcane straw availability, quality, recovery and energy use: A literature review. *Biomass and Bioenergy* **53**:11–19.

Luo, L., van der Voet, E., Huppes, G. (2009). Allocation issues in LCA methodology: A case study of corn stover-based fuel ethanol. *The International Journal of Life Cycle Assessment* **14**:529–539.

Luo, L., van der Voet, E., Huppes, G. (2010). Biorefining of lignocellulosic feedstock – Technical, economic and environmental considerations. *Bioresource Technology* **101**:5023–5032.

Macedo, I. C., Leal, M., Hassuani, S. (2001). Sugar cane residues for power generation in the sugar/ethanol mills in Brazil. *Energy for Sustainable Development* **5**:77–82.

Macedo, I.C., Seabra, J.E.A., Silva, J.E.A.R. (2008). Greenhouse gases emissions in the production and use of ethanol from sugarcane in Brazil: The 2005/2006 averages and a prediction for 2020. *Biomass and Bioenergy* **32**:582–595.

Mariano, A. P., Dias, M. O. S., Junqueira, T. L., Cunha, M. P., Bonomi, A., Maciel Filho, R. (2013). Utilization of pentoses from sugarcane biomass: techno-economics of biogas vs. butanol production. *Bioresource Technology* **142**:390–399.

Melamu, R., von Blottnitz, H. (2011). 2nd Generation biofuels a sure bet? A life cycle assessment of how things could go wrong. *Journal of Cleaner Production* **19**:138–144.

Mora-Pale, M., Meli, L., Doherty, T. V., Linhardt, R. J., Dordick, J. S. (2011). Room temperature ionic liquids as emerging solvents for the pretreatment of lignocellulosic biomass. *Biotechnology and bioengineering* **108**:1229–1245.

Mosqueira-Salazar, K. J., Palacios-Bereche, R., Chávez-Rodríguez, M., Seabra, J., Nebra, S. A. (2013). Reduction of water consumption in an integrated first- and second-generation ethanol plant. *Energy for Sustainable Development*. doi:10.1016/j.esd.2013.08.001

Ometto, A.R., Hauschild, M.Z., Roma, W.N.L. (2009). Lifecycle assessment of fuel ethanol from sugarcane in Brazil. *The International Journal of Life Cycle Assessment* **14**:236–247.

Palacios-Bereche, R., Mosqueira-Salazar, K. J., Modesto, M., Ensinas, A. V., Nebra, S. A., Serra, L. M., Lozano, M.-A. (2013). Exergetic analysis of the integrated first- and second-generation ethanol production from sugarcane. *Energy*. doi:10.1016/j.energy.2013.05.010

Pellegrini, L. F., Oliveira Júnior, S. (2011). Combined production of sugar, ethanol and electricity: Thermoeconomic and environmental analysis and optimization. *Energy* **36**:3704–3715.

Pereira, L. T. C., Pereira, L. T. C., Teixeira, R. S. S., Bon, E. P. S., Freitas, S. P. (2011). Sugarcane bagasse enzymatic hydrolysis: rheological data as criteria for impeller selection. *Journal of Industrial Microbiology and Biotechnology* **38**:901–907.

Rabelo, S. C., Carrere, H., Maciel Filho, R., Costa, A. C. (2011). Production of bioethanol, methane and heat from sugarcane bagasse in a biorefinery concept. *Bioresource Technology* **102**:7887–7895.

Renouf, M. A., Pagan R. J., Wegener, M. K. (2011). Life cycle assessment of Australian sugarcane products with a focus on cane processing. *The International Journal of Life Cycle Assessment* **16**:125–137.

Ruiz, H. A., Rodríguez-Jasso, R. M., Fernandes, B. D., Vicente, A. A., Teixeira, J. A. (2013). Hydrothermal processing, as an alternative for upgrading agriculture residues and marine biomass according to the biorefinery concept: A review. *Renewable and Sustainable Energy Reviews* **21**:35–51.

Sathitsuksanoh, N., George, A., Zhang, Y.-H. P. (2013). New lignocellulose pretreatments using cellulose solvents: a review. *Journal of Chemical Technology and Biotechnology* **88**:169–180.

Seabra, J. E. A., Tao, L., Chum, H. L., Macedo, I. C. (2010). A techno-economic evaluation of the effects of centralized cellulosic ethanol and co-products refinery options with sugarcane mill clustering. *Biomass and Bioenergy* **34**:1065–1078.

Silva, V. F. N., Arruda, P. V., Felipe, M. G. A., Gonçalves, A. R., Rocha, G. J. M. (2011). Fermentation of cellulosic hydrolysates obtained by enzymatic saccharification of sugarcane bagasse pretreated by hydrothermal processing. *Journal of Industrial Microbiology and Biotechnology* **38**:809–817.

Sørensen, A., Teller, P. J., Lübeck, P. S., Ahring, B. K. (2011). Onsite enzyme production during bioethanol production from biomass: screening for suitable fungal strains. *Applied Biochemistry and Biotechnology* **164**:1058–1070.

Srirangan, K., Akawi, L., Moo-Young, M., Chou, C. P. (2012). Towards sustainable production of clean energy carriers from biomass resources. *Applied Energy* **100**:172–186.

Stephen, J. D., Mabee, W., Saddler, J. N. (2012). Will second-generation ethanol be able to compete with first-generation ethanol? Opportunities for cost reduction. *Biofuels, Bioproducts Biorefining* **6**:159–176.

Tao, L., Aden, A., Elander, R. T., Pallapolu, V. R., Lee, Y. Y., Garlock, R. J., et al. (2011). Process and technoeconomic analysis of leading pretreatment technologies for lignocellulosic ethanol production using switchgrass. *Bioresource Technology* **102**:11105–11114.

Wu, L., Li, Y., Arakane, M., Ike, M., Wada, M., Terajima, Y., Ishikawa, S., Tokuyasu, K. (2011). Efficient conversion of sugarcane stalks into ethanol employing low temperature alkali pretreatment method. *Bioresource Technology* **102**:11183–11188.

Yu, Q., Zhuang, X., Lv, S., He, M., Zhang, Y., Yuan, Z., Qi, W., Wang, Q., Wang, W., Tan, X. (2013). Liquid hot water pretreatment of sugarcane bagasse and its comparison with chemical pretreatment methods for the sugar recovery and structural changes. *Bioresource Technology* **129**:592–598.

Zhao, X., Zhang, L., Liu, D., Chemistry, A. (2012). Biomass recalcitrance. Part II: Fundamentals of different pre-treatments to increase the enzymatic digestibility of lignocellulose. *Biofuels, Bioproducts Biorefining* **6**:561–579.

CHAPTER 13

Greenhouse gas abatement from sugarcane bioenergy, biofuels, and biomaterials

Marguerite A. Renouf

School of Geography, Planning and Environmental Management, Faculty of Science, University of Queensland, St. Lucia, Brisbane, Australia

13.1 Introduction

The depletion of fossil fuel resources, and the greenhouse gas (GHG) emissions from their use, have led to the pursuit of replacements for fossil fuels. Society is looking to biomass resources as replacement feedstock for energy, fuels, and materials (Ragauskas *et al.* 2006). There has been interest since the 1990s in bioproduction from agricultural crops, including sugarcane (Paturau 1989, Manohar Rao 1997), for curbing fossil fuel depletion (Sims *et al.* 2006). However, whether bioproduction can also abate GHG emissions is something that has been investigated over the last decade and is the focus of this chapter.

We know that GHG abatement can usually be achieved using crop-derived feedstock, but the scale varies depending on the type of crop and the land transformation that takes place. It can range from being considerable to very marginal, and some scenarios can actually be net GHG emitters. We also know that even when GHG abatement is possible, there can be other environmental stresses from the expansion or intensification of agricultural production (Miller *et al.* 2007, Goldemberg *et al.* 2008, Dornburg *et al.* 2010). Hence, it is recognized that careful assessment is needed to ensure that bioproduction from agricultural crops does achieve real GHG abatement, and that it does not compromise other environmental values and sustainability objectives (IEA 2009, Solomon 2010). As research and development in this field is now quite mature, it is timely to take stock of what conclusions can be drawn about GHG abatement from sugarcane bioenergy, biofuels, and biomaterials.

Sugarcane has been recognized for some time as a producer of substrates for bioproduction (sugar and cellulose) (Alexander 1985, Paturau 1989, Manohar Rao 1997, Nonato *et al.* 2001). The sucrose extracted from the stalk is a highly suitable building block chemical for producing fuels, plastics, solvents, and

Sugarcane-Based Biofuels and Bioproducts, First Edition. Edited by Ian M. O'Hara and Sagadevan G. Mundree.

organic acids. The fibrous component, either that contained in harvest residues or in bagasse, can be used to generate energy (thermal or electrical) or as cellulosic feedstock for fuel and chemical production.

Sugarcane also has attributes that make it a very resource-efficient feedstock. Being a C_4 plant, it has intrinsically high light, water, and nitrogen use efficiency. It has a high harvest index meaning that all components of the plant can be recovered in a single harvesting step. The processing of cane also requires very few resource inputs, since the fiber (bagasse) and water recovered from the cane can usually meet all energy and water needs of the process. This means that energy and resource inputs over the life cycle are low relative to outputs. Hence, when sugarcane products substitute fossil fuel products they can in theory displace more GHGs than they emit, a potential that has been noted since the late 1990s (Macedo 1998, Rosa and Ribeiro 1998, Beeharry 2001).

This chapter reviews what we know about the GHG abatement (or carbon abatement) from sugarcane bioproducts (fuels, electricity, and materials). It draws on a decade of published research (2001–2012) that has used environmental life cycle assessment (LCA) to quantify the GHG emissions intensity (and other environmental indicators) of these products. The GHG profiles of sugarcane products are first described and compared with those for their biomass and fossil fuel counterparts to ascertain the GHG abatement potential. The pathways and opportunities for optimizing GHG abatement are then discussed.

The review focuses on GHG abatement because GHG-induced climate change has attracted considerable interest and has been an emphasis of past research. However, bioproduction has implications, positive and negative, for other environmental values such as resource depletion (land, water, and minerals) and protection of environmental services (water quality, air quality, and biodiversity). To avoid emphasis on a single issue and provide a balanced perspective, observations about these other environmental considerations are also touched upon.

It is worth noting that the significance of GHG emissions from sugarcane product systems may be less than other impacts (Renouf *et al.* 2010). Land use, water use, or water quality impacts are likely to be more important depending on the regional context. Sugarcane industries would generally not expect to be high-priority targets for GHG *mitigation*. Instead, interest in sugarcane bioproduction relates to the opportunities for GHG *abatement* through their displacement of GHG-intensive products. Therefore, the flavor of this chapter is that of opportunities for sugarcane industries to abate GHG emissions, but moderated by an appreciation of other environmental risks and trade-offs.

Most of the review presented here is based on dedicated bioproduction from sugarcane, where all of the cane is directed to bioproduction, as opposed to coproducing bioproducts alongside sugar. It is unlikely that existing sugarcane production will be diverted away from sugar production, and therefore dedicated bioproduction (with expanded sugarcane production) is the most likely path.

Bioproduction from by-products, alongside the core business of sugar production, may be a path for many existing sugar industries, and so it is also discussed briefly.

13.2 Life cycle assessment (LCA) of sugarcane systems

13.2.1 Overview of LCA and carbon footprinting

Environmental LCA is a holistic assessment technique that considers the environmental impacts over the entire life of a product or service (Pennington *et al.* 2004, Rebitzer *et al.* 2004). The life cycle of sugarcane products (Figure 13.1) starts with the transformation of land and the extraction of natural resources (coal, oil, natural gas, minerals, metal ores, water, etc.) to produce inputs to the various life cycle stages (fuels, electricity, agrochemicals, processing chemicals, packaging materials, transport, etc.). It then proceeds through the agricultural production of sugarcane and primary processing (milling) to produce substrates for bioproduction (sucrose, bagasse, molasses). Subsequent downstream stages depend on the product pathway, but will generally include some form of secondary processing (e.g., hydrolysis, fermentation, cogeneration), packaging, distribution, retail, consumer use, and then end-of-life disposal of product and wastes.

Undertaken to its full extent, LCA aims to estimate environmental impact across the entire life cycle of a product ("cradle to grave"), and generate indicators for a wide range of environmental impact categories. However, the principles of LCA can be also applied at reduced scopes to assess partial life cycles up to the farm or factory gate ("cradle to gate"), and for single-impact categories such as GHG emissions (carbon footprints) or water use impacts (water footprints). This review focuses on the GHG accounting application of LCA, and draws on LCA studies conducted at both "cradle-to-grave" and "cradle-to-gate" scopes.

A carbon footprint (CF) is an indicator of the global warming potential resulting from anthropogenic GHGs emitted to atmosphere, expressed as kilograms of carbon dioxide equivalent (kg $CO_{2\text{-e}}$) per unit of product or product function. While the name would suggest it is an account of carbon-based substances, it actually accounts for a wide range of GHGs, some of which are not carbon based. The most significant GHGs are carbon dioxide (CO_2), methane (CH_4), nitrous oxide (N_2O), and chlorofluorocarbons (CFCs) (IPCC). As for CO_2, only long-term emissions such as those released from fossil fuel combustion (fossil-CO_2) or land use change (LUC) are accounted for. Short-term releases, such as those from bagasse combustion, are not counted because they are assumed to be assimilated by subsequent biomass growth (IPCC 2000).

The carbon footprinting process involves the identification of all processes in the product's life cycle, estimation of GHGs emitted from each process, aggregation of emissions across the life cycle, and finally characterization of the global warming impact using recognized 100-year global warming potentials (28 kg CO_{2eq}/kg CH_4; 265 kgCO_{2eq}/kg N_2O) (IPCC 2013).

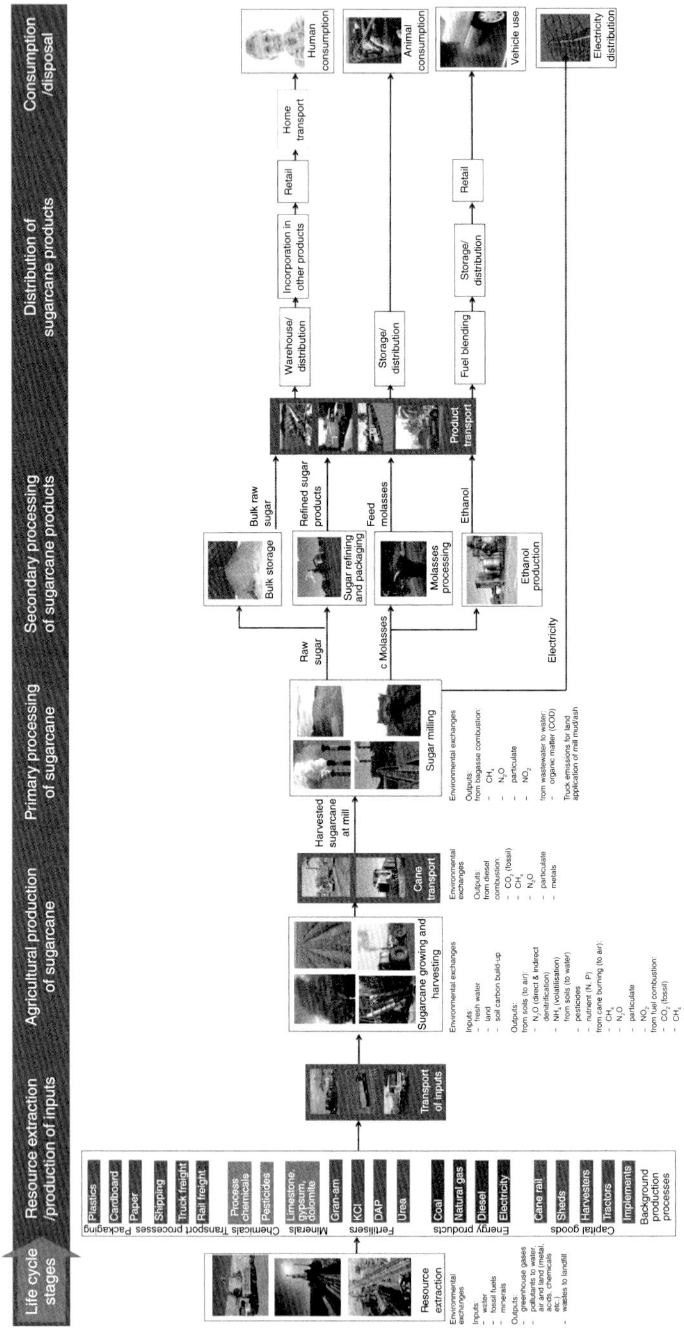

Figure 13.1 Life cycle of sugarcane products.

Carbon footprinting methods are well developed with international standards giving guidance for general application (ISO 2006a, 2006b, BSI 2008), more specific guidance on its application to sugarcane (Rein 2010) and biofuels (European Union 2009, RSB 2010b), and GHG calculators for verifying compliance with sustainability standards (Intelligent Energy Europe 2012, RSB 2012). The existence of standards suggests that it should be possible to compare carbon footprint results between studies and between products, confident that consistent methods have been applied. However, methods can still vary depending on the study objective making this difficult in practice. The aspects that vary include how the product system is modeled (attributionally or consequentially), the system boundaries (what is included), the functional unit (the reference product), and the method for apportioning impacts to multiple products (Box 13.1).

Box 13.1 Methodological considerations in LCA and carbon footprinting

A sugarcane product system can be modeled or represented using an *attributional or consequential approach*, depending on whether the study's aim relates to the attributes of a specific product system (*What is the CF of product X?, Does product X have a lower CF than product Y?*) or the implication of introducing alternative products to existing markets (*What are the GHG implications of producing product X from sugarcane vs. other feedstock?*). The former focuses on the existing attributes of the sugarcane product system, assumed to already be in place (the status quo). The latter considers changes that occur with the introduction of or change to a system, including how it influences other product systems.

Results can be expressed with different *functional units* depending on the question being asked – per unit of product or product function (t, MJ, L, kWh, km car travel), or per unit of production systems (t sugarcane processed or ha sugarcane).

Coproduction of multiple products is a feature of many bioproduction systems. Primary processing (milling) of sugarcane fractionates the harvested sugarcane stalk into its multiple components (sugar, bagasse, and molasses). Secondary processing (into fuels, energy, or materials) produces residues (dunder, biomass, steam), which can be important secondary products. The issue of how to apportion emissions to multiple products has been much-debated. An *allocation* approach allocates emissions based on physical or economic ratios. A *system expansion* approach assigns all emissions to the primary (determining) product and subtracts avoided emissions when coproducts substitute other products in the market. Therefore, the system is expanded to consider the wider implications of substitution effects in other production systems. This is argued to be a truer reflection of impacts (Weldema 2001). The international standard for LCA provides a decision hierarchy for selecting the most appropriate approach. However, the approach actually taken varies between studies, and has been shown to influence the results of biofuel studies (Luo *et al.* 2009b, Hoefnagels *et al.* 2010, Kaufman *et al.* 2010).

13.2.2 Past LCA and carbon footprint studies of sugarcane bioproducts

The existing body of LCA and carbon footprinting research related to sugarcane products (Table 13.1) is extensive enough to compile GHG profiles for a range of sugarcane products. Products that substitute fossil fuels and petrochemicals have

Table 13.1 LCA and carbon footprint studies of sugarcane products.

Product	Substrate	Purpose/theme	Region	References
Raw sugar	Cane juice	Hot spot identification	Mauritius	Ramjeawon (2004)
			Brazil	Seabra *et al.* (2011)
			Thailand	Yuttitham *et al.* (2011)
			Australia	Renouf *et al.* (2011)
		Alternative processes	Cuba	Contreras *et al.* (2009)
		Carbon footprint label	Brazil	Tate and Lyle (2011)
Electricity	Bagasse	GHG abatement	Mauritius	Beeharry (2001), Ramjeawon (2008)
			South Africa	Botha and von Blottnitz (2006)
			Brazil	Campbell and Block (2010)
			Australia	Renouf *et al.* (2011)
		Alternative processes	Thailand	Kiatkittipong *et al.* (2009)
			Cuba	Gil *et al.* (2013)
Ethanol	Cane juice	Hot spot identification	Brazil	Ometto *et al.* (2009), Lisboa et al. (2011)
		GHG abatement	Brazil	Macedo *et al.* (2008), Wang *et al.* (2012), Seabra *et al.* (2011), Campbell and Block (2010)
			Taiwan	Su and Lee (2009), Su and Tso (2011)
		Feedstock comparison	Brazil	Smeets *et al.* (2009), Rosa *et al.* (2009), Hoefnagels *et al.* (2010), de Vries *et al.* (2010)
		Alternative processes	Brazil	Luo *et al.* (2009a), Galdos *et al.* (2010), Souza *et al.* (2012), Acreche and Valeiro (2013)
			Mexico	Garcia *et al.* (2011)
			Argentina	Amores *et al.* (2013)
	Molasses	GHG abatement	Thailand	Nguyen and Gheewala (2008)
			Brazil	Gopal and Kammen (2009)
			Australia	Beer *et al.* (2002), Renouf, *et al.* (2011)
			Not specified	Nguyen and Hermansen (2012)
		Alternative processes	Argentina	Amores *et al.* (2013)
			Thailand	Silalertruksa and Gheewala (2011)
			Mexico	Garcia *et al.* (2011)
			Argentina	Acreche and Valeiro (2013)
	Bagasse	Alternative processes	South Africa	Botha and von Blottnitz (2006)
			Brazil	Luo *et al.* (2009a)
Methanol	Bagasse	GHG abatement	Brazil	Reno *et al.* (2011)
Plastics	Cane juice	GHG abatement	South Africa	PHB – Harding *et al.* (2007)
			Brazil	PHB – Pietrini *et al.* (2007)
				PE – Liptow and Tillman (2012), Kikuchi *et al.* (2013), Alvarenga and Dewulf (2013)
			Australia	PLA – Renouf *et al.* (2013)
			Thailand	PLA – Groot and Boren (2010)

been studied extensively, particularly those that are established or in advanced stages of development:

- fermentation of sucrose (in sugarcane juice or molasses) to ethanol (first-generation biofuel);
- hydrolysis/fermentation of bagasse to ethanol (second-generation lignocellulosic biofuel);
- combustion of bagasse with cogeneration of steam and electricity;
- fermentation of sucrose to organic compounds and subsequent polymerization to plastics-polylactic acid (PLA) and polyethylene (PE);
- microbial production of plastics e.g., poly-β-hydroxybutyric acid (PHB) polymers.

Most of the significant research findings to date have focused on biofuels (ethanol). However, the observations drawn from the ethanol work can also be relevant to the wider range of bioproducts. There have been fewer studies on less-established pathways such as the gasification, pyrolysis, or digestion of bagasse to syngas or biogas (Kiatkittipong *et al.* 2009, Kameyama *et al.* 2010, Reno *et al.* 2011). There is also a noticeable absence of studies on products that substitute non-fossil fuels products (paper, board, or furfural), and interestingly few studies examining sugar, the traditional product from sugarcane. The emphasis on fossil fuel substitutes shows that research has been driven by energy conservation and GHG abatement objectives.

Early research was exploratory aimed at identifying hotspots in the life cycle of existing sugarcane product systems. This led to comparison of the GHG intensities of sugarcane bioproducts with the fossil fuel products they displace, or those produced from competing biomass feedstock. More recent work has shifted away from the profiling to considering the optimization of production pathways for GHG abatement.

13.3 Greenhouse gas/carbon footprint profile of sugarcane bioproducts

The sources of GHG emissions over the life cycle of the sugarcane bioproducts are described here, highlighting the most influential ones. The land transformation and agricultural phases ('cradle to factory gate') are discussed at length, as these generally make the greatest contribution, or are the most significant variables. In contrast, less insight is available for the downstream stages of the life cycle ("factory gate to grave").

13.3.1 Land use change

If sugarcane bioproduction necessitates expanded agricultural production, there will be some form of land use change (LUC) from a prior use (or disuse) to intensive cropping (Volpi 2010), and there will commonly be associated GHG emissions.

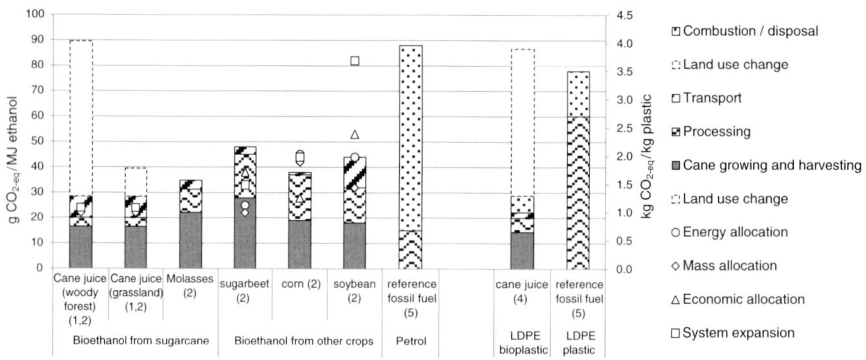

Figure 13.2 Indicative 'cradle to factory gate' greenhouse gas emissions for bioethanol and bioplastics relative to the reference fossil fuel, showing the influence of LUC and allocation method. Data derived from selected studies with "cradle-to-factory-gate" system boundaries, compatible functional units and sufficient transparency to present a contributional breakdown: (1) values for "cane growing, " "processing," and "transport" for ethanol from sugarcane juice are averages for a range of sugarcane growing regions, including Argentina (Amores *et al.* 2013), Mexico (Garcia *et al.* 2011), and Brazil (Hoefnagels *et al.* 2010). (2) Values for "land-use change" and variation due to alternative allocation approaches (energy, mass, and economic allocation, and system expansion) for ethanol derived from Hoefnagels *et al.* (2010) for Brazil. (3) Values for ethanol from molasses are averages across a range of sugarcane growing regions, including Argentina (Amores *et al.* 2013), Mexico (Garcia *et al.* 2011), and Australia (Renouf *et al.* 2011). (4) Values for low-density polyethylene (LDPE) plastic from sugarcane juice derived from Liptow and Tillman (2012).

Carbon dioxide (CO_2) is released when vegetation is cleared and when soil carbon stocks decline under intensive cropping (Anderson-Teixeira *et al.* 2009). We now appreciate the importance of natural landscapes in storing carbon (Fargione *et al.* 2008) and the need to account for releases of CO_2 from LUC in GHG assessments/carbon footprints (Gibbs *et al.* 2008). LUC can occur directly when vegetated land is cleared for crop production, or indirectly when an existing agricultural activity is displaced, and vegetation is cleared elsewhere to fill the gap in the supply of the displaced commodity.

LUC was not included in the scope of many early LCA/CF studies as their purposes were commonly focused on the attributes of existing sugarcane systems rather than the consequences of new or expanded production. However, it has been considered in more recent studies of sugarcane biofuels (Hoefnagels *et al.* 2010, Plassmann *et al.* 2010, Nguyen and Hermansen 2012, Wang *et al.* 2012) and bioplastics (Liptow and Tillman 2012), and found to greatly influence whether real reductions in GHG emissions can be achieved (see Figure 13.2).

13.3.2 Sugarcane production

The agricultural activities of growing and harvesting sugarcane are typically responsible for around half of the 'cradle to factory gate' GHG emissions from

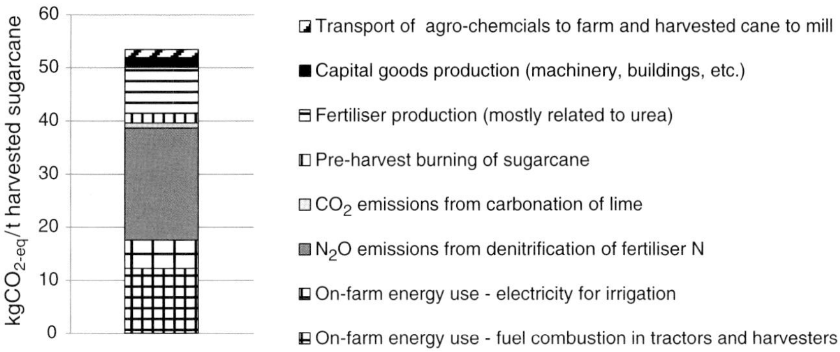

Figure 13.3 Indicative 'cradle to farm gate' greenhouse gas emissions for harvested sugarcane. This example represented a production system involving supplementary irrigation, mechanical cultivation and harvesting, with a portion of the crop being burnt before harvesting.

sugarcane bioproducts, excluding LUC (Figure 13.2). The most significant sources are on-farm energy use (fuel and electricity), fertilizer production, nitrous oxide (N_2O) from denitrification of applied *N*-fertilizers, and sugarcane burning where it occurs (Figure 13.3). Other sources, namely the production of other agrochemicals (pesticides, soil ameliorants), production of capital goods (machinery, farm buildings, sugarcane railways), carbonation of applied lime, and transport, are less significant. Environmental conditions (climate, soil type, topography) and production practices significantly influence emissions from the agricultural phase (Renouf *et al.* 2010), and so the GHG profile of sugarcane production will vary considerably between countries and between regions.

On-farm energy use is significant for highly mechanized farming systems, but less so for those that employ human labor. The need for supplementary irrigation also increases emissions considerably, particularly if high-pressure systems are used and if it is powered by emissions-intensive electricity (say from coal).

Cane-burning emissions (CH_4 and N_2O) have been noted as significant in regions where the crop is burnt at harvest, such as in Brazil (Galdos *et al.* 2010, Bordonal *et al.* 2012), and the cessation of burning has been flagged as a key mitigation strategy.

Nitrous oxide (N_2O) emitted from agricultural soils due to the denitrification of applied nitrogen has been identified as one of the most significant variables (Smeets *et al.* 2009, Renouf *et al.* 2010). The drivers for N_2O emissions are the presence of mineral nitrogen in the soil, high temperature, high soil water content (Allen *et al.* 2010), and also high soil organic content such as when harvest residues are retained as a mulch (Thorburn *et al.* 2010). So the hot, wet, tropical, and subtropical climates in which sugarcane is commonly grown are conducive to denitrification, and N_2O emission rates (N_2O-N emitted/N applied) are expected to be higher for sugarcane than for other crops (Box 13.2).

The rate of nitrogen application (or nitrogen use efficiency) is a key influencing factor and a primary target for N_2O mitigation (Dalal et al. 2003, Thorburn et al. 2010). Nitrogen use efficiency (i.e., the ratio of N taken up by the crop to the amount of N applied) has been estimated to range from 10% in less N-efficient countries to 60% in N-efficient countries (Lisboa et al. 2011, Robinson et al. 2011), with the average across sugarcane growing countries of around 47% (Robinson et al. 2011). This means that surplus N can accumulate in the soil and be prone to loss to the environment (to air as nitrous oxide and to water as nitrate).

While sugarcane is intrinsically an efficient user of N in theory (de Vries *et al.* 2010, Miller 2010), the tendency for generous application of N in some regions means its N-efficient status can be compromised (Robinson *et al.* 2011). The impacts of nitrogen use in agriculture generally (UNEP 2007) and for bioproduction (Crutzen *et al.* 2008, Erisman *et al.* 2010) have been acknowledged as important for both GHG mitigation and water quality.

Box 13.2 Nitrous oxide emission factors (N_2O EF)

LCA studies to date have commonly applied a default N_2O emission factor (EF) of 0.01 kgN_2O-N/kg applied N (1.0%) or similar, based on the IPCC Tier 1 approach for fertilizer-induced emission (FIE) (IPCC 2006). This EF is appropriate for global inventories, but does not consider the crop type and environmental conditions that influence N_2O emissions (Smeets *et al.* 2009).

Field measurements and modeling have found direct N_2O emissions from sugarcane soils to be in the order of 3–5% (Davidson *et al.* 1996, Allen *et al.* 2010, Thorburn *et al.* 2010, Lisboa *et al.* 2011). However, FIEs represent the direct emissions from a fertilized plot minus emissions from an unfertilized control plot (IPCC 2006). So for the purposes of deriving EF for use in carbon footprinting, these measured and modeled emissions estimates need to be considered against the reference situation.

Smeets *et al.* (2009) estimated N_2O emissions against various reference situations for different crops and geographical regions using a statistical estimation model. This work suggests that when unfertilized plots are the reference case EFs for sugarcane would be 2.0% for South Africa, 1.3% for South America, 1.8% for South Asia, and 2.3% for Australia. This means that carbon footprint estimated using a default value of around 1% may slightly under-report N_2O emissions.

13.3.3 Sugarcane biorefining

In most cases, the sugarcane biorefining stage contributes few GHG emissions to the carbon footprint of sugarcane products. This is because all or most of the energy requirements for milling and processing can be met by bagasse. Hence there is little or no fossil energy input, and there are also few material inputs. Sources of GHG emissions for this phase are commonly limited to methane and nitrous oxide released from bagasse combustion, and embodied impacts for producing inputs (processing chemicals, nutrients, lubricants, etc.). However, some biorefining products, such as aviation biofuels, can have high energy input, in which case the GHG emissions from this phase are significant (Cox *et al.* 2014).

In fact, this phase of the life cycle plays an important role in reducing the carbon footprint of sugarcane products because it transforms sugarcane components into the products that substitute emissions-intensive products from fossil fuels, which offsets emissions.

13.3.4 Downstream phases

The downstream phases of distribution, packaging, retail, use, and disposal have been less studied. The few studies of Brazilian ethanol that included transport to the end consumer (Smeets *et al.* 2009, Walter *et al.* 2011) found that this phase can have some significance. Land-based transport (in trucks) of ethanol to the port can add 5–15 $gCO_{2\text{-eq}}$/MJ fuel depending on the distance, but subsequent overseas shipping and distribution adds only around 2 $gCO_{2\text{-eq}}$/MJ fuel (Walter *et al.* 2011).

13.4 Greenhouse gas (GHG) abatement from sugarcane products

GHG abatement occurs when GHG-intensive products, such as those derived from or dependent on fossil fuels, are replaced with products with lower emissions intensity. We can judge the GHG abatement potential of sugarcane bioproducts by comparing their GHG emissions (carbon footprints) against these reference products. We can also compare the GHG abatement of sugarcane-derived bioproducts with those derived from other biomass feedstock.

13.4.1 Comparing sugarcane products with fossil fuel products

If comparisons do not consider land use change (LUC), then sugarcane-derived electricity, ethanol, and plastics are consistently shown to have lower carbon footprints than their fossil fuel equivalents (Figure 13.2). The main reason is avoided emission of fossil-CO_2 when bioproducts are combusted or decomposed. This means that substitution of fossil fuel products with sugarcane products will result in GHG abatement or "carbon offsets."

The scale of GHG savings/abatement expected from bioproducts is specified in standards, directives, and certification schemes. To date, these schemes have only covered biofuels but this is expanding to include biomaterials (RSB 2013). Renewable energy directives in the European Union, United Kingdom, United States, and Brazil govern the production or importation of biofuels in these regions (Desplechin 2010, Yeh and Sperling 2010), whereas the Roundtable for Sustainable Biofuels (RSB) provides certification globally (RSB 2010a). They all define both GHG abatement and land-use criteria for biofuels. The GHG accounting methods that support them are based on similar methodological principles but with some slight differences in assumptions (Hennecke *et al.* 2013).

GHG savings/abatements are expressed as the percentage (%) reduction in GHG emissions relative to the reference fossil fuel. For biofuels, the minimum GHG abatement (excluding LUC) stipulated by the RSB is 50% (RSB 2010a), and for the EU Directive it is currently 35% but will increase to 50% in 2017 (European Union 2009). For bioproducts, producers can make claims when GHG abatement is greater than 10% (RSB 2013). As can be seen in Figure 13.2, sugarcane ethanol and plastics meet these standards. If there are LUC-induced emissions, the carbon footprint can rise to be equivalent or greater than the fossil fuel equivalent. The standards deal with this by placing limitations on the type of land that can be converted.

The approach taken for assigning emissions to multiple products (Box 13.1) is known to influence carbon footprint results (Luo *et al.* 2009b, Hoefnagels *et al.* 2010, Kaufman *et al.* 2010, Renouf *et al.* 2011), as shown in Figure 13.2. However, this variable does not appear to compromise the favorable outcome for sugarcane bioproducts relative to fossil fuels.

Nitrous oxide (N_2O) emissions are a potentially influential variable, but uncertainty about N_2O emission factors (EF) (Box 13.2) means it is difficult to be sure of its significance (Erisman *et al.* 2010). When emissions of N_2O-N are assumed to be around 1% of applied N, which has been the case in many past studies, the carbon footprint of sugarcane ethanol can be up to 80–90% lower than petrol for Brazilian ethanol (Smeets *et al.* 2009). However, where it is higher (2.3% or more) this advantage falls to 70% (Smeets *et al.* 2009). So this variable, while important, does not appear to compromise the favorable status of sugarcane ethanol, and by inference other bioproducts. The exception may be in extreme cases where specific conditions such as extended water logging or acid sulfate soils lead to very high emissions (Denmead *et al.* 2010).

In summary, sugarcane bioproducts are robust GHG savers, meaning that abatement can be achieved irrespective of assessment methodology, the N_2O EF, sugarcane yield variability, transportation distances, and so on.

13.4.2 Influence of land-use change

The variable that does influence GHG abatement is land use change (LUC). Njakou Djomo and Ceulemans (2012) reviewed LUC-induced CO_2 fluxes reported in biofuel studies and determined it to range from −52 to 34 g CO_2/MJ for direct LUC and 0 to 327 g CO_2/MJ for indirect LUC. Direct LUC can result in emissions of CO_2 (+ve flux) when a cropping system displaces a carbon-rich ecosystem (forests, grasslands), but sequestration of CO_2 (−ve flux) when a perennial crop with relatively high carbon stocks displaces degraded cropland with low carbon stocks. Indirect LUC can have no (zero) net emissions when displaced activities move to marginal or degraded land, or large net emissions when relocated into newly opened up land, typically carbon-rich forests (Njakou Djomo and Ceulemans 2012). The outcome depends on the yield of the biomass feedstock (i.e., the amount of land converted), the type of land used or displaced (i.e., the carbon stocks displaced), and period over which the CO_2 flux is amortized (Fargione *et al.* 2008, Gibbs *et al.* 2008, Njakou Djomo and Ceulemans 2012).

So if LUC removes existing carbon stocks it can result in more CO_2 being emitted than is saved by the carbon offsets of bioproducts from each year's crop. This is referred to as the "carbon debt"(Fargione *et al.* 2008). In an extreme example of sugarcane expansion for biofuel production based on the conversion of Cerrado woodland in Brazil, Fargione *et al.* (2008) estimated that it would leave a carbon debt of 134 tCO_2/ha. With annual carbon offsets from displaced fossil fuel use of 9.8 $tCO_{2\text{-}e}$/ha/year, the carbon payback time would be 17 years. However, Walter *et al.* (2011) assessed that expansion of sugarcane in Brazil for ethanol had to that point mostly been at the expense of pasturelands and temporary crops, resulting in the net carbon sequestration. In another example, Liptow and Tillman (2012) estimated that indirect LUC can more than double the carbon footprint of bioplastics, when a worst-case scenario was assumed, making it comparable to that of fossil-based LDPE (Figure 13.2).

Gibbs *et al.* (2008) estimated the "ecosystem carbon payback time" (ECPT) (Fargione *et al.* 2008) for different biofuels LUC pathways based on tropical crops. The ECPT for first-generation sugarcane biofuel (from sugarcane juice) is less than 1 year when sugarcane is grown on degraded or existing cropland areas, between 5 and 9 years when grown on grassland (depending on yields and technology), and between 10 and 100 years when it replaces woody savannah, degraded forest or forest. Even if second-generation cellulosic technology were to double ethanol yields, more than three decades would still be required to replace lost forest carbon. Gibbs (*ibid*) concludes that no foreseeable technologies can make tropical deforestation a carbon beneficial enterprise.

In summary, sugarcane-derived energy, fuels, and materials are potentially very robust GHG savers, meaning that the savings are fairly resilient to most variables. However, LUC is the variable that determines whether real GHG abatement occurs. Abatement is only achievable when land with low-carbon stocks is used or displaced, such as degraded land.

13.4.3 Comparing sugarcane with other biomass feedstock

Since fossil fuel combustion is the main source of GHG emissions over the life cycle of bioproducts, fossil energy ratios (FER) (MJ out/MJ in) can be a

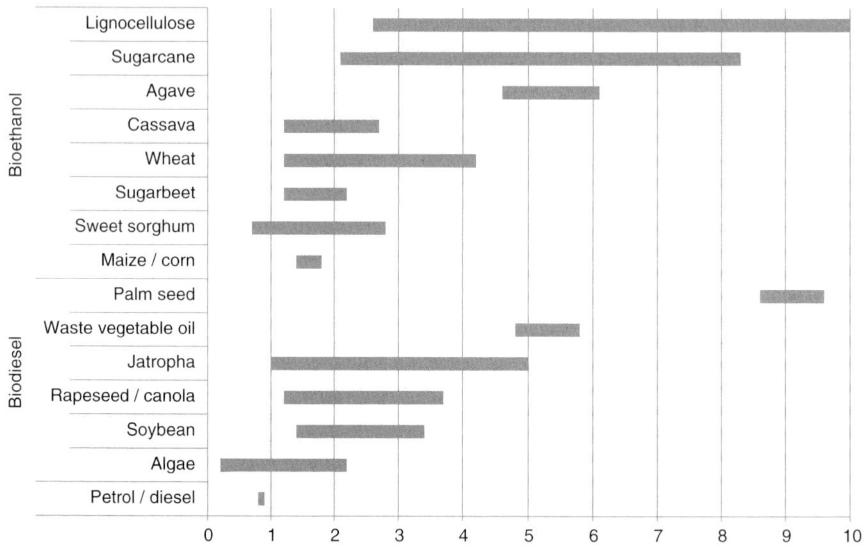

Figure 13.4 Fossil energy ratio (FER) of biofuels (MJ bioethanol output/MJ fossil energy input). Reproduced from FAO (2008) and updated.

reasonable proxy for GHG abatement. FERs are often reported fairly consistently in biofuel studies, and so are used here to compare the GHG abatement potential of sugarcane bioproducts with other biomass feedstocks (Figure 13.4). FER does not account for the fertilizer-induced N_2O emissions and CO_2 fluxes from LUC noted earlier, so an FER comparison only provides an initial screening tool.

Sugar crops (sugarcane, sugar beet, sweet sorghum, agave) have the potential for better FER than starch crops (maize, wheat, cassava, grain sorghum), because they produce a substrate (sucrose) that can be used as a direct input in biorefining with minimal energy, whereas starches require additional energy for prehydrolysis. Sugarcane has the combined advantage of very high sugar yields (10–17 t/ha) and energy self-sufficiency when bagasse is used as a fuel in processing. Sugar beet in comparison has lower sugar yields (7–9 t/ha) and processing requires fossil energy input. Sweet sorghum has similar characteristics to sugarcane but with sugar yields comparable to sugar beet (8–10 t/ha). However, it can have good FER when processed in sugar mills with energy recovery from its bagasse (Hoefnagels *et al.* 2010). Similarly, agave holds promise because of reasonable sugar yields (4–8 t/ha), low energy input for cultivation, and the potential for whole plant harvesting to provide harvest residues as fuel (Yan *et al.* 2011).

Of the conventional oil seed crops, palm seed has the highest FER due to its high oil yields (6–7 t/ha), very low energy input to cultivation as it is a perennial crop, and less energy for processing. Soybean and rapeseed/canola have a similar FER to the starch crops.

Much hope rests on second-generation feedstocks such as lignocellulose, algae, agave, and leguminous oil seed tree crops as they can avoid the undesirable competition with agricultural land for food production and fresh water resources. The FER of lignocellulosic ethanol from agricultural residues and perennial energy crops such as switchgrass and *Miscanthus* are very promising (Hoefnagels *et al.* 2010, Wang *et al.* 2012). Microalgae is promoted as a promising pathway due to its potential for high productivity, ability to be cultivated on unproductive lands, and potential to utilize carbon dioxide (CO_2) from industrial flue gas. However, the energy-intensive cultivation and harvesting means a relatively low FER (Sills *et al.* 2013, Zaimes and Khanna 2013), flagging the need for further developments in this technology before its promise can be realized. Legume oil-seed tree crops such as *Jatropha* and *Pongamia* are also promising due to their perennial nature and low nitrogen and water requirements (Prueksakorn *et al.* 2010, Cox *et al.* 2014).

Findings from studies that evaluate and compare GHG abatement from different feedstocks do not stray too far from this initial screening based on FER (de Vries *et al.* 2010, Hoefnagels *et al.* 2010). As the bulk of bioethanol is currently produced from sugarcane (in Brazil) or corn (in the United States), the GHG abatement and environmental impacts of these two pathways have been assessed and compared extensively. It is now well established that sugarcane-ethanol provides greater GHG abatement than corn-ethanol (Renouf *et al.* 2008, Rosa *et al.* 2009, Garoma *et al.* 2012), including when LUC is considered (Wang *et al.* 2012). Studies that have compared a wider range of first-generation biofuel crops rank sugarcane as having one of the best GHG saving benefits (Gibbs *et al.* 2008, de Vries *et al.* 2010, Hoefnagels *et al.* 2010, Bos *et al.* 2012).

When second-generation feedstocks are added to the mix, then lignocellulose generally outranks all other feedstocks (Hoefnagels *et al.* 2010). Sugarcane is a feedstock for both first-generation and second-generation bioproduction, so offers opportunities in both avenues. Under the European Union's directive on the promotion and use of energy from renewable sources (European Union 2009), typical and default GHG savings have been established for a range of biofuel feedstocks (Table 13.2), which nicely summarize the rankings.

When LUC is considered, the top-ranking feedstocks (lignocellulose, sugarcane, palm oil) can also have the lowest ecosystem carbon payback times if grown on land that already has low carbon stocks (Gibbs *et al.* 2008). However, even these crops would have extremely long payback times if their production displaces forests (Fargione *et al.* 2008, Gibbs *et al.* 2008).

In summary, sugarcane possesses attributes that make it one of the preferred agricultural feedstocks for bioproduction, and can be utilized for either first-generation or second-generation pathways. So if GHG abatement becomes

Table 13.2 Default life cycle GHG abatement for different feedstocks emissions from land-use change. GHG savings are expressed as the percentage (%) reduction in GHG emissions relative to the reference fossil fuel.

Biofuel		Feedstock	GHG abatement (%)
Bioethanol	Second generation	Wheat straw (lignocellulosic)	85
		Waste wood (lignocellulosic)	74
	First generation	Sugarcane	71
		Wheat (straw as fuel in CHP plant)	69
		Sugar beet	52
		Corn (natural gas as fuel in CHP plant)	49
		Wheat (natural gas as fuel in CHP plant)	34
Biodiesel	Second generation	Wood waste (Fischer–Tropsch)	95
		Farmed wood (Fischer–Tropsch)	93
	First generation	Waste vegetable or animal oil	83
		Palm oil (with methane capture at mill)	56
		Sunflower oil	51
		Canola/rapeseed oil	38
		Soybean oil	31
		Palm oil (process not specified)	19
		Soybean oil	31

Source: Extract from European Union (2009) Annex V.

a driver in agricultural policy there will be justification for producing sugarcane over other crops in tropical and subtropical regions.

13.4.4 Attributes for GHG abatement

The preceding analysis allows us to define the attributes of bioproduction that ensure and optimize GHG abatement.

In relation to the LUC pathway, preference is for:

- use of land with low initial carbon stocks such as degraded land, and that will not displace an existing activity to land with high carbon stocks;
- use of land free of conditions and soil types that are prone to high nitrous oxide (N_2O) emissions.

In relation to crop characteristics, preference is for:

- crops that can deliver good yields of substrate (i.e., sucrose, cellulose, oil, etc.) on the aforementioned marginal lands to minimize the energy inputs per unit of substrate;
- crops with perennial growing characteristics enabling multiple harvests so that cultivation energy inputs are reduced.

In relation to crop management, preference is for:

- low demand for nitrogen (good nitrogen use efficiency) to minimize the embodied emissions of fertilizer production and losses of N_2O;

- low demand for supplementary water supply that necessitates energy-intensive irrigation; and
- avoided burning of crop residues to eliminate unnecessary combustion-related GHG emissions.

In relation to the processing pathway, preference is for:

- recovery of multiple crop components (crop plus residues) that can be converted into multiple product streams, thereby maximizing product outputs per hectare;
- processing crop components into commodities that displace emissions-intensive products thereby maximizing carbon offsets; and
- maximizing recovery of processing residues from which energy can be recovered to enable energy self-sufficiency.

13.5 Environmental trade-offs

Miller (2012) generalizes two major consequences of bioproduction, which hold true for all production based on agriculture; changes in elemental cycles (carbon, nitrogen, and phosphorus), and the consumption of limited resources (land use, water use, fossil fuels). Up to this point, we have explored the fossil fuel conservation and GHG abatement implications. Here, we touch on the other consequences.

13.5.1 Land use and associated environmental services

Land is a limited resource, and its efficient utilization and protection of environmental services and biodiversity is a key sustainability priority for bioproduction. The amount of land occupied over the lifecycle of a production system (ha) has been used to date as a proxy for land use related impacts. Land-use demand for bioproduction systems is concentrated in the agricultural phase, and mostly hinges on the agricultural yield of substrate. So sugarcane, with its high yields of both sucrose and cellulose, is an efficient user of land compared with other crops. Only algae (theoretical potential for 0.3 ha/TJ raw energy) outranks sugarcane (2.3 ha/TJ), and both are considerably more efficient than corn (4.9 ha/TJ) and soybeans (20.2 ha/TJ), for instance (Miller 2010).

However, the quantity of land occupied alone does not capture the impacts of land occupation, which include degradation of environmental services and biodiversity (Wiens *et al.* 2011). These hinge more on the qualities of land sacrificed and the compensatory ecosystem services that can be provided by the feedstock crops. LCA has not been a useful tool for assessing these impacts to date. They are best assessed at regional or landscape scales using assessment frameworks that employ spatially linked vegetation and habitat characterization (Miyake *et al.* 2015). Broad-acre crops grown for bioproduction do not offer compensatory environmental services and biodiversity. The exceptions may be perennial tree crops such as *Pongamia*, eucalypt, *Jatropha* or semi-perennial cellulosic crops such as switchgrass or *Miscanthus*, which offer some form of woody biomass habitat. Avoiding or mitigating the degradation of environmental services and biodiversity will be a significant challenge for large-scale expansion of sugarcane for bioproduction.

A number of authors have predicted that producing fuel from sugarcane will not draw on the world's available agricultural land as significantly as might be expected (Pacca and Moreira 2011, Nunez *et al.* 2013), but may produce significant local impacts (Leal *et al.* 2013). Dornburg *et al.* (2010) estimate that the limitations imposed by water, biodiversity protection, and food demand will be a more significant constraint to global biomass potentials, suggesting that only one-third of the projected global energy demand could be met in 2050.

13.5.2 Water use

Depletion of water resources from bioproduction will be a key constraint, but has only recently been examined at length (Harto *et al.* 2010, Stone *et al.* 2010, Gerbens-Leenes *et al.* 2012). Gerbens-Leenes *et al.* (2012) estimate that by 2030 water demand for biofuel production will grow to 5.5% of available fresh water, putting extra pressure on water resources, and that further expansion will be constrained by water availability after that.

Sugarcane is intrinsically water-efficient, producing more biomass per unit of water than most other crops (de Vries *et al.* 2010). However, it does need good water availability to achieve high yields, which is why it has been traditionally suited to high rainfall regions of the world. In such areas, supplementary water is rarely required. However, expansion of sugarcane into areas with lower rainfall has necessitated supplementary irrigation, in regions such as southern Africa and parts of Australia. Thus, water availability and competition with other users will be a major limitation for sugarcane expansion for bioproduction.

13.5.3 Water quality

Sugarcane production has the potential to adversely impact water quality. High rainfall in tropical regions where sugarcane is commonly grown can mobilize and transport nitrogen, phosphorus, pesticides, and sediment to waterways via surface water runoff and leaching, with the potential for eutrophication and ecotoxicity impacts. These potential water quality risks have been recognized in northern Australian sugarcane growing regions, where it is grown in catchments that drain to the Great Barrier Reef marine lagoon (Kroon *et al.* 2013). The particular sensitivity of the receiving marine environment has been recognized and necessitated management measures for agricultural activities in this region. Sugarcane growing in other regions may present similar risks.

The intrinsic N- and pesticide-use efficiency of sugarcane (de Vries *et al.* 2010) can in theory translate into low losses of nutrients and chemical agents to the environment and low risk of water quality impacts. However, management practices are far more influential and the tendency for generous application rates in some regions means that in practice waterborne pollutants can be a risk.

13.5.4 Phosphorus depletion

Depletion of phosphorous (P) resources is a concern for all agricultural systems because P is essential for all agricultural production, it cannot be substituted, and the world's stocks are finite. Biofuels currently consume around 2% of the global inorganic P fertilizer production (Hein and Leemans 2012), and this will increase with any bioproduction expansion. Hein and Leemans (2012) estimated that for biofuels generally, the negative impacts of P depletion will exceed their positive contribution to climate change mitigation. However, sugarcane is the only feedstock where, under some conditions, GHG mitigation benefits can exceed P depletion impacts by virtue of its favorable yields (Hein and Leemans 2012).

13.5.5 Balancing the GHG abatement benefits with the environmental trade-offs

LCA studies that have considered environmental impact categories other than GHG emissions, have found sugarcane bioproducts to have higher land use, water use, water quality, and air quality impacts than their fossil fuel equivalents (Harding *et al.* 2007, Ramjeawon 2008, Groot and Boren 2010, Renouf *et al.* 2011, Cavalett *et al.* 2013, Renouf *et al.* 2013). Therefore, sugarcane bioproduction will involve a trade-off between mitigating GHG emissions on the one hand, and further depleting ecosystem services, water resources, and water quality on the other. Such trade-offs are a feature of all bioproduction based on agricultural crops. However, the scale of trade-offs are expected to be lower for sugarcane than for other crops across a range of agroecological sustainability indicators (de Vries *et al.* 2010).

The LCA process can apply weighting factors to multiple impact categories to generate a single score that represents the overall environmental impact. When this has been done for sugarcane ethanol the result can be either an overall better or worse environmental score for ethanol than petrol depending on the weighting factors applied (Cavalett *et al.* 2013). The scientific community is still some way off being able to consistently and confidently weight impacts, and draw overall conclusions about whether the GHG abatement benefit of sugarcane bioproducts outweigh the other environmental trade-offs.

Irrespective, it will be crucial for adverse environmental impacts of sugarcane cultivation to be mitigated with consideration of local environmental sensitivities so that GHG abatement can be harnessed sustainably.

13.6 Production pathways that optimize GHG abatement

Optimising the GHG abatement of different pathways requires consideration of (i) the production basis (i.e., dedicated bioproduction vs. coproduction), (ii) the type of bioproducts produced, and (iii) the land used.

13.6.1 Production basis (dedicated vs. coproduction)

Thus far GHG abatement has been discussed in the context of dedicated bioproduction presumed to require expanded sugarcane production. However, bioproducts can be coproduced alongside sugar by utilizing the by-products of sugar milling. Common coproduction pathways are the generation of electricity from surplus bagasse or fermentation of molasses to ethanol. Carbon footprints of coproduced bioproducts can be expected to be lower than dedicated bioproducts because there are generally no LUC-induced CO_2 emissions from agricultural expansion. LUC can, however, occur indirectly if the by-product feedstock is diverted away from an existing use. For instance, if low-grade molasses whose existing use is livestock feed is diverted to ethanol, there will be increased production of feed grain to compensate for the absence of molasses in the animal feed market (Beer *et al.* 2002, Nguyen and Hermansen 2012, Renouf *et al.* 2013). Therefore, increased grain production with presumably some LUC may be a consequence. GHG abatement from by-product utilization will depend on whether the offsets from the bioproduct pathway are greater than the offsets from the existing pathway.

In reality, the quantity of by-products available for bioproduction is constrained by sugar production, necessitating expansion into dedicated bioproduction to meet future demands for fuels and materials. So while coproduction of bioproducts can be a more carbon-efficient route, it may play only a small role. Furthermore, sugar milling by-products are already beneficially utilized in most sugarcane industries in various application (Paturau 1989, Manohar Rao 1997). There may be value in examining the optimal GHG abatement pathways for their utilization.

13.6.2 Product outputs

Bioproducts from sugarcane biorefining can be energy (electricity, steam), fuels (e.g., ethanol) or materials (e.g., plastics). The scale of GHG abatement offered by these different product categories varies, and some observations can be made about those that give the best GHG abatement.

The product options considered for sugarcane fiber are bioelectricity and cellulosic ethanol. Sugarcane fiber most commonly refers to bagasse, but it could also include crop residues (trash) recovered from the field or through whole-cane harvesting. Producing bioelectricity from fibre to offset conventional electricity can achieve greater GHG abatement than producing cellulosic-ethanol to offset petrol. The abatement benefits from electricity generation are greater when the displaced electricity is from coal (Botha and von Blottnitz 2006) than from natural gas (Campbell and Block 2010), due to differences in the GHG intensity of the displaced electricity (~1 $kgCO_{2\text{-}e}$/kWh for coal vs. ~0.7 $kgCO_{2\text{-}e}$/kWh for gas). Bioelectricity is a better GHG saver than bioethanol when the GHG-intensity of the displaced electricity is more than 0.45 $kgCO_{2\text{-}e}$/kWh (Sun *et al.* 2013). In regions with an electricity mix predominantly based on fossil fuels (>50%) electricity production from sugarcane fiber would be favored. Whereas ethanol production would be favored in countries with a low-emission electricity mix (nuclear or hydro). When bioelectricity from sugarcane is used to power vehicles Campbell *et al.* (2009) found that this pathway outperforms its use in biofuel production, producing on average 108% more emissions offsets and 81% more transportation kilometers per unit area of cropland.

When sucrose (in sugarcane juice) is the feedstock of interest, the alternative products considered here are ethanol and biodegradable bioplastics (PLA, PE, or PHB). GHG savings from ethanol and PLA have been found to be similar when they are produced in a biorefinery scenario with all energy needs met by bagasse (Alvarenga and Dewulf 2013, Renouf *et al.* 2013). This makes sense as both are principally based on a fermentation process, and in the final life cycle stages both break down by combustion or decomposition to biogenic CO_2, which is not counted. This assumes that when bioplastics degrade (presumably in landfill) most of the methane (biogas) produced during anaerobic breakdown is captured for energy generation. Noncapture of methane would increase the carbon footprint of PLA relative to ethanol.

If bioplastics were produced to be nonbiodegradable (i.e., for use in furniture, building materials, etc.) with the carbon "captured" in the products for at least 100 years, then the plastic pathway would lead to a net sequestration of carbon making it eminently preferable over ethanol.

Bioplastics also have advantages over ethanol in terms of other environmental impacts. Fossil energy conservation from sugarcane bioplastics is particularly large because of the fossil fuel–intensive nature of petrochemical plastics, which consume fossil fuels for both feedstock and production energy. For every 100 t of sugarcane processed, PLA production has been estimated to avoid 740 GJ of fossil energy use compared with 315 GJ for ethanol (Renouf *et al.* 2013). The replacement of petrochemical plastics with sugarcane-derived bioplastic also avoids air emissions associated with petrochemical production, which contribute to acidification and human health impacts (respiratory organics and inorganics).

It can be seen that the GHG savings offered by sugarcane bioproducts partly depends on the emissions intensity of the products displaced (*ibid*).

As the product portfolio chosen for biorefineries will ultimately depend on economic returns, GHG abatement may be a consideration when there is a price on carbon and mechanisms for claiming financial credits for abatement. In anticipation of this studies have considered optimal biorefinery configurations for joint production of plastic, ethanol, and electricity that maximize both economic and GHG abatement outcomes (Moncada *et al.* 2013).

13.6.3 Land used

In the case of dedicated bioproduction, we have seen that net GHG abatement is only achievable when expanded sugarcane production uses or displaces land with low-carbon stocks. Acceptable LUC pathways would be those that convert already degraded land, existing cropping land, or possibly grassland. Converting lands with existing carbon stocks, such as woody savannah, degraded forests, or forest would not be acceptable pathways.

In addition to avoiding the removal of existing carbon stocks, standards for sustainable bioproduction also aim to guarantee that biodiverse areas are not converted, including both forest and grassland habitat (European Union 2009).

13.7 Opportunities for further optimizing GHG abatement

Further optimization of GHG abatement is possible by reducing GHG emissions in the sugarcane growing phase, through more ecoefficient practices, soil carbon sequestration, utilization of harvest residues, and new sugarcane varieties each of which are discussed here.

13.7.1 Ecoefficient sugarcane growing

In Australian studies, the GHG abatement from more ecoefficient sugarcane growing was found to be comparable to that achieved from electricity cogeneration and ethanol production from molasses, and offers benefits across the full range of impact categories, not just GHG mitigation (Renouf *et al.* 2013). Ecoefficient practices (such as those listed below) are likely to be the most cost-effective route for GHG abatement:

- Improved nitrogen use efficiency. This includes the better matching of application rates to crop needs and the incorporation of legume break crops and intercropping to reduce demand for synthetic N fertilizer;
- On-farm energy efficiency. For highly mechanized sugarcane growing, energy efficiencies include refinements to irrigation energy efficiency (up to 30%) and changes in practices such as minimum tillage (up to 20%) (Baillie and Chen 2012);
- Cessation of sugarcane burning. Brazilian studies recommend the cessations of preharvest burning not only for avoiding combustion emissions (of CH_4 and

N_2O), but also for avoided health impacts of "black carbon" (Galdos *et al.* 2010) and soil carbon sequestration from retention of harvest residues (Pinheiro *et al.* 2010, Bordonal *et al.* 2012).

13.7.2 Utilization of harvest residues

The previous sections noted the soil organic carbon (SOC) sequestration benefits of retaining harvest residues in the field. The flip-side of this is the recovery of post-harvest residues or whole sugarcane harvesting to boost quantities of feedstock available for bioproduction (Suramaythangkoor and Li 2012).

Soil scientists recommended caution in relation to residue removal so as not to compromise soil productivity (Huggins *et al.* 2011). The optimal situation for GHG abatement will be a balance between retaining sufficient residues to build and maintain good SOC, but using any surplus for bioproduction to generate carbon offsets. Research suggests that SOC effects of residue removal vary depending on the type of crop and site characteristics. So threshold retention levels will need to be assessed on a case-by-case basis (Huggins *et al.* 2011).

13.7.3 New sugarcane varieties

Continued efforts in plant breeding to reduce the yield gap will be important, as yield improvements are known to be a key variable in reducing GHG intensities per unit of end-product (Johnston *et al.* 2011). Sugarcane varieties that enable good yields of fiber and/or sugar in marginal conditions, such as some "energy sugarcane" varieties (Alexander 1985, Viator *et al.* 2010), will be crucial for ensuring that expanded sugarcane production does not sacrifice existing carbon stocks and compete with food production. Genetic modification of sugarcane varieties that synthesize biomaterials within the sugarcane plant (plastics and pharmaceutical ingredients, for instance) is also expected to be promising (Petrasovits *et al.* 2012), as it removes the need for energy-consuming processing at the biorefinery.

13.8 Summary

Sugarcane bioproducts have lower carbon footprints than their fossil fuel equivalents, typically 60–70% lower. So substitution of fossil fuels with sugarcane-derived energy, fuels, and materials can result in significant GHG abatement, which easily meet international GHG abatement standards. Sugarcane also ranks very highly as a feedstock for bioproduction compared with other agricultural feedstock, in terms of GHG abatement. Only waste-derived feedstocks and second-generation lignocellulosic feedstocks based on agricultural residues outrank it.

Sugarcane bioproducts are robust GHG savers meaning the savings are fairly resilient to many variables. Land use change however, is an important variable that influences whether real GHG abatement occurs. Abatement with acceptable "ecosystem carbon payback time" (ECPT) is only achievable when land with low-carbon stocks is used or displaced, such as degraded land.

However, the GHG abatement benefits of sugarcane bioproduction need to be balanced against the environmental downsides of expanded sugarcane production, which are further erosion of environmental services and biodiversity, phosphorus depletion, and the potential for water resources depletion and water quality impacts. It will be crucial for these to be mitigated with consideration of local environmental sensitivities so GHG abatement can be sustainable.

Water and phosphorus conservation, and water and air quality impacts can be mitigated through ecoefficient sugarcane growing practices. However, it will be more difficult to mitigate the degradation of environmental services and biodiversity from expansion of agricultural production. At this point, the judgment to be made will be whether the stabilization of climate change through bioproduction is valuable enough to justify the other impacts. This will need to be judged on a case-by-case basis.

There are opportunities for optimizing and enhancing the GHG abatement potential of sugarcane bioproducts through the strategic selection of biorefinery products for a given region, reducing the GHG intensity of sugarcane growing, developing new sugarcane varieties that are productive in marginal conditions, and partially utilizing harvest residues.

References

Acreche, M. M. and Valeiro, A. H. (2013). Greenhouse gasses emissions and energy balances of a non-vertically integrated sugar and ethanol supply chain: a case study in Argentina. *Energy* **54**: 146–154.

Alexander, A. G. (1985). *The Energy Cane Alternative*. Elsevier Science Publishers: Amsterdam.

Allen, D. E., Kingston, G., Rennenberg, H., Dalal, R. C. and Schmidt, S. (2010). Effect of nitrogen fertilizer management and waterlogging on nitrous oxide emission from subtropical sugarcane soils. *Agriculture Ecosystems & Environment* **136**(3–4): 209–217.

Alvarenga, R. A. F. and Dewulf, J. (2013). Plastic vs. fuel: which use of the Brazilian ethanol Can bring more environmental gains? *Renewable Energy* **59**: 49–52.

Amores, M. J., Mele, F. D., Jimenez, L. and Castells, F. (2013). Life cycle assessment of fuel ethanol from sugarcane in Argentina. *International Journal of Life Cycle Assessment* **18**(7): 1344–1357.

Anderson-Teixeira, K. J., Davis, S. C., Masters, M. D. and Delucia, E. H. (2009). Changes in soil organic carbon under biofuel crops. *Global Change Biology Bioenergy* **1**(1): 75–96.

Baillie, C. and Chen, G. N. (2012). Opportunities for reducing on-farm energy use and greenhouse gas emissions in sugarcane. *International Sugar Journal* **114**(1366): 725–730.

Beeharry, R. P. (2001). Carbon balance of sugarcane bioenergy systems. *Biomass & Bioenergy* **20**(5): 361–370.

Beer, T., Grant, T., Williams, D. and Watson, H. (2002). Fuel-cycle greenhouse gas emissions from alternative fuels in Australian heavy vehicles. *Atmospheric environment* **36**: 753–763.

Bordonal, R. D., de Figueiredo, E. B. and La Scala, N. (2012). Greenhouse gas balance due to the conversion of sugarcane areas from burned to green harvest, considering other conservationist management practices. *Global Change Biology Bioenergy* **4**(6): 846–858.

Bos, H. L., Meesters, K. P. H., Conijn, S. G., Corre, W. J. and Patel, M. K. (2012). Accounting for the constrained availability of land: a comparison of bio-based ethanol, polyethylene, and PLA with regard to non-renewable energy use and land use. *Biofuels Bioproducts & Biorefining-Biofpr* **6**(2): 146–158.

Botha, T. and von Blottnitz, H. (2006). A comparison of the environmental benefits of bagasse-derived electricity and fuel ethanol on a life-cycle basis. *Energy Policy* **34**(17): 2654–2661.

BSI (2008). *PAS 2050:2008 Specification for the assessment of the life cycle greenhouse gas emissions of goods and services*, BSI Standards: UK.

Campbell, J. E. and Block, E. (2010). Land-use and alternative bioenergy pathways for waste biomass. *Environmental Science & Technology* **44**(22): 8665–8669.

Campbell, J. E., Lobell, D. B. and Field, C. B. (2009). Greater transportation energy and GHG offsets from bioelectricity than ethanol. *Science* **324**(5930): 1055–1057.

Cavalett, O., Chagas, M. F., Seabra, J. E. A. and Bonomi, A. (2013). Comparative LCA of ethanol versus gasoline in Brazil using different LCIA methods. *International Journal of Life Cycle Assessment* **18**(3): 647–658.

Contreras, A. M., Rosa, E., Perez, M., Van Langenhove, H. and Dewulf, J. (2009). Comparative life cycle assessment of four alternatives for using by-products of cane sugar production. *Journal of Cleaner Production* **17**(8): 772–779.

Cox, K., Renouf, M. A., Dargan, A., Klein-Marcuschamer, D. and Turner, C. (2014). Environmental life cycle assessment (LCA) of aviation biofuel from microalgae, *Pongamia pinnata*, and sugarcane molasses. *Biofuels, Bioproducts and Biorefining* **8**(4): 579–593.

Crutzen, P. J., Mosier, A. R., Smith, K. A. and Winiwarter, W. (2008). N_2O release from agro-biofuel production negates global warming reduction by replacing fossil fuels. *Atmospheric Chemistry and Physics* **8**(2): 389–395.

Dalal, R. C., Wang, W. J., Robertson, G. P. and Parton, W. J. (2003). Nitrous oxide emission from Australian agricultural lands and mitigation options: a review. *Australian Journal of Soil Research* **41**(2): 165–195.

Davidson, E. A., Matson, P. A. and Brooks, P. D. (1996). Nitrous oxide emission controls and inorganic nitrogen dynamics in fertilized tropical agricultural soils. *Soil Science Society of America Journal* **60**(4): 1145–1152.

de Vries, S. C., van de Ven, G. W. J., van Ittersum, M. K. and Giller, K. E. (2010). Resource use efficiency and environmental performance of nine major biofuel crops, processed by first-generation conversion techniques. *Biomass & Bioenergy* **34**(5): 588–601.

Denmead, O. T., Macdonald, B. C. T., Bryant, G., Naylor, T., Wilson, S., Griffith, D. W. T., Wang, W. J., Salter, B., White, I. and Moody, P. W. (2010). Emissions of methane and nitrous oxide from Australian sugarcane soils. *Agricultural and Forest Meteorology* **150**(6): 748–756.

Desplechin, E. (2010). The Brazilian sugarcane ethanol industry's approach to certification. *Zuckerindustrie* **135**(4): 222–225.

Dornburg, V., van Vuuren, D., van de Ven, G., Langeveld, H., Meeusen, M., Banse, M., van Oorschot, M., Ros, J., van den Born, G. J., Aiking, H., Londo, M., Mozaffarian, H., Verweij, P., Lysen, E. and Faaij, A. (2010). Bioenergy revisited: key factors in global potentials of bioenergy. *Energy & Environmental Science* **3**(3): 258–267.

Erisman, J., Grinsven, H., Leip, A., Mosier, A. and Bleeker, A. (2010). Nitrogen and biofuels; an overview of the current state of knowledge. *Nutrient Cycling in Agroecosystems* **86**(2): 211–223.

European Union (2009). Promotion of the use of energy from renewable sources and amending and subsequently repealing Directives 2001/77/EC and 2003/30/EC. Official Journal of the European Union 2009/28/EC (23 April).

FAO (2008). *The State of Food and Agriculture. Chapter 2*, p. 17. Food and Agriculture Organisation of the United Nations: Rome.

Fargione, J., Jason, H., Tilman, D., Polasky, S. and Hawthorne, P. (2008). Land Clearing and the biofuel carbon debt. *Science* **319**(5867): 1235–1238.

Galdos, M. V., Cerri, C. C., Lal, R., Bernoux, M., Feigl, B. and Cerri, C. E. P. (2010). Net greenhouse gas fluxes in Brazilian ethanol production systems. *Global Change Biology Bioenergy* **2**(1): 37–44.

Garcia, C. A., Fuentes, A., Hennecke, A., Riegelhaupt, E., Manzini, F. and Masera, O. (2011). Life-cycle greenhouse gas emissions and energy balances of sugarcane ethanol production in Mexico. *Applied Energy* **88**(6): 2088–2097.

Garoma, T., Ben-Khaled, M. and Beyene, A. (2012). Comparative resource analyses for ethanol produced from corn and sugarcane in different climatic zones. *International Journal of Energy Research* **36**(10): 1065–1076.

Gerbens-Leenes, P. W., van Lienden, A. R., Hoekstra, A. Y. and van der Meer, T. H. (2012). Biofuel scenarios in a water perspective: the global blue and green water footprint of road transport in 2030. *Global Environmental Change-Human and Policy Dimensions* **22**(3): 764–775.

Gibbs, H. K., Johnston, M., Foley, J. A., Holloway, T., Monfreda, C., Ramankutty, N. and Zaks, D. (2008). Carbon payback times for crop-based biofuel expansion in the tropics: the effects of changing yield and technology. *Environmental Research Letters* **3**(3): 034001.

Gil, M. P., Moya, A. M. C. and Dominguez, E. R. (2013). Life cycle assessment of the cogeneration processes in the Cuban sugar industry. *Journal of Cleaner Production* **41**: 222–231.

Goldemberg, J., Coelho, S. T. and Guardabassi, P. (2008). The sustainability of ethanol production from sugarcane. *Energy Policy* **36**(6): 2086–2097.

Gopal, A. R. and Kammen, D. M. (2009). Molasses for ethanol: the economic and environmental impacts of a new pathway for the lifecycle greenhouse gas analysis of sugarcane ethanol. *Environmental Research Letters* **4**(4): 044005.

Groot, W. J. and Boren, T. (2010). Life cycle assessment of the manufacture of lactide and PLA biopolymers from sugarcane in Thailand. *International Journal of Life Cycle Assessment* **15**(9): 970–984.

Harding, K. G., Dennis, J. S., von Blottnitz, H. and Harrison, S. T. L. (2007). Environmental analysis of plastic production processes: Comparing petroleum-based polypropylene and polyethylene with biologically-based poly-β-hydroxybutyric acid using life cycle analysis. *Journal of Biotechnology* **130**(1): 57–66.

Harto, C., Meyers, R. and Williams, E. (2010). Life cycle water use of low-carbon transport fuels. *Energy Policy* **38**(9): 4933–4944.

Hein, L. and Leemans, R. (2012). The impact of first-generation biofuels on the depletion of the global phosphorus reserve. *Ambio* **41**(4): 341–349.

Hennecke, A. M., Faist, M., Reinhardt, J., Junquera, V., Neeft, J. and Fehrenbach, H. (2013). Biofuel greenhouse gas calculations under the European Renewable Energy Directive – a comparison of the BioGrace tool vs. the tool of the Roundtable on Sustainable Biofuels. *Applied Energy* **102**: 55–62.

Hoefnagels, R., Smeets, E. and Faaij, A. (2010). Greenhouse gas footprints of different biofuel production systems. *Renewable & Sustainable Energy Reviews* **14**(7): 1661–1694.

Huggins, D. R., Karow, R. S., Collins, H. P. and Ransom, J. K. (2011). Introduction: evaluating long-term impacts of harvesting crop residues on soil quality. *Agronomy Journal* **103**(1): 230–233.

IEA (2009). *Bioenergy – A Sustainable and Reliable Energy Source: A Review of Status and Prospects.* IEA Bioenergy Executive Committee: Paris.

Intelligent Energy Europe (2012). Biograce GHG Calculation Tool Version 4C. Harmonised Calculations of Biofuel Greenhouse Gas Emissions in Europe. Retrieved 2-12-2012, from www.biograce.net.

IPCC (2000). *Good Practice Guidance and Uncertainty Management in National Greenhouse Gas Inventories*. Intergovernmental Panel on Climate Change/OECD: Japan.

IPCC (2006). 2006 IPCC guidelines for national greenhouse gas inventories. In *Agriculture, Forestry and Other Land Uses*. **4**. H. S. Eggleston, L. Buendia, K. Miwa, T. Ngara and K. Tanabe, eds. IGES: Japan.

IPCC (2013). Fifth Assessment Report (AR4). Climate Change 2013, International Panel on Climate Change.

ISO (2006a). *Greenhouse gas management and related activities. ISO 14064:2006*. International Organization for Standardization: Geneva, Switzerland.

ISO (2006b). *ISO 14040:2006. Environmental management – Life Cycle Assessment – Principles and Framework*. International Organization for Standardization: Geneva, Switzerland.

Johnston, M., Licker, R., Foley, J., Holloway, T., Mueller, N. D., Barford, C. and Kucharik, C. (2011). Closing the gap: global potential for increasing biofuel production through agricultural intensification. *Environmental Research Letters* **6**(3): 034028.

Kameyama, K., Shinogi, Y., Miyamoto, T. and Agarie, K. (2010). Estimation of net carbon sequestration potential with farmland application of bagasse charcoal: life cycle inventory analysis through a pilot sugarcane bagasse carbonisation plant. *Australian Journal of Soil Research* **48**(6–7): 586–592.

Kaufman, A. S., Meier, P. J., Sinistore, J. C. and Reinemann, D. J. (2010). Applying life-cycle assessment to low carbon fuel standards – how allocation choices influence carbon intensity for renewable transportation fuels. *Energy Policy* **38**(9): 5229–5241.

Kiatkittipong, W., Wongsuchoto, P. and Pavasant, P. (2009). Life cycle assessment of bagasse waste management options. *Waste Management* **29**(5): 1628–1633.

Kikuchi, Y., Hirao, M., Narita, K., Sugiyama, E., Oliveira, S., Chapman, S., Arakaki, M. M. and Cappra, C. M. (2013). Environmental performance of biomass-derived chemical production: a case study on sugarcane-derived polyethylene. *Journal of Chemical Engineering of Japan* **46**(4): 319–325.

Kroon, F., Turner, R., Smith, R., Warne, M., Hunter, H., Bartley, R., Wilkinson, S., Lewis, S., Waters, D. and Carroll,C. (2013). *Sources of Sediments, Nutrients, Pesticides and Other Pollutants in the Great Barrier Reef catchment. 2013 Scientific Consensus Statement Chapter 4*. Queensland Government, Reef Water Quality Protection Plan Secretariat: Brisbane.

Leal, M., Nogueira, L. A. H. and Cortez, L. A. B. (2013). Land demand for ethanol production. *Applied Energy* **102**: 266–271.

Liptow, C. and Tillman, A. M. (2012). A comparative life cycle assessment study of polyethylene based on sugarcane and crude oil. *Journal of Industrial Ecology* **16**(3): 420–435.

Lisboa, C. C., Butterbach-Bahl, K., Mauder, M. and Kiese, R. (2011). Bioethanol production from sugarcane and emissions of greenhouse gases – known and unknowns. *Global Change Biology Bioenergy* **3**(4): 277–292.

Luo, L., van der Voet, E. and Huppes, G. (2009a). Life cycle assessment and life cycle costing of bioethanol from sugarcane in Brazil. *Renewable & Sustainable Energy Reviews* **13**(6–7): 1613–1619.

Luo, L., van der Voet, E., Huppes, G. and de Haes, H. A. U. (2009b). Allocation issues in LCA methodology: a case study of corn stover-based fuel ethanol. *International Journal of Life Cycle Assessment* **14**(6): 529–539.

Macedo, I. C., Seabra, J. E. A. and Silva, J. E. A. R. (2008). Greenhouse gases emissions in the production and use of ethanol from sugarcane in Brazil: the 2005/2006 averages and a prediction for 2020. *Biomass and Bioenergy* **32**(7): 582–595.

Macedo, I. D. (1998). Greenhouse gas emissions and energy balances in bio-ethanol production and utilization in Brazil (1996). *Biomass & Bioenergy* **14**(1): 77–81.

Manohar Rao, P. J. (1997). *Industrial Utilization of Sugar Cane and Its By-Products*. ISPCK Publishers and Distributors: New Delhi, India.

Miller, S. A. (2010). Minimizing land use and nitrogen intensity of bioenergy. *Environmental Science & Technology* **44**(10): 3932–3939.

Miller, S. A. (2012). Avoiding the unintended consequences of bioenergy, perspectives on biofuels: Potential benefits and possible pitfalls. *American Chemical Society.* **1116**: 87–100.

Miller, S. A., Landis, A. E. and Theis, T. L. (2007). Environmental trade-offs of biobased production. *Environmental Science & Technology* **41**(15): 5176–5182.

Miyake, S., Smith, C., Waters, D., Peterson, A., McAlpine, C. and Renouf, M. (2015). *Environmental sustainability of using 'underutilised agricultural land' for future bioenergy crop production.* Agricultural Systems **139**: 180–195.

Moncada, J., Matallana, L. G. and Cardona, C. A. (2013). Selection of process pathways for biorefinery design using optimization tools: a Colombian case for conversion of sugarcane bagasse to ethanol, poly-3-hydroxybutyrate (PHB), and energy. *Industrial & Engineering Chemistry Research* **52**(11): 4132–4145.

Nguyen, T. L. T. and Gheewala, S. H. (2008). Life cycle assessment of fuel ethanol from cane molasses in Thailand. *International Journal of Life Cycle Assessment* **13**(4): 301–311.

Nguyen, T. L. T. and Hermansen, J. E. (2012). System expansion for handling co-products in LCA of sugar cane bio-energy systems: GHG consequences of using molasses for ethanol production. *Applied Energy* **89**(1): 254–261.

Njakou Djomo, S. and Ceulemans, R. (2012). A comparative analysis of the carbon intensity of biofuels caused by land use changes. *GCB Bioenergy* **4**(4): 392–407.

Nonato, R. V., Mantelatto, P. E. and Rossell, C. E. V. (2001). Integrated production of biodegradable plastic, sugar and ethanol. *Applied Microbiology and Biotechnology* **57**(1–2): 1–5.

Nunez, H. M., Onal, H. and Khanna, M. (2013). Land use and economic effects of alternative biofuel policies in Brazil and the United States. *Agricultural Economics* **44**(4–5): 487–499.

Ometto, A. R., Hauschild, M. Z. and Roma, W. N. L. (2009). Lifecycle assessment of fuel ethanol from sugarcane in Brazil. *International Journal of Life Cycle Assessment* **14**(3): 236–247.

Pacca, S. and Moreira, J. R. (2011). A biorefinery for mobility? *Environmental Science & Technology* **45**(22): 9498–9505.

Paturau, J. M. (1989). *By-Products of the Cane Sugar Industry. An Introduction to Their Industrial Utilization.* Elsevier Science Publishers: Amsterdam.

Pennington, D. W., Potting, J., Finnveden, G., Lindeijer, E., Jolliet, O., Rydberg, T. and Rebitzer, G. (2004). Life cycle assessment Part 2: Current impact assessment practice. *Environment International* **30**(5): 721–739.

Petrasovits, L. A., Zhao, L., McQualter, R. B., Snell, K. D., Somleva, M. N., Patterson, N. A., Nielsen, L. K. and Brumbley, S. M. (2012). Enhanced polyhydroxybutyrate production in transgenic sugarcane. *Plant Biotechnology Journal* **10**(5): 569–578.

Pietrini, M., Roes, L., Patel, M. K. and Chiellini, E. (2007). Comparative life cycle studies on poly(3-hydroxybutyrate)-based composites as potential replacement for conventional petrochemical plastics. *Biomacromolecules* **8**(7): 2210–2218.

Pinheiro, E. F. M., Lima, E., Ceddia, M. B., Urquiaga, S., Alves, B. J. R. and Boddey, R. M. (2010). Impact of pre-harvest burning versus trash conservation on soil carbon and nitrogen stocks on a sugarcane plantation in the Brazilian Atlantic forest region. *Plant and Soil* **333**(1–2): 71–80.

Plassmann, K., Norton, A., Attarzadeh, N., Jensen, M. P., Brenton, P. and Edwards-Jones, G. (2010). Methodological complexities of product carbon footprinting: a sensitivity analysis of key variables in a developing country context. *Environmental Science & Policy* **13**(5): 393–404.

Prueksakorn, K., Gheewala, S. H., Malakul, P. and Bonnet, S. (2010). Energy analysis of *Jatropha* plantation systems for biodiesel production in Thailand. *Energy for Sustainable Development* **14**(1): 1–5.

Ragauskas, A. J., Williams, C. K., Davisons, B. H., Britovsek, G., Cairney, J., Eckert, C. A., Frederick, W. J., Hallett, J. P., Leak, D. J., Liotta, C. L., Mielenz, J. R., Murphy, R., Templer, R. and Tschaplinski, T. (2006). The path forward for biofuels and biomaterials. *Science* **311**: 484–489.

Ramjeawon, T. (2004). Life cycle assessment of cane-sugar on the island of Mauritius. *International Journal of Life Cycle Assessment* **9**(4): 254–260.

Ramjeawon, T. (2008). Life cycle assessment of electricity generation from bagasse in Mauritius. *Journal of Cleaner Production* **16**(16): 1727–1734.

Rebitzer, G., Ekvall, T., Frischknecht, R., Hunkeler, D., Norris, G., Rydberg, T., Schmidt, W. P., Suh, S., Weidema, B. P. and Pennington, D. W. (2004). Life cycle assessment: Part 1: Framework, goal and scope definition, inventory analysis, and applications. *Environment International* **30**(5): 701–720.

Rein, P. W. (2010). The carbon footprint of sugar. *Zuckerindustrie* **135**(7): 427–434.

Reno, M. L. G., Lora, E. S., Palacio, J. C. E., Venturini, O. J., Buchgeister, J. and Almazan, O. (2011). A LCA (life cycle assessment) of the methanol production from sugarcane bagasse. *Energy* **36**(6): 3716–3726.

Renouf, M. A., Pagan, R. J. and Wegener, M. K. (2011). Life cycle assessment of Australian sugarcane products with a focus on cane processing. *International Journal of Life Cycle Assessment* **16**(2): 125–137.

Renouf, M. A., Pagan, R. J. and Wegener, M. K. (2013). Bio-production from Australian sugarcane: an environmental investigation of product diversification in an agro-industry. *Journal of Cleaner Production* **39**: 87–96.

Renouf, M. A., Wegener, M. K. and Nielsen, L. K. (2008). An environmental life cycle assessment comparing Australian sugarcane with US corn and UK sugar beet as producers of sugars for fermentation. *Biomass and Bioenergy* **32**(12): 1144–1155.

Renouf, M. A., Wegener, M. K. and Pagan, R. J. (2010). Life cycle assessment of Australian sugarcane production with a focus on sugarcane growing. *International Journal of Life Cycle Assessment* **15**(9): 927–937.

Robinson, N., Brackin, R. and Vinall, K. (2011). Nitrate paradigm does not hold up for sugarcane. *PLOS One* **6**(4).

Rosa, L. P., de Campos, C. P. and de Araujo, M. S. M. (2009). Biofuel contribution to mitigate fossil fuel CO2 emissions: Comparing sugar cane ethanol in Brazil with corn ethanol and discussing land use for food production and deforestation. *Journal of Renewable and Sustainable Energy* **1**(3).

Rosa, L. P. and Ribeiro, S. K. (1998). Avoiding emissions of carbon dioxide through the use of fuels derived from sugar cane. *Ambio* **27**(6): 465–470.

RSB (2010a). Principles and criteria for sustainable biofuel production. RSB-STD-01-001. Version 2. Retrieved 05-11-2010, from http://rsb.org.

RSB (2010b). RSB GHG calculation methodology. Retrieved 15-12-2010, from http://rsb.org.

RSB (2012). RSB guidelines and GHG calculator. Version 1. Retrieved 2-12-2014, from http://rsb.org.

RSB (2013). RBS standard for certification of bio-products. RSB-STD-02-001. Version 1. Retrieved 05-11-2010, from http://rsb.org.

Seabra, J. E. A., Macedo, I. C., Chum, H. L., Faroni, C. E. and Sarto, C. A. (2011). Life cycle assessment of Brazilian sugarcane products: GHG emissions and energy use. *Biofuels Bioproducts & Biorefining-Biofpr* **5**(5): 519–532.

Silalertruksa, T. and Gheewala, S. H. (2011). Long-term bioethanol system and its implications on GHG emissions: a case study of Thailand. *Environmental Science & Technology* **45**(11): 4920–4928.

Sills, D. L., Paramita, V., Franke, M. J., Johnson, M. C., Akabas, T. M., Greene, C. H. and Testert, J. W. (2013). Quantitative uncertainty analysis of life cycle assessment for algal biofuel production. *Environmental Science & Technology* **47**(2): 687–694.

Sims, R. E. H., Hastings, A., Schlamadinger, B., Taylor, G. and Smith, P. (2006). Energy crops: current status and future prospects. *Global Change Biology* **12**(11): 2054–2076.

Smeets, E. M. W., Bouwmanw, L. F., Stehfest, E., van Vuuren, D. P. and Posthuma, A. (2009). Contribution of N2O to the greenhouse gas balance of first-generation biofuels. *Global Change Biology* **15**(1): 1–23.

Solomon, B. D. (2010). Biofuels and sustainability. *Ecological Economics Reviews* **1185**: 119–134. K. Limburg and R. Costanza.

Souza, S. P., de Avila, M. T. and Pacca, S. (2012). Life cycle assessment of sugarcane ethanol and palm oil biodiesel joint production. *Biomass & Bioenergy* **44**: 70–79.

Stone, K. C., Hunt, P. G., Cantrell, K. B. and Ro, K. S. (2010). The potential impacts of biomass feedstock production on water resource availability. *Bioresource Technology* **101**(6): 2014–2025.

Su, C. L. and Lee, Y. M. (2009). Development status and life cycle inventory analysis of biofuels in Taiwan. *Energy Policy* **37**(2): 754–758.

Su, M. H. and Tso, C. T. (2011). Life cycle assessment of environmental and economic benefits of bio-ethanol in Taiwan. *International Journal of Global Warming* **3**(4): 339–354.

Sun, X. Z., Fujimoto, S. and Minowa, T. (2013). A comparison of power generation and ethanol production using sugarcane bagasse from the perspective of mitigating GHG emissions. *Energy Policy* **57**: 624–629.

Suramaythangkoor, T. and Li, Z. G. (2012). Energy policy tools for agricultural residues utilization for heat and power generation: a case study of sugarcane trash in Thailand. *Renewable & Sustainable Energy Reviews* **16**(6): 4343–4351.

Tate and Lyle (2011). Tate and Lyle's environmental performance report. Retrieved 30-11-2015, from www.tateandlyle.com.

Thorburn, P. J., Biggs, J. S., Collins, K. and Probert, M. E. (2010). Using the APSIM model to estimate nitrous oxide emissions from diverse Australian sugarcane production systems. *Agriculture, Ecosystems and Environment* **136**(3–4): 343–350.

UNEP (2007). *Reactive Nitrogen in the Environment: Too Much or Too Little of a Good Thing*. United Nations Environment Programme: Paris.

Viator, R., White, P. and Richard, E. (2010). Sustainable production of energycane for bio-energy in the southeastern United States. In *Sustainability of the Sugar and Sugar-Ethanol Industries*, pp. 147–161. G. Eggleston. **1058**. American Chemical Society, Washington, DC.

Volpi, G. (2010). Biofuels in Brazil and land use change. *Journal of Biobased Materials and Bioenergy* **4**(3): 211–220.

Walter, A., Dolzan, P., Quilodran, O., de Oliveira, J. G., da Silva, C., Piacente, F. and Segerstedt, A. (2011). Sustainability assessment of bio-ethanol production in Brazil considering land use change, GHG emissions and socio-economic aspects. *Energy Policy* **39**(10): 5703–5716.

Wang, M., Han, J., Dunn, J. B., Cai, H. and Elgowainy, A. (2012). Well-to-wheels energy use and greenhouse gas emissions of ethanol from corn, sugarcane and cellulosic biomass for US use. *Environmental Research Letters* **7**(4): 045905.

Weidema, B. P. (2001). Avoiding co-product allocation in life-cycle assessment. *Journal of Industrial Ecology* **4**(3): 11–33.

Wiens, J., Fargione, J. and Hill, J. (2011). Biofuels and biodiversity. *Ecological Applications* **21**(4): 1085–1095.

Yan, X. Y., Tan, D. K. Y., Inderwildi, O. R., Smith, J. A. C. and King, D. A. (2011). Life cycle energy and greenhouse gas analysis for agave-derived bioethanol. *Energy & Environmental Science* **4**(9): 3110–3121.

Yeh, S. and Sperling, D. (2010). Low carbon fuel standards: implementation scenarios and challenges. *Energy Policy* **38**(11): 6955–6965.

Yuttitham, M., Gheewala, S. H. and Chidthaisong, A. (2011). Carbon footprint of sugar produced from sugarcane in eastern Thailand. *Journal of Cleaner Production* **19**(17–18): 2119–2127.

Zaimes, G. G. and Khanna, V. (2013). Microalgal biomass production pathways: evaluation of life cycle environmental impacts. *Biotechnology for Biofuels* **6**: 88.

CHAPTER 14
Environmental sustainability assessment of sugarcane bioenergy

Shabbir H. Gheewala[1,2], Sébastien Bonnet[1,2] and Thapat Silalertruksa[1,2]

[1] *Life Cycle Sustainability Assessment Laboratory, The Joint Graduate School of Energy and Environment (JGSEE), King Mongkut's University of Technology Thonburi (KMUTT), Bangkok, Thailand*

[2] *Center of Excellence on Energy Technology and Environment, PERDO, Bangkok, Thailand*

14.1 Bioenergy and the sustainability challenge

In response to energy security concerns resulting from fossil energy supply shortage and climate change issues, the production, conversion, and use of biomass materials for energy has been the subject of much development, particularly over the last two decades. The importance of the contribution of biomass for sustainable energy production is now recognized worldwide.

It is in this context that biomass-based liquid transportation fuels, commonly referred to as biofuels, have gained recognition as one of the most important alternative fuels for transport. The global production of biofuels has been increasing from 16 billion liters in 2000 to more than 100 billion liters in 2011. Biofuels provide around 3% of total road transport fuel use globally (on an energy basis), and it is anticipated that by 2050 biofuels could provide 27% of total transportation fuel use (IEA 2011). At present, biodiesel and ethanol produced from indigenous biomass feedstocks are the two most popular forms of transportation biofuels promoted worldwide as substitutes to conventional fossil fuels, that is, diesel and gasoline. This is because of the benefits they are expected to provide with regard to stability of supply (enhanced energy security) and supporting the rural economy (Silalertruksa and Gheewala 2011a, b, Bell *et al.* 2011).

However, there is still much debate concerning the sustainability of biofuels. On one hand, biofuels are seen as a means to help mitigate climate change via reduced greenhouse gas (GHG) emissions, improved air quality in cities via lower tailpipe emissions, and enhanced economic growth in rural areas and the grass root economy via stabilization of farmers' income and employment generation in local communities (Silalertruksa and Gheewala 2009). On the other hand, there are concerns that the expansion of biofuels may create several drawbacks including *inter alia* increased GHG emissions as a result of direct and indirect land-use

Sugarcane-Based Biofuels and Bioproducts, First Edition. Edited by Ian M. O'Hara and Sagadevan G. Mundree.

change (especially if native forests are replaced with energy crops), competition with food and energy production (food vs. fuel), increased water scarcity, and biodiversity loss (spread of monocultures) (Silalertruksa and Gheewala 2012, Gheewala *et al.* 2013a).

14.2 Prospect of sugarcane bioenergy

The intensive promotion of bioenergy worldwide has been accompanied in the transportation sector with an increasing demand for biofuels, notably ethanol. Among a variety of energy crops promoted for this purpose, sugarcane stands out positively considering its relatively high biomass yield even under low fossil energy inputs (Macedo *et al.* 2008). Apart from a high biomass yield, sugarcane produces a range of useful coproducts including sugar, molasses, and bagasse, all contributing positively to its economic competitiveness (FAO 2013). Downstream sugar processing is remarkable for its almost zero-fossil-energy inputs due to the use of bagasse as a major energy source. In addition, there can be surplus electricity production from the combustion of the bagasse produced at the sugar mill, which can be supplied to the grid. Molasses is also conventionally used as an animal feed additive as well as to produce other value-added chemicals such as monosodium glutamate.

To enhance profitability in a sustainable way, sugarcane offers the possibility of using molasses and sugar juice for biofuels (ethanol) production in situations where sugar production has priority over biofuels. This is particularly so in modern sugar mills acting as biorefineries in which molasses or sugarcane juice is converted to ethanol and diverse coproducts, for example, energy (methane), biofertilizers, and CO_2. Many studies have shown that large GHG reduction benefits can be achieved by substituting sugarcane juice ethanol and molasses ethanol for gasoline in transportation (Macedo *et al.* 2008, Prakash *et al.* 1998).

However, one concern arising from the cultivation of sugarcane for expanded ethanol production is water availability as the crop grows better when there is sufficient rainfall and there is lesser requirement for irrigation (as is the case in tropical countries such as Brazil and Thailand). Another key concern is the effect of land-use change from such an expanded cultivation. In this case, for instance, when forest or grassland is cleared for new cropland, organic matter in vegetation and soil is released to the atmosphere as CO_2. This CO_2 release may make the GHG performance of biofuels become worse than that of conventional fossil fuels (Fargione *et al.* 2008).

To assure the long-term viability and continued public support for such feedstocks for bioenergy, these challenges need to be addressed to ensure that biofuels can be produced, traded, and consumed sustainably.

14.3 Environmental sustainability assessment tools

As detailed earlier, ethanol as a biofuel, similar to any other biomass-based fuels, faces the issue of whether or not it can help reduce a country's dependency on fossil energy and is environmentally friendly. To address this issue, net energy balance (NEB) and life cycle assessment (LCA) are useful tools to evaluate environmental sustainability and identify opportunities for improvement of the environmental efficiency of biofuels. In general, NEB is used to measure energy efficiency and effectiveness in fossil resources reduction from using fuels by looking at the overall fuel cycle. However, LCA can be used to identify and evaluate potential environmental implications of biofuels in order to improve environmental performance. A variety of ethanol systems have been assessed by these two methods in the past decades (Stout 1990, Prakash *et al.* 1998, Shapouri *et al.* 2006, Macedo *et al.* 2008, Silalertruksa and Gheewala 2009, Gheewala *et al.* 2011, Gheewala 2013).

The energy efficiency and environmental performance of biofuels are dependent on many factors, such as farming practices, industrial operation and technology, and residues management. For assessing the energy efficiency of biofuels, the NEB for the entire production system must be assessed in terms of the difference in energy content of the biofuel produced and fossil fuels and other energy sources required to produce it. However, in the case of biofuels, the energy required for growing the feedstock is mainly from solar energy. The solar energy captured by biomass is considered energetically free and excluded in the NEB analysis thus making a positive energy balance possible (Stout 1990). The net energy ratio (NER), which is the ratio of the amount of usable energy output to the amount of primary energy input to obtain that energy, and renewability are two indicators obtained from the NEB calculations that enable an identification of net energy efficiency and the net replaced fossil energy of biofuels.

For assessing environmental performance of bioenergy, LCA is a useful tool. The crucial advantage of a life cycle approach is that all burdens from raw material extraction through production, to use and disposal will be accounted for in this assessment. This approach is useful for evaluating the environmental performance of transportation biofuels as compared with conventional fossil fuels as it focuses on the entire life cycle of the biofuels rather than just the combustion step in vehicles (Silalertruksa and Gheewala 2011b). The International Organization for Standardization (ISO) describes the framework for conducting LCA in the ISO 14040 standard, which consists of four steps: (i) definition of the goal and scope of the study; (ii) identification and quantification of environmental loads involved (inventory); (iii) evaluation of the potential environmental impacts of these loads (impact assessment); and (iv) assessment of available options for reducing these environmental impacts (interpretation) (ISO 2006, Guinée *et al.* 2002).

14.4 Environmental sustainability assessment of sugarcane bioenergy: Case of Thailand

14.4.1 Background and policy context

Over the last decade, liquid biofuels have been intensively promoted in Thailand for transportation, especially to reduce oil imports and spur rural development. Ethanol derived from sugarcane (molasses) and also cassava has been receiving strong attention to partially substitute for the use of conventional gasoline. Based on its Alternative Energy Development Plan, the Ministry of Energy has set an ambitious production target for ethanol of 9 ml/day by 2021 (DEDE 2013). Electricity from biomass is also promoted with a target of 3630 MW by 2021; sugarcane residues being one of the contributors.

Ethanol is predominantly produced from sugarcane molasses although production from cassava is being increasingly promoted in Thailand. Ethanol production directly from sugarcane juice is also planned for the future. The promotion of ethanol is mainly for domestic consumption; however, there is also a great potential for export as 91 ml of surplus ethanol were exported to countries including the EU, Australia, and the Philippines during 2007–2009 (DEDE 2009).

At present, three types of gasohol blends are available in Thailand, that is, E10, E20, and E85. Gasohol blend E10 (a 10% blend of ethanol with 90% gasoline) was first introduced in 2004 while E20 (a 20% ethanol blend) was introduced in 2008. E85 gasohol became available later in 2008 slowly expanding over the years from Bangkok where it was first introduced to other regions, although vehicles that can use this blend are not yet readily available in Thailand (Gheewala 2012).

14.4.2 Sugarcane farming and production system

Thailand is the world's second largest sugar exporter after Brazil, which is the outstanding sugarcane producer globally (OAE 2008). Sugar is mainly produced from sugarcane, which grows well in the tropical and subtropical climate of Thailand. In general, sugarcane cultivation areas are located in the nonirrigated areas with less farm management requirements. Sugarcane cultivating areas are widely distributed in the northeastern, central, and northern regions of Thailand contributing 38%, 35%, and 27% of the total planted areas, respectively (Silalertruksa and Gheewala 2010). The crop is usually planted either before or after the rainy season and can be harvested around 10–12 months after cultivation. The harvest season of sugarcane is typically short – only 4–5 months each year during December–March.

There are four main stages leading to ethanol production: (i) sugarcane farming and harvesting, (ii) sugar milling, (iii) ethanol conversion, and (iv) transportation. The details of each stage are presented in the following sections. The

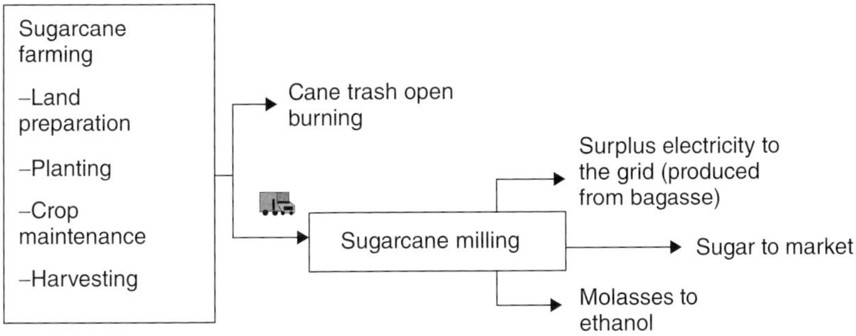

Figure 14.1 Value chain of sugarcane in Thailand.

overall stages leading to gasohol production and distribution are illustrated in Figure 14.1.

14.4.3 Sugarcane farming and harvesting

Sugarcane farming includes land preparation, planting, crop treatment, and harvesting. Most steps are labor intensive except for planting, which is fully mechanized (Gheewala *et al.* 2011). Crop maintenance after planting is a key factor influencing crop yield. It includes weeding as well as pesticide and fertilizer applications. A cycle of sugarcane planting and harvesting is about 12 months.

Sugarcane is initially grown from short sections of cane (plant cane) and the crops can be harvested 2–4 times before replanting with new sugarcane stems, which is required as a result of reduced yield (Gheewala *et al.* 2011). The average sugarcane yield from four cycles is about 60 tons per ha per year (Silalertruksa and Gheewala 2009).

Fertilizers and diesel are used for the cultivation and harvesting of sugarcane. Human labor is generally required for all farming activities including land preparation, planting, crop maintenance, and harvesting (Silalertruksa and Gheewala 2009).

In general, open burning is practiced before and after harvesting to ease the harvesting process and for removal of sugarcane trash residues (tops and leaves) from the field. Such burning activities result in air pollution and losses of valuable biomass materials that could otherwise be used for energy as well as for enhancing soil management practices, particularly, to maintain soil quality (organic content and moisture) and avoid erosion (Gheewala *et al.* 2011).

14.4.4 Sugarcane milling

Following harvesting, milling of the sugarcane takes place within 24 h to avoid sugar losses. This operation produces sugar (raw sugar, refined sugar, super refined sugar) and two major by-products – bagasse and molasses. The sugar milling process involves crushing sugarcane to extract sugarcane juice. This juice is clarified to remove any impurities and concentrated into syrup by boiling off excess water, seeded with raw sugar crystals in a vacuum pan and boiled until sugar crystals have formed and grown. The crystals are separated from the syrup by centrifugal processes before more crystals are grown in the syrup. Molasses is the syrup remaining after the sugar has passed through the centrifuge for the last time in a mill or refinery.

Bagasse is a biomass material that can be burnt in boilers at the mill for electricity and steam generation. Part of the electricity and steam produced can be used within the sugar mill while the excess electricity can be sold to the grid. This enables the mill not only to generate income from the biopower sold to the grid but also to gain GHG emission credits from the corresponding amount of fossil-based grid electricity that has been displaced. In Thailand, excess electricity generation as reported by Gheewala *et al.* (2011) is in the range of 14–15 kWh per ton of sugarcane. This is quite low due the use of low pressure boilers. Surplus electricity production levels of 60–180 kWh per ton of sugarcane are reported in the literature, which can be further increased to several hundred kilowatt-hour through optimization of sugarcane field residues utilization and combustion in dedicated power plants with high boiler pressures (Gheewala *et al.* 2011).

Molasses is an internationally traded commodity with total sugar content as the key quality criterion. Sugar in the molasses is converted to ethanol during the fermentation process.

14.4.5 Ethanol conversion

Ethanol production consists of the fermentation by yeast of sugars in molasses to produce a dilute alcohol solution, which is followed by distillation and dehydration to produce fuel-grade ethanol 99.5%. The three main by-products from this process are CO_2, stillage, and wastewater. The CO_2 can be collected, purified, and transformed for use in the coolant, soft drink, soda, dry ice, and fire extinguisher industries. The solids contained in the stillage supernatant can be separated from the remainder of the stillage and utilized for energy via biogas generation, whereas the settled portion from the whole process can be used as manure/soil conditioner for sugarcane plantations (Gheewala *et al.* 2011). Such utilization of by-products can contribute to optimized environmental performance (e.g., reduced emissions of pollutants including GHGs).

14.4.6 Transport

The whole cycle of ethanol production from molasses involves transportation of sugarcane to the sugar mill and transportation of molasses to the ethanol plant. The most widely used truck for transporting sugarcane to sugar mills in Thailand ranges between 18 and 25 tons per trip. Trucks are generally used for transporting molasses to the commercial ethanol manufacturers; however, some

Table 14.1 Key characteristics of the molasses-based ethanol plant investigated.

System characteristics	Molasses ethanol plant
Type of ethanol plant	Sugarcane milling and ethanol production
Steam source	Purchased from sugar mill
Electricity source	Purchased from sugar mill
End use of stillage	Organic fertilizer
Transportation of molasses	Pipeline from sugar mill

Source: Adapted from Silalertruksa and Gheewala (2009).

molasses ethanol plants receive molasses from the sugar mills through pipelines (Silalertruksa and Gheewala 2009).

14.5 Net energy balance and net energy ratio

The energy analysis presented here is based on an existing molasses ethanol plant in Thailand that is associated with a sugar mill and whose characteristics are as detailed in Table 14.1. The results of the net NER and renewability of this plant are presented later (see Table 14.1). An NER higher than 1 indicates that the fuel ethanol system has a net energy gain, while a renewability higher than 1 is a minimum requirement to indicate that the biofuel system can help reduce dependency on fossil energy.

The net energy output from molasses ethanol production includes the energy content of the ethanol and surplus electricity from sugar milling. The results (see Table 14.2) show that the NER of the molasses ethanol plant is slightly higher than 1 which is favorable. It is also observed that among all subsystems in the molasses ethanol cycle, ethanol conversion is the most energy-consuming stage, contributing two-thirds of the net energy demand. In this plant, because of the integrated approach to sugar milling and ethanol production, molasses was transported through pipes to the ethanol conversion plant resulting in reduced fossil energy use. The renewability factor of the molasses ethanol plant is 3.21 indicating that about 69% of the final fuel energy is obtained from renewable resources. This means that fuel ethanol produced from this plant can be considered renewable and the system approaches complete renewability (100%).

14.6 Life cycle environmental impacts

The life cycle environmental impacts associated with molasses ethanol production are discussed here (see Figure 14.2). These are presented for each step of production from sugarcane farming to ethanol conversion and for five major impact categories, that is, global warming potential, photo-oxidant formation,

Table 14.2 Energy balance (MJ) for production of 1000 l molasses-based ethanol[a].

Items	Molasses-based ethanol plant	
	Total energy	**Fossil energy**
1) Sugarcane farming		
NPK fertilizers	3,228	3,069
Herbicide	662	626
Diesel (farm machinery)	3,968	3,968
Labor	429	
2) Sugar milling		
Surplus electricity during normal operation (to grid)	(−2,608)	
Surplus bagasse (converted to electricity to grid)	(−7,047)	
3) Ethanol conversion		
Steam and electricity	17,378	
4) Transport		
Sugarcane	1,955	1,955
Molasses	238	
5) Energy values		
Net energy inputs	27,858	9,618
Net energy outputs (ethanol and surplus electricity)	30,855	
NEV[b]	2,997	
NRnEV[c]		21,237
6) Energy results		
NER[d]	1.11	
Renewability[e]		3.21

Source: Silalertruksa and Gheewala (2009).
[a]Energy content of ethanol: 21,200 MJ/1000 l ethanol.
[b]NEV: Energy content of ethanol (including surplus electricity) – net energy inputs
[c]NRnEV: Energy content of ethanol (including surplus electricity) – fossil energy inputs.
[d]NER: Net energy outputs/net energy inputs.
[e]Renewability: Net bioenergy outputs/net fossil energy inputs.

acidification, human toxicity, and eutrophication. Global warming potential refers to the warming of the atmosphere from the trapping of sunlight by GHGs, including, CO_2, N_2O, and CH_4 (in general calculated over 100-year horizon). Photo-oxidant formation refers to the production of ozone in the troposphere from the emissions of ozone precursors, including, NO_x, CO, CH_4, and other VOCs. Acidification is an impact resulting from the release of acidifying substances in the environment, notably, NO_x and SO_x. Human toxicity relates to impacts on human health from toxic pollutants such as PM (particulate matter), SO_x, NO_x, and VOCs. Eutrophication is the enrichment of nutrients (nitrogen and phosphorus) in surface waters.

These impacts were calculated on an averaged basis using information from three molasses-based ethanol plants in Thailand representing a range of characteristic systems for ethanol production in the country.

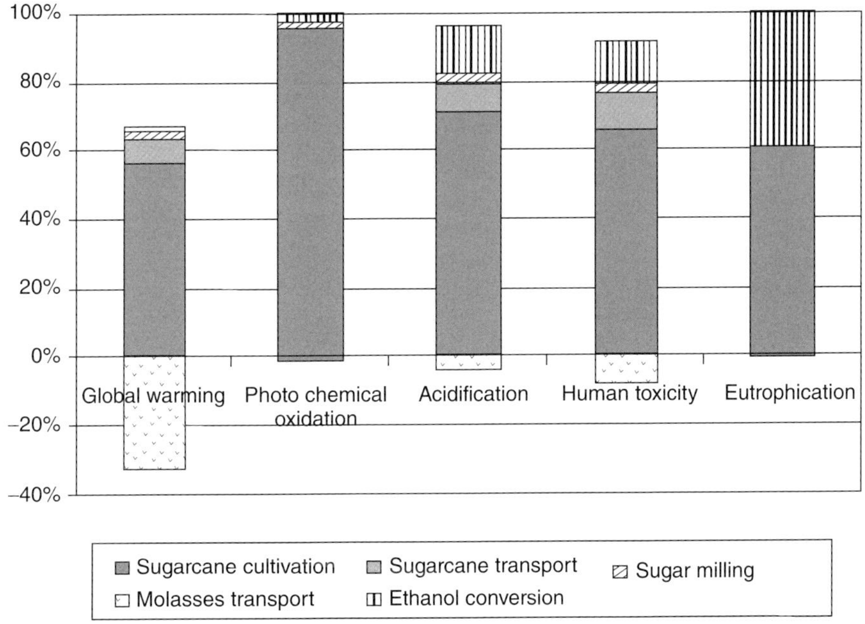

Figure 14.2 Average impact contribution from cradle to gate of molasses-based ethanol production. (Silalertruksa and Gheewala 2009, Figure 14.2. Reproduced with permission of Elsevier.)

The results indicate that the farming stage of sugarcane contributes the major impact for all environmental categories considered, in the range of 60–90%, over the life cycle of molasses-based ethanol.

Global warming appears to be negative for the sugar milling phase. This is because of the use of bagasse to produce steam and electricity, a part of which is recycled within the sugar mill and the surplus electricity sold to the grid. This brings important benefits to the mill in the forms of avoided fossil-based CO_2 emissions for grid electricity production (more than 30% avoided global warming impact). However, the farming stage is observed to contribute the largest share of the overall life cycle GHG emissions associated with the production of ethanol. This is because of open burning of sugarcane residues in the field at the time of harvesting. This activity also contributes heavily to photochemical ozone formation, human toxicity, and acidification, although, for this case, the contribution is also from SO_2 and NO_x emissions for feedstock production. In the case of eutrophication, this impact is contributed by sugarcane cultivation with almost 60% of the impact from fertilizer application and another 40% from the COD (chemical oxygen demand) of the spent wash of the molasses ethanol process (in the range of 100,000–150,000 mg/l), which is typically treated via open pond systems (leading to leaching of nutrients in the surrounding environment, including surface waters).

Table 14.3 Life cycle GHG emissions of molasses-based ethanol and gasoline 95.

	Molasses-based ethanol (kg CO_2eq)		Gasoline (kg CO_2eq)
	Base-case scenario with 70% cane trash open burnt	Alternative scenario with 35% cane trash open burnt	
Production	13.50	8.38	5.04
Use	—		21.66
Total	13.50	8.38	26.70

Source: Gheewala *et al.* (2011)

Results are based on a reference flow of 14.95 l ethanol that is equivalent to 9.89 l of gasoline 95.

To improve the environmental performance associated with molasses-based ethanol production, apart from the internal recycling of energy produced out of the coproducts generated along the chain leading to ethanol production, it is also imperative to minimize open burning of sugarcane residues. This common practice followed in Thailand to eliminate sugarcane trash from the field can be an important contributor to the overall environmental impacts associated with sugarcane ethanol production. For illustration purposes, the magnitude of the contribution of such a practice on global warming is shown here (see Table 14.3). This example is based on a biorefinery system in Thailand and considers a base-case scenario (actual situation) in which 70% of the residues are burnt in the field and an alternative scenario in which sugarcane trash burning would be reduced to 35%. The life cycle GHG emissions associated with gasoline production and use (from cradle to wheel) are also reported to illustrate the environmental performance of ethanol as compared with an equivalent amount of gasoline substituted.

In the base-case scenario, it is observed that the life cycle GHG emissions associated with molasses-based ethanol production are about 50% lower than those from an equivalent amount of gasoline. It is observed that a reduction in open burning of sugarcane residues from 70% to 35% would enable an additional 20% reduction in life cycle GHG emissions. Hence, if alternative uses for the sugarcane residues could be identified, for instance, as a source of energy, additional GHG emission credits could be earned, further benefiting the environmental performance. Optimization of farming practices may, therefore, also contribute an important role in climate change mitigation and should, therefore, be encouraged.

14.7 Key environmental considerations for promoting sugarcane bioenergy

Brazil is well known for being one of the pioneers in promoting the use of sugarcane as an indigenous energy feedstock for ethanol production to satisfy its own transportation sector's rising energy demand and to reduce dependence on fossil energy import. Several decades later, other tropical countries, including Thailand as illustrated in the earlier sections, have followed in the footsteps of Brazil, to promote sugarcane as an energy feedstock. The development of the associated biofuel industry was made possible thanks to strong commitments from national and international agencies in promoting renewable energy, the setting of international quality standards as well as national production policy targets, and the development of technologies, infrastructure, and financial mechanisms/incentives to meet these international and national targets.

As illustrated for the case of Thailand, sugarcane molasses-based ethanol shows a good potential as a substitute to fossil fuels, its energy ratio being higher than one (about 3.2 for the case of Thailand), which falls within the range of values reported in the literature, that is, between 2 and 8, as indicated by FAO (2013). In most of the studies associated with ethanol production, GHG benefits have also been highlighted (Macedo *et al.* 2008, Silalertruksa *et al.* 2009, Nguyen *et al.* 2010, Gheewala *et al.* 2011). However, in more recent years, some controversies have arisen regarding the scale of benefits depending on which elements are included within the system boundary.

One of the concerns associated with the intensification of sugarcane farming for expanded ethanol production is the associated land-use change in satisfying policy targets. Of particular concern is direct land-use change that may involve, for instance, the conversion of a forest or grassland to cropland. This assessment is commonly included in most life cycle carbon accounting procedure (FAO 2013). In this case, when for instance a forest or grassland is cleared for a new cropland, organic matter in vegetation and soil is released to the atmosphere as CO_2. This CO_2 release, if taken into account, may make the GHG performance of biofuels worse than that of conventional fossil fuels.

A study from Fargione *et al.* (2008) showed, for instance, that net carbon benefits associated with using ethanol for transportation if produced from a sugarcane plantation grown on a land that was initially occupied by a cerrado wood ecosystem would only accrue after 17 years of production. This analysis accounts for the time it takes for the ethanol to repay the carbon debt associated with such a direct land-use change when compared with using the equivalent amount of gasoline displaced.

Another similar study in Thailand from Nguyen *et al.* (2010) showed that 4.5–7 years of production would be required if the land-use change concerned grassland. These impacts when included in the assessment of the GHG contribution of sugarcane molasses ethanol can completely change the picture.

It is, therefore, vital that, for any policies related to the intensification of sugarcane cultivation for bioenergy, the land-use component be considered within the boundary of the biofuel production system being assessed. This is necessary to avoid compromising the environmental sustainability benefit of bioenergy with regard to its contribution to global warming mitigation.

Another aspect more recently incorporated in the assessment of impacts from land-use change is indirect land-use change (Silalertruksa *et al.* 2009). Indirect land-use change is the ripple effect that results in unintended consequences of using land for a new activity, which may result in geographically remote impacts including the diversion of land currently used for food for fuel production (Gnansounou *et al.* 2008). Indirect land-use change refers, therefore, to second, third, and higher degrees of land substitutions. The impacts of indirect land-use change are much harder to estimate, and no widespread agreement has been reached on a common methodology for analysis.

There is currently a debate about the measurement of GHG emissions resulting from indirect land-use change that may occur when increased demand for biofuel crops displaces other crops in new areas. Because natural lands such as rainforests and grasslands store and sequester carbon in their soil and biomass, as plants grow each year clearance of such lands for new sugarcane farms in other regions or countries translates into a net increase in GHG emissions (FAO 2013). Because of this change in the carbon stock of the soil and the biomass, indirect land-use change would have also consequences in the GHG balance of sugarcane molasses-based ethanol if considered within the system boundaries.

Fresh water consumption to grow energy crop such as sugarcane is another environmental sustainability challenge, which over the past few years has been the subject of much scrutiny among the scientific community. As is well known, water is a limited resource with only 3% freshwater on earth (97% is salty water) of which 1% is available as ground water and a smaller fraction as surface water. Almost 70% of surface freshwater is used for irrigation. This fact alone emphasizes the importance that the agricultural sector plays in terms of water consumption. An assessment of water management revealed that a fifth of the world's population or around 1.2 billion people live in areas of physical water scarcity and a further 500 million people are approaching this situation (IWMI 2007). Therefore, water security is another important challenge requiring urgent and effective management measures both at the global and local level (FAO 2013).

Because of climate change, it is also foreseen that disparities and uncertainties in the availability of water resources will increase in the future. To assess the water impact of bioenergy crops such as sugarcane, several studies have been conducted using the concept of water footprint (WF) to determine the volumetric demand in fresh water for the production and consumption of bioenergy (Gerbens-Leenes *et al.* 2009a, b, Mekonnen and Hoekstra 2011, Gheewala *et al.* 2013b). For sugarcane molasses-based ethanol production, Gheewala *et al.* (2013a, b) showed that feedstock cultivation is the major consumer of freshwater (irrigation water) at about 96%, followed by sugar milling at about 3% (process water), and mixing and distillation, 1% (process water as well). Improvement of farming practices and crop land zoning policies and planning based on the knowledge of local conditions of climate (amount of rainfall) and fresh water resources availability are, therefore, of utmost importance to optimize crop-related fresh water use efficiency and minimize competition with other users. According to the International Water Management Institute (IWMI 2007), 1% of global irrigation water supply is currently used for biofuels. A continued strong biofuel expansion could raise this percentage. In the case of India, estimations show that the amount of sugarcane needed to replace fossil fuels by 10% by 2030 would mean an increase in irrigation water of 22,000 billion liters (De Fraiture 2007).

It is also important to consider that not only water quantity but also water quality is affected by bioenergy production. An important contributor to such pollution is from the field applications of pesticides and fertilizers. Under rainfall conditions, chemicals rich in nitrogen, phosphorus, and potassium may infiltrate the soil and pollute underground water. These nutrients may also be washed away from the field as runoff and contaminate nearby surface waters such as streams and rivers. In the case of small streams, such nutrient inputs may result in eutrophication problems.

Biodiversity, defined as the abundance of species in a habitat, is essential to maintain the good health of an ecosystem. Biodiversity loss is identified as one of the major threats of the 21st century. Agricultural intensification and urban expansion have been important drivers to such losses. Preserving biodiversity or avoiding biodiversity loss from biofuels production is, therefore, another critical criterion for environmental sustainability. However, there are no standard ways to measure which systems to promote, except in general terms (such as use of rotations). Most current production systems do not indicate or assess stability or maintenance of biodiversity. Biomass production under intensive monoculture systems can have negative impacts on biodiversity including habitat loss, the expansion of invasive species and contamination from fertilizers and herbicides. However, native and perennial plantations including crop rotation are likely to be less damaging as well as the deployment of biomass on previously degraded or abandoned land. However, for the latter, strong incentives are needed, such as payments for environment services and internalization of externalities (FAO 2013).

Biofuels are an integral part of agriculture and energy crops, expected to play a major role for bioenergy in the future. By 2020–2030, bioenergy feedstocks, including sugarcane, are predicted to represent about 15% of the total area occupied by arable land, land under permanent crops, and land under permanent pasture (based on the levels of year 2000) (Berndes 2002, Berndes *et al.* 2003). Hence, the challenges of sustainability highlighted for biofuels are also true for all other biomass-based products, notably food and fibers. It is, therefore, important to recognize that the increased demand for such biomass-based products is a major challenge for agriculture to address and not solely for bioenergy, which is but a fraction of the whole system.

References

Bell, D.R., Silalertruksa, T. Gheewala S.H. and Kamens, R. (2011). The net cost of biofuels in Thailand: An economic analysis. *Energy Policy* **39**(2):834–843.

Berndes, G. (2002). Bioenergy and water: The implications of large-scale bioenergy production for water use and supply. *Global Environmental Change* **12**(4):253–271.

Berndes, G., Hoogwijk, M. and Van Den Broek R. (2003). The contribution of biomass in the future global energy supply: A review of 17 studies. *Biomass and Bioenergy* **25**(1):1–28.

DEDE (2009). Monthly exported ethanol (January 2007–August 2009). Retrieved 09-12-2015, from www.dede.go.th.

DEDE (2013). The renewable and alternative energy development plan for 25% in 10 years (AEDP 2012–2021). Department of Alternative Energy Development and Efficiency (DEDE), Ministry of Energy, Thailand. Retrieved 09-12-2015, from www.dede.go.th/dede/images/stories/dede&uscore;aedp&uscore;2012&uscore;2021.pdf.

De Fraiture, C. (2007). Biofuel crops could drain developing world dry. Science and Development Network. Retrieved 09-12-2015, from www.scidev.net/en/opinions.biofuel-crops-could-drain-developing-world-dry.html.

FAO (2013). *Biofuels and the Sustainability Challenge: A Global Assessment of Sustainability Issues, Trends and Policies for Biofuels and Related Feedstocks*. Trade and Markets Division, Food and Agriculture Organization (FAO), United Nations, Rome.

Fargione, J., Hill, J., Tilman, D., Polasky, S. and Hawthorne, P. (2008). Land clearing and the biofuel carbon debt. *Science* **319**(5867):1235–1238.

Gerbens-Leenes, P.W., Hoekstra A.Y. and Van Der Meer, T.H. (2009a). The water footprint of energy from biomass: A quantitative assessment and consequences of an increasing share of bioenergy in energy supply. *Ecological Economics* **68**(4):1052–1060.

Gerbens-Leenes, P.W., Hoekstra A.Y. and Van Der Meer, T.H. (2009b). The Water Footprint of Bioenergy. *Proceedings of the National Academy of Sciences* **2009**:0812619106v1.

Gheewala, S.H., Bonnet, S., Prueksakorn K. and Nilsalab P. (2011). Sustainability assessment of a biorefinery complex in Thailand. *Sustainability* **3**:518–530.

Gheewala, S.H. (2012). Food, fuel and climate change: Policy performance and prospects for biofuels in Thailand. In *RAP Publication 2012/14 on: Sustainable Bioenergy in Asia: Improving Resistance to High Food Prices and Climate Change*. B. Damen and S. Tvinnereim, eds. FAO, ISBN 978-92-5-107291-2.

Gheewala, S.H. (2013). Environmental sustainability assessment of ethanol from cassava and sugarcane molasses in a life cycle perspective, In *Life Cycle Assessment of Renewable Energy Sources*, A. Singh , S.L. Olsen , and D. Pant , eds. Springer.

Gheewala, S.H., Damen, B., Shi, X. (2013a). Biofuels: economic, environmental and social benefits and costs for developing countries in Asia, *WIREs Climate Change* **4**(6): 497–511.

Gheewala, S.H., Silalertruksa, T., Nilsalab, P., Mungkung, R., Perret, S.R. and Chaiyawannakarn, N. (2013b). Implications of the biofuels policy mandate in Thailand on water: The case of bioethanol. *Bioresource Technology* **150**: 457–465.

Gnansounou E., Panichelli L., Dauriat A., Villegas J.D. (2008). *Accounting for Indirect Landuse Changes in GHG Balances of Biofuels: Review of Current Approaches*. Ecole Polytechnique Federale De Lausanne.

Guinée J.B., Gorre M., Heijungs R., Huppes G., Kleijn R., Koning A., et al. (2002). *Handbook on Life Cycle Assessment: Operational Guide to the ISO Standards*. Dordrecht: Kluwer Academic Publishers.

IEA (2011). Technology roadmap – Biofuels for transport. International Energy Agency (IEA). Retrieved 09-12-2015, from www.iea.org.

ISO (2006). *ISO 14040:2006. Environmental Management Life Cycle Assessment Principles and Framework*. Geneva: International Organization for Standardization (ISO).

IWMI (2007). *Water for Food, Water for Life: A Comprehensive Assessment of Water Management in Agriculture*. Earthscan, London.

Macedo, I.C., Seabra, J.E.A., Silva, J.E.A.R. (2008). Greenhouse gases emissions in the production and use of ethanol from sugarcane in Brazil: The 2005/2006 averages and a prediction for 2020. *Biomass and Bioenergy* **32**(7):582–595.

Mekonnen, M.M. and Hoekstra, A.Y. (2011). The green, blue and grey water footprint of crops and derived crop products. *Hydrology and Earth System Sciences* **15**:1577–1600.

Nguyen, T.L.T., Gheewala, S.H. Sagisaka, M. (2010). Greenhouse gas savings potential of sugar cane bioenergy systems. *Journal of Cleaner Production* **18**:412–418.

OAE (2008). *Background Information of Agricultural Economics, Center for Agricultural Information, Office of Agri-cultural Economics (OAE)*, Ministry of Agriculture and Co-operatives, Bangkok.

Prakash, R., Henham, A. and Bhat, I.K. (1998). Net energy and gross pollution from bioethanol production. *Fuel* **77**:1629–1633.

Shapouri, H., Wang, M. and Duffield, J.A. (2006). Net energy balancing and fuel-cycle analysis. In J. Dewulf and H.van Langenhove, eds. *Renewables-Based Technology: Sustainability Assessment*, John Wiley and Sons Ltd., West Sussex

Silalertruksa, T. and Gheewala, S.H. (2009). Environmental sustainability assessment of bio-ethanol production in Thailand. *Energy* **34**:1933–1946.

Silalertruksa, T., Gheewala, S.H. and Sagisaka, M. (2009). Impacts of Thai bio-ethanol policy target on land use and greenhouse gas emissions. *Applied Energy* **86**:S170–S177.

Silalertruksa, T. and Gheewala, S.H. (2010). Security of feedstocks supply for future bio-ethanol production in Thailand. *Energy Policy* **38**:7476–7486.

Silalertruksa, T. and Gheewala, S.H. (2011a). Long-term bioethanol system and its implications on GHG emissions: A case study of Thailand. *Environmental Science and Technology* **45**(11):4920–4928.

Silalertruksa, T. and Gheewala, S.H. (2011b). The environmental and socio-economic: Impacts of bio-ethanol production in Thailand. *Energy Procedia* **9**:35–43.

Silalertruksa, T. and Gheewala, S.H. (2012). Food, fuel and climate change: Is palm-based biodiesel a sustainable option for Thailand? *Journal of Industrial Ecology* **16**(4):541–551.

Stout B.A. (1990). *Handbook of Energy for World Agriculture*. London: Elsevier Applied Science.

Index

Sugarcane-Based Biofuels and Bioproducts, First Edition. Edited by Ian M. O'Hara and Sagadevan G. Mundree.